6th Baltic-Nordic Workshop on Algebra, Geometry, and Mathematical Physics 2010

(AGMP-6)

AA002873

Journal of Physics: Conference Series Volume 346

Tjarno, Sweden
25-30 October 2010

Editors:

V. Abramov
J. Fuchs
E. Paal
Y. Shestopalov
S. Silvestrov
A. Stolin

ISBN: 978-1-62276-692-5
ISSN: 1742-6588

Printed from e-media with permission by:

Curran Associates, Inc.
57 Morehouse Lane
Red Hook, NY 12571

Some format issues inherent in the e-media version may also appear in this print version.

Printed by Curran Associates, Inc. (2013)

For permission requests, please contact the Institute of Physics
at the address below.

Institute of Physics
Dirac House, Temple Back
Bristol BS1 6BE UK

Phone: 44 1 17 929 7481
Fax: 44 1 17 920 0979

techtracking@iop.org

Additional copies of this publication are available from:

Curran Associates, Inc.
57 Morehouse Lane
Red Hook, NY 12571 USA
Phone: 845-758-0400
Fax: 845-758-2634
Email: curran@proceedings.com
Web: www.proceedings.com

TABLE OF CONTENTS

Journal of Physics: Conference Series **346** (2012) 011001 doi:10.1088/1742-6596/346/1/011001

Preface

Introduction

This proceedings volume presents results obtained by the participants of the 6th Baltic-Nordic workshop Algebra, Geometry, Mathematical Physics (AGMP-6) held at the Sven Lovén Centre for Marine Sciences in Tjärnö, Sweden on October 25-30, 2010.

The Baltic-Nordic Network AGMP (http://www.agmp.eu), "Algebra, Geometry, Mathematical Physics" was initiated in 2005 on the initiative of two Estonian universities, and two Swedish Universities: Tallinn University of Technology represented by Eugen Paal (coordinator of the network), Tartu University represented by Viktor Abramov, Lund University represented by Sergei Silvestrov and Chalmers University of Technology and University of Gothenburg represented by Alexander Stolin with the goal of promoting international and interdisciplinary cooperation between scientists and research groups in the countries of Baltic-Nordic region, in Mathematics and Mathematical Physics with the special emphasize on important role played by algebra and geometry in modern physics, engineering and technologies. The main activities of the AGMP network consist of a series of regular annual international workshops, conferences and research schools. The AGMP network constitutes also an important educational forum for scientific exchange and dissimilation of research results for PhD students and Postdocs. The network has expanded since its creation, and nowadays its activities extend beyond countries in Baltic-Nordic region to universities in other European countries and participants from elsewhere in the world. As one of the important research dissimilation outcomes of activities, the network has a tradition of producing high quality research Proceedings volumes after the network events, publishing them with various international publishers.

AGMP-6 and other AGMP meetings

The international workshop at Tjärnö (AGMP-6) was the 6th workshop in AGMP network workshop series:

- AGMP-7 University of Haute Alsace, Mulhouse, France, October 24-26, 2011 (Organizing Committee: A. Makhlouf (Chair), V. Abramov, M. Bordemann, E. Paal, S. Silvestrov, A. Stolin)
- AGMP-6 Sven Lovén Centre for Marine Sciences in Tjärnö, Sweden, October 25-30, 2010 (Organizing Committee: A. Stolin (Chair), V. Abramov, E. Paal, S. Silvestrov)
- AGMP-5 Stefan Banach International Mathematical Center, October 12-16, 2009 (Organizing Committee: A. Tralle (Chair), V. Abramov, J. Grabovski, J. Fuchs, E. Paal, A. Stolin, P. Urbanski)
- AGMP-4 University of Tartu, October 9-11, 2008 (Organizing Committee: V. Abramov (Chair), E. Paal, S. Silvestrov, A. Stolin)
- AGMP-3 Chalmers University of Technology and University of Gothenburg, October 11-13, 2007 (Organizing Committee: A. Stolin (Chair),V. Abramov, E. Paal, S. Silvestrov)
- AGMP-2 Lund University, October 12-14, 2006 (Organizing Committee: S. Silvestrov (Chair), V. Abramov, E. Paal, A. Stolin)

- AGMP-1 Tallinn University of Technology, October 8, 2005 (Organizers: E. Paal, S. Silvestrov)

Among other major events organized or to be organized by the AGMP Network recently are:

- AGMP-8, Brno, 12-14 September 2012 (Organizing Committee: M. Kures (Chair), J. Hrdina, E. Paal, P. Sehnalova, A. Stolin, P. Vasik, J. Zatocilova)
- The international conference "3Quantum: Algebra, Geometry, Information" (QQQ), Tallinn, 10-14 July 2012 (Organizing Committee: E. Paal (Chair), A. Fialowski, J. Fuchs, M. Schlichenmaier, V. Shumeiko, A. Stolin, P. Zusmanovich)
- AGMP&MP2 Summer School, Tjärnö, Sweden, September 05-09, 2011 (Organizing Committee: A. Stolin (Chair), J. Fuchs, H. Johannesson, E. Paal)

Main topics of the the AGMP-6 workshop in Tjärnö included, but were not limited to Algebra, Geometry, Lie theory, Representation theory, Deformation theory, Integrable systems, Mathematical methods of high energy physics, Quantum groups, Hopf algebras, Quantum information and applications in physics. The scientific programme of the AGMP-6 workshop in Tjärnö contained more then 63 talks including 11 plenary talks in the following special thematic sessions:

- Algebraic deformation theory (Organized by A. Siqveland)
- Integrable systems (Organized by H. Johannesson and P. Kulish)
- Noncommutative algebra (Organized by N. Iyudu and A. Laudal)
- Quantum algebra (Organized by S. Caenepeel and H. Rosengren)
- Categorical methods in physics (Organized by I. Burban)
- Operator algebras, dynamics, applications (Organized by S. Eilers, S. Silvestrov and L. Turowska)
- Number theoretical methods in string theory (Organized by B. E. W. Nilsson)
- Quantum information (Organized by T. Dorlas)
- Generalized Lie theory (Organized by A. Makhlouf and S. Silvestrov)
- Geometry (Organized by M. Kures)
- Spectral problems in quantum mechanics (Organized by G. Rozenblioum)
- Spectral and comp methods for Maxwell equations (Organized by L. Beilina and Y. Shestopalov)

In total, 79 scientists from Belgium, Brazil, Czech Republic, Denmark, Estonia, France, Germany, Hungary, Iran, Ireland, Italy, Japan, Norway, Portugal, Spain, Romania, Russia, Sweden, Uganda, UK, Ukraine, USA, Switzerland have taken part in the conference together making the conference a highly successful event.

Review of AGMP-6 Proceedings

The research articles selected for this volume cover a wide range of topics representing the main directions of the workshop and the topics of the special thematic sessions. All article submitted for this volume were refereed, and the Editors take this opportunity to thank all the referees for their work. Among contributors to the volume there are 42 researchers from different countries around the world making this proceedings volume into an important international interdisciplinary publication.

In the article "Weil jets theory, connections and curvature tensor", R. J. Alonso-Blanco and J. Muñoz-Díaz describe connections in the framework of Weil jets theory which consist of considering jets as ideals and refer all the operations to the ring of functions of the base manifold. With this approach they obtain a new perspective of the curvature tensor and point out as an application, that the Ricci tensor is obtained in a way somewhat more direct than usual, which could be interesting for a better understanding of this important object. M. Bacovský, M. Havlíček and S. Pošta, in their article "Central dependencies in quantum algebras", analyze the structure of the center of the quantum algebra $U_q(sl_2)$ in detail for both cases when deformation parameter q is or is not a root of unity, obtaining description of the center and its generators in a purely algebraic way without considering representation theory. In the article "Categorical Non-commutative Geometry", P. Bertozzini, R. Conti and W. Lewkeeratiyutkul outline the current status of some recent research programs aiming at a categorification of parts of A.Connes non-commutative geometry and provide an outlook on some possible future developments in categorical non-commutative geometry. S. Burciu, in his article "Normal coideal subalgebras of semisimple Hopf algebras", shows that the restriction functor from the category of representations of a semisimple Hopf algebra to the category of representations of a normal coideal subalgebra has a similar behavior to the restriction functor to the category of representations of a normal Hopf subalgebra. Commutator subalgebras as normal left (right) coideal subalgebras are also studied. R. Conti, J. H. Hong and W. Szymański, in their expository article "Analysis of endomorphisms", discuss the recent progress in the study of endomorphisms and automorphisms of the Cuntz algebras and, more generally graph C^*-algebras (or Cuntz-Krieger algebras), and in particular, the definition and properties of both the full and the restricted Weyl group of such an algebra. The authors outline a powerful combinatorial approach to analysis of endomorphisms arising from permutation unitaries, emphasize that the restricted Weyl group consists of automorphisms of this type, discuss the action of the restricted Weyl group on the diagonal MASA and its relationship with the automorphism group of the full two-sided n-shift and present several open problems. In the article "Representations of posets: Linear versus Unitary", V. Futorny, Y. Samoilenko and K. Yusenko, consider the representation theory of partially ordered sets in unitary spaces with the so called orthoscalar relation. Such theory generalizes the classical theory which studies the representations of partially ordered sets in linear spaces. In the article "Remarks on rational solutions of Yang-Baxter equations", T. Henrich studies the unitary rational solutions of the associative Yang-Baxter equation with three spectral parameters, explains how such solutions arise from the geometry of vector bundles on a cuspidal cubic curve, and moreover investigates how these solutions are related to the quantum and classical Yang-Baxter equations. S. Hohenegger, in his article "Automorphic products, generalized Kac-Moody algebras and string amplitudes", reviews automorphic products and generalized Kac-Moody algebras from a physics point of view, discuss the appearance of automorphic products in BPS-saturated oneloop quantities in heterotic string theory and review particular examples of this extension for compactifications with $N = 2$ and $N = 4$ space-time supersymmetry. J. Hrdina, in his article "Geometric properties of Frobenius algebras" discusses the properties of algebras and their modules proving that the study of generic rank reduces to the study of Frobenius algebras. In his article "Subring depth below an ideal" by L. Kadison, a minimum depth is assigned to a ring homomorphism and a bimodule over its codomain; when the homomorphism is an inclusion and the bimodule is the codomain, the recent notion of depth of a subring by Boltje-Danz-Külshammer is recovered, and it is pointed out that subring depth below an ideal gives a lower bound for BDK's subring depth of a group algebra pair or a semisimple complex algebra pair. In the article "The Belavin-Drinfeld theorem on non-degenerate solutions of the classical Yang-Baxter equation", L. Kierans and B. Kreussler give a coordinate free proof of Belavin and Drinfeld's Theorem about the classification of non-degenerate solutions of the classical Yang-Baxter equation, and the equivalence of different characterizations of non-degeneracy is also shown in such a way. In the article "Tensor powers for non-simply laced Lie algebras B_2-case", P. P. Kulish, V. D. Lyakhovsky and O. V. Postnova study the decomposition problem for tensor powers of B_2-fundamental modules. To solve this problem singular weight technique and injection fan algorithms are applied. Properties of multiplicity coefficients are formulated in terms of multiplicity functions. These functions are constructed showing explicitly the dependence of multiplicity coefficients on the highest weight coordinates and the tensor power parameter. It is thus possible to study general properties of multiplicity coefficients for powers of the fundamental B_2-modules. In the article "Initial-boundary value problems for linear equations of electrodynamics with nonlinear boundary conditions" by K. Lukin, a new class of electromagnetic (EM) problems is introduced, initial-boundary value problem for linear wave equation with nonlinear boundary conditions, and methods for their solution are elaborated. Theory of nonlinear 1D maps and difference-differential equations have been applied to solve the problem, and it has been shown that spatiotemporal dynamics of EM field in a resonator with nonlinear reflecting surfaces may be either regular or chaotic in spite of the major part of the system is formed by linear media. The author points out that the approach may be used for analysis of chaotic processes in many electromagnetic and electronic devices. In the article "On $(\varepsilon;\delta)$-Freudenthal Kantor triple systems and anti-structurable algebras with certain conditions", N. Kamiya, D. Mondoc and S. Okubo discuss a characterization of anti-structurable algebras in connection with their relation with $(-1;-1)$-Freudenthal Kantor triple systems. In the article "Mal'tsev algebras and triality" by E. Paal, a concept of the Malt'tsev pair is presented based on the generalized Maurer-Cartan equations of a local analytic Moufang loop. The triality can be seen as a fundamental property of such pairs. Based on triality, the Yamagutian is constructed and properties of the Yamagutian are studied. In the article "Automorphic Instanton Partition Functions on Calabi-Yau Threefolds", D. Persson surveys recent results on quantum corrections to the hypermultiplet moduli space in type IIA/B string theory on a compact Calabi-Yau threefold, or, equivalently, the vector multiplet moduli space in type IIB/A on $X \times S^1$. In the article "Four ways across the wall" B. Pioline considers an important question in the study of $N = 2$ supersymmetric string or field theories is to compute the jump of the BPS spectrum across walls of marginal stability in the space of parameters or vacua. The author surveys four apparently different answers for this problem, two of which are based on the mathematics of generalized Donaldson-Thomas invariants (the Kontsevich-Soibelman and the Joyce-Song formulae), while the other two are based on the physics of multi-centered black hole solutions (the Coulomb branch and the Higgs branch formulae, discovered in joint work with Jan Manschot and Ashoke Sen). The author points out also that explicit computations indicate that these formulae are equivalent, though a combinatorial proof is currently lacking. In the article "New approach for solving Maxwell equations with strong singularity", V. A. Rukavishnikov and A. O. Mosolapov introduce a notion of R_ν-generalized solution to Maxwell equations with strong singularity in a $2D$ nonconvex polygonal domain, develop a new weighted edge finite element method (FEM) based on the conception of R_ν-generalized solution of the Maxwell equations with strong singularity due to a reentrant corner on the boundary, and point out that the results of numerical experiments prove the efficiency of this method. Y. Shestopalov and Y. Smirnov, in their article "Inverse scattering in guides", present statements and a method of solution to the inverse scattering problem of reconstructing permittivity of a dielectric inclusion in a $2D$ or $3D$ waveguide from the transmission characteristics, establish the unique solvability of a volume singular integral equation (VSIE), demonstrate that the inverse problem is solved by the method of iterations applied to VSIE and prove the convergence of the method. In the article "Some formulas for Legendre functions related to the Poisson transform and Lorentz group representation", I. A. Shilin and A. I. Nizhnikov consider the representation of Lorentz group in the spaces of homogeneous functions on cone and hyperboloid. Using the Poisson transform, which maps a function from the first space into the corresponding function from the second space, the authors derive some new formulas containing Legendre functions, and show that using the restriction of the representation in the first space on a one-parameter subgroup of hyperbolic rotations another new formula for Legendre functions can be obtained. In the article

"Burchnall-Chaundy annihilating polynomials for commuting elements in Ore extension rings" by J. Richter and S. D. Silvestrov, further progress is made in extending the Burchnall-Chaundy type determinant construction of annihilating polynomial for commuting elements to broader classes of rings and algebras by deducing an explicit general formula for the coefficients of the annihilating polynomial obtained by the Burchnall-Chaundy type determinant construction in Ore extension rings. It is also demonstrated how this formula can be used to compute the annihilating polynomials in several examples of commuting elements in Ore extensions. Also it is demonstrated that additional properties which may be possessed by the endomorphism, such as for example injectivity, may influence strongly the annihilating polynomial. J. Vajakas, in his article "Some Generalizations of the Hewitt Extension", introduces two extensions of a Tychonoff space that generalize the notions of the Hewitt extension and the generalized Hewitt extension introduced by Bachman, Beckenstein, Narici and Warner. The extensions are determined by sets of continuous functions. In particular, the author studies the extensions determined by all continuous functions and the extensions determined by all continuous bounded functions and shows that these can be regarded as epireflective functors. The author also discusses the possibility of using these extensions to describe the characters of function algebras. In the article "Tangent structures: sector-forms, jets and connections", M. Rahula, P. Vašík and N. Voicu point out that while to define a higher order connection on a fibered manifold one can use the sections of nonholonomic jet prolongations, a more natural approach seems to be the one assuming the structure of a higher-order tangent bundle and using White's sector-forms on these bundles.

This international interdisciplinary volume will be a source of inspiration for a broad spectrum of researchers and for research students, as the contributions lie at the intersection of the research directions and interests of several large research communities and research groups on modern Mathematics and Physics.

Acknowledgments
We gratefully acknowledge the support of sponsors, which provided support for the organization of the workshop as a whole and for the special thematic sessions. These include the Estonian Science Foundation, the ITGP Network of the European Science Foundation, Network Swedish Research Council, NordForsk, the Swedish Foundation for international cooperation in research and higher education (STINT), the Royal Swedish Academy of Sciences, the Crafoord Foundation, and a number of other European and International foundations and universities. We are also grateful for support to the hosting organizations, the University of Gothenburg and Chalmers University of Technology and the the Sven Lovén Centre for Marine Sciences in Tjärnö for the hospitality which provided excellent environment for making this event into an international research success. Finally, we acknowledge Astralgo Ltd in Tallinn for the web hosting and IOP Publishing for the collaboration during the preparation of the proceedings.

V. Abramov, J. Fuchs, E. Paal, Y. Shestopalov, S. Silvestrov, A. Stolin, Eds.

Journal of Physics: Conference Series **346** (2012) 011001 doi:10.1088/1742-6596/346/1/011001

List of Participants

- Viktor Abramov, University of Tartu, Estonia
- Ricardo Alonso Blanco, University of Salamanca, Spain
- Joakim Arnlind, Max Planck Institute for Gravitational Physics, Germany
- Mohammad Asadzadeh, Chalmers, Sweden
- Martin Bacovský, Czech Technical University, Czech Republic
- Larisa Beilina, Chalmers and University of Gothenburg, Sweden
- Igor Burban, University of Bonn, Germany
- Sebastian Burciu, Institute of Mathematics "Simion Stoilow", Romania
- Stefaan Caenepeel, Vrije Universiteit Brussel, Belgium,
- Roberto Conti, Universita' G. D'Annunzio di Chieti-Pescara, Italy
- Nuno Costa Dias, Universidade Lusófona, Portugal
- Ahmad Doosthoseini, Islamic Azad University, Karaj, Iran
- Tony (Teunis) Dorlas, Dublin Institute for Advanced Studies, Ireland
- Edward Effros, UCLA, USA
- Eivind Eriksen, BI Norwegian School of Managemennt, Norway
- Gunnar Fløystad, University of Bergen, Norway
- Jürgen Fuchs, Karlstads universitet, Sweden
- Vyacheslav Futorny, University Sao Paulo, Brazil
- Magnus Goffeng, Chalmers and University of Gothenburg, Sweden
- Trond Stølen Gustavsen, Buskerud University College, Norway
- Bo Hellsing, University of Gothenburg, Sweden
- Thilo Henrich, University of Bonn, Germany
- Stefan Hohenegger, ETH, Switzerland
- Alexander Holevo, Steklov Mathematical Institute, Russia
- Jaroslav Hrdina, Brno University of Technology, Czech Republic
- Hilja Lisa Huru, Finnmark University College, Norway
- Radu Iordanescu, Institute of Mathematics of the Romanian Academy, Romania
- Henrik Johannesson, University of Gothenburg, Sweden
- Lars Kadison, Universidade do Porto, Portugal
- Noriaki Kamiya, Aizu University, Japan
- Johannes Kleppe, Høgskolen i Buskerud, Norway
- Jan Oddvar Kleppe, Oslo University College, Norway
- Pavel Krajča, Brno University of Technology, Czech Republic
- Bernd Kreussler, Mary Immaculate College, Ireland
- Miroslav Kures, Brno University of Technology, Czech Republic
- Par Kurlberg, KTH, Sweden
- Olav Arnfinn Laudal, Universitetet i Oslo, Norway
- Stefan Lemurell, Chalmers and University of Gothenburg, Sweden
- Konstantin Lukin, Institute for Radiophysics and Electronics of the National Academy of Sciences of Ukraine, Ukraine
- Sara Madariaga, University of La Rioja, Spain

- Jan Manschot, IPhT, France
- Aleksandar Mikovic, Lusofona University and GFMUL, Portugal
- Daniel Mondoc, Lund University, Sweden
- Akos Nagy, Budapest University of Technology and Economics, Hungary
- Bengt E. W. Nilsson, Chalmers, Sweden
- Roman Novikov, CNRS, France
- Johan Oinert, University of Copenhagen, Denmark
- Eugen Paal, Tallinn University of Technology, Estonia
- Daniel Persson, ETH Zrich, Switzerland
- Boris Pioline, LPTHE, CNRS & Universit Pierre et Marie Curie, France
- Severin Posta, Czech Technical University, Czech Republic
- João Prata, Universidade Lusófona, Portugal
- Johan Richter, Lund University, Sweden
- Hjalmar Rosengren, Chalmers, Sweden
- Grigori Rozenblioum, Chalmers and University of Gothenburg, Sweden
- Victor Rukavishnikov, Russian Academy of Sciences, Computing Center of Far-Eastern Branch, Russia
- Helena Rukavishnikova, Russian Academy of Sciences, Computing Center of Far-Eastern Branch, Russia
- Yury Shestopalov, Karlstad University, Sweden
- Ilya Shilin, M. Scholokhov Moscow State University for the Humanities and Moscow Aviation Institute, Russia
- Sergey Shitov, University of Tartu, Estonia
- Tatiana Shulman, University of Copenhagen, Denmark
- Sergei Silvestrov, Lund University, Sweden
- Arvid Siqveland, Buskerud University College, Norway
- Arne B Sletsjøe, University of Oslo, Norway
- Agata Smoktunowicz, University of Edinburgh, UK
- Wojciech Szymanski, University of Southern Denmark, Denmark
- Carl Stigner, Karlstad University, Sweden
- Alexander Stolin, Chalmers and University of Gothenburg, Sweden
- Pavel Stovicek, Czech Technical University, Czech Republic
- Fatemeh Zarezadeh, University of Teheran, Iran
- Natalia Zhukavets, Czech Technical University, Czech Republic
- Valeriy Tolstoy, Lomonosov Moscow State University, Russia
- Alex Tumwesigye, Mekerere University, Uganda
- Lyudmila Turowska, Chalmers, Sweden
- Jaan Vajakas, University of Tartu, Estonia
- Petr Vasik, Brno University of Technology, Czech Republic
- Jan Vondra, Masaryk University, Czech Republic
- Shinji Yamashita, Kyushu University, Japan

All papers published in this volume of *Journal of Physics: Conference Series* have been peer reviewed through processes administered by the proceedings Editors. Reviews were conducted by expert referees to the professional and scientific standards expected of a proceedings journal published by IOP Publishing.

Algebra, Geometry, and Mathematical Physics 2010 IOP Publishing
Journal of Physics: Conference Series **346** (2012) 012001 doi:10.1088/1742-6596/346/1/012001

Weil jets theory, connections and curvature tensor

R J Alonso-Blanco and J Muñoz-Díaz

Departamento de Matemáticas, Universidad de Salamanca, Plaza de la Merced 1-4, E-37008 Salamanca, Spain.

E-mail: `ricardo@usal.es, clint@usal.es`

Abstract. We will describe connections in the framework of Weil jets theory which consist of considering jets as ideals and refer all the operations to the ring of functions of the base manifold. With this approach we gain some clarity on certain aspects of the formalism. Moreover, we will obtain, based on it, a slightly new perspective of the curvature tensor. As an application, the Ricci tensor is obtained in a way somewhat more direct than usual, which could be interesting for a better understanding of this important object.

MSC 2010: 58A20, 58A30, 53B05, 53B15

1. Introduction

The theory of connections is an issue that has been extensively covered because of its important applications in geometry and physics. What we bring here is a small contribution to the understanding of its fundamentals, based on the approach of the theory of jets of Weil (see [7, 19, 1]). In essence, the aforementioned theory of jets (in part by continuing the ideas of [21]) is the development of a perspective in which jets (or contact elements, as they are sometimes called) are understood as ideals of the ring of differentiable functions of the manifold of departure and, as much as possible, all the operations are referred to that ring, a bit in the spirit of algebraic geometry and continuing a tradition originated in Plücker and Lie [11]: are points of a manifold, not only the ordinary ones but also the elements of line, surface, etc. This framework has allowed simplify the presentations (for instance, the prolongation of a tangent vector field to jet spaces is, in a sense, the field itself properly understood) and obtain important applications to the theory of PDE, formal integrability, pseudogroups, differential invariants, etc. (see [2, 3, 8, 9, 12, 13, 14, 15, 16, 17]).

On the other hand, the most primitive concept of connection is that of tangent distribution (of a given dimension) and this is identified exactly with the notion of (first order) jet-field. For a given function, the taking of quotient class by the ideal associated with the jet corresponding at each point, define, already here, a notion of covariant differential of that function. If on the distribution we further require transversallity to a given fibred structure or linearity, if we handle a vector bundle, we recover the familiar concepts of connection in a bundle and linear connection in a vector bundle (the same could be done, although we do not develop this possibility here, in the case of principal bundles if we require equivariance under the action of a Lie group, etc.). Definitions of covariant differential or derivative fit in this frame naturally, providing us with a good point of departure.

The description of jets of tangent fields on a manifold M, based on the theory of jets of Weil, allows us to refer all concepts and manipulations to the ring $\mathcal{C}^\infty(M)$ (and its quotient by the powers of the maximal ideal of a given point). For instance, the 1-jet of a tangent field at a point $x \in M$ is thus a derivation of the ring $\mathcal{C}^\infty(M)$ with values in the first order Taylor expansions $\mathcal{C}^\infty(M)/\mathfrak{m}_x^2$. In this way, a connection on the tangent bundle, associate with each tangent vector a derivation of $\mathcal{C}^\infty(M)$ with values in the ring of first order expansions. As an application, we will see that the torsion tensor when applied on two tangent vectors is given by the commutator of the infinitesimal extensions that the connection determines as above mentioned. Also, we will see that the geodesic field appears quite naturally within this framework.

We will discuss the curvature of a connection within the scheme outlined above by changing the usual point of view. This goes as follows. Each general connection (distribution) defines a certain submanifold of the first order jet space (of a fixed dimension m): the image $\nabla(V)$ of the associated jet-field. Two "fields" of tangent m-planes can be naturally defined along that submanifold. First, the distribution itself, once transported to $\nabla(V)$. Second, that produced when we take the "value" of sections of the distribution in the corresponding jet. We will see that, in general, both fields of planes no match and we will prove that they are equal only when the connection is integrable. Therefore, the difference between these fields measures the integrability of ∇; for this purpose we will proceed as follows: given a tangent vector D_v in the distribution at the point v, we consider the difference between $\nabla_* D_v$ and the "value" $D_{\nabla(v)}$ of a section D of the distribution (extending the given vector); the result is a tangent vector $R(D_v)$ at the point $\nabla(v)$, which we will call curvature of the connection at D_v because the vanishing of R implies the integrability of ∇. In the case of a connection on a bundle we consider the operator of two variables: given a tangent vector X on a point x of the base manifold and a point v belonging to the fiber of x, we just will take the value of R on the horizontal lift of X to v.

When a linear connection on the tangent bundle TM is considered, the curvature we have defined above gives an operator depending on two tangent vectors and taking values on the homomorphisms bundle $T^*M \otimes TM$ (from which the Riemann-Christoffel tensor follows). If, then, we take the trace, we obtain the Ricci tensor without the need to first "twist" the order of the indices as in the usual approach. Perhaps this little advantage could suggest a way for better understanding what the Ricci tensor really means.

The structure of the paper is as follows. In Section 2 we give a short account of the main facts about Weil jets theory in the first order case, the only one that we will need for working with connections; in Section 3 we consider general connections as synonymous of tangent distributions and introduce the covariant differential of a function; then, we specialize the above notions to the case of connections on fiber bundles and linear connections on vector bundles and define the covariant differential of a section; in Section 4 we first describe jets of tangent fields in terms of the ring of functions, and then we take advantage of this description in order to consider connections on the tangent bundle; finally, Section 5 is devoted to our approach to the curvature in all the several cases above considered joint with the Ricci tensor of an affine connection.

2. Weil jets

In this section we will give without proofs the more basic ideas and results about Weil jets theory. In addition, we restrict ourselves to the first order case because this will be sufficient for the purposes of this article. For details we refer to [14, 19, 1, 4].

2.1. Weil bundles

A commutative $\mathbb{R}$-algebra is named a Weil algebra if it is finite, local and rational. The most interesting examples for applications are the rings of truncated polynomials

$$\mathbb{R}_m^k := \mathbb{R}[\epsilon_1, \dots, \epsilon_m]/(\epsilon_1, \dots, \epsilon_m)^{k+1},$$

where $\epsilon_1, \ldots, \epsilon_m$ are undetermined variables. Algebras $\mathbb{R}^k_m$ are the models for the rings of Taylor expansions of order k at a point of an m-dimensional manifold. For instance, given an smooth manifold V, the class of a parameterized submanifold of dimensión m at the order k in the point $x \in V$ is given by an $\mathbb{R}$-algebra morphism

$$\mathcal{C}^\infty(V) \to \mathbb{R}^k_m.$$

In general, for a Weil algebra A and a smooth manifold V, the set

$$V^A := \mathrm{Hom}_{\mathbb{R}\text{-alg}}(\mathcal{C}^\infty(V), A),$$

called the *manifold of A-near points* (or simply, *A-points*) of V, is naturally endowed with the structure of a bundle $V^A \to V$ called *Weil bundle* (see [21, 10]).

When $A = \mathbb{R}^k_m$ we will use the notation V^k_m instead of $V^{\mathbb{R}^k_m}$. A particular instance is $A = \mathbb{R}^1_1$ (*dual numbers*); in this case, we recover the tangent bundle: $TV = V^1_1$.

Weil bundles enjoy of a nice property: if A and B are Weil algebras, then $V^{A \otimes B} = (V^A)^B = (V^B)^A$; for instance, if we take $B = \mathbb{R}^1_1$ we get $T(V^A) = (TV)^A$ or, what is the same, for each $\phi \in V^A$ we have

$$T_\phi V^A = \mathrm{Der}_{\mathbb{R}}(\mathcal{C}^\infty(V), A) \tag{2.1}$$

where A is an $\mathcal{C}^\infty(V)$-module via ϕ.

2.2. *Jets of submanifolds*

Let $S \subset V$ be an m-dimensional submanifold of an n-dimensional smooth manifold V. The class of the submanifolds having a contact of order 1 with S at a point $v \in V$ is naturally identified with the ideal

$$j^1_v S := I_S + \mathfrak{m}^2_v \subset \mathcal{C}^\infty(V)$$

where $I_S \subset \mathcal{C}^\infty(V)$ (resp. $\mathfrak{m}_v \subset \mathcal{C}^\infty(V)$) is the ideal attached with S (resp. with v). For this reason we give the following

Definition 2.1. An *1-jet of m-dimensional submanifold* of V is, by definition, an ideal $\mathfrak{p} \subset \mathcal{C}^\infty(V)$ such that

$$\mathcal{C}^\infty(V)/\mathfrak{p} \simeq \mathbb{R}^1_m;$$

equivalently, an 1-jet is the kernel of a surjective (="regular") $\mathbb{R}^1_m$-point. The set of such ideals will be called *space of 1-jets of m-dimensional submanifolds* of V and denoted by $J^1_m V$.

The space $J^1_m V$ can be also seen as the quotient of the set of regular $\mathbb{R}^1_m$-points $\check{V}^1_m$ under the action of the Lie group $G^1_m := \mathrm{Aut}_{\mathbb{R}-\mathrm{alg}}\mathbb{R}^1_m$; so that $\check{V}^1_m \to J^1_m V$ is a principal fiber bundle.

Later on, because there will be no risk of confusion, we will denote by the same letter, say $\mathfrak{p}$, as the ideal $\mathfrak{p} \subset \mathcal{C}^\infty(V)$ as the induced quotient map $\mathfrak{p} \colon \mathcal{C}^\infty(V) \to \mathcal{C}^\infty(V)/\mathfrak{p}$.

For each 1-jet $\mathfrak{p} \in J^1_m V$, there is a unique point $v \in V$, called *source* of $\mathfrak{p}$, such that $\mathfrak{p} \subset \mathfrak{m}_v$; in this way, the space of jets is endowed with a natural projection

$$\pi_1 \colon J^1_m V \to V, \quad \mathfrak{p} \mapsto v \tag{2.2}$$

Coordinates. Let $(U; x_1, \ldots, x_n)$ be a local chart of V. Now, let us consider the open subset $\underline{J}^1_m U$ comprised by those jets $\mathfrak{p} \in J^1_m V$ such that $\mathbb{R}[x_1, \ldots, x_m]/\mathfrak{p} \cap \mathbb{R}[x_1, \ldots, x_m] \simeq \mathbb{R}^1_m$. In addition, put $y_j = x_{m+j}$, $j = 1, \ldots, n-m$. Then, for each $\mathfrak{p} \in \underline{J}^1_m U$, there are numbers $y_{ji}(\mathfrak{p})$ such that $\mathfrak{p}$ is generated by functions

$$y_j - y_j(v) - y_{ji}(\mathfrak{p})(x_i - x_i(v), \quad j = 1, \ldots, n-m, \tag{2.3}$$

joint with $\mathfrak{m}_v^2$. Functions x_i, y_j, y_{ji} thus defined, give a local chart on $J^1_m U$. It follows easily how the whole of $J^1_m V$ may be covered by local charts.

Finally observe that a first order jet $\mathfrak{p} \in J^1_m V$, with source $v \in V$, defines (and is defined by) the m-plane

$$L_\mathfrak{p} := \{D_v \in T_v V \mid D_v \text{ annihilates } \mathfrak{p}\} = \ker(d_v \mathfrak{p}) \subset T_v V \qquad (2.4)$$

2.3. The tangent space at a jet

The following statement is a basic property and allows us to identify tangent vectors at a jet in $J^1_m V$ as (classes of) derivations of $\mathcal{C}^\infty(V)$ with values in $\mathcal{C}^\infty(V)/\mathfrak{p}$. This enforces the idea of considering jets as being also "points" of V besides the ordinary ones.

Theorem 2.2 ([14]). *For each $\mathfrak{p} \in J^1_m V$ the following isomorphism holds,*

$$T_\mathfrak{p} J^1_m V \simeq Der_\mathbb{R}(\mathcal{C}^\infty(V), \mathcal{C}^\infty(V)/\mathfrak{p})/\sim$$

where two derivations $D, \overline{D} \in Der_\mathbb{R}(\mathcal{C}^\infty(V), \mathcal{C}^\infty(V)/\mathfrak{p})$ are considered as equivalents if and only if they coincide when applied on the ideal $\mathfrak{p} \subset \mathcal{C}^\infty(V)$. In particular, $T_\mathfrak{p} J^1_m V$ is a $\mathcal{C}^\infty(V)/\mathfrak{p}$-module.

The above statement follows from (2.1) by taking into account that $J^1_m V$ is the quotient of $\check{V}^1_m$ under the action of the Lie group $\mathrm{Aut}_{\mathbb{R}\text{-alg}}\mathbb{R}^1_m$, by identifying $\mathcal{C}^\infty(V)/\mathfrak{p}$ with $\mathbb{R}^1_m$ and, finally, by computing the tangent space to the orbits. For details, see [1, 14, 19].

In terms of coordinates, the correspondence established in the above theorem is locally given by

$$\left(\frac{\partial}{\partial x_i}\right)_\mathfrak{p} = \left[\frac{\partial}{\partial x_i}\right]_\mathfrak{p}, \quad \left(\frac{\partial}{\partial y_j}\right)_\mathfrak{p} = \left[\frac{\partial}{\partial y_j}\right]_\mathfrak{p}, \quad \left(\frac{\partial}{\partial y_{ji}}\right)_\mathfrak{p} = \left[(x_i - x_i(v))\frac{\partial}{\partial y_j}\right]_\mathfrak{p}, \qquad (2.5)$$

where $[\ \ ]_\mathfrak{p}$ denotes the class modulo equivalence relation $\sim$.

2.4. Value of a derivation at a jet. Prolongation of a tangent field

Given a tangent field on V or, what is the same, if we dispose of a derivation D of the ring $\mathcal{C}^\infty(V)$, for each jet $\mathfrak{p} \in J^1_m V$ we can carry out the composition

$$\mathcal{C}^\infty(V) \xrightarrow{D} \mathcal{C}^\infty(V) \xrightarrow{\mathfrak{p}} \mathcal{C}^\infty(V)/\mathfrak{p} \qquad (2.6)$$

whose class, according Theorem 2.2, defines a tangent vector

$$D_\mathfrak{p} := [\mathfrak{p} \circ D]_\mathfrak{p} \in T_\mathfrak{p} J^1_m(V) \qquad (2.7)$$

We will call $D_\mathfrak{p}$ the *value of D at $\mathfrak{p}$*.

It is not difficult to check that tangent field $\mathfrak{p} \mapsto D_\mathfrak{p}$ thus defined is the usual 1-jet *prolongation*, say D^1, of the tangent field D from V to $J^1_m V$. In this way, prolongation of a field D is D itself but considered as "acting" on another points of V.

2.5. The contact system

We associate to each jet $\mathfrak{p}$ the set of differentials of functions that belong to the ideal $\mathfrak{p}$: The distribution of covectors on $J^1_m V$ defined as

$$\mathfrak{p} \mapsto \Omega_\mathfrak{p} := \pi_1^* d_v \mathfrak{p} = \pi_1^*\{d_v f \in T^*_v V \mid f \in \mathfrak{p}\} \subset T^*_\mathfrak{p} J^1_m V, \quad v = \pi_1(\mathfrak{p}), \qquad (2.8)$$

is called the *contact system* (or *contact distribution*). In a sense, we could say that jet $\mathfrak{p}$ and contact system $\Omega_\mathfrak{p}$ are "the same object". If we take local coordinates x_i, y_j, y_{ji}, as in (2.3), we obtain that the contact system is generated by 1-forms $dy_j - y_{ji}dx_i$, $j = 1, \ldots, n-m$.

Dually, the tangent vectors annihilated by the contact system define the so called *Cartan distribution*

$$\mathfrak{p} \mapsto \mathcal{C}_{\mathfrak{p}} := \{X_{\mathfrak{p}} \in T_{\mathfrak{p}} J^1_m V \mid \pi_{1*} X_{\mathfrak{p}}(f) = 0, \forall f \in \mathfrak{p}\} \subset T_{\mathfrak{p}} J^1_m V \tag{2.9}$$

(so that $\mathcal{C}_{\mathfrak{p}}$ is exactly the set of those tangent vectors $X_{\mathfrak{p}}$ such that $\pi_{1*} X_{\mathfrak{p}} \in L_{\mathfrak{p}}$).

2.6. *Jets of sections*

When V is endowed with an structure of fiber bundle $\pi\colon V \to M$, with $\dim M = m$, we consider the open subset

$$J^1(M, V) := \{\mathfrak{p} \in J^1_m V \mid \text{composition } \mathcal{C}^\infty(M) \hookrightarrow \mathcal{C}^\infty(V) \to \mathcal{C}^\infty(V)/\mathfrak{p} \text{ is surjective}\}$$

called the *space of 1-jets of sections* of $\pi\colon V \to M$.

By the very definition, if $\mathfrak{p} \in J^1(M, V)$ and x is its projection on M, it easily follows the isomorphism

$$\mathcal{C}^\infty(M)/\mathfrak{m}_x^2 \simeq \mathcal{C}^\infty(V)/\mathfrak{p}. \tag{2.10}$$

We see that space $J^1(M, V)$, by means of the above identification, equals the set of surjective maps

$$\mathcal{C}^\infty(V) \to \mathcal{C}^\infty(M)/\mathfrak{m}_x^2,$$

such that, when restricted to $\mathcal{C}^\infty(M)$ became the factor map $\mathcal{C}^\infty(M) \to \mathcal{C}^\infty(M)/\mathfrak{m}_x^2$. Conversely, if σ is a local section of π defined around $x \in M$, the induced morphism from $\mathcal{C}^\infty(V)$ to $\mathcal{C}^\infty(M)$, modulo $\mathfrak{m}_x^2$, gives a jet which is denoted by $j^1_x \sigma$.

The translation of Theorem 2.2 is

$$T_{\mathfrak{p}} J^1(M, V) = \mathrm{Der}_{\mathbb{R}}(\mathcal{C}^\infty(V), \mathcal{C}^\infty(M)/\mathfrak{m}_x^2)/\sim,$$

(recall that $D \sim \overline{D}$ iff $D - \overline{D}$ kills $\mathfrak{p}$). Further analysis proves that the subspace formed by the tangent vectors in $T_{\mathfrak{p}} J^1(M, V)$ which are vertical with respect to projection $\pi_1\colon J^1(M, V) \to V$ are identified with the module of derivations

$$\mathrm{Der}_{\mathcal{C}^\infty(M)}(\mathcal{C}^\infty(V), \mathfrak{m}_x/\mathfrak{m}_x^2).$$

2.7. *Affine structure*

When two jets $\mathfrak{p}, \mathfrak{q} \in J^1(M, V)$ project onto the same point $v \in V$, it can be checked that their difference

$$Z := \mathfrak{q} - \mathfrak{p}\colon \mathcal{C}^\infty(V) \to \mathcal{C}^\infty(M)/\mathfrak{m}_x^2, \quad x := \pi(v), \tag{2.11}$$

is a derivation. Development of this fact leads to the following conclusion: the projection $J^1(M, V) \to V$ is an affine bundle modeled on the the vector bundle $Q \to V$ whose fiber at each $v \in V$ is given by $Q_v = \mathrm{Der}_{\mathcal{C}^\infty(M)}(\mathcal{C}^\infty(V), \mathfrak{m}_x/\mathfrak{m}_x^2)$, where $\mathfrak{m}_x/\mathfrak{m}_x^2$ is considered as a $\mathcal{C}^\infty(V)$-module via an arbitrary jet $\mathfrak{p}$ with $\pi_1(\mathfrak{p}) = v$.

When $\pi\colon V \to M$ is a vector bundle, a section of V^* define a function on V; from this it can be derived that $Q_v = V_x \otimes T^*_x M$. Hence, $J^1(M, V)$ is an affine bundle modeled on the vector bundle $\pi^*(V \otimes T^*M) \to V$.

3. Connections

In this section we discuss the concept of connection in the context of Weil jets theory.

3.1. General connections

We give a quite general definition of (first order) connection by considering it as a synonymous of distribution:

Definition 3.1. A *connection* (of dimension m) on V is a distribution of m-dimensional tangent planes; this is to say, a section

$$V \xrightarrow{\nabla} J^1_m V$$

of the jet bundle projection $J^1_m V \to V$.

For each $v \in V$ we have three equivalent objects:

(i) The 1-jet $\nabla(v)$ which we identify with an **ideal** $\subset \mathcal{C}^\infty(V)$.

(ii) The **quotient map** $\mathcal{C}^\infty(V) \xrightarrow{\nabla(v)} \mathcal{C}^\infty(V)/\nabla(v)$ (which we denote by the same symbol).

(iii) The **m-plane** $L_v = \{D_v \in T_vV \mid D_v\nabla(v) = 0\}$ which is the value at v of the distribution L defined by ∇ (this is the $L_{\nabla(v)}$ according (2.4)).

So that, for each $v \in V$, $\nabla(v)$ is given in local coordinates as $x_i(\nabla(v)) = x_i(v)$, $y_j(\nabla(v)) = y_j(v)$, and

$$y_{ji}(\nabla(v)) = \Gamma^j_i(v),$$

for some locally defined functions $\Gamma^j_i \in \mathcal{C}^\infty(V)$. In this way, $\nabla(v) \subset \mathcal{C}^\infty(V)$ is the ideal

$$\Big(y_j - y_j(v) - \Gamma^j_i(v)(x_i - x_i(v))\Big)_{j=1,\ldots,n-m} + \mathfrak{m}^2_v; \tag{3.1}$$

the quotient map is

$$x_i \longmapsto x_i, \quad y_j \longmapsto y_j(v) + \Gamma^j_i(v)(x_i - x_i(v)), \quad \operatorname{mod} \mathfrak{m}^2_v$$

and the associated plane L_v is the kernel of $d_vy_j - \Gamma^j_i(v)d_vx_i$, $j = 1, \ldots, n-m$; this is to say,

$$L_v = \langle D_{1v}, \ldots, D_{mv}\rangle \quad \text{where} \quad D_{iv} := \left(\frac{\partial}{\partial x_i}\right)_v + \Gamma^j_i(v)\left(\frac{\partial}{\partial y_j}\right)_v . \tag{3.2}$$

On the other hand, as $\nabla(v) \supset \mathfrak{m}^2_v$, we have the factor map

$$\mathfrak{m}_v/\mathfrak{m}^2_v \to \mathfrak{m}_v/\nabla(v).$$

That allows us to give the following

Definition 3.2. The *covariant differential* of a function $F \in \mathcal{C}^\infty(V)$ at a point $v \in V$ is, by definition, the class of $d_vF \in \mathfrak{m}_v/\mathfrak{m}^2_v$ modulo $\nabla(v)/\mathfrak{m}^2_x$ and we will denote it by

$$d^\nabla_v F \in \mathfrak{m}_v/\nabla(v) \simeq L^*_v.$$

Coordinates. Given $F \in \mathcal{C}^\infty(V)$, we have

$$F \equiv F(v) + \left(\frac{\partial F}{\partial x_i}\right)_v (x_i - x_i(v)) + \left(\frac{\partial F}{\partial y_j}\right)_p (y_j - y_j(v)) \quad \operatorname{mod} \mathfrak{m}^2_v.$$

Since the jet $\nabla(v) = (x_i(v), y_j(v), y_{ji} = \Gamma^j_i(v))$ represents the ideal (3.1), we get

$$y_j - y_j(v) \equiv \Gamma^j_i(v)(x_i - x_i(v)) \quad \operatorname{mod}\ \nabla(v),$$

and then, according to definitions,

$$d^\nabla_v F \equiv \left[\left(\frac{\partial F}{\partial x_i}\right)_v + \Gamma^j_i(v)\left(\frac{\partial F}{\partial y_j}\right)_v\right](x_i - x_i(v)) = (D_{iv}F)(x_i - x_i(v)) \quad \operatorname{mod} \nabla(v),$$

where vectors D_{iv} are given as in (3.2).

3.2. Connections on fiber bundles

When V is endowed with an structure of fiber bundle $\pi\colon V\to M$, we will consider distributions

$$\nabla\colon V\to J^1(M,V),$$

so that the distribution defined is transversal to the fibers of π. This kind of object is also known as *jet field* (see, e.g. [20], p. 145).

In this situation, if $v\in V$ is in the fiber of $x\in M$, we have a natural identification $\mathfrak{m}_p/\nabla(p)\simeq\mathfrak{m}_x/\mathfrak{m}_x^2$ (like in (2.10) for $\mathfrak{p}=\nabla(v)$), which allow us to interpret the covariant differential of each function $F\in\mathcal{C}^\infty(V)$ as an ordinary differential form at x:

$$d_v^\nabla F\in\mathfrak{m}_v/\nabla(v)\simeq\mathfrak{m}_x/\mathfrak{m}_x^2=T_x^*M. \tag{3.3}$$

In fact, we can consider the covariant differential of a function restricted to each fiber $V_x:=\pi^{-1}(x)$, for $x\in M$:

$$d_x^\nabla F\colon V_x\longrightarrow T_x^*M;\qquad v\mapsto d_v^\nabla F \tag{3.4}$$

On the other hand, for each $X_x\in T_xM$ and each $v\in V_x$, we will denote by $X_v^\nabla\in T_vV$ the tangent vector defined by composition

$$\mathcal{C}^\infty(V)\xrightarrow{\nabla(v)}\mathcal{C}^\infty(V)/\nabla(v)\simeq\mathcal{C}^\infty(M)/\mathfrak{m}_x^2\xrightarrow{X_x}\mathbb{R}. \qquad \text{(composite arrow labelled } X_v^\nabla\text{)} \tag{3.5}$$

It is easily checked that, 1) By the very definition, $\pi_*X_v^\nabla=X_x$ and 2) $X_v^\nabla\in L_v$, because map $\nabla(v)$ annihilates ideal $\nabla(v)$. Vector X_v^∇ is the only one fulfilling these two conditions. In terms of that "lift", we have

$$\langle d_v^\nabla F,X_x\rangle=\langle d_vF,X_v^\nabla\rangle=X_x(\nabla(v)F). \tag{3.6}$$

3.3. Connections on vector bundles

When $\pi\colon V\to M$ is a vector bundle, we will impose on $\nabla\colon V\to J^1(M,V)$ the condition of being linear. So, for each linear function $F\in\mathcal{C}^\infty(V)$ and each point $x\in M$, the map defined in (3.4) is also linear and then,

$$d_x^\nabla F\in V_x^*\otimes T_x^*M.$$

On the other hand, each section ω of the dual bundle $V^*\to M$, defines a linear function, say F_ω, on V; that allow us to form its covariant differential

$$d_x^\nabla\omega:=d_x^\nabla F_\omega\in V_x^*\otimes T_x^*X,\qquad x\in X.$$

For each tangent vector $D_x\in T_xM$, we define the covariant derivative as usual by the coupling

$$D_x^\nabla\omega:=\langle D_x,d_x^\nabla\omega\rangle\in V_x^*. \tag{3.7}$$

Coordinates. The linearity of a connection is reflected on the coordinate expressions in the following way. Assume that the y_j's are the coordinates defined by a collection of sections α_j of the dual bundle $V^*\to M$; in this case,

$$\Gamma_i^j(x,y)=-\,\Gamma_{ik}^j(x)\,y_k,$$

for suitable local functions $\Gamma_{ik}^j\in\mathcal{C}^\infty(M)$ (the sign minus is traditional).

Journal of Physics: Conference Series **346** (2012) 012001 doi:10.1088/1742-6596/346/1/012001

If $\omega = \sum_k f_k\,\alpha_k$, with $f_k \in \mathcal{C}^\infty(M)$, when is considered as a function on V equals $F_\omega = \sum_k f_k\,y_k$ and then,

$$\left(\frac{\partial^\nabla \omega}{\partial x_i}\right)_x = \left(\left(\frac{\partial f_k}{\partial x_i}\right)_x - \Gamma^j_{ik}(x) f_j(x)\right) y_k;$$

so that

$$d^\nabla_x \omega = \left(\left(\frac{\partial f_k}{\partial x_i}\right)_x - \Gamma^j_{ik}(x) f_j(x)\right) (\alpha_k)_x \otimes d_x x_i.$$

3.4. Covariant differential of a section

Let $\sigma\colon M \to V$ be a section of $\pi\colon V \to M$. For each $x \in M$ we have two jets: $j^1_x\sigma$ and $\nabla(\sigma(x))$. The *covariant differential* of section σ at x is, by definition, the difference

$$d^\nabla_x \sigma := j^1_x\sigma - \nabla(\sigma(x)) \tag{3.8}$$

as a map $\mathcal{C}^\infty(V) \to \mathcal{C}^\infty(M)/\mathfrak{m}^2_x$.

Because both jets, $j^1_x\sigma$ and $\nabla(\sigma(x))$, project onto $\sigma(x)$, we can consider $d^\nabla_x\sigma$ as an element of $V_x \otimes_{\mathbb{R}} T^*_x M$ (see Subsection 2.7). Also, we can define the covariant derivative of σ with respect to a vector $D_x \in T_x M$ as

$$D^\nabla_x \sigma = \langle D_x, d^\nabla_x \sigma\rangle \in V_x \tag{3.9}$$

On the other hand, a section σ is a *solution* of (distribution) ∇ if for each $x \in M$, we have $\sigma_* T_x M = L_{\sigma(x)}$; by taking incident on both members we get

$$j^1_x\sigma = \nabla(\sigma(x));$$

in words: "in an infinitesimal neighborhood of each point $v = \sigma(x)$, σ coincides with the extension $\nabla(\sigma(x))$". So that, almost by the very definition of covariant differential:

Proposition 3.3. *Section σ is a solution of distribution ∇ if and only if $d^\nabla\sigma = 0$.*

Coordinates. Let us suppose that section σ is locally given by $y_j = \sigma_j(x)$; in this case,

$$j^1_x\sigma = (x_i, \sigma_j(x), (\partial\sigma_j/\partial x_i)_x)$$

and

$$\nabla(\sigma(x)) = (x_i, \sigma_j(x), -\sum_k \Gamma^j_{ik}(x)\sigma_k(x)).$$

It follows that

$$d^\nabla_x \sigma = \sum \left(\left(\frac{\partial \sigma_j}{\partial x_i}\right)_x + \Gamma^j_{ik}(x)\sigma_k(x)\right) d_x x_i \otimes \alpha_j.$$

3.5. Relationship between both covariant derivatives

Given a linear connection ∇ on $V \to M$ we have defined covariant derivatives of sections of both V and V^* (see equations (3.9) and (3.7), respectively). They are not unrelated: let σ be a section of $V \to M$, ω a section of $V^* \to M$ and D_x a tangent vector at $x \in M$; according to the definitions, we have

$$\begin{aligned}\langle D^\nabla_x\sigma, \omega(x)\rangle &= D_x((j^1_x\sigma - \nabla(\sigma(x)))F_\omega) = D_x(\sigma^* F_\omega) - D_x(\nabla(\sigma(x))F_\omega) \\ &= D_x\langle\sigma,\omega\rangle - D^\nabla_{\sigma(x)} F_\omega = D_x\langle\sigma,\omega\rangle - \langle\sigma(x), D^\nabla_x\omega\rangle\end{aligned}$$

or

$$D_x\langle\sigma,\omega\rangle = \langle D^\nabla_x\sigma, \omega(x)\rangle + \langle\sigma(x), D^\nabla_x\omega\rangle \tag{3.10}$$

which is the usual relation between the couple of covariant derivatives defined on V and V^*.

Journal of Physics: Conference Series **346** (2012) 012001 doi:10.1088/1742-6596/346/1/012001

4. Connections on tangent bundles

4.1. Jets of tangent fields

Prior to introduce connections on the tangent bundle, we will describe jets $\mathfrak{p} \in J^1(M, TM)$ in a convenient way for our purposes.

We know that such a jet $\mathfrak{p}$ is equivalent to give an $\mathbb{R}$-algebra morphism

$$\mathfrak{p}\colon \mathcal{C}^\infty(TM) \to \mathcal{C}^\infty(M)/\mathfrak{m}_x^2$$

whose restriction to $\mathcal{C}^\infty(M)$ is the natural quotient map.

On the other hand we have the map "differential":

$$\delta\colon \mathcal{C}^\infty(M) \to \mathcal{C}^\infty(TM), \quad f \mapsto \delta f$$

defined by the rule

$$\delta f(D_x) := D_x f, \quad \text{for each } f \in \mathcal{C}^\infty(M), D_x \in TM.$$

Each jet $\mathfrak{p} \in J^1(M, TM)$ is completely determined by its composition with δ, because its restriction to $\mathcal{C}^\infty(M)$ becomes always the quotient map modulo $\mathfrak{m}_x^2$; in this way, we can identify $\mathfrak{p}$ with the derivation

$$\mathfrak{p} \circ \delta\colon \mathcal{C}^\infty(M) \to \mathcal{C}^\infty(M)/\mathfrak{m}_x^2.$$

Conversely, if we have a (local) tangent field D on M, then its jet $j_x^1 D$ is the only jet (thought as a map) such that the following square commutes,

$$\begin{array}{ccc} \mathcal{C}^\infty(M) & \xrightarrow{D} & \mathcal{C}^\infty(M) \\ {\scriptstyle\delta}\downarrow & & \downarrow \\ \mathcal{C}^\infty(TM) & \xrightarrow[j_x^1 D]{} & \mathcal{C}^\infty(M)/\mathfrak{m}_x^2. \end{array}$$

In other words, we have a natural identification

$$J^1(M, TM)_x \simeq \mathrm{Der}_{\mathbb{R}}\left(\mathcal{C}^\infty(M), \mathcal{C}^\infty(M)/\mathfrak{m}_x^2\right) \tag{4.1}$$

(or also, $\simeq \mathrm{Der}_{\mathbb{R}}\left(\mathcal{C}^\infty(M)/\mathfrak{m}_x^3, \mathcal{C}^\infty(M)/\mathfrak{m}_x^2\right)$).

4.2. Affine connections

We now consider connections on the tangent bundle $V = TM \to M$; they are the so known *affine connections* (or sometimes also *linear connections*).

According with the interpretation of jets in $J^1(M, TM)$, a linear connection associates with each vector $D_x \in T_x M$ a derivation $\widetilde{D}_x$ which makes commutative the following diagram:

$$\begin{array}{ccc} \mathcal{C}^\infty(M)/\mathfrak{m}_x^3 & \xrightarrow{\widetilde{D}_x} & \mathcal{C}^\infty(M)/\mathfrak{m}_x^2 \\ \downarrow & & \downarrow \\ \mathcal{C}^\infty(M)/\mathfrak{m}_x^2 & \xrightarrow[D_x]{} & \mathcal{C}^\infty(M)/\mathfrak{m}_x \end{array} \tag{4.2}$$

(vertical arrows represent natural quotient maps).

4.3. An example: Levi-Civita connection

Let g be a non degenerated metric on M. Let us fix a point $x \in M$ and let $0_x \in T_xM$ be the null tangent vector at x. The value of g at x defines a metric g_0 on the vector space T_xM and, then, also on the manifold T_xM. It is possible to determine univocally (up to second order) an isometry around x between (M, g) and (T_xM, g_0) which reduces to identity at the level of tangent spaces: if x_i are local coordinates around x, with $x_i(x) = 0$ and denote by t_i the linear coordinates that they induce on T_xM, a calculation gives

$$\mathcal{C}^\infty(M)/\mathfrak{m}_x^3 \to \mathcal{C}^\infty(T_xM)/\mathfrak{m}_{0_x}^3, \quad x_i \mapsto t_i - \frac{1}{2}\Gamma^i_{hk}t_ht_k \tag{4.3}$$

where Γ^i_{uv} are the well known Christoffel symbols associated to g (at the point x); this is to say: $\partial g_{ik}/\partial x_h - g_{lk}\Gamma^l_{ih} - g_{il}\Gamma^l_{kh} = 0$. Hence, by means of the above identification, both metric are osculating in the sense of Cartan (see [6], p. 90, or [22], p. 268).

Once identification (4.3) is obtained, we can transfer the trivial connection defined on the vector space T_xM to the manifold M: for each $D_x = a_i(\partial/\partial x_i)_x$, $a_i \in \mathbb{R}$, we take its image (under (4.3)) $\overline{D}_{0_x} = a_i(\partial/\partial t_i)_{0_x} \in T_{0_x}(T_xM)$; the trivial connection consist of associating to vector $\overline{D}_{0_x}$ the derivation $\widetilde{\overline{D}_{0_x}} = a_i(\partial/\partial t_i)$ (this is to say, maintaining constant the coefficients). Finally, the derivation that we associate with D_x is the composition

$$\underbrace{\mathcal{C}^\infty(M)/\mathfrak{m}_x^3 \xrightarrow{(4.3)} \mathcal{C}^\infty(T_xM)/\mathfrak{m}_{0_x}^3 \xrightarrow{\widetilde{\overline{D}_{0_x}}} \mathcal{C}^\infty(T_xM)/\mathfrak{m}_{0_x}^2 \xrightarrow{(4.3)^{-1}} \mathcal{C}^\infty(M)/\mathfrak{m}_x^2}_{\widetilde{D}_x},$$

so that $\widetilde{D}_x(x_i) = \widetilde{\overline{D}_{0_x}}(t_i - \frac{1}{2}\Gamma^i_{hk}t_ht_k) = a_i - \Gamma^i_{hk}a_hx_k$.

4.4. Torsion tensor and covariant derivatives

Let $D_{1,x}$ a tangent vector at x and D_2 a vector field (both on M). According the precedent definitions, we have the covariant derivative

$$D^\nabla_{1,x}D_2 = D_{1,x} \circ (j^1_xD_2 - \nabla(D_{2,x})) \tag{4.4}$$

If we follow the definitions and denote by $\widetilde{D}_{2,x}$ the derivation representing the jet $\nabla(D_{2,x})$, then

$$D^\nabla_{1,x}D_2 = D_{1,x} \circ D_2 - D_{1,x} \circ \widetilde{D}_{2,x}. \tag{4.5}$$

On the other hand, following the usual definition, the torsion of ∇ can be computed as

$$\begin{aligned}
\mathrm{Tor}_\nabla(D_{1,x}, D_{2,x}) &= D^\nabla_{1,x}D_2 - D^\nabla_{2,x}D_1 - [D_1, D_2]_x \\
&= D_{1,x} \circ D_2 - D_{1,x} \circ \widetilde{D}_{2,x} - (D_{2,x} \circ D_1 - D_{2,x} \circ \widetilde{D}_{1,x}) - [D_1, D_2]_x \\
&= -D_{1,x} \circ \widetilde{D}_{2,x} + D_{2,x} \circ \widetilde{D}_{1,x} \\
&= -[\widetilde{D}_{1,x}, \widetilde{D}_{2,x}]
\end{aligned}$$

As a consequence, we get a nice interpretation of torsion tensor.

Proposition 4.1. *The value of the torsion when applied on two tangent vectors D_{1x}, D_{2x}, is the commutator of the corresponding jet extensions defined by the connection. This is to say, it* *is the obstruction to the square*

$$\begin{array}{ccc} \mathcal{C}^\infty(M)/\mathfrak{m}_x^3 & \xrightarrow{\widetilde{D}_{1,x}} & \mathcal{C}^\infty(M)/\mathfrak{m}_x^2 \\ {\scriptstyle \widetilde{D}_{2,x}}\downarrow & & \downarrow{\scriptstyle D_{2,x}} \\ \mathcal{C}^\infty(M)/\mathfrak{m}_x^2 & \xrightarrow[D_{1,x}]{} & \mathcal{C}^\infty(M)/\mathfrak{m}_x \simeq \mathbb{R} \end{array} \tag{4.6}$$

to be commutative.

Besides (4.5) there is another covariant derivative, also associated with ∇, which seems to be quite 'natural',

$$D^{\overline{\nabla}}_{1,x} D_2 := [\widetilde{D}_{1,x}, D_2] \tag{4.7}$$

Covariant derivative $D^{\overline{\nabla}}_{1,x} D_2$ is the Lie derivative of D_2 along the jet extension of $D_{1,x}$ defined by the connection.

Covariant derivatives (4.5) and (4.7) differ by the torsion tensor so that they define the *conjugate connection* one of each other (see, for instance, [5], p. 232 or [18], p. 78). From this point of view, we can say that *"Covariant derivatives are also Lie derivatives"*.

4.5. Geodesic field

As another example, we will show here how easily the second order equation of geodesic curves of a connection on TM appears.

With each tangent vector $D_x \in T_xM \subset TM$, connection ∇ associates a jet $\nabla(D_x) \in J^1(M, TM)$; hence, we can perform the composition

$$\mathcal{C}^\infty(TM) \xrightarrow{\nabla(D_x)} \mathcal{C}^\infty(M)/\mathfrak{m}_x^2 \xrightarrow{D_x} \mathbb{R}_1^1 \tag{4.8}$$

(here we consider D_x as a point of Weil bundle $M_1^1 = TM$) which, according to 2.1 is a tangent vector, say $G(D_x)$, in $T_{D_x}TM$. In this way, a tangent field

$$G\colon TM \longrightarrow TTM, \quad D_x \mapsto G(D_x),$$

known as the *geodesic field* associated with connection ∇, is defined (although it depends only on the symmetrized connection).

It is not too difficult to formally check that integral curves of G are the geodesic curves of ∇ (curves with null acceleration). Nevertheless, a calculation in coordinates is an alternative way to show that the assertion in this paragraph is correct.

5. Curvature

We come back to consider general (m-dimensional) connections $\nabla\colon V \to J^1_m V$. Later, we will need the following result.

Lemma 5.1. *Let $D \in Der_{\mathbb{R}}(\mathcal{C}^\infty(V), \mathcal{C}^\infty(V))$ be a section of the distribution L (induced by ∇); then the value $D_{\nabla(v)}$ of D at a jet $\nabla(v) \in J^1_m(V)$ (see Equation (2.7)), depends uniquely on the vector D_v.*

Proof. Let suppose $D_v = 0$ and let, locally, $L = \langle D_1, \ldots, D_m \rangle$, for some vector fields D_i. Each section of L, by construction, kills the ideal $\nabla(v)$ at the point v, so that

$$D_i(\nabla(v)) \subset \mathfrak{m}_v, \quad i = 1, \ldots, m.$$

On the other hand, $D = \sum_i f_i D_i$, for appropriate functions $f_i \in \mathfrak{m}_v$, because D vanishes at v by assumption.

As a consequence, $D(\nabla(v)) \subset \mathfrak{m}_v^2 \subset \nabla(v)$, and, therefore,

$$D_{\nabla(v)} := [\nabla(v) \circ D] = 0 \in T_{\nabla(v)} J^1_m V,$$

(see Theorem 2.2). □

We have accordingly two canonical fields of tangent m-planes along $\nabla(V)$: the distribution L itself transported to $\nabla(V)$ and the distribution H obtained by taking values as in the lema above. Concretely, to each point $\nabla(v) \in \nabla(V) \subset J^1_m V$ we can attach two well defined m-planes contained within $T_{\nabla(v)} J^1_m V$:

(i) $\nabla_* L_v \subseteq T_{\nabla(v)} \nabla(V)$, where L_v is the distribution L at v.

(ii) $H_{\nabla(v)} := \{\text{Set of values at } \nabla(v) \text{ of the sections of } L\}$.

These two m-plans are different. In fact, as we will see, $H_{\nabla(v)}$ can be not tangent to the submanifold $\nabla(V) \subset J^1_m V$.

Computations. Let x_i, y_j local coordinates such that the tangent fields

$$D_i := \frac{\partial}{\partial x_i} + \Gamma^j_i \frac{\partial}{\partial y_j}, \quad i = 1, \ldots, m,$$

span distribution L. A calculation gives

$$\nabla_* (D_i)_v = \left(\frac{\partial}{\partial x_i}\right)_{\nabla(v)} + \Gamma^j_i(v) \left(\frac{\partial}{\partial y_j}\right)_{\nabla(v)} + (D_i)_v (\Gamma^h_r) \left(\frac{\partial}{\partial y_{hr}}\right)_{\nabla(v)}$$

$$(D_i)_{\nabla(v)} = \left(\frac{\partial}{\partial x_i}\right)_{\nabla(v)} + \Gamma^j_i(v) \left(\frac{\partial}{\partial y_j}\right)_{\nabla(v)} + (D_r)_v (\Gamma^h_i) \left(\frac{\partial}{\partial y_{hr}}\right)_{\nabla(v)},$$

where we can see that all is identical except by a permutation on indexes r and i in the last coefficients of both expressions.

In the sequel we will proceed to analyze what this difference intrinsically is.

Lemma 5.2. *Let L be the distribution induced by ∇ and let $\mathfrak{p}$ be a jet in $J^1_m V$; then*

$$\mathfrak{p} \in \nabla(V) \quad \Leftrightarrow \quad \langle \omega_\mathfrak{p}, D_\mathfrak{p} \rangle = 0, \text{ for all } D \in L, \ \omega_\mathfrak{p} \in \Omega_\mathfrak{p},$$

where $\Omega_\mathfrak{p}$ is the contact system at $\mathfrak{p}$ (see Subsection 2.5).

Proof. Let $v \in V$ be the source point of the jet $\mathfrak{p}$ and $D \in L$. Since $\Omega_\mathfrak{p} = \pi_1^* d_v \mathfrak{p}$, we have $\omega_\mathfrak{p} = \pi_1^* d_v f$ for some function $f \in \mathfrak{p}$:

$$\langle \omega_\mathfrak{p}, D_\mathfrak{p} \rangle = \langle d_v f \,, \pi_{1*} D_\mathfrak{p} \rangle = \langle d_v f \,, D_v \rangle = D_v f.$$

From which we derive the equivalence

$$\langle \omega_\mathfrak{p}, D_\mathfrak{p} \rangle = 0, \quad \forall D \in L, \ \omega_\mathfrak{p} \in \Omega_\mathfrak{p} \quad \Leftrightarrow \quad d_v \mathfrak{p} = d_v \nabla(v) \quad \Leftrightarrow \quad \mathfrak{p} = \nabla(v)$$

□

As a consequence we get intrinsic and symbolically the *equations of* $\nabla(V) \subset J^1_m V$:

$$\langle \Omega,\, L^1 \rangle = 0 \tag{5.1}$$

where we denote by L^1 the set of prolongations of vector fields in L to $J^1_m V$, so that $L^1_{\mathfrak{p}} = \{D^1_{\mathfrak{p}} := D_{\mathfrak{p}} \mid D \in L\}$ (here, $D^1_{\mathfrak{p}}$ denotes the tangent vector that vector field D^1 on $J^1_m V$ gives at point $\mathfrak{p}$; instead, $D_{\mathfrak{p}}$ is the value at $\mathfrak{p}$ of a vector field D on the base manifold V according Subsection 2.4).

The previous lemma going to be used to obtain a geometric characterization of the symmetries of the distribution L. The statement is, moreover, that one would expect.

Proposition 5.3. *Let $\overline{D}$ be a tangent vector field on V. Then,*

$$\overline{D} \text{ is a symmetry of } \nabla \quad \Leftrightarrow \quad \overline{D}_{\mathfrak{p}} \in T_{\mathfrak{p}}\nabla(V), \text{ for all } \mathfrak{p} \in \nabla(V).$$

(this is to say, if the prolongation of $\overline{D}$ to $J^1_m V$ is tangent to submanifold $\nabla(V)$).

Proof. Let $D \in L$, then:

$$\overline{D}^1 \langle \Omega,\, D^1 \rangle = \langle L_{\overline{D}^1}\Omega,\, D^1 \rangle + \langle \Omega,\, [\overline{D}^1, D^1] \rangle$$

But, it is well known that prolongation of vector fields to jet spaces holds always the following properties: $L_{\overline{D}^1}\Omega \subset \Omega$ (preserves the contact system) and $[\overline{D}^1, D^1] = [\overline{D}, D]^1$ (defines a Lie algebra morphism); in this way, taking into account Equations (5.1), we have that

$$\overline{D}^1 \langle \Omega,\, D^1 \rangle = \langle \Omega,\, [\overline{D}, D]^1 \rangle$$

holds on $\nabla(V)$. Hence, for every point $\mathfrak{p} = \nabla(v)$ we get

$$\overline{D}^1_{\mathfrak{p}} \langle \Omega,\, D^1 \rangle = \langle \pi_1^* d_v \mathfrak{p},\, [\overline{D}, D]^1_{\mathfrak{p}} \rangle = \langle d_v \nabla(v),\, [\overline{D}, D]_v \rangle.$$

□

Corollary 5.4. *The distribution L induced by ∇ on V is involutive (or completely integrable) if and only if for all $D \in L$ and $\mathfrak{p} \in \nabla(V)$ we have $D_{\mathfrak{p}} \in T_{\mathfrak{p}}\nabla(V)$; that is to say, the jet prolongation of each section of L is tangent to the submanifold $\nabla(V)$.*

Consequently, we can measure the integrability of L (or ∇) by comparing vectors in $\nabla_* L$ and H in the following way,

Definition 5.5. For each $D_v \in L_v$ let us define the *curvature of* ∇ *at* D_v by

$$R(D_v) := \nabla_* D_v - D_{\nabla(v)} \in Q_{\nabla(v)},$$

where $D \in L$ is an arbitrary section extending D_v and $Q \subset TJ^1_m V$ denotes the vertical tangent sub-bundle w.r.t. projection $\pi^1 \colon J^1_m V \to V$.

According to the above corollary, the map of vector bundles $R \colon L \to Q_{|\nabla(V)}$ (on $V \simeq \nabla(V)$) thus defined holds, as the name of curvature given to it suggests, that

$$R = 0 \text{ if and only if } \nabla \text{ is completely integrable.}$$

Local computation. Let x_i, y_j, local coordinates on V such that $D_i := \frac{\partial}{\partial x_i} + \Gamma^j_i \frac{\partial}{\partial y_j}$, $i = 1, \dots, m$, span L. Then

$$R(D_{iv}) = \left(D_{iv}\Gamma^h_r - D_{rv}\Gamma^h_i\right)\left(\frac{\partial}{\partial y_{hr}}\right)_{\nabla(v)} = \left(\frac{\partial \Gamma^h_r}{\partial x_i} + \Gamma^j_i \frac{\partial \Gamma^h_r}{\partial y_j} - \frac{\partial \Gamma^h_i}{\partial x_r} - \Gamma^j_r \frac{\partial \Gamma^h_i}{\partial y_j}\right)_v \left(\frac{\partial}{\partial y_{hr}}\right)_{\nabla(v)}$$

5.1. Case of a fiber bundle
Let ∇ be a connection on the fiber bundle $\pi\colon V \to M$. Now we can take a tangent vector $X \in T_xM$, lift it horizontally up to a point $v \in V_x$ (see (3.5) and then take the curvature; so, we define

$$R(X;v) := R(X_v^{\nabla})$$

If, in addition, π is a vector bundle and ∇ is linear, the same is applied but, now, $R(X;v)$ depends linearly on both arguments, $X \in T_xM$ and $v \in V_x$.

Local computations. Case of a fiber bundle. In a local chart as above,

$$R(\partial/\partial x_i;v) = \left(\left(\frac{\partial}{\partial x_r}\right)_v^{\nabla}\Gamma_i^h - \left(\frac{\partial}{\partial x_i}\right)_v^{\nabla}\Gamma_r^h\right)\left(\frac{\partial}{\partial y_{hr}}\right)_{\nabla(v)}$$

If the connection ∇ depends linearly on the fiber coordinates y_j, then

$$\Gamma_i^j(x,y) = -\Gamma_{ik}^j(x)\, y_k$$

for some functions $\Gamma_{ik}^j \in \mathcal{C}^\infty(M)$ (we continue the tradition with a minus sign); we get

$$R((\partial/\partial x_i)_x;v) = \left(\frac{\partial\Gamma_{rl}^h}{\partial x_i} - \frac{\partial\Gamma_{il}^h}{\partial x_r} + \Gamma_{ij}^h\,\Gamma_{rl}^j - \Gamma_{rj}^h\,\Gamma_{il}^j\right)_x y_l \left(\frac{\partial}{\partial y_{hr}}\right)_{\nabla(v)}$$

whose coefficients contain the familiar expressions of the curvature tensor.

5.2. Case of affine connections. The Ricci tensor
When working with connections on the tangent bundle, the vertical bundle $Q_\mathfrak{p} \subset T_\mathfrak{p}J^1(M,TM)$ is identified with

$$V_x \otimes T_x^*M = T_xM \otimes T_x^*M.$$

So,

$$R(X_x;Y_x) = R(X_{Y_x}^{\nabla}) \in T_xM \otimes T_x^*M,$$

from which we easily could recover the usual Riemann-Christoffel (3,1)-tensor (homomorphism $R(X_x;Y_x)$ must not be confused with the usual commutator of covariant derivatives).

Now, we can consider the natural coupling:

$$c : TM \otimes T^*M \to \mathbb{R}, \quad Z_x \otimes \omega_x \to \langle Z_x\,, \omega_x\rangle.$$

In this way, *Ricci tensor* is directly obtained by taking trace,

$$\mathrm{Ric} := c \circ R, \quad \text{or} \quad \mathrm{Ric}\,(X_x, Y_x) = c\,(R(X_x;Y_x)),$$

as a little calculation would show.

Acknowledgments
The first author thanks the organizers of AGMP 2010, specially to chair A. Stolin and vice-chair E. Paal for their attentions during these days at Tjärno making the stay confortable and fruitful. Also he is highly indebted to professor M. Kureš, head of the section to which this work belongs, for the invitation to participate in the conference, for inspiring conversations and for all the kindness spent with him.

References

[1] Alonso-Blanco R J 2000 Jet manifold associated to a Weil bundle *Arch. Math. (Brno)* **36** 195–9
[2] Alonso-Blanco R J 2003 A transformation of PDE systems related to Drach theorem *J. Math. Anal. Appl.* **288** 530–9
[3] Alonso-Blanco R J, Jiménez S and Rodríguez J 2009 Some canonical structures of Cartan planes in jet spaces and applications *Differential Geometry, Symmetries and Integrability, 5th Abel Symp., Tromø, Norway 2008* ed B Kruglikov et al (Berlin: Springer) 1–20
[4] Alonso-Blanco R J and Muñoz J 2004 The contact system for the A-jet manifolds *Arch. Math. (Brno)* **40** (3) 233–48
[5] Bishop R L and Goldberg S I 1968 *Tensor Analysis on Manifolds* (New York: The Macmillan Co.)
[6] Cartan E 1928 *Leçons sur la Géométrie des Espaces de Riemann* (Paris: Gauthier-Villars)
[7] Jiménez S, Muñoz J and Rodríguez J 2002 On the reduction of some systems of partial differential equations to first order systems with only one unknown function *Proc. of the 8th Int. Conf. Differential Geometry and its Applications, Opava, Czech Republic 2001* ed O Kowalski et al (Opava: Silesian University) pp 187–95
[8] Jiménez S 2011 Symmetries of PDE systems and correspondences between jet spaces *J. Math. Anal. Appl.* **378** 23–41
[9] Jiménez S, Muñoz J and Rodríguez J 2005 Correspondences between jet spaces and PDE systems *J. Lie Theory*, **15** 197–218
[10] Kolář I, Michor P W and Slovák J 1993 *Natural Operations in Differential Geometry* (Berlin: Springer-Verlag)
[11] Lie S 1888 *Theorie der Transformationsgruppen* (Leipzig: Teubner)
[12] Muñoz J, Muriel J and Rodríguez J 1999 The canonical isomorphism between the prolongation of the symbols of a nonlinear Lie equation and its attached linear Lie equation *Proc. of the 7th Int. Conf. Differential Geometry and its Applications, Brno, Czech Republic, 1998* ed I Kolář et al (Brno: Masaryk Univ.) pp 255–61
[13] Muñoz J, Muriel J and Rodríguez J 2000 Integrability of Lie equations and pseudogroups *J. Math. Anal. Appl.* **252** 32–49
[14] Muñoz J, Muriel J and Rodríguez J 2000 Weil bundles and jet spaces *Czechoslovak Math. J.* **50** (125) 721–48
[15] Muñoz J, Muriel J and Rodríguez J 2001 A remark on Goldschmidt's theorem on formal integrability *J. Math. Anal. Appl.* **254** 275–90
[16] Muñoz J, Muriel J and Rodríguez J 2001 The contact system on the (m,l)-jet spaces *Arch. Math.(Brno)* **37** 291–300
[17] Muñoz J, Muriel J and Rodríguez J 2003 On the finiteness of differential invariants *J. Math. Anal. Appl.* **284** 266–82
[18] Nomizu K 1956 *Lie Groups and Differential Geometry* (Tokyo: Math. Soc. of Japan)
[19] Rodríguez J 1990 Sobre los espacios de jets y los fundamentos de la teoría de los sistemas de ecuaciones en derivadas parciales Ph. D. Thesis Universidad de Salamanca
[20] Saunders D J 1989 *The Geometry of Jet Bundles* London Math. Soc. Lecture Note Ser. **142** (Cambridge: Cambridge Univ. Press)
[21] Weil A 1953 Théorie des points proches sur les variétés différentiables *Colloque de Géométrie Différentielle, C.N.R.S.* 111–7
[22] Willmore T J 1959 *An Introduction to Differential Geometry* (New York: Oxford Univ. Press)

Algebra, Geometry, and Mathematical Physics 2010 IOP Publishing
Journal of Physics: Conference Series **346** (2012) 012002 doi:10.1088/1742-6596/346/1/012002

Central dependencies in quantum algebras

Martin Bacovský, Miloslav Havlíček and Severin Pošta

Department of mathematics, FNSPE, Czech Technical University in Prague, Trojanova 13, CZ-120 00 Prague, Czech Republic

E-mail: bacovmar@fjfi.cvut.cz

E-mail: miloslav.havlicek@fjfi.cvut.cz

E-mail: severin.posta@fjfi.cvut.cz

Abstract. Structure of the center of the quantum algebra $U_q(\mathrm{sl}_2)$ is analyzed in detail for both cases when deformation parameter q is/is not a root of unity. It is shown that the result can be obtained in a pure algebraic way without considering representation theory.

1. Introduction

In the past several decades many types of quantum deformations and other similar types of algebras appeared in the literature and shown its importance for various physics fields. For being able to take advantage of such particular structure, one often first needs to develop representation theory of the structure. Detailed knowledge about the properties of the center is important for developing the representation theory. On the example of the simplest quantum group $U_q(\mathrm{sl}_2)$ we show how one can obtain information about the center having no preliminary knowledge about the representation theory. Similar approach can be successful in the case of the other types of algebras (see e. g. [1]).

$U_q(\mathrm{sl}_2)$ is complex associative algebra with generators E, F, K, K^{-1} and relations [2]

$$KK^{-1} = K^{-1}K = 1, \quad KEK^{-1} = q^2E, \quad KFK^{-1} = q^{-2}F, \quad [E,F] = \frac{K-K^{-1}}{q-q^{-1}}. \tag{1}$$

Poincaré-Birkhoff-Witt theorem is valid for this algebra. That means that the elements $E^kF^lK^m$, where k, $l \in \{0,1,...\}$ and $m \in \mathbb{Z}$ constitute vector space basis of the algebra. It can be shown by induction that formulas (1) imply

$$\begin{aligned}
&[E, F^m] = [m]_q F^{m-1}[K; 1-m], \\
&[E^n, F] = [n]_q E^{n-1}[K; n-1], \\
&E^nF^m = \sum_{r=0}^{n} \frac{[n]_q![m]_q!}{[r]_q![n-r]_q![m-r]_q!} F^{m-r}E^{n-r} \times \\
&\qquad \times [K; n-m][K; n-m-1]\dots[K; n-m-r+1], \quad n \le m.
\end{aligned} \tag{2}$$

where $[a]_q = \frac{q^a - q^{-a}}{q - q^{-1}}$, $[m]_q! = [1]_q[2]_q \cdot \ldots \cdot [m]_q$ and $[K;n] = \frac{Kq^n - K^{-1}q^{-n}}{q - q^{-1}}$. Using commutation relations one can easily see that the Casimir element

$$C_q = EF + \frac{q^{-1}K + qK^{-1}}{(q - q^{-1})^2} = FE + \frac{qK + q^{-1}K^{-1}}{(q - q^{-1})^2} \tag{3}$$

lies in the center of the algebra, i. e. it commutes with E, F, K and K^{-1}.

2. Center of $U_q(\mathrm{sl}_2)$ when q is not a root of unity

When q is not a root of unity, the center of the algebra $U_q(\mathrm{sl}_2)$ is generated by the element C_q (see also theorem 45' in sect. 6.3.4 of [2]). We will prove now this fact in purely algebraic way. First let us choose the different vector space basis of $U_q(\mathrm{sl}_2)$, namely the set

$$\{C_q^\alpha E^k K^m \mid \alpha,\, k \geq 0,\ m \in \mathbb{Z}\} \cup \{C_q^\alpha F^k K^m \mid \alpha\ \geq 0, k \geq 1,\ m \in \mathbb{Z}\}. \tag{4}$$

Each element X from the algebra can be then written as a linear combination of the elements from (4), that is

$$X = \sum_{\alpha,k,m} c_{\alpha,k,m} C_q^\alpha E^k K^m + \sum_{\substack{\alpha,k,m \\ k>0}} d_{\alpha,k,m} C_q^\alpha F^k K^m, \tag{5}$$

$c_{\alpha,k,m}$ and $d_{\alpha,k,m}$ being arbitrary complex coefficients. For any element X of the form (5) and for any $s \in \{0, 1, \ldots\}$ let us define

$$\deg_s X = \sum_{\substack{\alpha,k,m \\ 2\alpha+k+|m|=s}} c_{\alpha,k,m} C_q^\alpha E^k K^m + \sum_{\substack{\alpha,k,m \\ 2\alpha+k+|m|=s \\ k>0}} d_{\alpha,k,m} C_q^\alpha F^k K^m$$

and also

$$\operatorname{maxdeg} X = \max\{2\alpha + k + |m| \mid c_{\alpha,k,m} \neq 0 \text{ or } d_{\alpha,k,m} \neq 0\}.$$

We have, for example,

$$\begin{aligned}
&\operatorname{maxdeg} 0 = -\infty, \\
&\operatorname{maxdeg} 1 = 0, \\
&\operatorname{maxdeg} E = \operatorname{maxdeg} F = \operatorname{maxdeg} K = \operatorname{maxdeg} K^{-1} = 1, \\
&\operatorname{maxdeg} C_q = 2, \\
&\operatorname{maxdeg}(E^2 + 2EK^{-3} - C_q F^2) = 4,
\end{aligned}$$

etc. Now let X belong to the center of the algebra. We want to show that the commutation relations

$$[X, E] = 0, \quad [X, F] = 0, \quad [X, K] = 0$$

imply

$$c_{\alpha,k,m} = 0 \quad \text{for all } (k, m) \neq (0, 0) \quad \text{and} \quad d_{\alpha,k,m} = 0 \quad \text{for all } k > 0.$$

Let us remove from X the terms for which $(k, m) = (0, 0)$, denote

$$X' = \sum_{\substack{\alpha,k,m \\ (k,m)\neq(0,0)}} c_{\alpha,k,m} C_q^\alpha E^k K^m + \sum_{\substack{\alpha,k,m \\ k>0}} d_{\alpha,k,m} C_q^\alpha F^k K^m. \tag{6}$$

We want to show $X' = 0$. We have

$$[X', K] = \sum_{\substack{\alpha,k,m \\ (k,m)\neq(0,0)}} c_{\alpha,k,m} C_q^\alpha [E^k, K] K^m + \sum_{\substack{\alpha,k,m \\ k>0}} d_{\alpha,k,m} C_q^\alpha [F^k, K] K^m.$$

Let us assume $s = \operatorname{maxdeg} X' > 0$. It is sufficient to show that $\deg_s X' = 0$ (contradiction). By commuting with generators E, F, $K^{\pm 1}$ the maximum degree of X' can rise up to $s+1$. By induction we have from (1) for any $m \in \mathbb{Z}$

$$K^m E = q^{2m} E K^m, \quad K^m F = q^{-2m} F K^m.$$

Similarly for all $k \geq 0$ we have

$$KE^k = q^{2k} E^k K, \quad KF^k = q^{-2k} F^k K. \tag{7}$$

Hence

$$\begin{aligned} &[K^m, E] = (q^{2m} - 1) E K^m, && [K^m, F] = (q^{-2m} - 1) F K^m, \\ &[E^k, K] = (1 - q^{2k}) E^k K, && [F^k, K] = (1 - q^{-2k}) F^k K. \end{aligned} \tag{8}$$

Now using (8) we have

$$\begin{aligned} \deg_{s+1}[X', K] &= \deg_{s+1} \sum_{\substack{\alpha,k,m \\ (k,m)\neq(0,0)}} c_{\alpha,k,m} C_q^\alpha [E^k, K] K^m + \sum_{\substack{\alpha,k,m \\ k>0}} d_{\alpha,k,m} C_q^\alpha [F^k, K] K^m \\ &= \sum_{\substack{\alpha,k,m \\ 2\alpha+k+|m|=s \\ k>0}} c_{\alpha,k,m} (1 - q^{2k}) C_q^\alpha E^k K^{m+1} + \\ &\quad \sum_{\substack{\alpha,k,m \\ 2\alpha+k+|m|=s \\ k>0}} d_{\alpha,k,m} (1 - q^{-2k}) C_q^\alpha F^k K^{m+1}. \end{aligned} \tag{9}$$

Because the elements in the sum (9) are linear independent, from the condition $\deg_{s+1}[X', K] = 0$ we get the system of linear equations

$$c_{\alpha,k,m} = d_{\alpha,k,m} = 0 \text{ for } 2\alpha + k + |m| = s, \; k > 0. \tag{10}$$

Applying the conditions (10) to (6) we get

$$\deg_s X' = \sum_{\substack{\alpha,m \\ 2\alpha+|m|=s \\ m>0}} c_{\alpha,0,m} C_q^\alpha K^m. \tag{11}$$

To eliminate the remaining coefficients in (11) we commute X' with the generator E. Using (7) we obtain

$$\deg_{s+1}[X', E] = \deg_{s+1} \sum_{\substack{\alpha,m \\ 2\alpha+|m|=s \\ m>0}} c_{\alpha,0,m} C_q^\alpha [K^m, E] = \sum_{\substack{\alpha,m \\ 2\alpha+|m|=s \\ m>0}} c_{\alpha,0,m} (q^{2m} - 1) C_q^\alpha E K^m.$$

This way we obtain several new equations:

$$c_{\alpha,0,m} = 0 \text{ for } 2\alpha + |m| = s, \; m > 0. \tag{12}$$

Putting the conditions (12) into (11) we get $\deg_s X' = 0$. as desired. Therefore $\operatorname{maxdeg} X' < s$ which is contradiction to the assumption. Hence $X' = 0$. This leads to the following theorem.

Theorem 1. Let $q \in \mathbb{C}$, $q \neq 0$, q is not root of unity. The center of the algebra $U_q(\mathrm{sl}_2)$ defined by the relations (1) is a free polynomial ring in variable C_q defined by (3).

3. Center of $U_q(\mathrm{sl}_2)$ when q is a root of unity

When q is a primitive root of unity (see also [4]), $q^n = 1$, $n > 2$, $q^j \neq 1$ for all $j \in \{1, 2, ..., n-1\}$, one can easily see from (2) and (8) that center of the algebra $U_q(\mathrm{sl}_2)$ contains besides the element C_q additional four Casimir elements, namely

$$C_E = E^{n'}, \ C_F = F^{n'}, \ C_K^{\pm 1} = K^{\pm n'}, \tag{13}$$

where $n' = \frac{n}{2}$ when n is even and $n' = n$ when n is odd. The set

$$\{C_E^\alpha C_F^\beta C_K^\gamma E^k F^l K^m \mid \alpha,\ \beta \geq 0,\ \gamma \in \mathbb{Z},\ 0 \leq k,\ l,\ m \leq n'-1\}$$

clearly constitutes new vector space basis of $U_q(\mathrm{sl}_2)$. Let us denote

$$Z = \mathbb{C}[C_E, C_F, C_K^{\pm 1}]$$

commutative polynomial ring in variables C_E, C_F, $C_K^{\pm 1}$. $U_q(\mathrm{sl}_2)$ can then be seen as a finite-dimensional module over the ring Z with the dimension equal to n'^3 and a basis

$$\{E^k F^l K^m \mid 0 \leq k,\ l,\ m \leq n'-1\}.$$

Trivial corollary of this fact is that any irreducible representation of $U_q(\mathrm{sl}_2)$ is finite dimensional (see also corollary 16, p. 67 in [2]).

This module is also Z-spanned by the following n'^3 elements:

$$\begin{array}{ll} C_q^\delta E^k K^m, & 0 \leq \delta \leq n'-1,\ 0 \leq k \leq n'-1-\delta,\ 0 \leq m \leq n'-1, \\ C_q^\delta F^l K^m, & 0 \leq \delta \leq n'-1,\ 1 \leq l \leq n'-1-\delta,\ 0 \leq m \leq n'-1. \end{array} \tag{14}$$

That means that we have also new vector space basis of the whole algebra $U_q(\mathrm{sl}_2)$ consisting of the following elements:

$$\begin{array}{lr} C_E^\alpha C_F^\beta C_K^\gamma C_q^\delta E^k K^m, & \alpha,\ \beta \geq 0,\ \gamma \in \mathbb{Z},\ 0 \leq \delta \leq n'-1, \\ & 0 \leq k \leq n'-1-\delta,\ 0 \leq m \leq n'-1, \\ C_E^\alpha C_F^\beta C_K^\gamma C_q^\delta F^l K^m, & \alpha,\ \beta \geq 0,\ \gamma \in \mathbb{Z},\ 0 \leq \delta \leq n'-1, \\ & 1 \leq l \leq n'-1-\delta,\ 0 \leq m \leq n'-1. \end{array} \tag{15}$$

Because the element $C_q^{n'}$ does not belong to the basis (14) and it is of course element from the center of the algebra $U_q(\mathrm{sl}_2)$, this implies algebraic dependency between C_q and the elements C_E, C_F and $C_K^{\pm 1}$. For each $k = 1, 2, ...$ we have

$$E^k F^k = \prod_{m=0}^{k-1} \left(C_q - \frac{q^{-2m-1}K + q^{2m+1}K^{-1}}{(q-q^{-1})^2} \right), \tag{16}$$

which is easily shown by induction: with the help of (8) one has

$$E^{k+1}F^{k+1} = E^k E F F^k = E^k \left(C_q - \frac{q^{-1}K + qK^{-1}}{(q-q^{-1})^2} \right) F^k = E^k F^k \left(C_q - \frac{q^{-1-2k}K + q^{1+2k}K^{-1}}{(q-q^{-1})^2} \right).$$

When we put $k = n'$ in (16) we obtain

$$C_E C_F = \prod_{m=0}^{n'-1} \left(C_q - \frac{q^{-2m-1}K + q^{2m+1}K^{-1}}{(q-q^{-1})^2} \right). \tag{17}$$

When we expand righthand side of (17), we find it is free of any powers of K^j, $j = -n'+1,\dots,-1,1,\dots,n'-1$. Really, for any complex x we have

$$\prod_{m=0}^{n'-1}(q^{2m+1}-x) = 1+(-1)^n x^{n'}, \tag{18}$$

because the polynomials on both sides of (18) have the same roots q, q^3, ..., q^{n-1}.

For any nonzero complex a, b we have

$$\begin{aligned} a-\frac{q^{-2m-1}b+q^{2m+1}b^{-1}}{(q-q^{-1})^2} &= -aq^{-2m-1}\left(q^{2(2m+1)}\tilde{a}b^{-1}-q^{2m+1}+\tilde{a}b\right)\\ &= -aq^{-2m-1}\tilde{a}b^{-1}\left(q^{2(2m+1)}-\tilde{a}^{-1}bq^{2m+1}+b^2\right)\\ &= -aq^{-2m-1}\tilde{a}b^{-1}\left(q^{2m+1}-b\frac{1-\sqrt{1-4\tilde{a}^2}}{2\tilde{a}}\right)\\ &\times\left(q^{2m+1}-b\frac{1+\sqrt{1-4\tilde{a}^2}}{2\tilde{a}}\right) \end{aligned} \tag{19}$$

where

$$\tilde{a} = \frac{1}{a(q-q^{-1})^2}.$$

Multiplying (19) with the help of (18) and binomial theorem we obtain

$$\begin{aligned} &\prod_{m=0}^{n'-1}\left(a-\frac{q^{-2m-1}b+q^{2m+1}b^{-1}}{(q-q^{-1})^2}\right) = \\ &(-1)^{n'}a^{n'}\prod_{m=0}^{n'-1}\left(q^{-2m-1}\right)\tilde{a}^{n'}b^{-n'}\left(1+(-1)^n\left(b\frac{1-\sqrt{1-4\tilde{a}^2}}{2\tilde{a}}\right)^{n'}\right)\times \\ &\left(1+(-1)^n\left(b\frac{1+\sqrt{1-4\tilde{a}^2}}{2\tilde{a}}\right)^{n'}\right) = \\ &a^{n'}\left(b^{-n'}\tilde{a}^{n'}+\frac{(-1)^n}{2^{n'}}\left(\left(1-\sqrt{1-4\tilde{a}^2}\right)^{n'}+\left(1+\sqrt{1-4\tilde{a}^2}\right)^{n'}\right)+b^{n'}\tilde{a}^{n'}\right) = \\ &\frac{1}{b^{n'}(q-q^{-1})^{2n'}}+\frac{(-1)^n}{2^{n'-1}}\sum_{m=0}^{[n'/2]}\binom{n'}{2m}a^{n'-2m}\frac{(a^2(q-q^{-1})^4-4)^m}{(q-q^{-1})^{4m}}+\frac{b^{n'}}{(q-q^{-1})^{2n'}}. \end{aligned}$$

Therefore, the relation (17) one can rewrite as

$$C_E C_F = \frac{C_K+C_K^{-1}}{(q-q^{-1})^{2n'}}+\frac{(-1)^n}{2^{n'-1}}\sum_{m=0}^{[n'/2]}\binom{n'}{2m}C_q^{n'-2m}\frac{(C_q^2(q-q^{-1})^4-4)^m}{(q-q^{-1})^{4m}} \tag{20}$$

Using the combinatorial identity

$$\sum_{m=l}^{[n/2]}\binom{n}{2m}\binom{m}{l} = \frac{m\cdot 2^{n-2l-1}}{n-l}\binom{n-l}{n-2l},$$

we can rewrite (20) to the form

$$C_E C_F = \frac{C_K + C_K^{-1}}{(q-q^{-1})^{2n'}} + (-1)^n n' \sum_{l=0}^{[n'/2]} \frac{(-1)^l (q-q^{-1})^{-4l}}{n'-l} \binom{n'-l}{n'-2l} C_q^{n'-2l}.$$

As a result we obtain there exists a complex polynomial P in five variables such that

$$P(C_E, C_F, C_K, C_K^{-1}, C_q) = 0,$$

which explicit form is

$$P(x,y,z,\tilde{z},w) = -xy + \frac{z+\tilde{z}}{(q-q^{-1})^{2n'}} + (-1)^n n' \sum_{l=0}^{[n'/2]} \frac{(-1)^l (q-q^{-1})^{-4l}}{n'-l} \binom{n'-l}{n'-2l} w^{n'-2l}. \quad (21)$$

Any other polynomial relation Q between elements C_E, C_F, $C_K^{\pm 1}$ and C_q can be written in the form

$$Q(x,y,z,\tilde{z},w) = \alpha(x,y,z,\tilde{z},w) P(x,y,z,\tilde{z},w) + R(x,y,z,\tilde{z},w),$$

where α and R are polynomials and the degree of R is less than n' with respect to variable w. The condition

$$Q(C_E, C_F, C_K, C_K^{-1}, C_q) = 0,$$

implies

$$R(C_E, C_F, C_K, C_K^{-1}, C_q) = 0.$$

This together with linear independence of C_q^j for $j = 0, 1, \ldots, n'-1$ gives

$$R(x,y,z,\tilde{z},w) \equiv 0.$$

Any other relation between Casimir elements is then equal to the P multiplied by a polynomial.

Our final task is to show that the center of $U_q(\mathrm{sl}_2)$ has a basis

$$\{C_E^\alpha C_F^\beta C_K^\gamma C_q^\delta \mid \alpha, \beta \geq 0,\ \gamma \in \mathbb{Z},\ 0 \leq \delta \leq n'-1\}, \quad (22)$$

i. e. there is no other algebraically independent Casimir element. Let us write an element X from $U_q(\mathrm{sl}_2)$ that is not in the span of (22) in the basis (15)

$$X = \sum_{\delta,k,m} \alpha_{\delta,k,m} C_q^\delta E^k K^m + \sum_{\delta,l,m} \beta_{\delta,l,m} C_q^\delta F^l K^m,$$

where we sum through

$$0 \leq \delta \leq n'-1,\ 0 \leq k \leq n'-\delta-1,\ 1 \leq l \leq n'-\delta-1,\ 0 \leq m \leq n'-1, \quad (23)$$

and where $\alpha_{\delta,k,m}$ and $\beta_{\delta,l,m}$ are arbitrary polynomials in elements C_E, C_F and $C_K^{\pm 1}$ except $\alpha_{\delta,0,0}$ which are identically zero. Commuting X with the element K we obtain

$$[X, K] = \sum_{\delta,k,m} \alpha_{\delta,k,m} (1-q^{2k}) C_q^\delta E^k K^{m+1} + \sum_{\delta,l,m} \beta_{\delta,l,m} (1-q^{-2l}) C_q^\delta F^l K^{m+1}.$$

Considering the bounds (23) only those coefficients $1-q^{2k}$ can be equal to zero for which $k=0$. Thus all polynomials $\alpha_{\delta,k,m}$ and $\beta_{\delta,l,m}$ must be identically zero except $\alpha_{\delta,0,m}$. X has now the form

$$X = \sum_{\delta,m} \alpha_{\delta,0,m} C_q^\delta K^m.$$

Commuting with E we obtain

$$[X, E] = \sum_{\delta,m} \alpha_{\delta,0,m}(1 - q^{2m}) C_q^\delta E K^m = 0,$$

which implies

$$\alpha_{\delta,0,m} = 0 \quad \text{for all } 0 \le \delta \le n' - 1,\ 1 \le m \le n' - 1.$$

This yields $X = 0$, which proves that (22) is the vector space basis of the center of the algebra $U_q(\mathrm{sl}_2)$. Thus we have shown the following.

Theorem 2. Let $q \in \mathbb{C}$ be a primitive root of unity, i. e. $q^n = 1$, $n > 2$, $q^j \neq 1$ for all $j \in \{1, 2, ..., n-1\}$. The center of the algebra $U_q(\mathrm{sl}_2)$ defined by the relations (1) is generated by the elements C_q defined by (3) and C_E, C_F and $C_K^{\pm 1}$ defined by (13) and satisfy the relation $P(C_E, C_F, C_K, C_K^{-1}, C_q) = 0$ where P is given by (21).

Acknowledgments

We acknowledge financial support provided by Czech Technical University grant SGS10/210/OHK4/2T/14.

References

[1] Havlicek M and Posta S 2010 *Proc. Int. Conf. XXIX Workshop on Geometric Methods in Physics* ed V Buchstaber, A Odzijewicz et al (New York: AIP) pp 125-130

[2] Klimyk A U and Schmüdgen K 1997 *Quantum groups and their representations* (Berlin: Springer)

[3] Bergman G M 1978 *Adv. Math.* **29** 178

[4] Concini C and Kac V G 1990 *Operator algebras, unitary representations, enveloping algebras, and invariant theory* (Boston: Birkhäuser)

Categorical Non-commutative Geometry

Paolo Bertozzini[1], **Roberto Conti**[2], **Wicharn Lewkeeratiyutkul**[3]

[1] Department of Mathematics and Statistics, Faculty of Science and Technology, Thammasat University, Pathumthani 12121, Thailand
[2] Dipartimento di Scienze, Università di Chieti-Pescara "G. D'Annunzio", Viale Pindaro 42, I-65127 Pescara, Italy
[3] Department of Mathematics, Faculty of Science, Chulalongkorn University, Bangkok 10330, Thailand

E-mail: paolo.th@gmail.com, conti@sci.unich.it, Wicharn.L@chula.ac.th

Abstract. The purpose of this short note is to outline the current status of some recent research programs aiming at a categorification of parts of A.Connes non-commutative geometry and to provide an outlook on some possible future developments in categorical non-commutative geometry.

1. Introduction

Category theory has been applied to operator algebraic settings since many years ago, probably starting around 1980 with the pioneering work of John Roberts in algebraic quantum field theory [20] and since then it has been constantly used in the theory of superselection sectors.

As far as we know, apart from a couple of independent proposals by A.Sitarz [28, section 3.2] and Y.Manin [23], the study of "categorical non-commutative geometry", in the setting of A.Connes' spectral triples, started around 2002-2003 as a by-product of our research project "Modular Spectral Triples in Non-commutative Geometry and Physics" (Thai Research Fund Grant: RSA4580030). There, in order to identify a non-commutative configuration space from a non-commutative phase-space obtained by Tomita-Takesaki modular theory, a kind of polarization was necessary and for this purpose we were motivated to introduce a definition of sub-object and study the most elementary notion of morphism of spectral triples [1].

The research on categorical non-commutative geometry soon started to become one of the main areas of our activity as documented in the survey paper [2] that can still be considered a fairly good introduction to the subject.

Apart from the study of alternative simple notions of morphism of spectral triples [3], we started a project of "categorification" of A.Connes non-commutative geometry. Around 2006 we introduced the terms "horizontal categorification" and "vertical categorification" in order to distinguish the categorical "many-object version" of usual mathematical concepts from the more demanding "higher-morphism" counterpart and we concentrated in proving a horizontal categorified version of Gel'fand-Naĭmark duality for commutative full C*-categories, a result that has recently appeared in [5]. The spectrum of such a C*-category consists of a specific Fell line-bundle that we call "topological spaceoid".

The horizontal categorification of Krein-C*-algebras (essentially categories of bounded linear operators between complete semi-definite linear spaces) has also been investigated in [8].

In the meantime categorical non-commutative geometry (in A.Connes' sense) has been the subject of more and more investigations at different levels of technical sophistication by different authors:

- A. Connes, C. Consani, M. Marcolli in the fist part of their paper [13] describe two possible general approaches for the definition of categories of spectral triples, suggestions that are carried further by A. Connes, M. Marcolli, in their recent book [14],
- B. Mesland [25] introduced what is in our opinion the most sophisticated notion of category of spectral triples, based on "smooth correspondences" (certain specific KK-bimodules),
- R. Dawe Martins in [15, 16, 17] proposed other generalizations of the notion of spectral triple that are based on Fell bundles.

Research in categorical non-commutative geometry (as already claimed in our original motivation) seems to be of particular interest in all the attempts to provide a formulation of algebraic quantum gravity (see [4] and also [14, 19, 24, 17]).

2. Morphisms of Spectral Triples

Recall[1] that a (naive) spectral triple $(\mathcal{A}, \mathcal{H}, D)$ is given by a pre-C*-algebra $\mathcal{A}$ faithfully represented via bounded operators on a Hilbert space $\mathcal{H}$ and a possibly unbounded self-adjoint "Dirac" operator D on $\mathcal{H}$ that has compact resolvent and commutators $[D, \pi(x)]_-$, for all $x \in \mathcal{A}$, bounded on a common dense domain in $\mathcal{H}$. Typical examples of spectral triples originating in differential geometry are the Atiyah-Singer spectral triples of a compact orientable Riemannian spinorial manifold M, where $\mathcal{A} := C^\infty(M)$ is the algebra of smooth complex-valued functions on M represented by left multiplication on the Hilbert space $\mathcal{H} := L^2(S(M))$ of square integrable section of a spinor bundle on M with D given by the usual Pauli-Atiyah-Singer Dirac operator. Compact orientable Riemannian manifolds can also be described by spectral triples taking $\mathcal{A} := C^\infty(M)$ represented by left multiplication on the space $\mathcal{H} := L^2(\Lambda_\bullet(M))$ of sections of the Grassmann bundle of M with Dirac operator $D := d + d^*$.

2.1. Totally Geodesic Morphisms

In our first paper on this subject [1] we proposed this notion of morphism: given two spectral triples $(\mathcal{A}_j, \mathcal{H}_j, D_j)$, with $j = 1, 2$, a **morphism of spectral triples** is a pair (ϕ, Φ), where $\phi : \mathcal{A}_1 \to \mathcal{A}_2$ is a $*$-homomorphism between the pre-C*-algebras $\mathcal{A}_1, \mathcal{A}_2$ and $\Phi : \mathcal{H}_1 \to \mathcal{H}_2$ is a bounded linear map that "intertwines" the representations $\pi_1, \pi_2 \circ \phi$ and the Dirac operators D_1, D_2 i.e.: $\pi_2(\phi(x)) \circ \Phi = \Phi \circ \pi_1(x)$, $\forall x \in \mathcal{A}_1$, $D_2 \circ \Phi(\xi) = \Phi \circ D_1(\xi)$, $\forall \xi \in \mathrm{Dom}\, D_1$.

This definition of morphism clearly implies a strong relationship between the spectra of the Dirac operators of the two spectral triples. Loosely speaking, for ϕ epi and Φ coisometric (respectively mono and isometric), in the case of Atiyah-Singer spectral triples, one should expect such definition to become relevant only for maps that "preserve the geodesic structures" (totally geodesic immersions and respectively totally geodesic submersions).[2] Furthermore these morphisms depend, at least in some sense, on the spin structures: this "spinorial rigidity" (at least in the case of morphisms of real even spectral triples, when we also impose intertwining conditions between Φ and the real structures J_j and the gradings Γ_j) requires that such morphisms between spectral triples of different dimensions might be possible only when the difference in dimension is a multiple of 8.

[1] For basic background on A.Connes' definition of spectral triples we refer to [10, 21, 31, 9].

[2] Bertozzini P, Conti R, Lewkeeratiyutkul W, Non-commutative Totally Geodesic Submanifolds and Quotient Manifolds, work in progress.

2.2. Metric Morphisms

A notion of morphism that is essentially blind to the spin structures has been proposed in [5] where it has been used to prove a refined version of Gel'fand duality for Atiyah-Singer spectral triples and metric isometries of spinorial manifolds. Given two spectral triples $(\mathcal{A}_j, \mathcal{H}_j, D_j)$, with $j = 1, 2$, denote by $d_{D_j}(\omega_1, \omega_2) := \sup\{|\omega_1(x) - \omega_2(x)| \mid x \in \mathcal{A}, \ \|[D_j, \pi(x)]_-\| \leq 1\}$ the quasi-distance induced on the sets $\mathcal{P}(\mathcal{A}_j)$ of pure states of $\mathcal{A}_j$. A **metric morphism** of spectral triples is a unital epimorphism[3] $\phi : \mathcal{A}_1 \to \mathcal{A}_2$ of pre-C*-algebras whose pull-back $\phi^\bullet : \mathcal{P}(\mathcal{A}_2) \to \mathcal{P}(\mathcal{A}_1)$, $\phi^\bullet(\omega) := \omega \circ \phi$ is an isometry, i.e. $d_{D_1}(\phi^\bullet(\omega_1), \phi^\bullet(\omega_2)) = d_{D_2}(\omega_1, \omega_2)$, for all $\omega_1, \omega_2 \in \mathcal{P}(\mathcal{A}_2)$.

2.3. Riemannian Morphisms

A weaker notion of metric morphisms (that in the case of isomorphisms reduces to the unitary maps considered in [27]) and that for Atiyah-Singer spectral triples should reproduce the usual situation of Riemannian immersions and submersions of spinorial manifolds is as follows:[4] given two spectral triples $(\mathcal{A}_j, \mathcal{H}_j, D_j)$, with $j = 1, 2$, a **Riemannian morphism** is a pair (ϕ, Φ) where $\phi : \mathcal{A}_1 \to \mathcal{A}_2$ is a $*$-homomorphism between the pre-C*-algebras $\mathcal{A}_1, \mathcal{A}_2$ and $\Phi : \mathcal{H}_1 \to \mathcal{H}_2$ is a bounded linear map that "intertwines" the representations $\pi_1, \pi_2 \circ \phi$ and the commutators of the Dirac operators D_1, D_2: $\pi_2(\phi(x)) \circ \Phi = \Phi \circ \pi_1(x)$, $\forall x \in \mathcal{A}_1$, $[D_2, \pi_2(\phi(x))]_- \circ \Phi = \Phi \circ [D_1, \pi_1(x)]_-$, $\forall x \in \mathcal{A}_1$. Note that the boundedness of Φ (here as well as in the case of totally geodesic morphisms) can actually be weakened, considering unbounded operators, the important property here being the fact that the adjoint action of Φ on the algebra $\Omega_D(\mathcal{A})$ generated by $\pi(\mathcal{A})$ and the commutators $[D, \pi(x)]_-$, $x \in \mathcal{A}$ (the "non-commutative Clifford algebra") is a $*$-homomorphism extending ϕ.

2.4. Morita Morphisms of Spectral Triples

All of the several definitions of morphisms considered above have been essentially modelled on the case of commutative algebras of functions, where $*$-homomorphisms are abundant, and although they still make sense in the non-commutative case, they correspond to quite special "maps" of non-commutative spaces. In a wider perspective, a morphism of spectral triples $(\mathcal{A}_j, \mathcal{H}_j, D_j)$, for $j = 1, 2$, should be formalized as a "suitable" functor $\mathcal{F} : {}_{\mathcal{A}_1}\mathcal{M} \to {}_{\mathcal{A}_2}\mathcal{M}$, between the categories ${}_{\mathcal{A}_j}\mathcal{M}$ of $\mathcal{A}_j$-modules, having "appropriate intertwining" properties with the Dirac operators D_j. Under some "mild" hypothesis, by Eilenberg-Gabriel-Watt theorem, any such functor is given by "tensorization" with a bimodule. These bimodules, suitably equipped with spectral data (as in the case of spectral triples), will provide the natural setting for a general theory of morphisms of non-commutative spaces. This "Morita morphism" point of view has been first advocated by Y.Manin [23], but it is had already been implicitly exploited in A.Connes' "transfer" of Dirac operators via Morita equivalence bimodules equipped with a connection [11, 12].

In [2] we also noticed the construction of a strictly related category of **Morita-Connes morphisms** of spectral triples (containing A.Connes' "transfers and inner deformations" as isomorphisms) based on the choice of a connection on a Morita morphism (that is not necessarily an imprimitivity bimodule) i.e.: a left-$\mathcal{A}_2$ right-$\mathcal{A}_1$ bimodule that is a Hilbert C*-module over $\mathcal{A}_1$, a Hermitian connection[5] $\nabla : X \to X \otimes_{\mathcal{A}_1} \Omega^1_{D_1}(\mathcal{A}_1)$ on the bimodule X (the Dirac operators on the spectral triples $(\mathcal{A}_j, \mathcal{H}_j, D_j)$, $j = 1, 2$, being related to the connection ∇ by the Connes' "transfer" formula $D_2(\xi \otimes h) = \xi \otimes D_1(h) + (\nabla\xi)(h)$ where $h \in \mathcal{H}_1$ and $\xi \in X$) and with composition given by the bimodule $X^3 := X^2 \otimes_{\mathcal{A}_2} X^1$ equipped with the connection: $\nabla^3(\xi_2 \otimes \xi_1)(h) := \xi_2 \otimes (\nabla^1\xi_1)(h) + (\nabla^2\xi_2)(\xi_1 \otimes h)$, where $\xi_1 \in X^1, \xi_2 \in X^2$, $h \in \mathcal{H}_1$.

[3] Note that if ϕ is an epimorphism, its pull-back $\phi^\bullet$ maps pure states into pure states.

[4] Bertozzini P, Conti R and Lewkeeratiyutkul W, Categories of Spectral Triples and Morita Equivalence, work in progress.

[5] Here $\Omega^1_D(\mathcal{A})$ denotes the $\mathcal{A}$-bimodule inside the algebra $\Omega_D(\mathcal{A})$ spanned by the commutators $[D, \pi(x)]_-$, $x \in \mathcal{A}$.

2.5. Mesland Morphisms

Morphism of spectral triples via Morita correspondences have been further developed in the works by A.Connes, M.Marcolli [14, chapter 8.4] and M.Marcolli, A.Z.al Yasri [24] were "spectral correspondences", defined as Hilbert C*-bimodules, are used to provide a "bivariant version" of spectral triples.

The most complete proposal in this direction comes from the work by B.Mesland [25] that has defined a category of (unitary equivalence classes of) smooth KK-bimodules that seems to be the best candidate for a non-commutative metric category of spectral triples. A **Mesland morphisms** from the spectral triple $(\mathcal{B}, \mathcal{H}', D')$ to the spectral triple $(\mathcal{A}, \mathcal{H}, D)$ is given by a unitary isomorphism class of an unbounded "smooth" $\mathcal{A}$-$\mathcal{B}$-bimodule $(\mathcal{E}, S)$ with "smooth connection" ∇ such that: $\mathcal{H}$ is isomorphic to $\mathcal{E}\overline{\otimes}_{\mathcal{B}}\mathcal{H}'$ (where here $\overline{\otimes}$ denotes the Haagerup tensor product); $[\nabla, S]$ is a completely bounded operator; $D = S\otimes \mathrm{Id} + \mathrm{Id}\otimes_\nabla D'$ with $\mathrm{Id}\otimes_\nabla D'(x\otimes h) := (-1)^{\partial x}(x\otimes D'h + (\nabla_{D'}x)h)$, where $x\in\mathcal{E}$, $h\in\mathcal{H}'$ (∂x denoting the degree of x in the graded module $\mathcal{E}$).

3. Categorification of Gel'fand Naĭmark Duality

3.1. Horizontal Categorification

In the same way as a category can be seen as a "many-objects" version of a monoid or a groupoid can be thought as a multi-objects version of a group, a C*-category is a (horizontal) categorification of a C*-algebra. Furthermore, in the same way as every category $\mathcal{C}$ induces a projection functor $\pi : \mathrm{Hom}_{\mathcal{C}} \to \mathrm{Ob}_{\mathcal{C}} \times \mathrm{Ob}_{\mathcal{C}}$, a C*-category $\mathcal{C}$ can be identified as a very special kind of Fell bundle where the base category is a "double groupoid" $\mathrm{Ob}_{\mathcal{C}} \times \mathrm{Ob}_{\mathcal{C}}$.

More precisely, given an **inverse involutive category** $\mathcal{X}$ (i.e. a category equipped with an object-preserving contravariant functor $x \mapsto x^*$ such that for all arrows $x\in\mathcal{X}$, $(x^*)^* = x$ and such that $x\circ x^*\circ x = x$ for all $x\in\mathcal{X}$) a **unital Fell bundle over** $\mathcal{X}$ is a Banach bundle[6] $\pi : \mathcal{E}\to\mathcal{X}$ with, a total space $\mathcal{E}$ that is an involutive category, a projection π that is a covariant $*$-functor and such that: the composition in $\mathcal{E}$ is fiberwise bilinear and norm submultiplicative; the involution in $\mathcal{E}$ is fiberwise conjugate linear with the C*-property $\|e^*\circ e\| = \|e\|^2$ and such that $e^*\circ e$ is a positive element in the C*-algebra $\pi^{-1}(\pi(e^*\circ e))$.[7] A Fell bundle is **saturated** whenever the Hilbert bimodules $\pi^{-1}(x)$ are full over the C*-algebras $\pi^{-1}(x^*\circ x)$ and $\pi^{-1}(x\circ x^*)$. A (small) **C*-category** can be identified as a unital Fell bundle over an involutive category of the form $\mathcal{O}\times\mathcal{O}$ for a certain set $\mathcal{O}$. It is said to be full if it is saturated as a Fell bundle and commutative if the C*-algebras $\pi^{-1}(x^*\circ x)$ are Abelian.

In the search for an appropriate notion of "spectrum of a commutative full small C*-category" we defined a **topological spaceoid** as a unital Fell bundle of rank-one (i.e. with one-dimensional fibers) whose base category is given by a direct product $\Delta_X\times\mathcal{R}_{\mathcal{O}}$, where X is a compact Hausdorff space, $\mathcal{O}$ is a discrete space, $\Delta_X := \{(x,x) \mid x\in X\}$ is the "diagonal of X" and $\mathcal{R}_{\mathcal{O}} := \mathcal{O}\times\mathcal{O}$ is the maximal equivalence relation on $\mathcal{O}$.

In [5], we provide a categorical extension of the usual Gel'fand-Naĭmark duality between the category of unital $*$-homomorphisms of Abelian unital C*-algebras and the category of continuous maps of compact Hausdorff spaces to a new duality between the category of object-preserving $*$-functors of small commutative full C*-categories and a category of suitable morphisms of spaceoids.

In [7] we further generalize the notion of Fell bundle introducing a definition of involutive categorical bundle (Fell bundle) enriched in an involutive monoidal category (or even in an involutive 2-fold category) and we made use of this concept to relate three equivalent ways to describe the spectrum of a full commutative small C*-category.

[6] We refer to J.Fell, R.Doran [18] for all the details on Banach bundles.

[7] Note that $\pi^{-1}(x^*\circ x)$ is always a unital C*-algebra and $\pi^{-1}(x)$ is always a C*-bimodule onto the C*-algebras $\pi^{-1}(x^*\circ x)$ and $\pi^{-1}(x\circ x^*)$.

Journal of Physics: Conference Series **346** (2012) 012003 doi:10.1088/1742-6596/346/1/012003

An interesting by-product of this investigation is an alternative direct proof of a spectral theorem for imprimitivity Hilbert C*-bimodules over Abelian C*-algebras [6] (i.e. a Hermitian version of Serre-Swan theorem) that is suitable to provide a "bivariant version" of A. Takahashi's duality between categories of Hilbert C*-modules and categories of Hilbert bundles [29, 30].

3.2. Non-full C-categories*

One further essential step is to extend our spectral theorem to the case of non-full small commutative C*-categories. In this case the spectrum of the C*-category $\mathcal{C}$ is no more a line-bundle and can be described as a Fell bundle with fibers of dimension less than or equal to one. The locus of base points supporting zero-dimensional fibers is given by a family of closed sets F_{AB} for all $A, B \in \mathrm{Ob}_{\mathcal{C}}$ with the properties $F_{AA} = \varnothing$, $F_{AB} = F^*_{BA}$ and $F_{AC} \subset F_{AB} \circ F_{BC}$ for all $A, B, C \in \mathrm{Ob}_{\mathcal{C}}$. These non-full categories correspond of course to special cases of closed "ideals" of full commutative C*-categories.

3.3. Vertical Categorification

In view of a further vertically categorified extension of Gel'fand duality, we are investigating the existence of reasonable notions of strict n-C*-categories.[8]

Recall [22, section 1.4] that a **globular n-set** $\mathcal{C}^0 \leftleftarrows \mathcal{C}^1 \leftleftarrows \cdots \mathcal{C}^{m-1} \leftleftarrows \mathcal{C}^m \leftleftarrows \cdots \leftleftarrows \mathcal{C}^n$, $n \in \mathbb{N}$, is given by: a collections of classes $\mathcal{C}^m$, for all $m = 0, \dots, n$, whose elements are called **m-arrows**, and a pair of **source, target** maps $s_m, t_m : \mathcal{C}^m \to \mathcal{C}^{m-1}$, for all $m = 1, \dots, n$, such that for all $m = 1, \dots, n-1$, we have $s_m \circ s_{m+1} = s_m \circ t_{m+1}$, and $t_m \circ s_{m+1} = t_m \circ t_{m+1}$.

A (globular) **strict n-category** (for example see T.Leinster [22, section 1.4]) has been defined as a globular n-set that for all $0 \leq p < m \leq n$, is equipped with a partial **p-composition** map $\circ^m_p : \mathcal{C}^m \times_{\mathcal{C}^p} \mathcal{C}^m \to \mathcal{C}^m$, $(x, y) \mapsto x \circ^m_p y$, defined on the set $\mathcal{C}^m \times_{\mathcal{C}^p} \mathcal{C}^m$ of p-composable m-arrows $(x, y) \in \mathcal{C}^m \times_{\mathcal{C}^p} \mathcal{C}^m \Leftrightarrow t_{p+1} \circ \cdots \circ t_m(y) = s_{p+1} \circ \cdots \circ s_m(x)$, such that, for all $m = 0, \dots, n-1$, there is an **identity map** $\iota_m : \mathcal{C}^m \to \mathcal{C}^{m+1}$, in such a way that the following axioms are satisfied:

- for all $m = 0, \dots, n$, for all $p = 0, \dots, m-1$, for all $(x, y) \in \mathcal{C}^m \times_{\mathcal{C}^p} \mathcal{C}^m$,
  $s_m(x \circ^m_p y) = s_m(y), \quad t_m(x \circ^m_p y) = t_m(x), \quad \text{if } p = m-1;$
  $s_m(x \circ^m_p y) = s_m(x) \circ^{m-1}_p s_m(y), \quad \text{if } p = 0, \dots, m-2,$
  $t_m(x \circ^m_p y) = t_m(x) \circ^{m-1}_p t_m(y), \quad \text{if } p = 0, \dots, m-2;$
- for all $x \in \mathcal{C}^m$, $s_{m+1}(\iota_m(x)) = x, \quad t_{m+1}(\iota_m(x)) = x;$
- for all $m = 1, \dots, n$ and $p = 0, \dots, m-1$ and for all $x, y, z \in \mathcal{C}^m$, $(x \circ^m_p y) \circ^m_p z = x \circ^m_p (y \circ^m_p z)$, whenever $(x, y), (y, z) \in \mathcal{C}^m \times_{\mathcal{C}^p} \mathcal{C}^m$ holds;
- for all $m = 1, \dots, n$, for all $p = 0, \dots, m-1$, for all $x \in \mathcal{C}^m$,
  $\Big(\iota_{m-1} \circ \cdots \circ \iota_p\big(t_{p+1} \circ \cdots \circ t_m(x)\big)\Big) \circ^m_p x = x, \quad x = x \circ^m_p \Big(\iota_{m-1} \circ \cdots \circ \iota_p\big(s_{p+1} \circ \cdots \circ s_m(x)\big)\Big);$
- for all $m = 2, \dots, n$, for all $p, q = 0, \dots, m-1$, with $q < p$, for all $w, x, y, z \in \mathcal{C}^m$ such that $(w, x), (y, z) \in \mathcal{C}^m \times_{\mathcal{C}^p} \mathcal{C}^m$ and $(w, y), (x, z) \in \mathcal{C}^m \times_{\mathcal{C}^q} \mathcal{C}^m$, we have the exchange property $(w \circ^m_p x) \circ^m_q (y \circ^m_p z) = (w \circ^m_q y) \circ^m_p (x \circ^m_q z)$;
- for all $m = 1, \dots, n-1$, for all $p = 0, \dots, m-1$, for all $(x, y) \in \mathcal{C}^m \times_{\mathcal{C}^p} \mathcal{C}^m$, we have $\iota_m(x \circ^m_p y) = \iota_m(x) \circ^{m+1}_p \iota_m(y)$.

It is reasonable to define a **strict involutive n-category** as a strict n-category that is equipped with a family of "involutions" $*^m : \mathcal{C}^m \to \mathcal{C}^m$, for $0 < m \leq n$, that satisfy the

[8] Bertozzini P, Conti R, Lewkeeratiyutkul W, Suthichitranont N, Strict Higher C*-categories, work in progress. See also the slides "Categories of Non-commutative Geometries at the second workshop "Categories, Logic and Physics" in Imperial College.

following properties:[9]

- $s_m(x^{*^m}) = t_m(x)$, $t_m(x^{*^m}) = s_m(x)$, for all $x \in \mathcal{C}^m$,
- $(x \circ_p^m y)^{*^m} = y^{*^m} \circ_p^m x^{*^m}$, for $p = m-1$, $(x \circ_p^m y)^{*^m} = x^{*^m} \circ_p^m y^{*^m}$, for $0 \leq p < m-1$, for all $x, y \in \mathcal{C}^m \times_{\mathcal{C}^p} \mathcal{C}^m$ with $m = 1, \dots, n$,
- $(x^{*^m})^{*^m} = x$, for all $x \in \mathcal{C}^m$.

Finally, one might try to define a **strict-n-C*-category** to be a strict involutive n-category such that:

- for all $m = 1, \dots, n$, and $x, y \in \mathcal{C}^{m-1}$, the sets $\mathcal{C}^m(x,y) := \{z \in \mathcal{C}^m \mid s_m(z) = y,\ t_m(z) = x\}$ are Banach spaces with norm denoted by $x \mapsto \|x\|_m$, for $0 \leq p < m$,
- for all $w, x, y, z \in \mathcal{C}^{m-1}$ such that $\mathcal{C}^m(w,x) \times \mathcal{C}^m(y,z) \subset \mathcal{C}^m \times_{\mathcal{C}^p} \mathcal{C}^m$, the composition maps $\circ_p^m : \mathcal{C}^m(w,x) \times \mathcal{C}^m(y,z) \to \mathcal{C}^m$ are bilinear,
- for all $m = 1, \dots, n$, for all $x, y \in \mathcal{C}^{m-1}$, the maps $*^m : \mathcal{C}^m(x,y) \to \mathcal{C}^m$ are conjugate linear;
- for all $m = 1, \dots, n$, for all $p = 0, \dots, m-1$, for all pairs $(x,y) \in \mathcal{C}^m \times_{\mathcal{C}^p} \mathcal{C}^m$, $\|x \circ_p^m y\|_m \leq \|x\|_m \cdot \|y\|_m$,
- for all $m = 1, \dots, n$ and $0 \leq p < m$, for all $(x^{*^m}, x) \in \mathcal{C}^m \times_{\mathcal{C}^p} \mathcal{C}^m$, $\|x^{*^m} \circ_p^m x\|_m = \|x\|_m^2$.

Note that the above properties already imply that, for all $m = 1, \dots, n$ and for all $x \in \mathcal{C}_{m-1}$, the set $\mathcal{C}^m(x,x)$ is a C*-algebra with multiplication $\circ_{m-1}^m$ and involution $*^m$ and hence the following final condition is meaningful:

- for all $m = 1, \dots, n$, for all $x \in \mathcal{C}^m(u,v)$, $x^{*^m} \circ_{m-1}^m x \in \mathcal{C}^m(u,u)_+$, i.e. $x^{*^m} \circ_p^m x$ is a positive element in the C*-algebra $\mathcal{C}^m(u,u)$.

A **left module $_{\mathcal{C}}\mathcal{M}$ over the n-category** $\mathcal{C}$ is given by

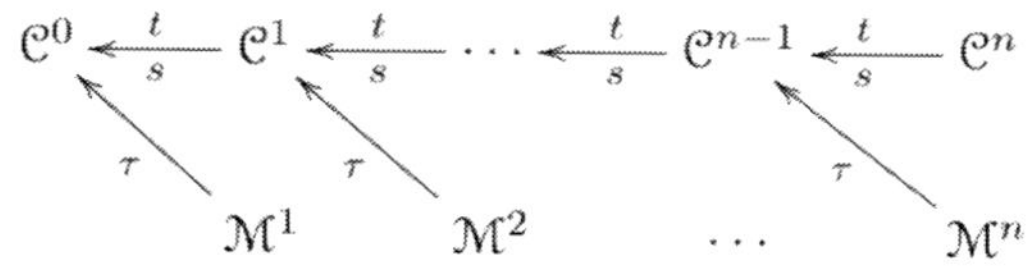

where for all $m = 0, \dots, n$, $\tau : \mathcal{M}^m \to \mathcal{C}^{m-1}$ is a fibered category over the $(m-1)$-category $\mathcal{C}^{m-1}$ and, for all $0 \leq p < m \leq n$, there is a left action $\mu_p^m : \mathcal{C}^m \times \mathcal{M}^m \to \mathcal{M}^m$ of the bi-fibered $(m-1)$-category $\mathcal{C}^m \rightrightarrows \mathcal{C}^{m-1} \times \mathcal{C}^{m-1}$ over $\mathcal{M}^m \to \mathcal{C}^{m-1}$ such that $\mu_p^m(\mathcal{C}^m(x,y) \times \mathcal{M}^m(z)) \subset \mathcal{M}^m(x)$ whenever $(y,z) \in \mathcal{C}^{m-1} \times_{\mathcal{C}^p} \mathcal{C}^{m-1}$ with $x = y \circ_p^{m-1} z$.[10] Similar definitions can be given for right modules $\mathcal{M}_{\mathcal{C}}$ and bimodules $_{\mathcal{C}}\mathcal{M}_{\mathcal{C}}$ over the n-category $\mathcal{C}$.

The notion of **left Hilbert C*-module $_{\mathcal{C}}\mathcal{M}$ over a strict n-C*-category** $\mathcal{C}$ should be given imposing that for all $m = 1, \dots n$, $\tau : \mathcal{M}^m \to \mathcal{C}^{m-1}$ is a "Fell bundle"(for all the compositions and involutions in $\mathcal{C}^{m-1}$) equipped with an inner product $\langle \cdot \mid \cdot \rangle_m : \mathcal{M}^m \times \mathcal{M}^m \to \mathcal{C}^m$ such that $\langle \mathcal{M}^m(x) \mid \mathcal{M}^m(y) \rangle_m \subset \mathcal{C}^m(y,x)$.[11]

Examples of rank-one strict-n-C*-categories i.e. strict-n-C*-categories such that for every $m = 1, \dots, n$, the Banach space $\mathcal{C}^m(x,y)$ is one-dimensional, can be constructed by hand recursively. In the theory of higher C*-categories they play the role of the scalar field $\mathbb{C}$.

[9] Actually it is perfectly possible to require the existence of involutions only for certain specific "arrow levels" so that, in the case of involution present only for the level $m = n$, previous definitions of 2-C*-categories can be recovered. In the opposite direction, it might also be possible to require further axioms for involutions $*_q^m : \mathcal{C}^m \to \mathcal{C}^m$ of depth q for $0 \leq q < m \leq n$, but we will not go into further details here.
[10] For $p = m-1$ we assume $\mathcal{C}^{m-1} \times_{\mathcal{C}^p} \mathcal{C}^{m-1} = \Delta_{\mathcal{C}^{m-1}}$.
[11] Again, corresponding definitions can be given for right Hilbert C*-modules and right/left bimodules over a strict n-C*-category, but it will be necessary to distinguish right and left structures also for bimodules.

Journal of Physics: Conference Series **346** (2012) 012003 doi:10.1088/1742-6596/346/1/012003

Hilbert C*-modules over rank-one strict n-C*-categories will play the role of n-Hilbert spaces. Examples of non-commutative strict-n-C*-categories are expected to arise as "categories of endomorphisms" of left Hilbert C*-modules over rank-one n-C*-categories.

A formulation of Gel'fand-Naĭmark duality in such higher C*-categorical context requires the usage of "iterated Fell line-bundles" and it is under investigation.

3.4. Horizontal Categorification of Spectral Triples

One of the main original motivations in the study of C*-categories comes from the realization that, since the "off-diagonal blocks" $\mathcal{C}_{AB} := \mathrm{Hom}_{\mathcal{C}}(B, A)$ are Hilbert C*-bimodules over the C*-algebras $\mathcal{C}_{AA}$ and $\mathcal{C}_{BB}$, the study of possible axiomatizations of spectral triples over C*-categories might provide some further light on the appropriate definition of "bivariant spectral triples" and more generally Morita morphisms of spectral triples. Of course in the case of full C*-categories, all the bimodules $\mathcal{C}_{AB}$ are imprimitivity bimodules (i.e. isomorphisms in the Morita-Rieffel category of Hilbert C*-bimodules) and so, in this special case, we are bound to obtain arrows in a groupoid of isomorphisms of spectral triples. Without entering into further details that will be developed elsewhere, we note that spectral triples over a C*-category can be simply defined as spectral triples over the enveloping C*-algebra of the C*-category. For example, in the attempt to generalize naive spectral triples to a categorified context[12] we can define a **categorical spectral geometry** as a triple $(\mathcal{C}, \mathcal{H}, \mathcal{D})$ given by:

- a pre-C*-category $\mathcal{C}$;
- a module $\mathcal{H}$ over $\mathcal{C}$ that is also a Hilbert C*-module over $\mathbb{C}$; in other terms a family of Hilbert spaces $\mathcal{H}$ equipped with an object bijective $*$-functor $\pi : \mathcal{C} \to \mathcal{B}(\mathcal{H})$ with values in the C*-category of bounded linear maps between the Hilbert spaces in the family $\mathcal{H}$;
- the generator $\mathcal{D}$ of a unitary one-parameter group on $\mathcal{H}$ (i.e. the generator of a one-parameter group whose adjoint action in the enveloping C*-algebra of $\mathcal{B}(\mathcal{H})$ leaves $\mathcal{B}(\mathcal{H})$ invariant) such that, for all $x \in \mathcal{C}$, $[\mathcal{D}, \pi(x)]_-$ is extendable to an operator in $\mathcal{B}(\mathcal{H})$.

In the case of C*-categories, the notion of bimodule over a C*-category is significantly different from that of left or right module (see for example P.Mitchener [26]) and this results in a further complication as can be seen in the following very tentative definition.

A **bivariant spectral geometry** over two pre-C*-categories (with the same objects) $\mathcal{A}$ and $\mathcal{B}$ is a quintuple $(\mathcal{A}, \mathcal{B}, \mathcal{H}, \mathcal{D}_{\mathcal{A}}, \mathcal{D}_{\mathcal{B}})$, where

- $\mathcal{H}$ is a bimodule over $\mathcal{A}$-$\mathcal{B}$ that is also a Hilbert C*-bimodule over $\mathbb{C}$ and hence it is equipped with two $*$-representations $\rho : \mathcal{A} \to \mathcal{B}_\rho(\mathcal{H})$ and $\lambda : \mathcal{B} \to \mathcal{B}_\lambda(\mathcal{H})$ into the right, and respectively the left, C*-category of the bimodule;
- $\mathcal{D}_{\mathcal{A}}$ (acting on the left) and $\mathcal{D}_{\mathcal{B}}$ (acting on the right) are two (generally unbounded) self-adjoint operators on $\mathcal{H}$ that generate on the enveloping algebras of $\mathcal{B}_\rho(\mathcal{H})$, and respectively of $\mathcal{B}_\lambda(\mathcal{H})$, one-parameter groups leaving $\mathcal{B}_\rho(\mathcal{H})$, and respectively $\mathcal{B}_\lambda(\mathcal{H})$, invariant and such that $[\mathcal{D}_{\mathcal{A}}, \rho(x)]_-$ and $[\mathcal{D}_{\mathcal{B}}, \lambda(y)]_-$ are extensible to bounded operators in $\mathcal{B}_\rho(\mathcal{H})$, $\mathcal{B}_\lambda(\mathcal{H})$, for all $x \in \mathcal{A}$ and $y \in \mathcal{B}$.

4. Outlook

A short-term objective of this line of research is to provide explicit examples of functors from suitable categories of geometrical spaces (such as for example oriented Riemannian or spinorial compact manifolds) to categories of spectral triples (such as the category described by B.Mesland [25] or possibly some variants of it).

[12] Bertozzini P, Conti R and Lewkeeratiyutkul W, Spectral Geometries over C*-categories and Morphisms of Spectral Geometries, work in progress.

Since spectral triples are a very sophisticated kind of mathematical tool where topological, measurable, smooth and metric structures are simultaneously present, it seems worth to spend some time investigating separately the categorical structures involved in the case of oriented spaces, measure spaces, (Riemannian/Hermitian) manifolds/bundles equipped with connections or with spinorial bundles and their "dual" categories of modules. Of particular interest is the case of "non-commutative measure spaces" and the study of the categorical structure implicit in Tomita-Takesaki modular theory and in Falcone-Takesaki non-commutative flow of weights.

Some more ambitious goals include:

- spectral reconstruction theorems for certain classes of morphism of spectral triples,
- extensions of our Gel'fand duality result to full non-commutative C*-categories,
- a "spectral theory" of spectral triples in terms of Fell line-bundles (along the lines envisaged by R. Martins) and its application to physics,
- the study of possible relations between (categorical) non-commutative geometry and Grothendieck's topoi.

Some applications of such mathematical structures in physics are also priorities:

- in the context of loop quantum gravity, we might provide (categorical) non-commutative geometries associated to the "quantum geometries" described by spin-networks,
- usage of (higher/modular) categorical structures to obtain a mathematical formulation of C.Rovelli's relational quantum mechanics,
- further progress in our modular algebraic quantum gravity proposal (see [4]).

References

[1] Bertozzini P, Conti R and Lewkeeratiyutkul W 2006 A category of spectral triples and discrete groups with length function *Osaka J. Math.* **43** 327 (Preprint math.OA/0502583v1)
[2] Bertozzini P, Conti R and Lewkeeratiyutkul W 2007 Non-commutative geometry categories and quantum physics *Contributions in Mathematics and Applications II, East-West J. Math. special volume 2007* 213 (Preprint 0801.2826v1)
[3] Bertozzini P, Conti R and Lewkeeratiyutkul W 2008 A remark on Gel'fand duality for spectral triples Preprint 0812.3584v1
[4] Bertozzini P, Conti R and Lewkeeratiyutkul W 2010 Modular theory, non-commutative geometry and quantum gravity *Symmetry, Integrability and Geometry: Methods and Applications SIGMA* **6** 067 (Preprint 1007.4094v2)
[5] Bertozzini P, Conti R and Lewkeeratiyutkul W 2011 A horizontal categorification of Gel'fand duality *Adv. Math.* **226** 584 (Preprint 0812.3601v2)
[6] Bertozzini P, Conti R and Lewkeeratiyutkul W 2008 A spectral theorem for imprimitivity C*-bimodules Preprint 0812.3596v1
[7] Bertozzini P, Conti R and Lewkeeratiyutkul W 2009 Enriched fell bundles and spaceoids (in preperation)
[8] Bertozzini P and Rutamorn K 2009 Krein C*-categories *Chamchuri J. Math.* **1** 61
[9] Carey A, Phillips J and Rennie A 2010 Spectral triples: examples and index theory Preprint 0812.3596v1)
[10] Connes A 1994 *Noncommutative Geometry* (San Diego CA: Academic Press)
[11] Connes A 1996 Gravity coupled with matter and the foundations of noncommutative geometry *Commun. Math. Phys.* **182** 155
[12] Connes A and Chamseddine A 2006 Inner fluctuation of the spectral action *J. Geom. Phys.* **57** 1-21 (Preprint hep-th/0605011v3)
[13] Connes A, Consani C and Marcolli M 2007 Noncommutative geometry and motives: the thermodynamics of endomotives *Adv. Math.* **214** 761 (Preprint math/0512138v2)
[14] Connes A and Marcolli M 2008 *Noncommutative Geometry, Quantum Fields and Motives* (American Mathematical Society Colloquium Publications vol 55) (Providence: AMS)
[15] Dawe Martins R 2007 Double fell bundles over discrete double groupoids with folding Preprint 0709.2972v3
[16] Dawe Martins R 2009 Categorified noncommutative manifolds *Int. J. Mod. Phys.* A **24** 2802 (Preprint 0811.1485v4)
[17] Dawe Martins R 2010 An outlook on quantum gravity from an algebraic perspective Preprint 1003.4434v1

[18] Fell J and Doran R 1998 *Representations of C^*-Algebras, Locally Compact Groups and Banach $*$-Algebraic Bundles* (San Diego CA: Academic Press) vol 1, 2
[19] Denicola D, Marcolli M and Zainy al-Yasry A 2010 Spin Foams and Noncommutative Geometry *Classical Quantum Gravity* **27** 205025 (Preprint 1005.1057v1)
[20] Ghez P, Lima R and Roberts J 1985 W*-categories *Pacific J. Math.* **120** 79
[21] Gracia-Bondia J, Varilly J and Figueroa H 2001 *Elements of Noncommutative Geometry* (Boston: Birkhäuser)
[22] Leinster T 2003 *Higher Operads, Higher Categories* (Cambridge: Cambridge University Press)
[23] Manin Y 2004 Real multiplication and noncommutative geometry *The Legacy of Niels Henrik Abel* (Berlin: Springer) pp 685–727 (Preprint math.AG/0202109)
[24] Marcolli M and Zainy al-Yasry A 2008 Coverings, correspondences, and noncommutative geometry *J. Geom. Phys.* **58** 1639 (Preprint 0807.2924v1)
[25] Mesland B 2009 Unbounded bivariant K-theory and correspondences in noncommutative geometry Preprint arXiv:0904.4383v1
[26] Mitchener P 2002 C*-categories *Proc. Lond. Math. Soc.* **84** 375
[27] Paschke M and Verch R 2004 Local covariant quantum field theory over spectral geometries *Classical Quantum Gravity* **21** 5299 (Preprint gr-qc/0405057)
[28] Sitarz A 2002 *The Basics of Noncommutative Geometry - Spectral Triples - Noncommutative Symmetries at Work* (Introduction of the Habilitation Thesis, Jagellonian University)
[29] Takahashi A 1979 Hilbert modules and their representation *Rev. Colombiana Mat.* **13** 1
[30] Takahashi A 1979 A duality between Hilbert Modules and fields of Hilbert Spaces *Rev. Colombiana Mat.* **13** 93
[31] Varilly J 2006 *An Introduction to Noncommutative Geometry* (Zürich: European Mathematical Society)

Normal coideal subalgebras of semisimple Hopf algebras

Sebastian Burciu

University of Bucharest, Faculty of Mathematics and Computer Science, 14 Academiei St., Bucharest, Romania
and
Institute of Mathematics " Simion Stoilow" of Romanian Academy, P.O. Box 1-764, RO-014700, Bucharest, Romania

E-mail: sebastian.burciu@imar.ro

Abstract. The restriction functor from the category of representations of a semisimple Hopf algebra to the category of representations of a normal coideal subalgebra is studied. It is shown that this functor has a similar behavior to the restriction functor to the category of representations of a normal Hopf subalgebra. Commutator subalgebras as normal left (right) coideal subalgebras are also studied.

1. Introduction

Semisimple Hopf algebras were intensively studied in the last ten years especially due to their connections with other fields in mathematics and physics such as quantum field theory, fusion categories and topological invariants of knots and manifolds.

In principle the structure of a Hopf algebra can be recovered from the structure of a normal Hopf subalgebra, its corresponding quotient Hopf algebra and some additional cohomological data (see [1]). For this reason, normal Hopf subalgebras of semisimple Hopf algebras are an important tool in the study of semisimple Hopf algebras, especially in their classification.

Recently it was proven in [2] that Hopf subalgebras are normal if and only if they are depth two subalgebras. This result was then extended by the author to coideal subalgebras in [3]. Moreover it was shown that in this situation, depth two and normality in the sense defined by Rieffel in [10] also coincide.

In this paper we study normal left coideal subalgebras L of semisimple Hopf algebras H. We show that the restriction functor from left H-modules to left L-modules has a similar behavior to the restriction functor to normal Hopf subalgebras. Using Rieffel's equivalence classes, formulae for this restriction functor are given in Section 4. Similar results for the induction functor are also proven in the same section.

The proof of these results is using the notion of double coset for semisimple finite dimensional Hopf algebras that was introduced in [4].

The left (right) commutator subalgebra of a semisimple Hopf algebra is defined as the smallest normal left (right) coideal subalgebra with the property that the corresponding quotient is a commutative Hopf algebra. We show that a semisimple Hopf algebra over an algebraically closed field of characteristic zero is of Frobenius type provided that its left (right) commutator algebra is of Frobenius type.

Recall that a finite dimensional algebra A over a field k is called of Frobenius type if the dimension of each irreducible A-module divides the dimension of A. One of Kaplansky's conjectures states that any semisimple Hopf algebra over an algebraically closed field is of Frobenius type.

The paper is organized as follows. In the second section we recall the basics on semisimple Hopf algebras that are needed for the rest of the paper. Section 3 proves some properties of the induction and restriction functors determined by a left coideal subalgebra of a semisimple Hopf algebra. Using the coset decomposition from [4] we prove in Section 4 the above mentioned formulae for induction and restriction. In Section 5 we define conjugate modules for normal coideal subalgebras, similar to the conjugate modules for normal subgroups. It is shown in Theorem 5.9 that if an irreducible module is a constituent of a restricted module then all the conjugate modules are also constituents of the restricted module. The left (right) commutator subalgebra of a semisimple Hopf algebra are defined in the last section. In Theorem 6.4 we show that if the commutator algebra of a semisimple Hopf algebra is of Frobenius type then the Hopf algebra itself is of Frobenius type.

We work over an algebraically closed field k of characteristic zero. For a k-vector space V we denote by $|V|$ the dimension $\dim_k(V)$ of V as vector space over k. All Hopf algebraic notations are those from [8] except the fact that we drop the sigma symbol in Sweedler's comultiplication.

2. Preliminaries

Throughout of this paper H will be a finite dimensional semisimple Hopf algebra over an algebraically closed field k of characteristic zero. Then H is also cosemisimple and $S^2 = \mathrm{Id}$ [7]. The set of irreducible characters of H is denoted by $\mathrm{Irr}(H)$. The Grothendieck group $\mathcal{G}(H)$ of the category of finite dimensional left H-modules is a ring under the tensor product of modules. Then $C(H) = \mathcal{G}(H) \otimes_{\mathbb{Z}} k$ is a semisimple subalgebra of H^* [9] and it has a basis given by the characters of the irreducible H-modules. By duality $C(H^*)$ is a semisimple subalgebra of H and coincides with the set of cocommutative elements of H.

Since H is also cosemisimple [7] the set of simple subcoalgebras of H is in bijection with the set of irreducible characters of H^* (see [6] for this correspondence).

For a vector subspace S of a Hopf algebra H we denote by $\omega(S)$ the vector subspace of H generated by $s - \epsilon(s)1$ for all $s \in S$.

2.1. Coideal subalgebras

Recall that a left coideal subalgebra S of H is a subalgebra S of H with $\Delta(S) \subset H \otimes S$. Then S is also a semisimple Hopf algebra by Lemma 4.0.2 of [3].

A left coideal subalgebra of H is called *normal* if L is closed under the left adjoint action of H on itself, i.e $a_1 L S(a_2) \subset L$ for all $a \in H$. If H is involuntary this is equivalent to $a_2 L S(a_1) \subset L$ for all $a \in H$.

It follows from [3] that S is also a semisimple algebra if H is a semisimple Hopf algebra. Denote by $C(S)$ the linear span of the trace functionals on S. It is the vector space with basis given by the set $\mathrm{Irr}(S)$ of irreducible characters of S.

2.2. The correspondence between Hopf ideals and normal left coideal subalgebras

Let H be a finite dimensional Hopf algebra. Based on the structure of relative Hopf modules, Takeuchi described in [13] a one to one correspondence between the Hopf ideals of H and the normal left coideal subalgebras of H. This correspondence can be summarized as follows. If I is a Hopf ideal of H then define $L(I) := H^{co\,\pi}$ where π is the usual Hopf map $\pi : H \to H/I$. Then it is easy to check that $L(I)$ is a normal left coideal subalgebra of H and $I = \omega(L(I))H$.

Conversely if L is a normal coideal subalgebra of H then $\omega(L)H$ is a Hopf ideal and $L(\omega(L)H) = L$. Denote by $H//L$ the quotient Hopf algebra $H/\omega(L)H$ and by π_L the canonical Hopf projection $\pi_L : H \to H//L$.

2.3. Coset decomposition
Let B be a Hopf subalgebra H. By Corollary 2.5 of [4] there is a coset decomposition for H

$$H = \oplus_{C/\sim} CB.$$

where $\sim$ is an equivalence relation on the set of simple subcoalgebras of H given by $C \sim C'$ if and only if $CB = C'B$. In [4] this equivalence relation is denoted by $r^H_{k,\,B}$.

In terms of characters it follows from Remark 2.9 of [4] that $d \sim d'$ if and only if

$$\frac{d\Lambda_B}{\epsilon(d)} = \frac{d'\Lambda_B}{\epsilon(d')} \tag{2.1}$$

where $\Lambda_B \in B$ is an integral of B.

2.4. Subcoalgebra associated to a comodule
Let W be a right H^*-module. Then W is a right H-comodule and one can associate to it a subcoalgebra of H denoted by C_W [6]. This is the minimal subcoalgebra of H with the property that $\rho(W) \subset C_W \otimes W$. If W is a simple comodule and $q = |W|$ then $|C_W| = q^2$ and the associated coalgebra C_W is a co-matrix coalgebra. It has a basis $\{x_{ij}\}_{1\leq i,j\leq q}$ such that $\Delta(x_{ij}) = \sum_{l=0}^{q} x_{il} \otimes x_{lj}$ for all $1 \leq i,j \leq q$. Moreover $W \cong k < x_{i1}|\ 1 \leq i \leq q >$ as right H-comodules where $\rho(x_{i1}) = \Delta(x_{i1}) = \sum_{l=0}^{q} x_{il} \otimes x_{l1}$ for all $1 \leq i \leq q$. The character of W as right H^*-module is $d \in C(H^*) \subset H$ and it is given by $d = \sum_{i=1}^{q} x_{ii}$. Then $\epsilon(d) = q$ and the simple subcoalgebra C_W is sometimes denoted by C_d.

3. Restriction to normal coideal subalgebras

In this section we prove some formulae for restriction of modules to arbitrary coideal subalgebras.

Let S be a left coideal subalgebra of a given Hopf algebra H. Then for any H-module M and any S-module V the comodule structure $\rho : S \to H \otimes S$ defines via pullback, an S-module structure on $M \otimes V$. Denote this module structure by $M \odot V$.

This makes the category of S-modules a left module category over the tensor category H-modules [14]. The following Proposition also appears in [3]. For the sake of completeness we sketch its proof here.

Proposition 3.1. *Let $S \subset H$ be a right H-comodule subalgebra of H. Then $M \otimes V \uparrow_S^H \cong (M \downarrow_S^H \odot V) \uparrow_S^H$ for any S-module V and any H-module M.*

Proof. The map $T : H \otimes_S (M \odot V) \to M \otimes (H \otimes_S V)$ given by $a \otimes_S (m \otimes v) \mapsto a_1 m \otimes (a_2 \otimes_S v)$ is a well defined map and a morphism of H-modules with inverse given by $m \otimes (a \otimes_S v) \mapsto a_2 \otimes_S (S^{-1}(a_1)m \otimes v)$. □

Corollary 3.2. *Let $S \subset H$ be a left coideal subalgebra of H. Then*

$$M \otimes \epsilon_S \uparrow_S^H = M \downarrow_S^H \uparrow_S^H \tag{3.3}$$

for any H-module S.

Proof. Put $V = k$, the trivial S-module in the above Proposition. Note that $M \odot k = M \downarrow_S^H$. □

The following Proposition is a straightforward computation. It shows that the restriction functor res : $H-\text{mod} \to S-\text{mod}$ is a morphism of $H-\text{mod}$ module categories.

Proposition 3.4. *Let S be a left coideal subalgebra of H and M, N be two H-modules. Then*

$$(M \otimes N) \downarrow_S^H = M \odot N \downarrow_S^H$$

Corollary 3.5. *Let S be a left coideal subalgebra of H and M, N be two H modules such that N is a trivial S-module. Then*

$$M \otimes N \cong M^{|N|}$$

as S-modules.

Proof. One has that $s(m \otimes n) = s_1 m \otimes s_2 n = s_1 m \otimes \epsilon(s_2) n = sm \otimes n$ for all $m \in M$, $n \in N$ and $s \in S$. □

3.1. On two endofunctors on $H-\text{mod}$ and respectively $S-\text{mod}$
Define the endofunctors:

$$\mathcal{T} : S-\text{mod} \to S-\text{mod} \quad \text{given by} \quad \mathcal{T}(V) = V \uparrow_S^H \downarrow_S^H \tag{3.6}$$

for any S-module V. Also define

$$\mathcal{V} : H-\text{mod} \to H-\text{mod} \quad \text{given by} \quad \mathcal{V}(M) = M \downarrow_S^H \uparrow_S^H \tag{3.7}$$

for any H-module M.

The following formulae for composition powers of these endofunctors were proven by induction on n in [3]:

Lemma 3.8 (Lemma 2.5.7 of [3]). *For any S-module V and any H-module M it follows that:*

(i)

$$\mathcal{V}^n(M) = M \otimes (\epsilon_S \uparrow_S^H)^n \tag{3.9}$$

(ii)

$$\mathcal{T}^{n+1}(V) = V \uparrow_S^H \odot \mathcal{T}^n(\epsilon_S) \tag{3.10}$$

for all $n \geq 1$.

3.2. Rieffel's equivalence relation for a coideal subalgebra
Let $S \subseteq H$ be a left coideal subalgebra of the semisimple Hopf algebra H. Then as above S is also semisimple algebra. For any V and W two S-modules one defines $V \sim W$ if and only if there is a simple H-module M such that V and W are both constituents of $M \downarrow_S^H$. The relation $\sim$ is reflexive and symmetric but not transitive in general. Its transitive closure is denoted by $\approx$. Thus we say that $V \approx W$ if and only if there is $m \geq 1$ and a sequence $V_{i_0}, V_{i_1}, \cdots, V_{i_{m-1}}, V_{i_m}$ of simple S-modules such that $V = V_{i_0} \sim V_{i_1} \sim V_{i_2} \sim \cdots \sim V_{i_{m-1}} \sim V_{i_m} = W$. As explained in [5] it follows that $V \approx W$ if and only if there is $n \geq 0$ such that V is a constituent to $\mathcal{T}^n(W)$. This is also equivalent to W to be a constituent of $\mathcal{T}^n(V)$. In the rest of the paper the above equivalence relation $\approx$ is denoted by d_S^H. This equivalence relation is considered in [10] in the context of extensions of semisimple rings.

Similarly one can define an equivalence relation u_S^H on the set of irreducible H-modules. For two irreducible$H$-modules M and N we say that $M \sim N$ if and only if their restriction to S have a common constituent. Then u_S^H is the transitive closure of $\sim$. Similarly, it can be shown that M and N are equivalent if and only if there is $n \geq 0$ such that M is a constituent to $\mathcal{V}^n(N)$. This is also equivalent to the fact that N is a constituent of $\mathcal{V}^n(M)$.

3.3. The formula for the induced trivial character
Let L be a normal left coideal subalgebra of H and $\epsilon_L := \epsilon_H|_L$ be the trivial character of L.

Proposition 3.11. *Let L be a normal left coideal subalgebra of a finite dimensional semisimple Hopf algebra H and $A = H//L$. If $t_A \in A^*$ is the integral on A with $t_A(1) = |A|$ then*

$$\epsilon_L \uparrow_L^H = t_A.$$

Moreover $t_A \downarrow_L^H = \frac{|H|}{|L|}\epsilon_L$.

Proof. One has from [7] that

$$t_A = \sum_{\chi \in \mathrm{Rep}(A//L)} \chi(1)\chi.$$

It is easy to see that $\chi \in \mathrm{Rep}(A//L)$ if and only if L acts trivially on the representation associated to χ. Therefore in this situation $\chi \downarrow_L = \chi(1)\epsilon_L$. By Frobenius reciprocity it follows that χ is a constituent of $\epsilon_L \uparrow_L^H$ with multiplicity $\chi(1)$. Thus $\epsilon_L \uparrow_L^H = t_A$. The second statement follows from Corollary 2.4.3 of [3]. □

4. A dual relation

Let L be a normal left coideal subalgebra of H and $A := H//L$. Then the natural projection $\pi : H \to A$ is a surjective Hopf map and then $\pi^* : A^* \to H^*$ is an injective Hopf map. We identify A^* with its image $\pi^*(A^*)$ in H^*. This is a Hopf subalgebra of H^*. In this section we will study the equivalence relation $r^{H^*}_{k,\,A^*}$ on $\mathrm{Irr}(H^{**}) = \mathrm{Irr}(H)$. Recall the definition of $r^{H^*}_{k,\,A^*}$ from subsection 2.3.

Proposition 4.1. *Let L be a normal left coideal subalgebra of a semisimple Hopf algebra H and $A = H//L$. Consider the equivalence relation $r^{H^*}_{k,\,A^*}$ on $\mathrm{Irr}(H)$. Then $\chi \sim \mu$ if and only if their restrictions to L have a common constituent.*

Proof. With the notations of subsection 2.3 note that $\Lambda_{A^*} = t_A$. Using equation 2.1 the equivalence relation $r^{H^*}_{k,\,A^*}$ on $\mathrm{Irr}(H)$ becomes $\chi \sim \mu$ if and only if $m_H(\chi,\, \mu t_A) > 0$. On the other hand, applying the previous Proposition it follows that:

$$\begin{aligned} m_H(\chi,\, \mu t_A) &= m_H(\chi,\, \mu\epsilon_L \uparrow_L^H) \\ &= m_H(\chi,\, \mu \downarrow_L^H \uparrow_L^H) \\ &= m_L(\chi \downarrow_L^H,\, \mu \downarrow_L^H) \end{aligned}$$

In the second equality we have applied formula 3.3. Thus $\chi \sim \mu$ if and only if their restriction to L have a common constituent. □

Corollary 4.2. *Let L be a normal left coideal subalgebra of a semisimple Hopf algebra H and $A = H//L$. With the above notations it follows that $u_L^A = r^{k,\,H^*}$.*

Theorem 4.3. *Let L be a normal left coideal subalgebra of a semisimple Hopf algebra H and $A = H//L$. Consider the equivalence relation $r^{H^*}_{k,\,A^*}$ on $\mathrm{Irr}(H)$. Then $\chi \sim \mu$ if and only if $\frac{\chi\downarrow_L}{\chi(1)} = \frac{\mu\downarrow_L}{\mu(1)}$.*

Proof. Let $\mathcal{C}_1, \mathcal{C}_2, \cdots, \mathcal{C}_s$ be the equivalence classes of $r^{H^*}_{k,\,A^*}$ on $\mathrm{Irr}(H)$ and let

$$a_i := \sum_{\chi \in \mathcal{C}_i} \chi(1)\chi \tag{4.4}$$

for $1 \leq i \leq l$. If $\mathcal{C}_1$ is the equivalence class of the trivial character ϵ_A of A then the definition of $r^{H^*}_{k,\, A^*}$ implies that $a_1 = t_A$. Formula 2.1 becomes

$$\frac{\chi}{\chi(1)} \frac{t_A}{|A|} = \frac{a_i}{a_i(1)} \tag{4.5}$$

for any irreducible character $\chi \in \mathcal{C}_i$.

By Proposition 3.11 one has that $t_A \downarrow^H_L = |A|\epsilon_L$. Thus by restricting to L both terms of the relation 4.5 and applying Lemma 3.5 one obtains that:

$$\frac{\chi \downarrow_L}{\chi(1)} = \frac{a_i \downarrow_L}{a_i(1)} \tag{4.6}$$

Therefore if $\chi \sim \mu$ then $\frac{\chi\downarrow_L}{\chi(1)} = \frac{\mu\downarrow_L}{\mu(1)}$. The converse follows from Proposition 4.1. □

Corollary 4.7. *With the above notations one has that* $u^H_L = r^{H^*}_{k,\,(H//L)^*}$ *as equivalence relations on* $\mathrm{Irr}(H)$.

4.1. Formulae for the restriction and induction functor

Let L be a normal left coideal subalgebra of A. Consider $\mathcal{L}_0, \cdots, \mathcal{L}_s$ the equivalence classes of d^H_L. By Proposition 4.3.1 of [3] one may assume $\mathcal{L}_0 = \{\epsilon_L\}$. Define

$$l_i := \sum_{\alpha \in \mathcal{L}_i} \alpha(1)\alpha,$$

for all $0 \leq i \leq s$. Also as before let $\mathcal{C}_0, \cdots, \mathcal{C}_s$ be the corresponding equivalence classes of u^H_L and let

$$a_i := \sum_{\chi \in \mathcal{C}_i} \chi(1)\chi.$$

The previous Theorem implies the following Corollary:

Corollary 4.8. *With the above notations, if* $\chi \in \mathcal{C}_i$ *and* $\alpha \in \mathcal{L}_i$ *then one has the following formulae:*

$$\chi \downarrow_L = \frac{\chi(1)}{a_i(1)} \frac{|H|}{|L|} l_i. \tag{4.9}$$

and

$$\alpha \uparrow^H_L = \frac{\alpha(1)}{a_i(1)} \frac{|H|}{|L|} a_i. \tag{4.10}$$

Proof. It follows form the previous Theorem that the restriction of two irreducible H-characters to L either have the same common constituents or they have no common constituents. Let t_H be the integral on H with $t_H(1) = |H|$. One has that $t_H = \sum_{i=1}^{l} a_i$ as t_H is the regular character of H. Since H is free as left L-module (see [12]) it follows that the restriction of t_H to L is the regular character r_L of L multiplied by $|H|/|L|$. Thus $t_H \downarrow^H_L = |H|/|L| r_L$. But $r_L = \sum_{\alpha \in \mathrm{Irr}(L)} \alpha(1)\alpha$. Since all the irreducible constituents of a_i are inside L_i it follows from Theorem 4.3 that

$$a_i \downarrow_L = \frac{|H|}{|L|} l_i$$

for any i with $0 \leq i \leq s$.

Then if $\chi \in \mathcal{C}_i$ formula 4.6 implies that

$$\chi \downarrow_L = \frac{\chi(1)}{a_i(1)} \frac{|H|}{|L|} l_i. \tag{4.11}$$

Evaluating at 1 the above equality one gets $a_i(1) = \frac{|H|}{|L|} l_i(1)$. By Frobenius reciprocity the above restriction formula implies that if $\alpha \in \mathcal{L}_i$ then

$$\alpha \uparrow_L^H = \frac{\alpha(1)}{a_i(1)} \frac{|H|}{|L|} a_i. \tag{4.12}$$

□

5. The restriction of modules to normal left coideal subalgebras

Let G be a finite group and H a normal subgroup of G. If M is an irreducible H-module then

$$M \uparrow_H^G \downarrow_H^G = \oplus_{i=1}^s {}^{g_i}M$$

where ${}^gN$ is a conjugate module of M and $\{g_i\}_{i=1,\,s}$ is a set of representatives for the left cosets of H in G. For $g \in G$ the H-module ${}^gM$ has the same underlying vector space as M and the multiplication with $h \in H$ is given by $h.m = (ghg^{-1})m$ for all $m \in M$. It is easy to see that ${}^gM \cong^{g'} M$ if $gH = g'H$.

Let L be a normal left coideal subalgebra of H and M be an irreducible L-module. In this section we will define the notion of a conjugate module to M similar to the above group situation. If $d \in \mathrm{Irr}(H^*)$ we define a conjugate module ${}^dM$. We will show that as in the group situation, the irreducible constituents of $M \uparrow_L^H \downarrow_L^H$ and $\oplus_{d \in \mathrm{Irr}(H^*)} {}^dM$ are the same for any irreducible L-module M.

5.1. *The conjugate modules*

Let M be an irreducible L-module with character $\alpha \in C(L)$. We consider the following notion of conjugate module similar to the one introduced in [4] for normal Hopf subalgebras. It was also previously considered in [11] in the cocommutative case.

If W is a right H^*-module then $W \otimes M$ becomes an L-module with the following structure:

$$l(w \otimes m) = w_0 \otimes (Sw_{-1} l w_{-2}))m \tag{5.1}$$

for all $l \in L$, $w \in W$ and $m \in M$. Note that $l(w \otimes m) \in W \otimes M$ since H is involutory. Here we used that any right H^*-module W is a left H-comodule with the structure map $\rho(w) = w_{-1} \otimes w_0$.

In order to check that $W \otimes M$ is an L-module one has that

$$\begin{aligned} l.(l'.(w \otimes m)) &= l(w_0 \otimes (S(w_{-1})l'w_{-2})m) \\ &= w_0 \otimes (S(w_{-1})lw_{-2}).(S(w_{-3})l'w_{-4}).m) \\ &= w_0 \otimes (S(w_{-1})ll'w_{-2})m \\ &= (ll').(w \otimes m) \end{aligned}$$

for all $l, l' \in L$, $w \in W$ and $m \in M$.

It can be easily checked that if $W \cong W'$ as right H^*-modules then $W \otimes M \cong W' \otimes M$ as L-modules. Thus for any irreducible character $d \in \mathrm{Irr}(H^*)$ associated to a simple H-comodule W one can define the L-module ${}^dM \cong W \otimes M$. Its character will be denoted with ${}^d\alpha$ where α is the character of M. Clearly $W \otimes (M \oplus N) = (W \otimes M) \oplus (W \otimes N)$ for any two L-modules M and N of L.

5.2. Formula for the conjugate character

Proposition 5.2. *Let L be a normal left coideal subalgebra of H and M be an irreducible L-module with character $\alpha \in C(L)$. Suppose that W is an irreducible right H^*-module with character $d \in \mathrm{Irr}(H^*)$. Then the character ${}^d\alpha$ of the L-module ${}^dM$ is given by the following formula:*

$$^d\alpha(x) = \alpha(S(d_2)xd_1) \tag{5.3}$$

for all $x \in L$.

Proof. Indeed as in subsection 2.4 one may suppose that $W \cong k < x_{i1} \mid 1 \leq i \leq q >$ where $C_d = k < x_{ij} \mid 1 \leq i, j \leq q >$ is the coalgebra associated to W and $q = \epsilon(d) = |W|$. Then formula 5.1 becomes $l(x_{i1} \otimes w) = \sum_{j,\, r=1}^{q} x_{r1} \otimes (S(x_{jr})lx_{ij})m$. Since $d = \sum_{i=1}^{q} x_{ii}$ one gets the the formula for the character ${}^d\alpha$. □

Proposition 5.4. *With the above notations one has that ${}^{dd'}\alpha = {}^d({}^{d'}\alpha)$ for all $d, d' \in \mathrm{Irr}(H^*)$ and $\alpha \in C(L)$. This shows that $C(L)$ is a left $C(H^*)$-module.*

Proof. Clearly ${}^1\alpha = \alpha$. The above formula can be checked directly using the formula for the character ${}^d\alpha$ from the previous Proposition. Indeed for any $x \in L$ one has that ${}^{dd'}\alpha(x) = \alpha(S(d_2d'_2)xd_1d'_1) = ({}^{d'}\alpha)(Sd_2xd_1) = ({}^d({}^{d'}\alpha))(x)$. □

5.3. Compatibility with restriction

Let N be a left H-module and W a right H^*-module. Then $W \otimes N$ becomes an H-module with the following structure:

$$h(w \otimes m) = w_0 \otimes (S(w_{-1})hw_{-2}))m \tag{5.5}$$

It can be checked that $W \otimes N \cong N^{|W|}$ as H-modules. Indeed the map $\phi : W \otimes N \to$ given by ${}_\epsilon W \otimes N$ $w \otimes n \mapsto w_0 \otimes w_{-1}n$ is an isomorphism of H-modules where ${}_\epsilon W$ is considered left H-module with the trivial action. Its inverse is given by $w \otimes n \mapsto w_0 \otimes S(w_{-1})n$. To check that ϕ is an H-module map one has that

$$\begin{aligned}
\phi(h.(w \otimes n)) &= \phi(w_0 \otimes (S(w_{-1})hw_{-2})n) \\
&= w_0 \otimes w_{-1}(S(w_{-2}hw_{-3})n \\
&= w_0 \otimes hw_{-1}n \\
&= h.(w_0 \otimes w_{-1}n) \\
&= h\phi(w \otimes n)
\end{aligned}$$

for all $w \in W$, $m \in M$ and $h \in H$.

Proposition 5.6. *Let L be a normal left coideal subalgebra of H and M be an irreducible L-module with character $\alpha \in C(L)$. If $d \in \mathrm{Irr}(H^*)$ then*

$$\frac{1}{\epsilon(d)}\, {}^d\alpha \uparrow_L^H = \alpha \uparrow_L^H .$$

Proof. Using the notations from subsection 4.1 let $\mathcal{L}_i$ be the subset of $\mathrm{Irr}(L)$ which contains α. It is enough to show that the irreducible constituents of ${}^d\alpha$ are contained in this set and then the induction formula 4.12 from the same subsection can be applied for each of these constituents.

For this, suppose N is an irreducible H-module and

$$N \downarrow_L = \oplus_{i=1}^{s} N_i \tag{5.7}$$

Journal of Physics: Conference Series **346** (2012) 012004 doi:10.1088/1742-6596/346/1/012004

where N_i are irreducible L-modules. The above result implies that $W \otimes N \cong N^{|W|}$ as H-modules. Therefore $(W \otimes N)\downarrow_L = (N\downarrow_L)^{|W|}$ as L-modules. But $(W \otimes N)\downarrow_L = \oplus_{i=1}^s (W \otimes N_i)$ where each $W \otimes N_i$ is an L-module by 5.1. Thus

$$\oplus_{i=1}^s {N_i}^{|W|} = \oplus_{i=1}^s (W \otimes N_i) \tag{5.8}$$

This shows that if N_i is a constituent of $N\downarrow_L$ then $W \otimes N_i$ has all the irreducible L- constituents among those of $N\downarrow_L$. □

Proposition 5.9. *Let L be a left normal coideal subalgebra of H and M be an irreducible L-module. Then $M \uparrow_L^H \downarrow_L^H$ and $\oplus_{d \in \mathrm{Irr}(H^*)}\, {}^d M$ have the same irreducible constituents.*

Proof. Let $\mathcal{S}$ be the set of simple subcoalgebras of H. Then $H = \oplus_{C \in \mathcal{S}} C$ and therefore $H = \sum_{C \in \mathcal{S}} CL$. It follows that the induced module $M \uparrow_L^H$ can be written as the sum:

$$M \uparrow_L^H = H \otimes_L M = \sum_{C \in \mathcal{S}} CL \otimes_L M.$$

Note that each CL is a free L-module by Theorem 6.1 of [12]. On the other hand each $CL \otimes_L M$ is an L-module by left multiplication with elements of L since

$$l.(cl' \otimes_L m) = c_3(Sc_2lc_1)l' \otimes_L m = c_3 \otimes_L (Sc_2lc_1)l'm$$

for all $l,\ l' \in L$, $c \in C$ and $m \in M$. Thus $M \uparrow_L^H$ restricted to L is the sum of the L-modules $CL \otimes_L M$ with $C \in \mathcal{S}$. On the other hand it can be checked directly that the composition of the canonical maps $C \otimes M \hookrightarrow CL \otimes M \to CL \otimes_L M$ is a surjective morphism of L-modules. Let d be the character associated to C. Since $C \otimes M \cong \epsilon(d)\ {}^d M$ as L-modules this implies that $CL \otimes_L M$ is a homeomorphic image of $\epsilon(d)$ copies of ${}^d M$. Therefore the irreducible constituents of $M \uparrow_L^H \downarrow_L^H$ are among those of $\oplus_{d \in \mathrm{Irr}(H^*)}\, {}^d M$. In the proof of the previous Proposition we have shown the other inclusion. Thus $M \uparrow_L^H \downarrow_L^H$ and $\oplus_{d \in \mathrm{Irr}(H^*)}\, {}^d M$ have the same irreducible constituents □

Remark 5.10. *It follows that the equivalence class of a character $\alpha \in \mathrm{Irr}(L)$ under the relation d_L^H is given by all the irreducible constituents of ${}^d\alpha$ as d runs through all irreducible characters of H^*.*

6. On the commutator subalgebra and the dimension of an irreducible A-module

Let $\mathcal{S}$ be the set of one dimensional modules of a semisimple Hopf algebra H and

$$I := \cap_{M \in \mathcal{S}} \mathrm{Ann}_H(M).$$

Since $\mathcal{S}$ is closed under tensor product it follows from [15] that I is a Hopf ideal of H. Let $\pi : H \to H/I$ be the canonical Hopf projection.

Definition 6.1. *The normal left coideal subalgebra $L := H^{\mathrm{co}\ \pi}$ of A is called the left commutator subalgebra of A.*

It follows that $H//L$ is a commutative Hopf algebra since $I = \omega(L)H$ and all the irreducible representations of H/I are one dimensional. Similarly $R :={}^{\mathrm{co}\ \pi} H$ is called the right commutator subalgebra of A.

Lemma 6.2. *Let L and L' be two normal left coideal subalgebras. If $\omega(L)H \subset \omega(L')H$ then $L \subset L'$*

Proof. Let $\phi : H//L \to H//L'$ be given by $\overline{h} \mapsto \hat{h}$. Then it is easy to check that ϕ is a well defined Hopf map and $\phi \circ \pi_L = \pi_{L'}$ where π_L and $\pi_{L'}$ are the Hopf projections of H onto $H//L$ respectively $H//L'$. Thus $L = H^{\mathrm{co}\,\pi_L} \subset H^{\mathrm{co}\,\pi_{L'}} = L'$. □

Next proposition shows that L is the smallest normal left coideal subalgebra of H with $H//L$ a commutative Hopf algebra.

Proposition 6.3. *Let H be a semisimple Hopf algebra, L its left commutator subalgebra and L' an arbitrary normal left coideal subalgebra of H. If $H//L'$ is a commutative Hopf algebra then $L \subset L'$.*

Proof. Since all the representations of $H//L'$ are one dimensional it follows that $\mathrm{Irr}(H//L') \subset \mathcal{S}$. Thus by taking annihilators one has $\omega(L')H \supset I = \omega(L)H$ and therefore $L' \supset L$ by Lemma 6.2. □

Recall that a finite dimensional algebra A is said to be of Frobenius type if the dimension of any irreducible representation of A divides the dimension of A.

Theorem 6.4. *Suppose that the left commutator algebra of a semisimple Hopf algebra H is of Frobenius type. Then H is also of Frobenius type.*

Proof. Let as above $\mathcal{S}$ be the set of one dimensional modules of H and L its left commutator algebra. Then $\mathrm{Irr}(H//L) = \mathcal{S}$. Let M be an irreducible H-module and let V be an irreducible constituent of $M\downarrow_L^H$. Then by Equation 4.5 it follows all the other constituents of $H \otimes_L V$ are of the type $M \otimes X$ with $X \in \mathcal{S}$. Thus all the irreducible constituents of $H \otimes_L V$ have the same dimension, namely the dimension $|M|$ of M. This implies that $|M|$ divides $\frac{|H||V|}{|L|}$. But since L is of Frobenius type it follows that $|V|$ divides dimension $|L|$ of L. Therefore $|M|$ divides the dimension $|H|$ of H and H is also of Frobenius type. □

Acknowledgments
This work was supported by the strategic grant POSDRU/89/1.5/S/58852, Project "Postdoctoral programme for training scientific researchers" cofinanced by the European Social Found within the Sectorial Operational Program Human Resources Development 2007 - 2013.

References
[1] Andruskiewitsch N and Devoto 1995 *Algebra i Analiz* **7** 22–69
[2] Boltje R and Külshammer B 2009 *J. Algebra* **323** 1783–96
[3] Burciu S 2010 Kernels of representations and coideal subalgebras of Hopf algebras Preprint 1012.3096v1
[4] Burciu S 2009 *Comm. Alg* **37** 3573–85
[5] Burciu S and Kadison L and Külshammer B 2011 *Int. Electr. J Alg.* **9** 133–66
[6] Larson R. G. 1971 *J Alg.* **17** 352–68
[7] Larson R. G. and Radford D. E. 1988 *J. Algebra* **117** 267–89
[8] Montgomery S 1997 *Hopf algebras and their actions on rings* (AMS: Providence)
[9] Zhu Y 1994 *Int. Math. Res. Not.* **1** 53–9
[10] M. Rieffel 1979 *J. Alg.* **24** 364–86
[11] Schneider H- J 1990 *Israel J. Math.* **72** 196–231
[12] Skryabin Y 2007 *Trans. Amer. Math. Soc* **359** 2597–623
[13] Takeuchi M 1995 *Comm. Alg.* **22** 2503–23
[14] Ostrik V 2003 *Transform. Groups* **26** 177–206
[15] Passman D. S. and Quinn D. 1995 *Proc. Am. Math. Soc* **123** 327–33

Journal of Physics: Conference Series **346** (2012) 012005 doi:10.1088/1742-6596/346/1/012005

Analysis of endomorphisms

Roberto Conti[1], Jeong Hee Hong[2] and Wojciech Szymański[3]

[1] Dipartimento di Scienze, Università di Chieti-Pescara "G. D'Annunzio" , Viale Pindaro 42, I-65127 Pescara, Italy

[2] Department of Data Informtion, Korea Maritime University, Busan 606-791, South Korea

[3] Department of Mathematics and Computer Science, The University of Southern Denmark, Campusvej 55, DK-5230 Odense M, Denmark

E-mail: conti@sci.unich.it, hongjh@hhu.ac.kr, szymanski@imada.sdu.dk

Abstract. In this expository article, we discuss the recent progress in the study of endomorphisms and automorphisms of the Cuntz algebras and, more generally graph C^*-algebras (or Cuntz-Krieger algebras). In particular, we discuss the definition and properties of both the full and the restricted Weyl group of such an algebra. Then we outline a powerful combinatorial approach to analysis of endomorphisms arising from permutation unitaries. The restricted Weyl group consists of automorphisms of this type. We also discuss the action of the restricted Weyl group on the diagonal MASA and its relationship with the automorphism group of the full two-sided n-shift. Finally, several open problems are presented.

1. Introduction

The purpose of this note is to summarize some recent results obtained in a series of papers concerning endomorphisms and especially automorphisms of the Cuntz algebras and, more generally, graph algebras, [47, 17, 14, 18, 16, 27, 11, 12, 13].

1.1. Cuntz algebras and their subalgebras

The Cuntz algebra $\mathcal{O}_n$ is defined as the universal C^*-algebra generated by n isometries S_i, $i = 1, \ldots, n$, whose range projections have sum equal to 1, [19]. It turns out that $\mathcal{O}_n$ is a unital, separable, nuclear, simple, purely infinite C^*-algebra with the K-groups $K_0 \cong \mathbb{Z}_{n-1}$ and $K_1 = 0$, [19, 22]. Being simple, $\mathcal{O}_n$ has no ideals, however it has interesting unital subalgebras, namely the canonical UHF subalgebra $\mathcal{F}_n$ (isomorphic to the CAR algebra, [6], for $n = 2$), generated by Wick ordered monomials $S_\alpha S_\beta^*$ where the lengths of the multi-indices α and β are the same, and the diagonal $\mathcal{D}_n$ generated by the range projections $P_\alpha = S_\alpha S_\alpha^*$, which is a MASA (maximal abelian subalgebra) in $\mathcal{O}_n$.

1.2. Endomorphisms of Cuntz algebras

It is of great interest to study the group of automorphisms of the Cuntz algebra $\mathrm{Aut}(\mathcal{O}_n)$ providing insight into internal symmetries of this complicated C^*-algebra. More generally, it is also of interest to study (unital, $*$-) endomorphisms of $\mathcal{O}_n$, as for instance they provide a direct link to subfactor and sector theory, [38], or wavelets, [5]. One nice feature of the Cuntz algebras is that it is possible to describe all endomorphisms in a fairly simple way, namely they can be written in the form λ_u, where u is a unitary in $\mathcal{O}_n$, [20]. Here, λ_u is the unique endomorphism of $\mathcal{O}_n$ which acts on generators as $\lambda_u(S_i) = uS_i, i = 1,\ldots,n$. For instance, for the so-called canonical endomorphism φ of $\mathcal{O}_n$, defined by $\varphi(x) = \sum_{i=1}^n S_i x S_i^*$, one has $\varphi = \lambda_\theta$, where θ is the flip unitary $\sum_{i,j=1}^n S_i S_j S_i^* S_j^*$. This φ may be considered a noncommutative analogue of the one-sided shift on the space X_n of infinite strings over the alphabet $\{1,\ldots,n\}$.

Notice that for $x \in \mathcal{F}_n$ one has

$$\lambda_u(x) = \lim_{k\to\infty} u\varphi(u)\ldots\varphi^{k-1}(u)x\varphi^{k-1}(u^*)\ldots\varphi(u^*)u^*.$$

1.3. State of the arts

Recently, a significant progress in understanding the general structure of the endomorphisms of the Cuntz algebras preserving both the UHF-subalgebra and the diagonal MASA has been achieved. The main motivation for this analysis was to provide an answer to a question posed by Cuntz about thirty years ago, [20], about the structure of certain Weyl groups as explained below. Perhaps one reason for this problem remaining open for such a long time was the high level of complexity of the involved combinatorial structure. At the same time, the solution to this problem provides also an interesting direct link to the theory of shifts as studied in the theory of classical dynamical systems, [34], and this link opens great new perspectives, [12]. It is an exciting possibility that it could even provide some new tools for attacking certain old open problems in symbolic dynamics.

It is worth pointing out that in this research theoretical progress often goes hand-in-hand with massive computer calculations. Due to the large scale of the involved data, pen and paper calculations albeit possible in principle are utterly impractical. On the other hand, symbolic computer calculation (performed on *Magma* software, [4]) allowed for locating interesting examples illuminating the abstract theory and for complete classification of automorphisms in lower dimensional cases, [17, 14].

1.4. Graph algebras

By now, this program has been partially extended to the much wider class of graph algebras. Graph C^*-algebras are generated by partial isometries with mutually commuting domain and range projections, satisfying certain natural relations, [36, 35, 43]. For finite graphs, their C^*-algebras coincide with the class of Cuntz-Krieger algebras, [23, 21, 44].

Recently, both the full and the restricted Weyl groups of a graph algebra $C^*(E)$ have been defined, [13], at least in the case of a finite graph E. A combinatorial approach to permutative endomorphisms of graph algebras has been developed as well. In this approach, we see many similarities to the simpler case of the Cuntz algebras. Nevertheless, a number of new conceptual phenomena and technical difficults arise.

2. Cuntz algebras and their Weyl groups

2.1. The structure of the Weyl group of a Cuntz algebra

For algebras $A \subseteq B$ we denote by $\mathcal{N}_B(A) = \{u \in \mathcal{U}(B) : uAu^* = A\}$ the normalizer of A in B, and by $A' \cap B = \{b \in B : (\forall a \in A)\ ab = ba\}$ the relative commutant of A in B. For algebras $A \subseteq B$, we denote by $\mathrm{Aut}(B, A)$ the collection of all those automorphisms α of B such that $\alpha(A) = A$, and by $\mathrm{Aut}_A(B)$ those automorphisms of B which fix A point-wise.

One can show that λ establishes an isomorphism between the group $\mathcal{U}(\mathcal{D}_n)$ and the maximal abelian subroup $\mathrm{Aut}_{\mathcal{D}_n}(\mathcal{O}_n)$ of $\mathrm{Aut}(\mathcal{O}_n)$, which is the limit of higher dimensional tori, [20]. Furthermore, $\mathcal{O}_n^{\mathrm{Aut}_{\mathcal{D}_n}(\mathcal{O}_n)} = \mathcal{D}_n$. The idea is now to define a Weyl group as the quotient of the normalizer of $\mathrm{Aut}_{\mathcal{D}_n}(\mathcal{O}_n)$ in $\mathrm{Aut}(\mathcal{O}_n)$.

Proposition 2.1. *Let $A \subset B$ be a unital inclusion of C^*-algebras such that $B^{\mathrm{Aut}_A(B)} = A$. Then*

$$\mathcal{N}_{\mathrm{Aut}_A(B)}(\mathrm{Aut}(B)) = \mathrm{Aut}(B, A)\ . \tag{1}$$

One then defines the Weyl group $\mathfrak{W}_n$ for $\mathrm{Aut}(\mathcal{O}_n)$ as

$$\mathfrak{W}_n := \mathrm{Aut}(\mathcal{O}_n, \mathcal{D}_n)/\mathrm{Aut}_{\mathcal{D}_n}(\mathcal{O}_n), \tag{2}$$

as well as its restricted version

$$\mathfrak{R}\mathfrak{W}_n := \mathrm{Aut}(\mathcal{O}_n, \mathcal{D}_n) \cap \mathrm{Aut}(\mathcal{O}_n, \mathcal{F}_n)/\mathrm{Aut}_{\mathcal{D}_n}(\mathcal{O}_n). \tag{3}$$

The Weyl group is countable, [20].

Next, it turns out that $\mathrm{Aut}(\mathcal{O}_n, \mathcal{D}_n) = \lambda(\mathcal{N}_{\mathcal{O}_n}(\mathcal{D}_n))^{-1}$ and $\mathrm{Aut}(\mathcal{O}_n, \mathcal{D}_n) \cap \mathrm{Aut}(\mathcal{O}_n, \mathcal{F}_n) = \lambda(\mathcal{N}_{\mathcal{F}_n}(\mathcal{D}_n))^{-1}$, where for a subset $E \subset \mathcal{U}(\mathcal{O}_n)$ we set $\lambda(E)^{-1} = \{\lambda_u \mid u \in E\} \cap \mathrm{Aut}(\mathcal{O}_n)$.

Let $\mathcal{S}_n$ denote the group of unitaries that can be written as finite words in the S_i's and their adjoints and $\mathcal{P}_n := \mathcal{S}_n \cap \mathcal{F}_n$. Then $\mathcal{P}_n = \cup \mathcal{P}_n^k$, with $\mathcal{P}_n^k$ isomorphic to the symmetric group on n^k elements. Using the fact that $\mathcal{N}_{\mathcal{O}_n}(\mathcal{D}_n) = \mathcal{U}(\mathcal{D}_n) \cdot \mathcal{S}_n$ and, similarly, $\mathcal{N}_{\mathcal{F}_n}(\mathcal{D}_n) = \mathcal{U}(\mathcal{D}_n) \cdot \mathcal{P}_n$, [42], one can show that $\mathrm{Aut}(\mathcal{O}_n, \mathcal{D}_n) = \mathrm{Aut}_{\mathcal{D}_n}(\mathcal{O}_n) \rtimes \lambda(\mathcal{S}_n)^{-1}$ and $\mathrm{Aut}(\mathcal{O}_n, \mathcal{D}_n) \cap \mathrm{Aut}(\mathcal{O}_n, \mathcal{F}_n) = \mathrm{Aut}_{\mathcal{D}_n}(\mathcal{O}_n) \rtimes \lambda(\mathcal{P}_n)^{-1}$, [17]. It follows that there is an isomorphism $\mathfrak{R}\mathfrak{W}_n \cong \lambda(\mathcal{P}_n)^{-1}$. Furthermore there exists a short exact sequence

$$1 \to \mathcal{P}_n \overset{\mathrm{ad}}{\to} \lambda(\mathcal{P}_n)^{-1} \to G_n \to 1$$

where G_n is the image of $\lambda(\mathcal{P}_n)^{-1}$ through the quotient map $q : \mathrm{Aut}(\mathcal{O}_n) \to \mathrm{Out}(\mathcal{O}_n)$. This is a first step towards a deeper understanding of the detailed structure of the automorphism group of $\mathcal{O}_n$. Several challenging questions remain open about the properties of these groups. For instance, is $\lambda(\mathcal{P}_n)^{-1}$ isomorphic to $\lambda(\mathcal{P}_m)^{-1}$ when $n \neq m$? It can be shown that these groups are non-amenable, [17, 12], but we do not know if they are exact or possess any other weak amenability-type property or what kind of operator algebras they generate. We will come back to this issue again later.

2.2. Combinatorial approach to permutative endomorphisms

In this line of research, along with investigating global properties of the aforementioned groups, it is important to be able to analyze endomorphisms (and automorphisms) corresponding to specific elements of the unitary group of $\mathcal{O}_n$. Results of this type have been appearing right from the beginnings of the theory, [1, 41, 25, 24, 32, 33]. To this end, efficient combinatorial algorithms have been developed for deciding invertibility of endomorphisms related to permutation unitaries in the matrix algebras of the canonical UHF subalgebra $\mathcal{F}_n$, [17]. Somewhat surprisingly, labeled rooted trees play an important role in this scheme and allow some significant simplification of the computational complexity of the problem. It is worth pointing out that (finite) trees continue to be used in modern analysis in a number of different contexts, sometimes closely related to operator algebras, see e.g. [2, 3, 37, 10, 26] for a small sample.

We have obtained a complete classification of permutative endomorphisms of $\mathcal{O}_n$ associated to unitaries in $\mathcal{P}_n^k$ for small values of n and k, namely $n+k \leq 6$, [17, 14]. This was achieved with help of large scale symbolic computer calculations (in principle, these could be performed by hand, save for the high impracticality of this approach). Larger values of the parameters require capabilities which are not available at present, even for fast modern computers. It appears useful and interesting to carry out a complete analysis in this vein of certain well-defined subclasses of permutative endomorphisms, for example of the induced permutations considered in [27]. Closely related to this classification is the problem of determining the subgroup of the outer automorphism group of $\mathcal{O}_n$ generated by the image of $\lambda(\mathcal{P}_n^k)^{-1}$, at least for some small values of the parameters n and k. So far this has been achieved only for $k = 1$ (and arbitrary n), and $n = 2$ and $k \leq 3$, [17]. Partial results in the case $n = 2$ and $k = 4$ have been obtained in [17].

2.3. *Interaction with symbolic dynamics*

The restriction map yields an embedding of the Weyl group $\mathfrak{W}_n$ of $\mathcal{O}_n$ into the automorphim group of the canonical diagonal MASA $\mathcal{D}_n$. By Gelfand's theorem, the latter is isomorphic to the algebra of complex-valued, continuous functions on the full one-sided shift space on n-letters $X_n := \{1, \ldots, n\}^{\mathbb{N}}$. A deeper analysis of the action of $\mathfrak{W}_n$ on X_n reveals a striking relation between Cuntz algebras and classical symbolic dynamical systems. Namely, it turns out that, modulo inner automorphisms of $\mathcal{O}_n$, the restricted Weyl group $\mathfrak{RW}_n$ admits a natural embedding into the group of automorphisms of the full two-sided n-shift Σ_n modulo its center $\langle\sigma\rangle$, generated by the shift itself, [12]. That is, we have the following:

$$\mathfrak{RW}_n \cong \lambda(\mathcal{P}_n)^{-1} \xrightarrow{q} G_n \hookrightarrow \mathrm{Aut}(\Sigma_n)/\langle\sigma\rangle.$$

A key step in arriving at this relationship is the observation that elements of the restricted Weyl group may be identified (by restriction to the diagonal) with those homeomorphisms of X_n which eventually commute with the one-sided shift, [12].

Furthermore, if n is prime then this embedding of G_n is surjective and hence an isomorphism. Immediate consequences of this embedding include residual finiteness of the restricted Weyl group of $\mathcal{O}_n$ (modulo inner automorphisms) and its non-amenability.

It is expected that this relation between Cuntz algebras and symbolic dynamics can be exploited further, in both ways. Indeed, despite the many available results, the precise structure of the automorphism group of the two-sided shift remains unknown and one of the most outstanding problems in this theory is the question of the isomorphism of these groups corresponding to different primes. It would also be interesting to analyze in greater detail the properties of those elements in the Weyl group corresponding to certain specific automorphisms of the shift spaces. For example, marker automorphisms provide a convenient set of generators for $\mathrm{Aut}(X_n)$ and it would be interesting to know exactly the types of trees arising from the associated permutative automorphisms of $\mathcal{O}_n$.

Currently, investigations of the action of the full Weyl group $\mathfrak{W}_n$ on the (spectrum of) the diagonal MASA $\mathcal{D}_n$ are under way.

2.4. *Exotic endomorphisms*

This circle of ideas eventually contributed also to the construction of endomorphisms of $\mathcal{O}_n$ with quite unexpected properties, [16]. These exotic endomorphisms are associated to unitaries u which are not in $\mathcal{F}_n$ but nonetheless they satisfy $\lambda_u(\mathcal{F}_n) \subset \mathcal{F}_n$. As one can show that automorphisms with these properties do not exist, this results was rather surprising.

2.5. *Index and entropy*

There is a complementary aspect to the study of automorphisms of $\mathcal{O}_n$. Since its birth, the theory of endomorphisms of $\mathcal{O}_n$ has revealed a close relationship with the theory of subfactors (and its parallel C^*-algebraic version), [31, 15]. For instance, let $\omega = \tau \circ \epsilon$ be the state on $\mathcal{O}_n$ given by composition of the unique trace on $\mathcal{F}_n$ with the canonical conditional expectation $\epsilon : \mathcal{O}_n \to \mathcal{F}_n$ obtained by averaging over the gauge action of $\mathbb{T}$. Moreover, let u be a unitary in $M_n \otimes M_n \otimes \cdots \otimes M_n$ (k factors) inside $\mathcal{F}_n$. Then it is well-known that $\pi_\omega(\lambda_u(\mathcal{O}_n))'' \subset \pi_\omega(\mathcal{O}_n)''$ is an inclusion of type $III_{1/n}$ factors with index bounded by $n^{2(k-1)}$. It is then a challenging problem to compute the index values obtained in this way and to study the properties of the associated inclusions of factors, in terms of the unitary u, [15]. Inclusions corresponding to endomorphisms λ_u are also important as a tool for realizing specific examples of subfactors with preassigned fusion rules (or principal graphs) and other combinatorial invariants, [28, 29, 30].

Similarly, one can compute the topological entropy of suitable classes of endomorphisms of $\mathcal{O}_n$ and, in the case of permutative endomorphisms, also the entropy of their restriction to $\mathcal{D}_n$, [9, 46, 45].

3. Graph algebras and their endomorphisms

Recently, a large part of the theory of localized endomorpisms of the Cuntz algebras has been extended to a much wider class of graph C^*-algebras (Cuntz-Krieger algebras), [13]. Among basic examples of such endomorphisms there are the quasi-free automorphisms, [48]. For most of the results, some mild hypotheses on the underlying graphs were imposed. In this way, the key concept of the Weyl group and the restricted Weyl group has been introduced in the context of graph algebras. The two Weyl groups were shown to possess a number of properties very similar to those which hold in the case of $\mathcal{O}_n$. Furthermore, criteria for deciding invertibility of localized endomorphisms have been found, and a combinatorial approach to the analysis of permutative endomorphisms has been developed.

The aforementioned progress notwithstanding, a number of interesting challenging problems remain open. firstly, it would be useful to try and extend these results to include larger classes of graphs (infinite, not row-finite, with sinks, etc.). In particular, it appears natural to ask how these results would be affected by relaxing the hypothesis that the canonical diagonal subalgebra be a MASA (cf. [8, Theorem 6]). Another natural question arises with regard to the combinatorics involved in the analysis of permutative endomorphisms. At this stage, it is not clear if labelled trees are sufficient in this process, and perhaps one should search for some objects of greater complexity. Another natural line of future investigations would be finding a closer relationship between the restricted Weyl group of a graph algebra and the automorphism groups (both one- and two-sided) of the corresponding subshift of finite type, cf. [39, 40, 7]. Certainly, building a larger collection of well understood examples will help in these investigations.

References

[1] Archbold R J 1979 On the flip-flop automorphism of $C^*(S_1, S_2)$ *Quart. J. Math. Oxford Ser. (2)* **30** 129
[2] Arnol'd V I 1994 Topological invariants of plane curves and caustics *Dean Jacqueline B. Lewis Memorial Lectures presented at Rutgers University, New Brunswick, New Jersey* University Lecture Series, 5 (Providence, RI: Amer. Math. Soc.)
[3] Bhattacharyya B 1998 Krishnan-Sunder subfactors and a new countable family of subfactors related to trees *Doctoral Dissertation* UC Berkeley
[4] Bosma W, Cannon J and Playoust C 1997 The Magma algebra system. I. The user language *J. Symbolic Comput.* **24** 235
[5] Bratteli O and Jørgensen P E T 1997 Isometries, shifts, Cuntz algebras and multiresolution wavelet analysis of scale N *Integral Equations & Operator Theory* **28** 382
[6] Bratteli O and Robinson D 1987 *Operator Algebras and Quantum Statistical Mechanics I. C^*- and W^*-Algebras, Symmetry Groups and Decomposition of States* Second edition. Texts and Monographs in Physics (New York: Springer-Verlag)
[7] Carlsen T M and Silvestrov S 2007 C^*-crossed products and shift spaces *Expo. Math.* **25** 275
[8] Carlsen T M and Silvestrov S 2009 On the Exel crossed product of topological covering maps *Acta Appl. Math.* **108** 573
[9] Choda M 1999 Entropy of Cuntz's canonical endomorphism *Pacific J. Math.* **190** 235
[10] Connes A and Kreimer D 1998 Hopf algebras, renormalization and noncommutative geometry *Commun. Math. Phys.* **199** 203
[11] Conti R 2010 Automorphisms of the UHF algebra that do not extend to the Cuntz algebra, *J. Aust. Math. Soc.* **89** 309
[12] Conti R, Hong J H and Szymański W 2010 The restricted Weyl group of the Cuntz algebra and shift endomorphisms *J. Reine Angew. Math.* in press (arXiv:1006.4791)
[13] Conti R, Hong J H and Szymański W Endomorphisms of graph algebras Preprint 1101.4210
[14] Conti R, Kimberley J and Szymański W 2010 More localized automorphisms of the Cuntz algebras *Proc. Edinburgh Math. Soc.* **53** 619
[15] Conti R and Pinzari C 1996 Remarks on the index of endomorphisms of Cuntz algebras *J. Funct. Anal.* **142** 369
[16] Conti R, Rørdam M and Szymański W 2010 Endomorphisms of $\mathcal{O}_n$ which preserve the canonical UHF-subalgebra *J. Funct. Anal.* **259** 602
[17] Conti R and Szymański W 2011 Labeled trees and localized automorphisms of the Cuntz algebras *Trans. Amer. Math. Soc.* **363** 5847

[18] Conti R and Szymański W 2009 Computing the Jones index of quadratic permutation endomorphisms of $\mathcal{O}_2$ *J. Math. Phys.* **50** 012705
[19] Cuntz J 1977 Simple C^*-algebras generated by isometries *Commun. Math. Phys.* **57** 173
[20] Cuntz J 1980 Automorphisms of certain simple C^*-algebras *Quantum Fields-Algebras-Processes* (Bielefield, 1978) ed. L Streit (Vienna: Springer) pp 187-96
[21] Cuntz J 1981 A class of C^*-algebras and topological Markov chains. II. Reducible chains and the Ext-functor for C^*-algebras *Invent. Math.* **63** 25
[22] Cuntz J 1981 K-theory for certain C^*-algebras *Ann. of Math.* **113** 181
[23] Cuntz J and Krieger W 1980 A class of C^*-algebras and topological Markov chains *Invent. Math.* **56** 251
[24] Enomoto M, Fujii M, Takehana H and Watatani Y 1979/80 Automorphisms on Cuntz algebras. II *Math. Japon.* **24** 463
[25] Enomoto M, Takehana H and Watatani Y 1979/80 Automorphisms on Cuntz algebras *Math. Japon.* **24** 231
[26] Hollands S and Olbermann H 2009 Perturbative quantum field theory via vertex algebras *J. Math. Phys.* **50** 112304
[27] Hong J H, Skalski A and Szymański W On invariant MASAs for endomorphisms of the Cuntz algebras Indiana Univ. Math. J. in press (Preprint 1001.1899)
[28] Izumi M 1993 Subalgebras of infinite C^*-algebras with finite Watatani indices I. Cuntz algebras *Commun. Math. Phys.* **155** 157
[29] Izumi M 1998 Subalgebras of infinite C^*-algebras with finite Watatani indices II: Cuntz-Krieger algebras *Duke Math. J.* **91** 409
[30] Izumi M 2001 The structure of sectors associated with Longo-Rehren inclusions. II. Examples *Rev. Math. Phys.* **13** 603
[31] Jones V F R 1994 On a family of almost commuting endomorphisms *J. Funct. Anal.* **122** 84
[32] Kawamura K 2005 Polynomial endomorphisms of the Cuntz algebras arising from permutations. I. General theory *Lett. Math. Phys.* **71** 149
[33] Kawamura K 2006 Branching laws for polynomial endomorphisms of Cuntz algebras arising from permutations *Lett. Math. Phys.* **77** 111
[34] Kitchens B P 1998 *Symbolic Dynamics: One-Sided, Two-Sided and Countable State Markov Shifts* (Berlin: Springer)
[35] Kumjian A, Pask D and Raeburn I 1998 Cuntz-Krieger algebras of directed graphs *Pacific J. Math.* **184** 161
[36] Kumjian A, Pask D, Raeburn I and Renault J 1997 Graphs, groupoids, and Cuntz-Krieger algebras *J. Funct. Anal.* **144** 505
[37] Loday J-L and Ronco M O 1998 Hopf algebra of the planar binary trees *Adv. Math.* **139** 293
[38] Longo R 1994 A duality for Hopf algebras and for subfactors. I *Commun. Math. Phys.* **159** 133
[39] Matsumoto K 2000 On automorphisms of C^*-algebras associated with subshifts *J. Operator Theory* **44** 91
[40] Matsumoto K 2010 Orbit equivalence of topological Markov shifts and Cuntz-Krieger algebras *Pacific J. Math.* **246** 199
[41] Matsumoto K and Tomiyama J 1993 Outer automorphisms of Cuntz algebras *Bull. London Math. Soc.* **25** 64
[42] Power S C 1998 Homology for operator algebras, III. Partial isometry homotopy and triangular algebras *New York J. Math.* **4** 35
[43] Raeburn I 2005 Graph Algebras *CBMS Regional Conf. Series in Math.* **103** (Providence: Amer. Math. Soc.)
[44] Rørdam M 1995 Classification of Cuntz-Krieger algebras *K-Theory* **9** 31
[45] Skalski A Noncommutative topological entropy of endomorphisms of Cuntz algebras II (*Preprint* arXiv:1002.2276)
[46] Skalski A and Zacharias J 2008 Noncommutative topological entropy of endomorphisms of Cuntz algebras *Lett. Math. Phys.* **86** 115
[47] Szymański W 2008 On localized automorphisms of the Cuntz algebras which preserve the diagonal subalgebra *R.I.M.S. Kôkyûroku* **1587** 109
[48] Zacharias J 2000 Quasi-free automorphisms of Cuntz-Krieger-Pimsner algebras *C^*-Algebras* (Münster, 1999) (Berlin: Springer) pp 262–72

Algebra, Geometry, and Mathematical Physics 2010 IOP Publishing
Journal of Physics: Conference Series **346** (2012) 012006 doi:10.1088/1742-6596/346/1/012006

Representations of posets: linear versus unitary

Vyacheslav Futorny[1], **Yurii Samoilenko**[2], **Kostyantyn Yusenko**[1,2]

[1] IME-USP, São-Paulo, Brasil
[2] Institute of Mathematics, Kiev, Ukraine

Abstract. A number of recent papers treated the representation theory of partially ordered sets in unitary spaces with the so called orthoscalar relation. Such theory generalizes the classical theory which studies the representations of partially ordered sets in linear spaces. It happens that the results in the unitary case are well-correlated with those in the linear case. The purpose of this article is to shed light on this phenomena.

MSC 2010: 16G20, 53D20, 14L24.

1. Introduction

In the second part of XX century the new results concerning to representation theory in linear spaces were obtained, in particular, the results about relative position of several subspaces and partially ordered collections of subspaces in linear spaces (see, for example, [7, 24]). Representations of partially ordered sets (*posets* in sequel) were first introduced by L.Nazarova and A.Roiter (see [26]), using the language of "matrix problems", as a tool to prove the second Brauer-Thrall conjecture. Later P.Gabriel gave an equivalent (up to a finite number of indecomposables) definition of representations using "subspace" language (see, for example, [29, Chapter 3]). In the sequel we use the latter terminology.

One can try to develop a similar theory for unitary spaces. Namely, one can transfer straightforward the definitions from linear to unitary case keeping in mind that the morphisms between two representations should preserve their unitary structure; that is, the morphisms are unitary maps. With this restriction the classification problem becomes extremely difficult already in very simple situations. For example the problem of classifying unitary representations of the poset $\mathcal{P} = \{a_1, a_2, a_3 \mid a_2 \prec a_3\}$ is hopeless, since by [20, 21] it contains the problem of classifying any system of operators in unitary spaces (see [27, Chapter 3] for more details).

The idea, hence, is to consider those representations which satisfy some additional conditions. In such a way there were defined orthoscalar representations of quivers (see [18, 19]) and orthoscalar representations of posets (see, for example, [1, 28]). Certainly the additional conditions can be chosen differently. Orthoscalarity relation appears in different areas of mathematics. For example, using [14, 16], one can show that every indecomposable orthoscalar representation of a poset generate a stable reflexible sheaf on certain toric variety. This is one of the motivation to study such representations.

Analyzing the results related to orthoscalar representations of quivers and posets (see [1, 18, 19, 28] and references therein) one can see the mysterious (on the first sight) connection with the classical results (more details in Section 1). The main goal of this paper is to explain this phenomena.

For a given finite poset $\mathcal{P}$ we consider the variety $R_{\mathcal{P},d}$ of all representations of $\mathcal{P}$ with dimension vector $d = (d_0; d_i)_{i\in\mathcal{P}}$. We show that the classes of unitary non-equivalent orthoscalar representations of $\mathcal{P}$ with dimension vector d can be viewed as the symplectic quotient of $R_{\mathcal{P},d}$. On the other side we consider the GIT quotient of $R_{\mathcal{P},d}$ with respect to some polarization of this variety. This quotient can be viewed as the set of isomorphism classes of polystable linear representations of $\mathcal{P}$ with dimension d. We show that the identification between these two quotients is a consequence of the fundamental Kirwan-Ness theorem (see, for example, [23]).

Let us mention that the idea of the symplectic reduction in similar contexts is not new. For example, it appeared in [10] (constructing moduli spaces for quivers) and in [15] (in connection with Horn's problem). In this text we apply it in the setting of partially ordered sets.

1.1. Structure of the paper.

In Section 2 we recall basic facts about the posets and their indecomposable representations in linear spaces and indecomposable orthoscalar representations in unitary spaces, in particular we compare some known results about the structure of representations in both cases. In Section 3 we give some basic facts about symplectic quotients. Section 4 is dedicated to the GIT quotients. We explain the construction and state the Kirwan-Ness theorem which connects symplectic and GIT quotients. In Section 5 we prove Theorem 3 which relates representations of posets in linear spaces with the ones in unitary spaces with orthoscalar condition. In Section 6 we estimate the dimension of the variety of uni-classes of representations of a given poset with a given dimension vector using the connection between posets and bound quivers.

2. Posets and their representations.

Let $\mathcal{P}$ be a finite poset with partial order relation $\prec$. Denote by $\mathcal{P}^*$ the extension of $\mathcal{P}$ by a unique maximal element $*$. To $\mathcal{P}$ we associate the Hasse quiver of $\mathcal{P}^*$ which is denoted by $Q^{\mathcal{P}}$. The *order* of $\mathcal{P}$ is the number of its elements. A poset is said to be *primitive* and denoted by $(n_1, \ldots, n_s)$ if it is a disjoint (cardinal) sum of linearly ordered sets with the orders n_i. A poset $\mathcal{P}$ is primitive if and only if the corresponding quiver $Q_{\mathcal{P}}$ has a star-shaped form.

Example 1. Let $\mathcal{P} = (1, 2)$ be the poset consisting of three elements a_1, a_2, a_3 with a unique relation $a_2 \prec a_3$. Then the associated quiver $Q^{\mathcal{P}}$ has the following form:

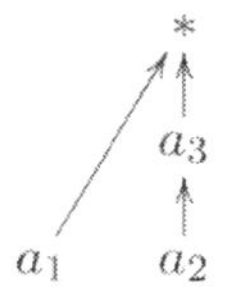

2.1. Linear representations

Following P.Gabriel (see [29, Chapter 3]) we define the category $\mathcal{P}-sp$ of representations of $\mathcal{P}$ in linear spaces. The objects of $\mathcal{P}-sp$ are systems $(V;V_i)_{i\in\mathcal{P}}$, in which V is a vector space, each V_i is a subspace of V and $V_i\subset V_j$ if $i\prec j$. The set of morphisms between two objects $(V;V_i)_{i\in\mathcal{P}}$ and $(\tilde{V};\tilde{V}_i)_{i\in\mathcal{P}}$ consists of the linear maps $f:V\to\tilde{V}$ such that $f(V_i)\subset\tilde{V}_i$ for all $i\in\mathcal{P}$. Two objects are *isomorphic* if there exists an invertible $f:V\to\tilde{V}$ with $f(V_i)=\tilde{V}_i$ for all $i\in\mathcal{P}$. The category $\mathcal{P}-sp$ is additive with usual direct sums. An object $(V;V_i)_{i\in\mathcal{P}}$ is said to be *indecomposable* it is not isomorphic to a direct sum of two non-zero objects.

Example 2. Let $\mathcal{P}=(1,1)$ be a poset consisting of two incomparable elements. Fixing $\lambda\in\mathbb{C}$ we build the following linear two-dimensional representation of $\mathcal{P}$

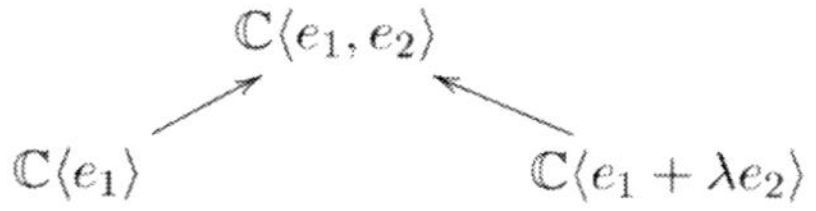

It is easy to see that any such representation with $\lambda\neq 0$ is equivalent to the one with $\lambda=1$ and hence it splits into one-dimensional representations:

One can show also that this poset has only finite number of isomorphism classes of indecomposable objects in $\mathcal{P}-sp$.

The *dimension vector* $d=(d_0;d_i)_{i\in\mathcal{P}}\in\mathbb{N}^{|\mathcal{P}|+1}$ of a given representation $(V;V_i)_{i\in\mathcal{P}}$ is defined as $d_0=\dim V$, $d_i=\dim V_i$. Fixing a dimension vector d we define the variety $R_{\mathcal{P},d}$ of representations of $\mathcal{P}$ with dimension d

$$R_{\mathcal{P},d}=\left\{(V_i)_{i\in\mathcal{P}}\in\prod_{i\in\mathcal{P}}\mathrm{Gr}(d_i,d_0)\;\middle|\;V_i\subset V_j,\ i\prec j\right\}.$$

The group $GL(d_0)$ acts on $R_{\mathcal{P},d}$ via simultaneous base change $g\cdot(V_i)_{i\in\mathcal{P}}=(g(V_i))_{i\in\mathcal{P}}$, so that the orbits of this action are in one-to-one correspondence with the isomorphism classes of representations of $\mathcal{P}$ with the dimension vector d.

Example 3. Consider the poset $\mathcal{P}=(1,1,1,1)$ consisting of four incomparable elements. L.Nazarova [24] and I.Gelfand with V.Ponomarev [7] completely classified indecomposable systems of four subspaces. Let us recall the description for the dimension vector $d=(2;1,1,1,1)$:

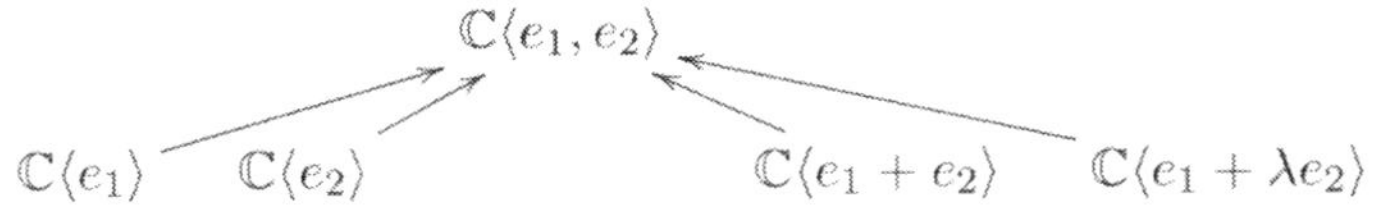

It was shown that non-isomorphic indecomposable representations are parametrized by the extended complex plane $\mathbb{C} \cup \{\infty\} \simeq S^2$ (when $\lambda = \infty$ the corresponding subspace $\mathbb{C}\langle e_1 + \lambda e_2\rangle$ is defined to be $\mathbb{C}\langle e_2\rangle$). One can see that when $\lambda = 0$, 1 or ∞ then the subspaces in corresponding representations are not in general position (this cases are exceptional). Therefore we have the sphere without three points of non-exceptional representations.

2.2. Unitary representations. Orthoscalar representations.
Define the category $\mathcal{P} - usp$ of unitary representations of a poset $\mathcal{P}$. The objects of $\mathcal{P} - usp$ are systems $(U; U_i)_{i\in\mathcal{P}}$, in which U is a unitary space, each U_i is a subspace of U and $U_i \subset U_j$ if $i \prec j$. The set of morphisms between two objects $(U; U_i)_{i\in\mathcal{P}}$ and $(\tilde{U}; \tilde{U}_i)_{i\in\mathcal{P}}$ consists of isometric maps $\varphi : U \mapsto \tilde{U}$ such that $\varphi(U_i) \subset U_j$. Two systems are said to be *unitary equivalent* if the morphism between them can be chosen to be a unitary map. An object $(U; U_i)_{i\in\mathcal{P}}$ is said to be *indecomposable* it is not unitary equivalent to an orthogonal sum of two non-zero objects in $\mathcal{P} - usp$.

Example 4. The poset $\mathcal{P} = (1, 1)$ contains a continuous family of unitary non-equivalent representations. Indeed, consider the representations of the poset $\mathcal{P}$ as in Example 2. Taking usual scalar product in $\mathbb{C}\langle e_1, e_2\rangle$ we view these representations as the objects in $\mathcal{P} - usp$. For different $\lambda \in (0, 1]$ they are indecomposable and unitary non-equivalent. Moreover, P.Halmosh [8] proved that the representations as in Example 2 gives a complete list of all indecomposable objects in $\mathcal{P} - usp$.

Example 5. Consider the poset $\mathcal{P} = (1, 2)$. There exists only a finite number of indecomposable objects in $\mathcal{P} - sp$ up to the isomorphism. At the same time it is an extremely hard problem to classify the indecomposable objects in $\mathcal{P} - usp$ up to the unitary equivalence. By S.Kruglyak and Yu.Samoilenko [20, 21], such problem is $*$-wild; that is, it contains the problem of classifying any system of operators in unitary spaces (see [27, Chapter 3] for more details).

Fix a weight $\chi = (\chi_0; \chi_i)_{i\in\mathcal{P}} \in \mathbb{Z}_+^{|\mathcal{P}|+1}$ and consider those objects $(U; U_i)_{i\in\mathcal{P}}$ in $\mathcal{P} - usp$ that satisfy

$$\sum_{i\in\mathcal{P}} \chi_i P_{U_i} = \chi_0 I,$$

where P_{U_i} is the orthogonal projection onto U_i. Such objects are called χ-*orthoscalar* ([1, 18, 19]) and form the subcategory denoted by $\mathcal{P}_\chi - usp$. One of the main motivation to study χ-orthoscalar objects is that any such object gives rise to a polystable reflexible sheaves on certain toric variety (see [14, 16] and references therein).

Example 6. Consider the poset $\mathcal{P} = (1, 1)$. The objects in $\mathcal{P}_\chi$-usp are those systems $(U; U_1, U_2)$ that satisfy

$$\chi_1 P_{U_1} + \chi_2 P_{U_2} = \chi_0 I.$$

One can see that the projections P_{U_1} and P_{U_2} commute. Hence the indecomposable objects are at most one-dimensional. For the category $\mathcal{P}_\chi$-usp to be non-empty χ_0 should be equal to $\chi_1+\chi_2$ (trace identity). Moreover, if this condition is satisfied then the indecomposables in $\mathcal{P}_\chi - usp$ are in one-to-one correspondence with the indecomposables in $\mathcal{P} - sp$.

Example 7. Consider the poset $\mathcal{P} = (1,1,1,1)$. As in Example 3 we consider the objects with the dimension vector $(2;1,1,1,1)$. Let us take the weight $\chi = (2;1,1,1,1)$. The description of unitary non-equivalent irreducible quadruples of projections that satisfy

$$P_{U_1} + P_{U_2} + P_{U_3} + P_{U_4} = 2I,$$

is the following (e.g. [27, Chapter 2.2]):

$$P_{U_1} = \frac{1}{2}\begin{pmatrix} 1+a & -b-ic \\ -b+ic & 1-a \end{pmatrix}, \quad P_{U_2} = \frac{1}{2}\begin{pmatrix} 1-a & b-ic \\ b+ic & 1+a \end{pmatrix},$$
$$P_{U_3} = \frac{1}{2}\begin{pmatrix} 1-a & -b+ic \\ -b-ic & 1+a \end{pmatrix}, \quad P_{U_4} = \frac{1}{2}\begin{pmatrix} 1+a & b+ic \\ b-ic & 1-a \end{pmatrix}.$$

Topologically the set of parameters a, b, c is a sphere S^2 without three points where representations split into one-dimensional. Moreover by [22], to each triple (a,b,c) one can associate $\lambda \neq 0, 1, \infty$ (and vice versa) such that the corresponding system $(U; U_i)_{i\in\mathcal{P}}$ is equivalent to a system $(V; V_i)_{i\in\mathcal{P}}$ which corresponds to λ as in Example 3. Hence, the uni-classes of indecomposable objects in $\mathcal{P}_{(2;1,1,1,1)} - usp$ with dimension vector $(2;1,1,1,1)$ are in one-to-one correspondence with the iso-classes of indecomposable non-exceptional objects in $\mathcal{P} - sp$ with dimension vector $(2;1,1,1,1)$.

Recall that $\mathcal{P}$ is said to be *representation-finite* (resp. *$*$-representation-finite*) if it has only a finite number of isomorphism classes of indecomposables in $\mathcal{P} - sp$ (resp. in $\mathcal{P}_\chi - usp$ for any weight χ). P.Gabriel classified all representation-finite quivers together with their indecomposable representations. Correspondingly, M.Kleiner [13] classified representation-finite posets, and also their indecomposable representations. There is a direct analogue of the Gabriel's classification obtained by S.Kruglyak and A.Roiter [19] for unitary orthoscalar representations of quivers (see [19] for the definition of orthoscalarity for quivers). An analogue of the Kleiner's classification was obtained in [28]. In particular it turned out that

$$\mathcal{P} \text{ is representation-finite} \Leftrightarrow \mathcal{P} \text{ is } *\text{-representation-finite}.$$

2.3. Non-commutative Hopf fibration

Consider the subset of $R_{\mathcal{P},d}$ consisting of the representations which satisfy χ-orthoscalar condition:

$$R^\chi_{\mathcal{P},d} = \left\{ (V_i)_{i\in\mathcal{P}} \in R_{\mathcal{P},d} \;\middle|\; \sum_{i\in\mathcal{P}} \chi_i P_{V_i} = \chi_0 I \right\},$$

where the projections P_{V_i} are taking with respect to the standard Hermitian metric on $\mathbb{C}^{d_0}$. The group $U(d_0)$ acts on $R_{\mathcal{P},d}$ as a subgroup of $GL(d_0)$. The orbits of this action on $R^{\chi}_{\mathcal{P},d}$ are in one-to-one correspondence with the uni-classes of χ-orthoscalar representations of $\mathcal{P}$ with dimension vector d.

We will see that the connection between orbit spaces $R_{\mathcal{P},d}/GL(d_0)$ and $R^{\chi}_{\mathcal{P},d}/U(d_0)$ is a generalization of the following commutative identification.

The group $\mathbb{C}^*$ (identified with $GL(1)$) acts on $\mathbb{C}^n$ by multiplication $c * z = (cz_1, \ldots, cz_n)$. The corresponding orbit space is not Hausdorff (the orbit of 0 lies in any neighbourhood of any other orbit), but $\mathbb{C}^n - \{0\}/\mathbb{C}^*$ is Hausdorff and homeomorphic to a projective space $\mathbb{CP}^{n-1}$. Consider the subset of $\mathbb{C}^n$ consisting of those points $(z_1, \ldots, z_n)$ that satisfy $z_1\overline{z}_1 + \ldots + z_n\overline{z}_n = 1$. This subset determines the sphere S^{2n-1}. The group S^1 (identified with the unitary group $U(1)$) acts on S^{2n-1} by rotations. The corresponding orbit space S^{2n-1}/S^1 (so called *Hopf-fibration*) is again a projective space $\mathbb{CP}^{n-1}$. So we trivially have

$$\{(z_1, \ldots, z_n) \in \mathbb{C}^n \mid z_1\overline{z}_1 + \ldots + z_n\overline{z}_n = 1\}/U(1) \cong \mathbb{C}^n - \{0\}/GL(1).$$

Let $\mathcal{P}$ be a poset with n non-comparable elements, d be dimension vector and χ be the weight. Formally substitute each z_i by the matrix $A_i \in M_{d_0 \times d_i}(\mathbb{C})$. The corresponding equation between z_i has the following form

$$A_1A_1^* + \ldots + A_nA_n^* = I.$$

Take $(\chi'_1, \ldots, \chi'_n) = (\chi_1\chi_0^{-1}, \ldots, \chi_n\chi_0^{-1})$. Viewing each χ'_i as $\|A_i\|^2$ we write the non-commutative version of $\overline{z}_iz_i = |z_i|^2$ as $A_i^*A_i = \chi'_iI_{d_i}$ From last equation we get $A_iA_i^* = \chi'_iP_{\mathrm{Im}(A_i)}$. We say that $(A_1, \ldots, A_n)$ and $(\tilde{A}_1, \ldots, \tilde{A}_n)$ lie in the same equivalence class under the action of $U(d_0)$ if there exists $\varphi \in U(d_0)$ such that $\varphi(\mathrm{Im}(A_i)) = \mathrm{Im}(\tilde{A}_i)$ for all $i \in \mathcal{P}$. Then the quotient

$$\left\{(A_i)_{i\in\mathcal{P}} \in (M_{d_0\times d_i}(\mathbb{C}))_{i\in\mathcal{P}} \;\middle|\; A_1A_1^* + \ldots + A_nA_n^* = I, \quad A_i^*A_i = \chi'_iI_{d_i} \right\} \Big/ U(d_0)$$

parametrizes the equivalence classes of χ-orthoscalar representations with dimension vector d of the poset $\mathcal{P}$ and can be seen as a *non-commutative Hopf fibration*. On the other hand the quotient

$$\left\{(A_i)_{i\in\mathcal{P}} \in (M_{d_0\times d_i}(\mathbb{C}))_{i\in\mathcal{P}} \;\middle|\; A_i^*A_i = \chi'_iI_{d_i} \right\} \Big/ GL(d_0)$$

can be identified with $R_{\mathcal{P},d}/GL(d_0)$. In what follows we show (see Section 4) that for any poset $\mathcal{P}$ these two quotients are connected similarly to commutative example.

3. Symplectic quotient.

3.1. Lie groups and algebras. Coadjoint representations

We briefly recall necessary information about Lie groups, Lie algebras and coadjoint representations of Lie groups (more information see, for example, in [12]).

By G we denote a Lie group (which is assumed to be finite-dimensional), $\mathfrak{g} = \mathfrak{Lie}(G)$ its Lie algebra (the tangent space to the identity element of G) and $\mathfrak{g}^*$ denotes the dual Lie algebra. A group G is called *complex matrix group* if G is a subgroup and at the

Journal of Physics: Conference Series **346** (2012) 012006 doi:10.1088/1742-6596/346/1/012006

same time a smooth submanifold of $GL(n)$. In this paper we mainly consider the case when $G = U(n)$; that is, the group of unitary matrices in n-dimensional complex space $\mathbb{C}^n$. The corresponding Lie algebra $\mathfrak{u}(n)$ consists of skew-Hermitian matrices.

A Lie group G acts on itself by inner automorphisms

$$A : G \to Aut(G), \quad A(g) : h \mapsto ghg^{-1}.$$

Differentiating we get the *adjoint representation* $d(A(g))_e = Ad_g : \mathfrak{g} \to \mathfrak{g}$. In the case when G is a matrix group this representations is given by

$$Ad_g(x) = gxg^{-1}, \quad g \in G, \quad x \in \mathfrak{g}.$$

Let $\langle \cdot, \cdot \rangle : \mathfrak{g}^* \times \mathfrak{g} \to \mathbb{K}$, $(\xi, x) \to \langle \xi, x \rangle = \xi(x)$ be natural pairing between $\mathfrak{g}$ and $\mathfrak{g}^*$. The *coadjoint representation* $Ad^* : G \to Aut(\mathfrak{g}^*)$ is defined by

$$\langle Ad_g^* \xi, x \rangle = \langle \xi, Ad_{g^{-1}} x \rangle, \quad x \in \mathfrak{g}, \quad \xi \in \mathfrak{g}^*.$$

In the case when G is a matrix group the coadjoint representations is given by

$$Ad_g^*(\xi) = g\xi g^{-1}, \quad g \in G, \quad \xi \in \mathfrak{g}^*.$$

Recall that if $\mathfrak{g}^*$ is semisimple then adjoint and coadjoint representations are equivalent. By the *coadjoint orbits* we understand the orbits of G on $\mathfrak{g}^*$.

Example 8. Let $G = U(n)$. Its dual Lie algebra $\mathfrak{u}(n)^*$ consists of the Hermitian matrices via the identification $\mathfrak{u}(n)^* \cong i\mathfrak{u}(n)$. Then each coadjoint orbit $\mathcal{O}_\lambda$ is the set Hermitian matrices that have the spectrum $\lambda = (\lambda_1, \ldots, \lambda_n) \in \mathbb{R}^n$.

We say that a Lie group $G^{\mathbb{C}}$ is a *complexification* of G, if G is a closed sub-Lie group of $G^{\mathbb{C}}$ and $\mathfrak{g}^{\mathbb{C}} = \mathfrak{Lie}(G^{\mathbb{C}})$ is a vector-space complexification of $\mathfrak{g} = \mathfrak{Lie}(G)$; that is, $\mathfrak{g}^{\mathbb{C}} = \mathfrak{g} \oplus i\mathfrak{g}$.

Example 9. The complexification of $G = SL(n, \mathbb{R})$ is the group $SL(n, \mathbb{R})^{\mathbb{C}} = SL(n, \mathbb{C})$, due to obvious identification $\mathfrak{sl}(n, \mathbb{R})^{\mathbb{C}} = \mathfrak{sl}(n, \mathbb{C})$. The complexification of $G = U(n)$ is $GL(n)$, because $\mathfrak{u}(n)^{\mathbb{C}}$ is the full matrix algebra.

3.2. Symplectic manifolds.

Let us recall some basic facts about symplectic manifolds (more details see, for example, [5]). Let M be a manifold, and ω be a closed 2-form on the tangent space which assumed to be non-degenerate; that is, for any tangent vector $x_1 \in T_pM$, there exists another vector $x_2 \in T_pM$ with $\omega(x_1, x_2)$ nonzero. Due to non-degeneracy of ω the manifold M has to be even dimensional.

Example 10. Consider the space $\mathbb{R}^{2n}$ with basis $\{x_1, \ldots, x_n, y_1, \ldots, y_n\}$ and the form ω acting by $\omega(x_i, x_j) = \omega(y_i, y_j) = 0$, and $\omega(x_i, y_j) = \omega(y_j, x_i) = \delta_{ij}$ (Kronecker delta). Such form ω is called *standard.*

If (M, ω) is locally isomorphic to $\mathbb{R}^{2n}$ with standard ω then M is called a *symplectic manifold*, and ω is its *symplectic form.*

Example 11. Complex projective space $\mathbb{CP}^n$ can be equipped with the Fubini-Study form which makes it into a symplectic manifold (see, for example, [5, Section 16]). We do not define this form, just note that one can think of it as the one realized on the quotient space after the identification $\mathbb{CP}^n = S^{2n+1}/S^1$.

The coadjoint orbits have a structure of symplectic manifold, moreover each coadjoint orbit possesses a G-invariant symplectic structure (see [11, Chapter 1]). In some cases the opposite is also true: a symplectic Hausdorff manifold (M, ω), with the action of a Lie group G which preserves ω, is isomorphic to a coadjoint orbit of G (see, for example, [17]).

Example 12. Describe the symplectic form on a coadjoint orbit of $U(n)$. It is enough to calculate it in a generic point $p \in \mathfrak{u}(n)^*$. Any tangent vector to p looks like $ad_p(x) = [p, x]$ for some $x \in \mathfrak{u}(n)$. Then the symplectic form is given by $\omega_p(ad_p(x_1), ad_p(x_2)) = tr(x_1 p x_2 - x_2 p x_1)$.

Example 13. Consider a coadjoint orbit $\mathcal{O}_\lambda$ of $U(n)$ with $\lambda_1 \neq \lambda_2 = \lambda_3 = \ldots = \lambda_n$. The eigenspaces corresponding to elements of $\mathcal{O}_\lambda$ are the line and the orthogonal hyperplane. Hence, $\mathcal{O}_\lambda$ is homeomorphic to $\mathbb{CP}^{n-1}$ and by the Kirillov's construction we have many symplectic forms on $\mathbb{CP}^{n-1}$: one for each distinct pair of real numbers. In particular, the Fubiny-Study form (Example 11) corresponds to the choice $\lambda_1 = 1$, $\lambda_2 = 0$. The corresponding coadjoint orbit is just the set of one-dimensional orthogonal projections.

3.3. Moment map

For a given function $f : M \to \mathbb{C}$ its *symplectic gradient* $X_f : M \to T_* M$ (called the *Hamiltonian vector field*) is defined by the following equation

$$d_x f = \omega(x, X_f(p)),$$

where x and $X_f(p)$ are tangent vectors to the point $p \in M$ and $d_x f$ is the derivation of the function f in the direction $x \in T_p M$. Since ω is non-degenerate this defines X_f uniquely.

Let G be a connected Lie group acting on some symplectic manifold M smoothly preserving the symplectic form. Then it generates a Lie algebra homomorphism from $\mathfrak{g}$ to the Lie algebra of smooth vector fields on M. Indeed, an element of G near the identity gives a diffeomorphism of M close to the identity; differentiating we get that each tangent vector to the identity generates a vector field on M. By $p \mapsto a_p$ we denote the vector field on M associated to $a \in \mathfrak{g}$.

We say that $\mu : M \to \mathfrak{g}^*$ is a *moment map* for the action of G on M if the following two conditions are satisfied:

(i) μ is G-equivariant with respect to the action of G on M and to the coaction of G on $\mathfrak{g}^*$; that is, the following holds

$$\mu(g \cdot p) = g\mu(p)g^{-1}, \quad p \in M,\ g \in G;$$

(ii) $d_x \langle \mu(p), a \rangle = \omega(x, a_p)$, $\xi \in T_x M$ and $a \in \mathfrak{g}$. This property means that the vector field associated to any $a \in \mathfrak{g}$ equals to the symplectic gradient of the function $f_a(p) = \langle \mu(p), a \rangle : M \to \mathbb{C}$.

Journal of Physics: Conference Series **346** (2012) 012006 doi:10.1088/1742-6596/346/1/012006

If an action has moment map it is said to be *Hamiltonian*, and the corresponding moment map is uniquely determined up to adding a constant in $\mathfrak{g}^*$. On the other hand given a moment map, one can recover the action of the Lie algebra and the Lie group on the manifold.

Example 14. Let $M = \mathbb{R}^2$ and the Lie group S^1 acts by rotations on M. It obviously preserves ω, and the corresponding moment map $\mu : \mathbb{R}^2 \to \mathbb{R}$ is defined as $\mu(x) = x_1^2 + x_2^2 + \text{const}$.

Example 15. Assume that M is a coadjoint orbit of G. There exists a unique symplectic structure on M (called Kirillov-Konstant-Souriau) such that the moment map is the embedding

$$\mu : M \to \mathfrak{u}(n)^*, \quad \mu : x \mapsto x.$$

Example 16. Assume the $M = M_{n\times k}(\mathbb{C})$ with the symplectic form $\omega(A, B) = \operatorname{tr}(A^*B) - \operatorname{tr}(B^*A)$, and with the natural action of $U(k)$ which obviously preserves the symplectic structure. Then $\mu : M \to \mathfrak{u}(k)^*$ is given by $\mu : A \mapsto \frac{i}{2}AA^* + \text{const} \cdot I$.

An important property of symplectic manifolds is that having two manifolds (M_1, ω_1) and (M_2, ω_2) one can form the product manifold $M_1 \times M_2$ with symplectic structure $\pi_1^*\omega_1 + \pi_2^*\omega_2$, where $\pi_i : M_1 \times M_2 \to M_i$ is the projection onto i-th factor, and π_i^* is the pull-back of π_i.

Assume that M_1 and M_2 are symplectic manifolds with the action of the same Lie group G and corresponding moment maps μ_1 and μ_2. Then the symplectic manifold $M_1 \times M_2$ possesses the diagonal action of G with the moment map

$$\mu : M_1 \times M_2 \to \mathfrak{g}^*, \qquad \mu : (x, y) \mapsto \mu_1(x) + \mu_2(y).$$

Example 17. Let $M = \mathbb{R}^{2n}$ with the action of S^1. Then $\mu : \mathbb{R}^{2n} \to \mathbb{R}$ is defined as $\mu(x) = x_1^2 + \ldots + x_{2n}^2 + \text{const}$.

Example 18. Consider the product $\mathcal{O}_{\lambda^{(1)}} \times \ldots \times \mathcal{O}_{\lambda^{(n)}}$ of coadjoint orbits of $U(n)$. Moment map takes the set of Hermitian matrices $(A_1, \ldots, A_n)$ with $\sigma(A_i) = \lambda^{(i)}$ to their sum:

$$\mu : (A_1, \ldots, A_n) \mapsto A_1 + \ldots + A_n + \text{const} \cdot I.$$

3.4. Symplectic quotient

Assume that M is a manifold with Hamiltonian action of a Lie group G and with the corresponding moment map $\mu : M \to g^*$. Then G acts on the fiber $\mu^{-1}(0) \subset M$. The following theorem says that the corresponding orbit space has the structure of a symplectic manifold.

Theorem 1. *(Marsden-Weinstein) The quotient space $\mu^{-1}(0)/G$ is a symplectic manifold. If the action of G is free on $\mu^{-1}(0)$ then $\mu^{-1}(0)/G$ has dimension* $\dim M - 2\dim G$.

The manifold $\mu^{-1}(0)/G$ is called the *symplectic quotient.*

Journal of Physics: Conference Series **346** (2012) 012006 doi:10.1088/1742-6596/346/1/012006

Example 19. Consider the manifold and moment map as in Example 17. The constant in the moment map can be chosen to be -1, hence $\mu^{-1}(0)$ is the sphere S^{2n-1} and the symplectic quotient is homeomorphic to $\mathbb{CP}^{n-1}$ (Hopf fibration).

Example 20. Let M be the manifold as in Example 16. We choose the constant in the moment map to be equal to $-\frac{i}{2}$. Then $\mu^{-1}(0) = \{A \in M_{n\times k}(\mathbb{C}) \mid AA^* = I\}$ and the quotient $\mu^{-1}(0)/U(k)$ is the set of k-dimensional subspaces in n-dimensional space; that is, $\mathrm{Gr}(k, n)$.

4. Quick GIT. Stability conditions and Kirwan-Ness theorem

Assume that M is a variety and G is a linear algebraic group acting on M. Group G is assumed to be *reductive*; that is, it is a complexification of some compact Lie group. Geometric Invariant Theory tries to build the quotient of M by G. The main problem is that the quotient M/G may not exist in the category of algebraic or projective varieties (especially when the group G is not finite). One of the possible solution is to remove some points from M, by taking an open subset M' of M as large as possible such that M'/G is a variety. The following elementary example explains the idea. The space $\mathbb{CP}^n$ can be described as the GIT quotient of the space $\mathbb{C}^{n+1}$ by $\mathbb{C}^*$, namely $\mathbb{CP}^n = (\mathbb{C}^{n+1} - 0)/\mathbb{C}^*$.

4.1. Projective spectrum and GIT quotients

Let $R = \bigoplus_{k\in\mathbb{N}} R_k$ be a *graded* ring (the product of an element from R_n with an element from R_m lies in R_{n+m}). The *projective spectrum* of the ring R (denoted by $\mathrm{Proj}(R)$) is defined as the set of maximal graded ideals of R. There is alternative definition in terms of ordinary spectrum of the ring. We have the projection $R \twoheadrightarrow R_0$, hence $\mathrm{Spec}(R_0) \hookrightarrow \mathrm{Spec}(R)$ ($\mathrm{Spec}(\cdot)$ is contravariant). The group $\mathbb{C}^*$ acts on R by rotating R_k's. Then

$$\mathrm{Proj}(R) = (\mathrm{Spec}(R) \setminus \mathrm{Spec}(R_0))/\mathbb{C}^*.$$

Example 21. Let $R = \mathbb{C}[x_1, \ldots, x_n]$. Then $R_0 = \mathbb{C}$. Hence, $\mathrm{Proj}(\mathbb{C}[x_1, \ldots, x_n]) = (\mathbb{C}^n \setminus \{0\})/\mathbb{C}^*$ which is identified with $\mathbb{CP}^{n-1}$.

A *polarization* L of M is a line bundle $L \to M$, equipped with the action of G. Graded ring $R(M)$ associated to M is defined as

$$R(M) = \bigoplus_{n\geq 0} \Gamma(M, L^n),$$

where L^n is the n-th tensor power of L and $\Gamma(M, L^n)$ is the set of G-invariant sections of L^n. The *GIT quotient* of M by G with respect to the line bundle L is defined as $\mathrm{Proj}(R(M))$ and denoted by $M//G$.

A point $p \in M$ is:

- *semistable* if $s(p) > 0$ for some section $s \in \Gamma(M, L^n)$, in which $n > 0$;
- *polystable* if p is semistable and the orbit $\{g \cdot p \mid g \in G\}$ is closed;
- *stable* if p is polystable with a finite stabilizer;

- *unstable* if m is not semistable.

Let M^{ss}, M^{ps}, M^s and M^{us} be semistable, polystable, stable and unstable locus respectively. D.Mumford gave a numerical criterion to decide the type of the point $m \in M$ and proved that the GIT quotient $M//G$ can be identified with M^{ps}/G (see [23, Chapter 2] for the details).

Example 22. ([23, Chapter 8]) Let $M = (\mathbb{CP}^1)^n$ with $G = SL(2)$ acting diagonally. Then we have:

$$\begin{aligned} M^s &= \{(m_1, \dots, m_n) \in (\mathbb{CP}^1)^n \mid \text{at most n/2 points equal}\}, \\ M^{ss} &= \{(m_1, \dots, m_n) \in (\mathbb{CP}^1)^n \mid \text{less then n/2 points equal}\}, \\ M^{ps} - M^s &= \{(m_1, \dots, m_n) \in (\mathbb{CP}^1)^n \mid \#\{m_1, \dots, m_n\} = 2\}. \end{aligned}$$

4.2. Kirwan-Ness Theorem

Assume that M is a symplectic manifold with an action of a compact group G, which preserves symplectic form and has the moment map μ. Then there exists a unique continuation of the action of G to the action of $G^{\mathbb{C}}$ (its complexification). By [23], there exists an inclusion of $\mu^{-1}(0)$ into M^{ps}. The following theorem is a fundamental fact (proved by Kirwan and Ness independently) connecting the symplectic and the GIT quotients.

Theorem 2. *(Kirwan-Ness) The inclusion $\mu^{-1}(0)$ into M^{ps} induces a homeomorphism*

$$\mu^{-1}(0)/G \cong M//G^{\mathbb{C}}.$$

Example 23. Let $M = \mathbb{R}^{2n}$, and let S^1 acts by rotations on M. The complexification of S^1 is $\mathbb{C}^*$ which acts by multiplication on M. The Kirwan-Ness theorem is just an identification between S^{2n-1}/S^1 and $\mathbb{CP}^{n-1}$.

Example 24. Consider the space $M = (\mathbb{CP}^1)^n$ with diagonal action of $G = SL(2)$. By Example 22, $M^s = M^{ss}$ if and only if n is odd. Take the maximal compact subgroup $SU(2)$ of G. Its dual Lie algebra can be identified with the real vector space $\mathbb{R}^3$. The corresponding moment map takes the n-tuple of points on unit sphere to their sum in $\mathbb{R}^3$. Then the fiber $\mu^{-1}(0)$ consists of the points with the center of gravity in the origin. It is not hard to check directly that $\mu^{-1}(0) \subset M^{ss}$. In the case when $n = 4$ it can be straightforwardly proved that $(\mathbb{CP}^1)^n//SL(2)$ and $\mu^{-1}(0)/SU(2)$ are isomorphic to a sphere $\mathbb{CP}^1$ without three removed points. We refer the reader to [23, Chapter 8] for the details.

Note that the correspondence in previous example is the same as the correspondence between Example 3 and Example 7.

5. Correspondence between unitary and linear representations of the posets

For a given poset $\mathcal{P}$ fix a dimension vector $d = (d_0; d_i)_{i \in \mathcal{P}}$ and a weight $\chi = (\chi_0; \chi_i)_{i \in \mathcal{P}} \in \mathbb{N}^{|\mathcal{P}|+1}$, such that $\sum_{i \in \mathcal{P}} \chi_i d_i = \chi_0 d_0$. Let $\lambda^{(i)} = (\underbrace{\chi_i, \dots, \chi_i}_{d_i}, \underbrace{0, \dots, 0}_{d_0 - d_i})$, $i \in \mathcal{P}$. The variety $R_{\mathcal{P},d}$ is a subset of the product of Grassmanians $\prod_{i \in \mathcal{P}} \mathrm{Gr}(d_i, d_0)$. Given an Hermitian

Journal of Physics: Conference Series **346** (2012) 012006 doi:10.1088/1742-6596/346/1/012006

metric on $\mathbb{C}^{d_0}$ one can view each $\mathrm{Gr}(d_i, d_0)$ as the set of $d_0 \times d_0$ matrices $\chi_i P_i$, where P_i is the d_i-dimensional orthoprojection. Therefore $\mathrm{Gr}(d_i, d_0)$ is identified with the coadjoint orbit $\mathcal{O}_{\lambda^{(i)}}$. Our first aim is to show how to embed the variety $R_{\mathcal{P},d}$ into some projective space so that its symplectic structure coincides with symplectic structure on the product of corresponding coadjoint orbits. We use slightly modified standard construction (see, for example, [6, Chapter 11] and [15]).

A standard way to embed $\mathrm{Gr}(d_i, d_0)$ into a projective space is via Plucker embedding; that is, for an element $V_i \in \mathrm{Gr}(d_i, d_0)$ we take its basis vectors a_j and wedge them together $a_1 \wedge \cdots \wedge a_{d_i}$ obtaining an element of $\mathbb{P}(\wedge^{d_1}\mathbb{C}^{d_0})$ (this is well-defined because if we change a basis then $a_1 \wedge \cdots \wedge a_{d_1}$ changes by a scalar).

If V is some vector space then one can form a symmetric tensor d-power of V denoted $Sym^d(V)$ (for $d < 0$ one takes $Sym^{-d}(V^*)$, for $d = 0$ we have $Sym^0(V) = \mathbb{C}$). We take the symmetric tensor χ_i-power of the space $\wedge^{d_i}\mathbb{C}^{d_0}$. By Veronese map, we embed the projective space $\mathbb{P}(V)$ into the space $\mathbb{P}(Sym^d(V))$. Hence, we have the following sequences of inclusions:

$$\mathrm{Gr}(d_i, d_0) \hookrightarrow \mathbb{P}(\wedge^{d_i}\mathbb{C}^{d_0}) \hookrightarrow \mathbb{P}(Sym^{\chi_i}(\wedge^{d_i}\mathbb{C}^{d_0})).$$

Now it is a routine to check that the symplectic form on $\mathrm{Gr}(d_i, d_0)$ (taken as the restriction of the Fubiny-Study form on $\mathbb{P}(Sym^{\chi_i}(\wedge^{d_i}\mathbb{C}^{d_0}))$) coincides with the symplectic form on the corresponding coadjoint orbit $\mathcal{O}_{\lambda^{(i)}}$.

Correspondingly for the product of Grassmanians $\prod_{i\in\mathcal{P}} \mathrm{Gr}(d_i, d_0)$ we have the embedding

$$\prod_{i\in\mathcal{P}} \mathrm{Gr}(d_i, d_0) \hookrightarrow \prod_{i\in\mathcal{P}} \mathbb{P}(Sym^{\chi_i}(\wedge^{d_i}\mathbb{C}^{d_0})).$$

Using the Segre map $\mathbb{P}^n \times \mathbb{P}^m \hookrightarrow \mathbb{P}^{(n+1)(m+1)-1}$ we embed the last product into

$$\mathbb{P}\left(\prod_{i\in\mathcal{P}} Sym^{\chi_i}(\wedge^{d_i}\mathbb{C}^{d_0})\right).$$

Consider the moment map of the action of $U(d_0)$ on $\prod_{i\in\mathcal{P}} \mathrm{Gr}(d_i, d_0)$ after this embedding. As we mentioned above the Fubiny-Study form on $\mathbb{P}\left(\prod_{i\in\mathcal{P}} Sym^{\chi_i}(\wedge^{d_i}\mathbb{C}^{d_0})\right)$ coincides with the symplectic form on the product of coadjoint orbits $\prod_{i\in\mathcal{P}} O_{\lambda^{(i)}}$. Knowing the form of the moment map $\mu : (P_i)_{i\in\mathcal{P}} \mapsto \mathfrak{u}(d_0)^*$ (see Example 18) and taking a constant in the moment map to be equal to $-\chi_0$ we have the following

$$\mu^{-1}(0) = \left\{ (P_i)_{i\in\mathcal{P}} \in (M_{d_0}(\mathbb{C}))_{i\in\mathcal{P}} \middle| \begin{array}{c} P_i = P_i^* = P_i^2, \mathrm{rank}(P_i) = d_i, \quad i \in \mathcal{P}, \\ P_i P_j = P_i P_j = P_i, \quad i \prec j, \quad \sum_{i\in\mathcal{P}} \chi_i P_i = \chi_0 I \end{array} \right\}.$$

The continuation of the action of $U(d_0)$ to its complexification $GL(d_0)$ coincides with the action of $GL(d_0)$ on $R_{\mathcal{P},d}$. Using Muformd's numerical criterion it was shown in [6, Chapter 11] and [9, Theorem 2.2] the set of stable representations with respect to induced action of $GL(d_0)$ consists of those representations $(V; V_i)_{i\in\mathcal{P}}$ that satisfy

$$\frac{1}{\dim K} \sum_{i\in\mathcal{P}} \chi_i \dim(V_i \cap K) < \frac{1}{\dim V} \sum_{i\in\mathcal{P}} \chi_i \dim V_i$$

Journal of Physics: Conference Series **346** (2012) 012006 doi:10.1088/1742-6596/346/1/012006

for each proper subspace $K \subset V$.

Straightforward calculations show that if $(P_i)_{i\in\mathcal{P}} \in \mu^{-1}(0)$ then the system of subspaces $(\mathbb{C}^{d_0}; \mathrm{Im}P_i)_{i\in\mathcal{P}}$ is *χ-polystable*; that is, it decomposes into a direct sum of stable representations $(\tilde{V}; \tilde{V}_i)_{i\in\mathcal{P}}$ which satisfy $\sum_{i\in\mathcal{P}} \chi_i \dim \tilde{V}_i = \chi_0 \dim \tilde{V}_0$. Denote by Φ the corresponding map from $Ob(\mathcal{P}_\chi - usp)$ to $Ob(\mathcal{P} - sp)$. As a consequence of Kirwan-Ness theorem we have the following:

Theorem 3. *Let $\mathcal{P}$ be a poset, $(d_0; d_i)_{i\in\mathcal{P}}$ be a dimension vector and $\chi = (\chi_0; \chi_i)_{i\in\mathcal{P}}$ be a weight such that $\sum_{i\in\mathcal{P}} \chi_i d_i = \chi_0 d_0$. Then Φ induces a homeomorphism (with respect to the usual orbit-space topology) between*

$$\left\{ (P_i)_{i\in\mathcal{P}} \in (M_{d_0}(\mathbb{C}))_{i\in\mathcal{P}} \;\middle|\; \begin{array}{c} P_i = P_i^* = P_i^2,\ rank(P_i) = d_i, \quad i \in \mathcal{P}, \\ P_i P_j = P_i P_j = P_i, \quad i \prec j, \quad \sum_{i\in\mathcal{P}} \chi_i P_i = \chi_0 I \end{array} \right\} \Big/ U(d_0)$$

and

$$\left\{ (V; V_i)_{i\in\mathcal{P}} \in R_{\mathcal{P},d} \;\middle|\; (V; V_i)_{i\in\mathcal{P}}\ is\ \chi - polystable \right\} \Big/ GL(d_0).$$

Let us show some consequences of this theorem.

(i) In [18] it was shown that if two objects in $\mathcal{P}_\chi - usp$ are equivalent as the objects in $\mathcal{P} - sp$ then they are equivalent in $\mathcal{P}_\chi - usp$. This follows as a consequence of Theorem 3.

(ii) In [18] it was shown that each indecomposable object in $\mathcal{P}_\chi - usp$ is Schurian; that is, it has trivial endomorphism ring. In fact this result can be deduced from the previous observations. Indeed, each indecomposable object in $\mathcal{P}_\chi - usp$ corresponds to some χ-stable object in $\mathcal{P} - sp$. But stable objects are Schurian.

6. Calculating the dimensions of the quotients

6.1. Bound quivers

For a finite quiver Q denote by Q_0 the set of its vertices and by Q_1 the set of its arrows denoted by $\rho : i \to j$ for $i, j \in Q_0$. In the following we only consider quivers without oriented cycles. A finite-dimensional *representation* of Q is given by a tuple

$$X = ((X_i)_{i\in Q_0}, (X_\rho)_{\rho\in Q_1} : X_i \to X_j)$$

of finite-dimensional vector spaces and linear maps between them. We say that X is *subspace representation* if all maps X_ρ are injective. The dimension vector $\underline{\dim} X \in \mathbb{N}^{|Q_0|}$ of X is defined by

$$\underline{\dim} X = (\dim X_i)_{i\in Q_0}.$$

The variety $R_{Q,d}$ of representations of Q with the dimension vector $d \in \mathbb{N}^{|Q_0|}$ is defined as the affine complex space

$$R_{Q,d} = \bigoplus_{\rho: i\to j} \mathrm{Hom}(\mathbb{C}^{d_i}, \mathbb{C}^{d_j}).$$

The algebraic group $GL_d = \prod_{i \in Q_0} GL(d_i)$ acts on $R_{Q,d}$ via simultaneous base change; that is,

$$(g_i)_{i \in Q_0} \cdot (X_\rho)_{\rho \in Q_1} = (g_j X_\rho g_i^{-1})_{\rho: i \to j}.$$

The orbits of this action are in a bijection with the isomorphism classes of representations of Q with the dimension vector d.

Let $\mathbb{C}Q$ be the path algebra of Q (see, for example, [2, Chapter 2] for the definition) and let RQ be the arrow ideal. A relation in Q is a $\mathbb{C}$-linear combination of paths of length at least two which have the same head and tail. For a set of relations $(r_j)_{j \in J}$ we can consider the admissible ideal I generated by these relations, that means that $RQ^m \subseteq I \subseteq RQ^2$ for some $m \geq 2$. We say that a representation X of Q is *bound* by I (or a representation of the *bound quiver* (Q, I)) if $X_{r_j} = 0$ for all $j \in J$. For every dimension vector this defines a closed subvariety of $R_{Q,d}$ denoted by $R_{(Q,I),d}$. Fixing a minimal set of relations generating I, we denote by $r(i, j, I)$ the number of relations with starting vertex i and terminating vertex j. Following [3], for the dimension of $R_{(Q,I),d}$ we have the following lower bound

$$\dim R_{(Q,I),d} \geq \dim R_{Q,d} - \sum_{(i,j) \in (Q_0)^2} r(i, j, I) d_i d_j.$$

Let $C_{(Q,I)}$ be the Cartan matrix of (Q, I); that is, $c_{j,i} = \dim e_i (\mathbb{C}Q/I) e_j$ where e_i denotes the primitive idempotent (resp. the trivial path) corresponding to the vertex i. On $\mathbb{Z}^{|Q_0|}$ a non-symmetric bilinear form is defined by

$$\langle d, \tilde{d} \rangle := d^t (C_{(Q,I)}^{-1})^t \tilde{d}.$$

For two representation X and Y we have the following homological interpretation of this form:

$$\langle X, Y \rangle := \langle \underline{\dim} X, \underline{\dim} Y \rangle = \sum_{i=0}^{\infty} (-1)^i \dim \operatorname{Ext}^i(X, Y).$$

If Q is unbound, for two representations X, Y of Q with $\underline{\dim} X = d$ and $\underline{\dim} Y = \tilde{d}$ we have $\operatorname{Ext}^i(X, Y) = 0$ for $i \geq 2$ and

$$\langle X, Y \rangle = \dim \operatorname{Hom}(X, Y) - \dim \operatorname{Ext}(X, Y) = \sum_{q \in Q_0} d_q \tilde{d}_q - \sum_{\rho: i \to j \in Q_1} d_i \tilde{d}_j.$$

6.2. Bound quiver and posets

Recall briefly the relation between representations of posets and representations of bound quivers.

Let $\mathcal{P}$ be a poset and $Q^{\mathcal{P}}$ its Hasse quiver. All arrows of $Q^{\mathcal{P}}$ are oriented to one vertex q_0 which is called the root. Let $d \in \mathbb{N}^{|\mathcal{P}|+1}$ be a dimension vector. By $S_{Q^{\mathcal{P}},d} \subset R_{Q^{\mathcal{P}},d}$ we denote the open subvariety of subspace representations. For every (non-oriented) cycle $\rho_1 \ldots \rho_n \tau_k^{-1} \ldots \tau_1^{-1}$ with ρ_i, $\tau_j \in Q_1$ and $\rho_i \neq \tau_j$ we define a relation

$$r = \rho_1 \ldots \rho_n - \tau_1 \ldots \tau_k.$$

Let I be the ideal generated by all such relations.

Let $\pi = (V; V_i)_{i\in\mathcal{P}}$ be a representation of $\mathcal{P}$ with the dimension vector d. This defines a representation $X(\pi) \in S_{(Q^{\mathcal{P}},I),d}$ satisfying the stated relations. Indeed, every inclusion $V_i \subset V_j$ defines an injective map $X(\pi)_{\rho:i\to j} : V_i \to V_j$. Thus it defines an element of $S_{(Q^{\mathcal{P}},I),d}$. For two arbitrary representations $\pi = (V; V_i)_{i\in\mathcal{P}}$ and $\pi' = (W; W_i)_{i\in\mathcal{P}}$ a morphism $f : \pi \to \pi'$ induces a morphism $X(f) : X(\pi) \to X(\pi')$, where $X(f)_i := f|_{V_i} : X(V)_i \to X(W)_i$.

Conversely, let $X \in S_{(Q^{\mathcal{P}},I),d}$. This gives rise to a representation $\pi(X)$ of $\mathcal{P}$ by defining $\pi(X)_q = X_{\rho_n^q} \circ \ldots \circ X_{\rho_1^q}(X_q)$ for some path $p_q = \rho_1^q \ldots \rho_n^q$ from q to q_0. This definition is independent of the chosen path. Moreover, every morphism $\varphi = (\varphi_q)_{q\in Q_0} : X \to Y$ defines a morphism $\pi(\varphi)$ which is induced by $\varphi_{q_0} : X_0 \to Y_0$.

Therefore we get an equivalence between the categories of representations of $\mathcal{P}$ and subspace representations of $Q^{\mathcal{P}}$ bound by I. This equivalence also preserves dimension vectors.

6.3. GIT quotient of quivers

A representation X of (Q, I) is *semistable* (resp. *stable*) with respect to a form $\Theta \in \mathrm{Hom}_{\mathbb{Z}}(\mathbb{Z}^{|Q_0|}, \mathbb{Z})$ if $\Theta(\underline{\dim} X) = 0$ and for all subrepresentations $Y \subset X$ (resp. all proper subrepresentations $0 \neq Y \subsetneq X$) we have:

$$\Theta(\underline{\dim} Y) \geq 0 \quad (\text{resp. } \Theta(\underline{\dim} Y) > 0).$$

Denote the set of semistable (resp. stable) points by $R^{ss}_{(Q,I),d}$ (resp. $R^{s}_{(Q,I),d}$). A.King in [10] built the GIT quotient $R_{(Q,I),d}//GL_d$ for quivers. This quotient parametrizes polystable representations. For a stable representation X we have that its orbit is of maximal possible dimension. The scalar matrices act trivially on $R_{(Q,I),d}$, hence the isotropy group is one-dimensional. Therefore, if the quotient $R_{(Q,I),d}//GL_d$ is not empty, for its dimension we have the lower bound

$$\begin{aligned} \dim R_{(Q,I),d}//GL_d &= \dim R_{(Q,I),d} - (\dim G_d - 1) \\ &\geq 1 - \sum_{i\in Q} d_i^2 + \sum_{\rho:i\to j\in Q_1} d_i d_j - \sum_{(i,j)\in(Q_0)^2} r(i,j,I) d_i d_j. \end{aligned}$$

In the case $I = 0$ we have $\dim R_{(Q,I),d}//GL_d = 1 - \langle d, d\rangle$.

6.4. Dimension of symplectic quotients of the posets

Assume that the representation $\pi = (V; V_i)_{i\in\mathcal{P}}$ is stable with the weight $(\chi_0; \chi_i)_{i\in\mathcal{P}} \in \mathbb{Z}^{|\mathcal{P}|+1}$; that is

$$\frac{1}{\dim K} \sum_{i\in\mathcal{P}} \chi_i \dim(V_i \cap K) < \frac{1}{\dim V} \sum_{i\in\mathcal{P}} \chi_i \dim V_i$$

for each proper subspace $K \subset V$ and $\sum_{i\in\mathcal{P}} \chi_i \dim V_i = \chi_0 \dim V$. Then the corresponding representation $X(\pi)$ of the bound quiver $Q^{\mathcal{P}}$ is stable with respect to the form $\Theta = (\chi_0; -\chi_i)_{i\in\mathcal{P}}$ and vise versa. Therefore, using Theorem 3 we have that if there exists at least one object in $\mathcal{P}_\chi - usp$ with dimension vector $(d_0; d_i)_{i\in\mathcal{P}}$ then the unitary classes of all objects with this dimension vector depend on at least

$$1 - \sum_{i\in Q_0^{\mathcal{P}}} d_i^2 + \sum_{\rho: i\to j\in Q_1^{\mathcal{P}}} d_i d_j - \sum_{(i,j)\in (Q_0^{\mathcal{P}})^2} r(i,j,I) d_i d_j.$$

complex parameters. When the poset is primitive we have that the number of parameters is $1 - \sum_{i\in Q_0^{\mathcal{P}}} d_i^2 + \sum_{\rho:i\to j\in Q_1^{\mathcal{P}}} d_i d_j = 1 - \langle d, d\rangle$.

Example 25. Consider the case when $\mathcal{P}$ is primitive and representation-tame. Hence the corresponding quiver $Q^{\mathcal{P}}$ is one of the affine Dynkin quivers; that is, $Q^{\mathcal{P}} = \tilde{D}_4$, $\tilde{E}_6$, $\tilde{E}_7$ or $\tilde{E}_8$. Let us take the dimension vector $d_{\mathcal{P}}$ to be the minimal imaginary root of $Q^{\mathcal{P}}$:

$$\begin{aligned} d_{(1,1,1,1)} &= (2;1,1,1,1), \\ d_{(2,2,2)} &= (3;1,2;1,2;1,2), \\ d_{(1,3,3)} &= (4;2;1,2,3;1,2,3), \\ d_{(1,2,5)} &= (6;3;2,4;1,2,3,4,5). \end{aligned}$$

Using Theorem 3 and the dimension formula we have at most $1-\langle d_{\mathcal{P}}, d_{\mathcal{P}}\rangle = 1$–parametric family of uni-classes of indecomposable objects in $\mathcal{P}_\chi - usp$ with dimension $d_{\mathcal{P}}$. One can check that it is always possible to choose the weight $\chi_{\mathcal{P}}$ such that there exists at least one stable representation, hence for these weights set of uni-classes of orthoscalar representations of $\mathcal{P}$ in dimension $d_{\mathcal{P}}$ is isomorphic to $\mathbb{CP}^1$ (because the GIT quotient is one-dimensional and rational).

Example 26. Consider representation-tame non-primitive critical poset $\mathcal{P} = (N, 4)$, which has the following Hasse quiver

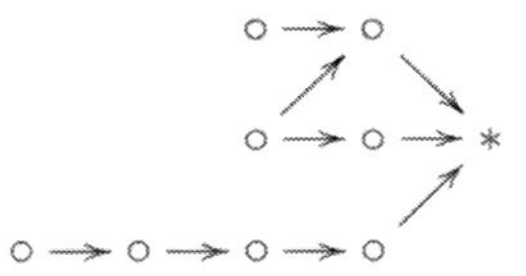

Fix the dimension vector $d_{(N,4)} = (5;2,4,3,2;1,2,3,4)$. One can check that with respect to the weight $\chi = (5;2,1,1,2;1,1,1,1)$ there exists at least one stable representation in dimension $d_{(N,4)}$, hence the corresponding GIT quotient is not empty. Due to the dimension formula we have at least $1-\langle d_{\mathcal{P}}, d_{\mathcal{P}}\rangle + 2\cdot 5 = 1$–parametric family of uni-classes of representation of $\mathcal{P}_\chi - usp$. Moreover, it is possible to show that this is a maximal number of parameters. Therefore, we have that for the weight $\chi = (5;2,1,1,2;1,1,1,1)$ the set of uni-classes of orthoscalar representations of $\mathcal{P}$ in dimension $d_{(N,4)}$ is isomorphic to $\mathbb{CP}^1$.

Acknowledgments

This work was done during the visit of the third author to the University of São Paulo as a postdoctoral fellow. This author is grateful to the University of São Paulo for hospitality and to Fapesp for financial support (2010/15781-0). The first author is supported in part by CNPq (301743/2007-0) and by Fapesp (2010/50347-9).

Journal of Physics: Conference Series **346** (2012) 012006 doi:10.1088/1742-6596/346/1/012006

References

[1] Albeverio S, Ostrovskyi V and Samoilenko Yu 2007 On functions of graphs and representations of a certain class of ∗-algebras *J. Algebra.* **308** 2 567–82

[2] Assem I, Simson D and Skowronski A 2007 *Elements of the Representation Theory of Associative Algebras* (New York: Cambridge University Press)

[3] Bongartz K 1983 Algebras and quadratic forms *J. London Math. Soc.* **28** 461–9

[4] Crawley-Boevey W 1999 Representations of quivers, preprojective algebras and deformations of quotient singularities *Preprint* Lectures from a DMV Seminar

[5] Cannas da Silva A 2001 *Lectures on Symplectic Geometry* (Berlin: Springer)

[6] Dolgachev I 2003 *Lectures on Invariant Theory* (Cambridge: Cambridge University Press)

[7] Gelfand I and Ponomarev V 1970 Problems of linear algebra and classification of quadruples of subspaces in a finitedimensional vector space *Coll. Math. Spc. Bolyai. Tihany* 5 163–237

[8] Halmosh P 1969 Two subspaces *Trans. of the Amer. Math. Soc.* **144** 381–9

[9] Hu Y 2005 Stable configurations of linear subspaces and quotient coherent sheaves *Q. J. Pure Appl. Math.* **1** 127–64

[10] King A 1994 Moduli of representations of finite-dimensional algebras *Quart. J. Math. Oxford Ser.* **45** 515–30

[11] Kirillov A 2004 *Lectures on Orbit Methods* (USA: AMS Graduate Studies in Mathematics)

[12] Kirillov A 2008 *An Introduction to Lie Groups and Lie Algebras* (Cambrige: Cambridge Studies in Advanced Mathematics) **113**

[13] Kleiner M 1975 Partially ordered sets of finite type *J. Soviet Math.* **3** 607–15.

[14] Klyachko A 1988 Stable bundles, representation theory and hermitian operators *Selecta Math.* **4** 419–45

[15] Knutson A 2000 The symplectic and algebraic geometry of Horn's problem *Linear Algebra Appl.* **319** 61–81

[16] Knutson A and Sharpe E 1998 Sheaves on toric varieties for physics *Adv.Theor.Math.Phys.* **2** 865–948

[17] Kostant B 1970 *Orbits and quantization theory* (Congrás international des mathámaticiens)

[18] Kruglyak S, Nazarova L and Roiter A 2010 Orthoscalar quiver representations corresponding to extended Dynkin graphs in the category of Hilbert spaces *Funct. Anal. Appl.* **44** 57–73

[19] Kruglyak S and Roiter A 2005 Locally Scalar Graph Representations in the Category of Hilbert Spaces *Funct. Anal. Appl.* **39** 13–30.

[20] Kruglyak S and Samoilenko Yu 1980 Unitary equivalence of sets of self-adjoint operators *Funct. Anal. Appl.* **14** 60–62

[21] Kruglyak S and Samoilenko Yu 2000 On the complexity of description of representations of ∗-algebras generated by idempotents *Proc. AMS* **128** 1655–64

[22] Moskaleva Yu and Samoilenko Yu 2006 Systems of n subspaces and representations of ∗- algebras generated by projections *Methods Funct. Anal. Topology* **12** 57–73

[23] Mumford D, Fogarty J and Kirwan F 1994 *Geometric Invariant Theory, 3rd Edition* (New York: Springer)

[24] Nazarova L 1967 Representations of a quadruple *Math. USSR Izv.* **31** 1361–77 (in Russian).

[25] Nazarova L 1975 Partially ordered sets of infinite type *Math. USSR Izv.* **9** 963–91

[26] Nazarova L and Roiter A 1975 Representations of the partially ordered sets *J. Soviet Math.* **3** 585–606

[27] Ostrovskyi V and Samoilenko Yu 1999 *Introduction to the theory of representations of finitely presented *-algebras. I. Representations by bounded operators* (London: Rev. Math. & Math. Phys.).

[28] Samoilenko Yu and Yusenko K 2011 Kleiner's theorem for unitary representations of the posets Preprint 1103.1085

[29] Simson D 1992 *Linear representations of partially ordered sets and vector space categories* (Amsterdam: Gordon and Breach Science Publishers)

Remarks on rational solutions of Yang-Baxter equations

Thilo Henrich

Mathematisches Institut, Universität Bonn, Endenicher Allee 60, 53115 Bonn

E-mail: henrich@math.uni-bonn.de

Abstract. In this article, we study unitary rational solutions of the associative Yang-Baxter equation with three spectral parameters. We explain how such solutions arise from the geometry of vector bundles on a cuspidal cubic curve. Moreover, we investigate how these solutions are related to the quantum and classical Yang-Baxter equations.

1. Introduction

In this article we study solutions of the associative Yang-Baxter equation (AYBE)

$$r^{12}(u;y_1,y_2)\,r^{23}(u+v;y_2,y_3) = r^{13}(u+v;y_1,y_3)\,r^{12}(-v;y_1,y_2) + r^{23}(v;y_2,y_3)\,r^{13}(u;y_1,y_3). \quad (1)$$

Here $r : (\mathbb{C}^3, 0) \to A \otimes A$ is the germ of a meromorphic function for $A = \mathrm{Mat}_{n\times n}(\mathbb{C})$. Moreover, for $i \neq j \in \{1,2,3\}$, we use the notation $r^{ij} = r \circ \rho_{ij}$ for the composition of r with the canonical embedding $\rho_{ij} : A^{\otimes 2} \to A^{\otimes 3}$, e.g. $\rho_{13}(a \otimes b) = a \otimes 1 \otimes b$. A solution r of (1) is called unitary if

$$r^{12}(v;y_1,y_2) = -r^{21}(-v;y_2,y_1)$$

and non-degenerate if the tensor $r(v;y_1,y_2) \in A \otimes A$ is non-degenerate for generic v, y_1, y_2. Non-degenerate unitary solutions of (1) have previously been studied by Polishchuk [2, 3] and Burban, Kreußler [1]. We will focus on solutions r of (1) satisfying the following Ansatz:

$$r(v;y_1,y_2) = \tfrac{1\otimes 1}{v} + r_0(y_1,y_2) + v\,r_1(y_1,y_2) + v^2\,r_2(y_1,y_2) + \ldots \quad (2)$$

This is motivated by the following fact. Let $\mathrm{pr} : A \to \mathfrak{sl}_n(\mathbb{C})$ denote the canonical projection $X \mapsto X - \frac{\mathrm{tr}X}{n}1$. It is not difficult to show, see [2, 1], that $\bar{r}_0(y_1,y_2) = (\mathrm{pr}\otimes\mathrm{pr})\,(r_0(y_1,y_2))$ satisfies the classical Yang-Baxter equation (CYBE)

$$\left[\bar{r}_0^{12}(y_1,y_2), \bar{r}_0^{23}(y_2,y_3)\right] + \left[\bar{r}_0^{12}(y_1,y_2), \bar{r}_0^{13}(y_1,y_3)\right] = -\left[\bar{r}_0^{13}(y_1,y_3), \bar{r}_0^{23}(y_2,y_3)\right]. \quad (3)$$

In section 2 we present an algorithm attaching to any pair of coprime integers (n,d) with $0 \leq d < n$ a non-degenerate unitary solution $r_{(n,d)}$ of the AYBE (1). The idea behind this algorithm is the computation of certain triple Massey products in the bounded derived category of coherent sheaves $\mathrm{D}^b(\mathrm{Coh}(E))$ for a cuspidal cubic curve $E = V(y^2z - x^3) \subset \mathbb{P}^2$. In all examples computed so far, solutions produced by this algorithm satisfy (2). Moreover, in all

Journal of Physics: Conference Series **346** (2012) 012007 doi:10.1088/1742-6596/346/1/012007

these examples $\bar{r}_0$ has no infinitesimal symmetries, i.e. there is no non-trivial $a \in \mathfrak{sl}_n(\mathbb{C})$ such that

$$\left[\bar{r}_0(y_1, y_2), a \otimes 1 + 1 \otimes a\right] = 0.$$

As observed by Polishchuk [3], certain unitary solutions of (1) are closely related with the quantum Yang-Baxter equation. The remaining sections of this paper are dedicated to the generalization of his results. Our main result is the following:

Theorem 1.1 *Let $r(v; y_1, y_2)$ be a non-degenerate unitary solution of the AYBE (1) of the form (2). If $\bar{r}_0(y_1, y_2) = (\mathrm{pr}\otimes\mathrm{pr})\,(r_0(y_1, y_2))$ has no infinitesimal symmetries, then the following hold.*

i) For fixed $v_0 \in \mathbb{C}^\times$, $\tilde{r}(y_1, y_2) = r(v_0; y_1, y_2)$ is a solution of the quantum Yang-Baxter equation (QYBE)

$$\tilde{r}^{12}(y_1, y_2)\,\tilde{r}^{13}(y_1, y_3)\,\tilde{r}^{23}(y_2, y_3) = \tilde{r}^{23}(y_2, y_3)\,\tilde{r}^{13}(y_1, y_3)\,\tilde{r}^{12}(y_1, y_2). \tag{4}$$

ii) Let $s(v; y_1, y_2)$ be another non-degenerate unitary solution of the AYBE (1) of the form (2) with $(\mathrm{pr}\otimes\mathrm{pr})\,(s_0) = \bar{r}_0$. Then there exists a meromorphic function $g : \mathbb{C} \to \mathbb{C}$ such that

$$s(v; y_1, y_2) = \exp\Big(v\,(g(y_2) - g(y_1))\Big)r(v; y_1, y_2).$$

If additionally both r and s can be obtained by the procedure described in section 2, then g is holomorphic. Thus the above equation states that r and s are gauge equivalent in that case.

This result is proved in two steps, which are theorem 5.1 respectively theorem 6.1 and corollary 6.3. Let us add the following remarks:

- Theorem 1.1 was shown by Polishchuk, see [3, Theorem 1.4] and [2, Theorem 6], in the case when $r(v; y_1, y_2)$ depends only on v and the difference $y = y_1 - y_2$. However, solutions obtained by the algorithm presented in section 2 do not have this property.
- The proof of theorem 1.1 is purely analytical.
- The solutions $\bar{r}_0$ of the CYBE (3) obtained by the procedure presented in section 2 belong to the class of *rational* solutions. Rational solutions of the CYBE (3) have been classified by Stolin [4]. As we shall show in a subsequent paper, the solutions $\bar{r}_0$ produced by our algorithm form a subclass of rational solutions which can be intrinsically described in terms of Stolin's classification. One of the benefits of our method is that theorem 1.1 gives an explicit way of lifting these solutions of the CYBE to solutions of the QYBE.
- There are solutions r of (1) which satisfy (4) even though $\bar{r}_0$ has infinitesimal symmetries. An example is given by the following function, depending only on the difference $y = y_1 - y_2$:

$$r(v, y) = \frac{1}{2v}1 \otimes 1 + \frac{1}{y}\left(e_{11} \otimes e_{11} + e_{22} \otimes e_{22} + e_{12} \otimes e_{21} + e_{21} \otimes e_{12}\right).$$

The corresponding solution of the CYBE (3) is the rational solution of Yang

$$\bar{r}_0(v) = \frac{1}{y}\left(\frac{1}{2}h \otimes h + e_{12} \otimes e_{21} + e_{21} \otimes e_{12}\right) \in \mathfrak{sl}_2(\mathbb{C}) \otimes \mathfrak{sl}_2(\mathbb{C})$$

whose infinitesimal symmetries are given by all non-zero elements of $\mathfrak{sl}_2(\mathbb{C})$. Hence, there might exist a generalization of theorem 1.1 i).

2. The algorithm

In this section we present an algorithm which takes as input a pair of coprime numbers (n, d) with $0 \leq d < n$ and produces a non-degenerate unitary solution $r_{(n,d)}$ of the AYBE (1) with values in $A \otimes A$ where $A = \mathrm{Mat}_{n\times n}(\mathbb{C})$. The algorithm is due to Burban and Kreußler [1, Section 10], see in particular [1, Algorithm 10.7]. The actual content of the described procedure is the computation of certain triple Massey products in the bounded derived category of coherent sheaves $\mathrm{D}^b(\mathrm{Coh}(E))$ for a cuspidal cubic curve $E = V(y^2z - x^3) \subset \mathbb{P}^2$.

$\star$ **Step 1: construction of the matrix $J = J(n-d, d)$.**
We introduce the following map defined on all tuples of coprime integers $(a, b) \neq (1, 1)$:

$$\epsilon(a,b) = \begin{cases} (a-b, b), & a > b \\ (a, b-a), & a < b \end{cases}$$

By assumption (n, d) is a tuple of coprime integers. Hence it induces a finite sequence of tuples ending with $(1, 1)$, defined as follows. We put $(a_0, b_0) = (n-d, d)$ and, as long as $(a_i, b_i) \neq (1, 1)$, we set $(a_{i+1}, b_{i+1}) = \epsilon(a_i, b_i)$. Next, let

$$J(1,1) = \left(\begin{array}{c|c} 0 & 1 \\ \hline 0 & 0 \end{array}\right) \in \mathrm{Mat}_{2\times 2}(\mathbb{C}).$$

Assuming

$$J(a,b) = \left(\begin{array}{c|c} J_1 & J_2 \\ \hline 0 & J_3 \end{array}\right)$$

with $J_1 \in \mathrm{Mat}_{a\times a}(\mathbb{C})$ and $J_3 \in \mathrm{Mat}_{b\times b}(\mathbb{C})$ has already been defined and that $(a, b) = \epsilon(p, q)$, we set

$$J(p,q) = \begin{cases} \left(\begin{array}{c|cc} 0 & 1 & 0 \\ \hline 0 & J_1 & J_2 \\ 0 & 0 & J_3 \end{array}\right), & p = a \\[2ex] \left(\begin{array}{cc|c} J_1 & J_2 & 0 \\ 0 & J_3 & 1 \\ \hline 0 & 0 & 0 \end{array}\right), & q = b. \end{cases}$$

Hence, to (n, d) we may associate the $n \times n$ matrix $J = J(n-d, d)$ that is obtained from the matrix $J(1, 1)$ and the sequence $\{(n-d, d), ..., (1, 1)\}$ by applying the recursive procedure described above.

Example 2.1 Let $(n, d) = (5, 2)$. We obtain an induced sequence $\{(3, 2), (1, 2), (1, 1)\}$ and $J = J(3, 2)$ is constructed as follows

$$\left(\begin{array}{c|c} 0 & 1 \\ \hline 0 & 0 \end{array}\right) \to \left(\begin{array}{c|cc} 0 & 1 & 0 \\ \hline 0 & 0 & 1 \\ 0 & 0 & 0 \end{array}\right) \to \left(\begin{array}{ccc|cc} 0 & 1 & 0 & 0 & 0 \\ 0 & 0 & 1 & 1 & 0 \\ 0 & 0 & 0 & 0 & 1 \\ \hline 0 & 0 & 0 & 0 & 0 \\ 0 & 0 & 0 & 0 & 0 \end{array}\right).$$

$\star$ **Step 2: definition of $\mathbf{Sol}_{n,d}^{v,y_1}$, $\mathbf{res}_{y_1}$ and $\mathbf{ev}_{y_2}$.**
For the partition of $A = \mathrm{Mat}_{n\times n}(\mathbb{C})$ induced by the partition of J as in step 1, we introduce the following subspace of the polynomial ring $A[z]$:

$$W_{n,d} = \left\{ F(z) = \left(\begin{array}{c|c} W & X \\ \hline Y & Z \end{array}\right) + \left(\begin{array}{c|c} W' & 0 \\ \hline Y' & Z' \end{array}\right) z + \left(\begin{array}{c|c} 0 & 0 \\ \hline Y'' & 0 \end{array}\right) z^2 \right\}.$$

Next, for $F(z) \in W_{n,d}$ we denote

$$F_0 = \left(\begin{array}{c|c} W' & X \\ \hline Y'' & Z' \end{array}\right) \text{ and } F_\epsilon = \left(\begin{array}{c|c} W & 0 \\ \hline Y' & Z \end{array}\right).$$

Then for $v, y_1 \in \mathbb{C}$, we define the following subspace of $W_{n,d}$:

$$\mathrm{Sol}_{n,d}^{v,y_1} = \left\{F(z) \in W_{n,d} \,\middle|\, [F_0, J] + (y_1 - v)F_0 + F_\epsilon = 0\right\}.$$

Proposition 2.2 [1, Section 10] *The vector space* $\mathrm{Sol}_{n,d}^{v,y_1}$ *has dimension* n^2 *and for* $y_1 \neq y_2 \in \mathbb{C}$, $\mathrm{ev}_{y_2} : \mathrm{Sol}_{n,d}^{v,y_1} \to A$ *defined by* $\mathrm{ev}_{y_2}(F(z)) = \frac{1}{y_2 - y_1} F(y_2)$ *is an isomorphism. For* $v \neq 0$, $\mathrm{res}_{y_1} : \mathrm{Sol}_{n,d}^{v,y_1} \to A$ *given by* $\mathrm{res}_{y_1}(F(z)) = F(y_1)$ *is an isomorphism as well.*

We will assume $v \neq 0$ and $y_1 \neq y_2$ for the rest of this section. Then we get a linear automorphism $\tilde{r}_{(n,d)}$ of the matrix algebra A given by the formula $\tilde{r}_{(n,d)}(v; y_1, y_2) = \mathrm{ev}_{y_2} \circ \mathrm{res}_{y_1}^{-1}$.

⋆ **Step 3: definition of the tensor** $r_{(n,d)}(v; y_1, y_2)$**.**
Note that we have a canonical isomorphism of vector spaces

$$\mathrm{can} : A \otimes A \to \mathrm{End}_{\mathbb{C}}(A),\ X \otimes Y \mapsto \big(Z \mapsto \mathrm{tr}(XZ)Y\big).$$

For fixed v, y_1, y_2, we set $r_{(n,d)}(v; y_1, y_2) = \mathrm{can}^{-1}\big(\tilde{r}_{(n,d)}(v; y_1, y_2)\big)$. The proof of the following theorem is contained in [1, Section 10].

Theorem 2.3 *The tensor-valued function* $r_{(n,d)} : \left(\mathbb{C}^3_{(v;y_1,y_2)}, 0\right) \to A \otimes A$ *is a non-degenerate unitary solution of (1). Moreover* $r_{(n,d)}(v; y_1, y_2)$ *is holomorphic on* $\big(\mathbb{C}^3 \setminus V(v(y_1 - y_2))\big)$.

Example 2.4 For any $n \in \mathbb{N}$, let $P = \sum_{1 \le i,j \le n} e_{ij} \otimes e_{ji} \in A \otimes A$.

i) Let $(n, d) = (2, 1)$. Then we have

$$\begin{aligned} r_{(2,1)}(v; y_1, y_2) =& \frac{1}{2v} 1 \otimes 1 + \frac{1}{y_2 - y_1} P \\ &+ (v - y_1)\, e_{21} \otimes \check{h} + (v + y_2)\, \check{h} \otimes e_{21} - \frac{v(v - y_1)(v + y_2)}{2} e_{21} \otimes e_{21} \end{aligned}$$

where $\check{h} = \mathrm{diag}(\frac{1}{2}, -\frac{1}{2})$.

ii) Let $(n, d) = (3, 1)$. Then we have

$$\begin{aligned} r_{(3,1)}(v; y_1, y_2) =& \frac{1}{3v} 1 \otimes 1 + \frac{1}{y_2 - y_1} P - e_{21} \otimes \check{h}_1 + \check{h}_1 \otimes e_{21} + e_{32} \otimes e_{12} - e_{12} \otimes e_{32} \\ &- y_1 e_{32} \otimes \check{h}_2 + y_2 \check{h}_2 \otimes e_{32} + (v - y_1) e_{31} \otimes e_{12} + (v + y_2) e_{12} \otimes e_{31} \\ &+ v e_{32} \otimes (e_{11} - e_{33}) + v(e_{11} - e_{33}) \otimes e_{32} \\ &+ \frac{1}{3} v(y_1 - 3v) e_{32} \otimes e_{21} + \frac{1}{3} v(y_2 + 3v) e_{21} \otimes e_{32} + v(v - y_1) e_{31} \otimes \check{h}_1 \\ &- v(v + y_2) \check{h}_1 \otimes e_{31} + \frac{2}{3} v^2 (y_1 - v) e_{31} \otimes e_{21} - \frac{2}{3} v^2 (y_2 + v) e_{21} \otimes e_{31} \\ &+ \frac{1}{3} v^2 (y_2 + v)(3v - y_1) e_{32} \otimes e_{31} + \frac{1}{3} v^2 (y_1 + v)(3v + y_2) e_{31} \otimes e_{32} \\ &+ \frac{2}{3} v e_{21} \otimes e_{21} + \frac{2}{3} v^3 (v - y_1)(v + y_2) e_{31} \otimes e_{31} \\ &+ \frac{1}{3} v(-6v^2 + 3v(y_1 - y_2) + 2y_1 y_2) e_{32} \otimes e_{32}, \end{aligned}$$

where $\check{h}_1 = \mathrm{diag}(\frac{2}{3}, -\frac{1}{3}, -\frac{1}{3})$ and $\check{h}_2 = \mathrm{diag}(\frac{1}{3}, \frac{1}{3}, -\frac{2}{3})$.

Next, recall [1, Lemma 2.11], which is based upon [2, Lemma 1.2]:

Lemma 2.5 *Let $r(v; y_1, y_2)$ be a unitary solution of (1) of the form (2). Then $\bar{r}_0(y_1, y_2) = (\mathrm{pr} \otimes \mathrm{pr})\,(r_0(y_1, y_2))$ is a unitary solution of the CYBE (3).*

Combining this result with example (2.4), we derive that $r_{(2,1)}$ and $r_{(3,1)}$ induce solutions of of the CYBE (3). Let us denote these by $c_{(2,1)}$ and $c_{(3,1)}$ respectively. Also, let Ω be the Casimir element of $\mathfrak{sl}_n(\mathbb{C}) \otimes \mathfrak{sl}_n(\mathbb{C})$ with respect to the trace form $(x, y) \mapsto \mathrm{tr}(x \cdot y)$

$$\Omega = \sum_{1 \le i \ne j \le n} e_{i,j} \otimes e_{j,i} + \sum_{1 \le l \le n-1} h_l \otimes \check{h}_l.$$

Observing that $(\mathrm{pr} \otimes \mathrm{pr})\,(P) = \Omega$, we derive the following formulae:

$$c_{(2,1)}(y_1, y_2) = \frac{\Omega}{y_2 - y_1} + y_2 \check{h} \otimes e_{21} - y_1 e_{21} \otimes \check{h} \in \mathfrak{sl}_2(\mathbb{C}) \otimes \mathfrak{sl}_2(\mathbb{C})$$

and

$$\begin{aligned} c_{(3,1)}(y_1, y_2) = & \frac{\Omega}{y_2 - y_1} + y_2 \check{h}_2 \otimes e_{32} - y_1 e_{32} \otimes \check{h}_2 + y_2 e_{12} \otimes e_{31} \\ & - y_1 e_{31} \otimes e_{12} - e_{21} \otimes \check{h}_1 + \check{h}_1 \otimes e_{21} + e_{32} \otimes e_{12} - e_{12} \otimes e_{32} \in \mathfrak{sl}_3(\mathbb{C}) \otimes \mathfrak{sl}_3(\mathbb{C}). \end{aligned}$$

It can be verified that neither $c_{(2,1)}$ nor $c_{(3,1)}$ has any infinitesimal symmetries. Thus theorem 1.1 1) yields that for fixed $v_0 \in \mathbb{C}^\times$, both $r_{(2,1)}(v_0; y_1, y_2)$ and $r_{(3,1)}(v_0; y_1, y_2)$ satisfy the QYBE (4).

3. Gauge equivalence and general results on the AYBE

In this section we explain the notion of gauge equivalence and collect some useful results for solutions of (1). In order to deal with gauge equivalences in a correct way, we have to consider a more general form of the AYBE in four variables

$$r^{12}(v_1, v_2; y_1, y_2)\, r^{23}(v_1, v_3; y_2, y_3) = r^{13}(v_1, v_3; y_1, y_3)\, r^{12}(v_3, v_2; y_1, y_2) \tag{5}$$
$$+ r^{23}(v_2, v_3; y_2, y_3)\, r^{13}(v_1, v_2; y_1, y_3) \tag{6}$$

where r now denotes the germ of a meromorphic function $r : (\mathbb{C}^4, 0) \to A \otimes A$. From the point of view of algebraic geometry, (5) is more natural than (1), see [1]. Note that solutions of (1) are just solutions of (5) depending on the difference of the first pair of spectral parameters $r(v_1, v_2; y_1, y_2) = r(v_1 - v_2; y_1, y_2) = r(v; y_1, y_2)$. As to the definition of gauge equivalence, this is given as follows:

Definition 3.1 Let $\phi : (\mathbb{C}^2, 0) \to \mathrm{GL}_n(\mathbb{C})$ be the germ of a holomorphic function and let $r(v_1, v_2; y_1, y_2)$ be a solution of (5). Then the tensor valued function

$$r'(v_1, v_2; y_1, y_2) = \Big(\phi(v_1; y_1) \otimes \phi(v_2; y_2)\Big)\, r(v_1, v_2; y_1, y_2)\, \Big(\phi^{-1}(v_2; y_1) \otimes \phi^{-1}(v_1; y_2)\Big)$$

is also a solution of (5). The solutions r and r' are said to be gauge equivalent and ϕ is called a gauge transformation.

Example 3.2 Let $r(v_1, v_2; y_1, y_2) \in A \otimes A$ be a solution of (5), $c \in \mathbb{C}$ and $\phi(v, y) = \exp(cvy) \cdot 1 : (\mathbb{C}^2, 0) \to \mathrm{GL}_n(\mathbb{C})$ be a gauge transformation. Then

$$\exp\Big(c\,(v_2 - v_1)\,(y_2 - y_1)\Big)\, r(v_1, v_2; y_1, y_2)$$

is a solution of (5), gauge equivalent to r.

Similarly, assume that $r(v_1, v_2; y_1, y_2) = r(v; y_1, y_2)$ is a solution of (5) which depends only on $v = v_1 - v_2, y_1, y_2$. Thus $r(v; y_1, y_2)$ is a solution of (1). Consider the gauge transformation $\phi(v, y) = \exp(vg(y)) \cdot 1 : (\mathbb{C}^2, 0) \to \mathrm{GL}_n(\mathbb{C})$ for some holomorphic function $g : \mathbb{C} \to \mathbb{C}$. Then $r'(v; y_1, y_2) = \exp(v\,(g(y_2) - g(y_1)))\, r(v; y_1, y_2)$ is a solution of (1) as well.

In the remainder of this section, we list some basic results on solutions of the AYBE which we shall need in the next sections.

Lemma 3.3 [1, Lemma 2.7] *Let $r(v_1, v_2; y_1, y_2)$ be a unitary solution of (5). Then writing $r^{ij}(v_1, v_2)$ as short-hand for $r^{ij}(v_1, v_2; y_i, y_j)$, r also satisfies the "dual equation"*

$$r^{23}(v_2, v_3)\, r^{12}(v_1, v_3) = r^{12}(v_1, v_2)\, r^{13}(v_2, v_3) + r^{13}(v_1, v_3)\, r^{23}(v_2, v_1).$$

Corollary 3.4 *If $r(v; y_1, y_2)$ is a unitary solution of (1), then we also have*

$$r^{23}(u+v; y_2, y_3)\, r^{12}(u; y_1, y_2) = r^{12}(-v; y_1, y_2)\, r^{13}(u+v; y_1, y_3) + r^{13}(u; y_1, y_3)\, r^{23}(v; y_2, y_3). \quad (7)$$

The proof of the next lemma is essentially contained in the proof of [2, Theorem 5].

Lemma 3.5 *Let $r(v; y_1, y_2)$ be a unitary solution of (1) of the form (2). Then r is uniquely determined by r_0 and r_1. Moreover, we have*

$$\begin{aligned} r_1^{12}(y_1, y_2) + r_1^{13}(y_1, y_3) + r_1^{23}(y_2, y_3) &= r_0^{12}(y_1, y_2)\, r_0^{13}(y_1, y_3) - r_0^{23}(y_2, y_3)\, r_0^{12}(y_1, y_2) \\ &\quad + r_0^{13}(y_1, y_3)\, r_0^{23}(y_2, y_3). \end{aligned} \quad (8)$$

Proof. First we show that r is uniquely determined by r_0, r_1 and r_2. To this end, we fix $k > 2$ and show how to construct r_k from $\{r_i\}_{0 \le i \le k-1}$. Let us insert the Laurent expansion (2) of r into (7) and examine the terms of total degree $k-1$ in the variables u and v. We derive the expression

$$r_k^{12}(y_1, y_2)\left[\frac{(-v)^k}{u+v} - \frac{u^k}{u+v}\right] + r_k^{13}(y_1, y_3)\left[\frac{u^k}{v} - \frac{(u+v)^k}{v}\right] + r_k^{23}(y_2, y_3)\left[\frac{v^k}{u} - \frac{(u+v)^k}{u}\right] \quad (9)$$

which equals an expression in terms of r_i with $i < k$ only. The polynomials in u and v are linearly independent for $k > 2$. Indeed, if we place everything over a common denominator and focus on the coefficients of u^k in the respective terms

$$u(-v)^{k+1} - vu^{k+1},\ u^{k+1}(u+v) - u(u+v)^{k+1},\ (u+v)v^{k+1} - v(u+v)^{k+1}$$

then these are $-vu$, $u(u+v) - \binom{k+1}{2}v^2$ and $-(k+1)\,v^2$ respectively. This proves our claim that r is determined by the r_k with $k \le 2$.

In the next step, we show that r_2 is already determined by r_0 and r_1. Indeed, for $k = 2$ the term (9) equals

$$(v-u)\ r_2^{12}(y_1, y_2)\ -\ (2u+v)\ r_2^{13}(y_1, y_3) - (u+2v)\, r_2^{23}(y_2, y_3)$$

that is

$$-u \cdot \left(r_2^{12}(y_1, y_2) + 2r_2^{13}(y_1, y_3) + r_2^{23}(y_2, y_3)\right) + v \cdot \left(r_2^{12}(y_1, y_2) - r_2^{13}(y_1, y_3) - 2r_2^{23}(y_2, y_3)\right).$$

As mentioned above, this equals an expression depending on r_0 and r_1 only. Let us denote the coefficient of $-u$ by a, that of v by b. Since a, b are determined by r_0 and r_1 only, so is $\frac{a+2b}{3} = r_2^{12}(y_1, y_2) - r_2^{23}(y_2, y_3)$. Thus, $r_2(y_1, y_2)$ is determined by r_0 and r_1. Putting $k = 1$ in (9) and writing down the equivalent expression in terms of r_i with $i < k$, we obtain (8). □

4. Poles of solutions of the AYBE

In this section, we study the poles of solutions of (1) along $y_1 = y_2$. We start with the following easy fact on $P = \sum_{1 \le i,j \le n} e_{i,j} \otimes e_{j,i} \in A \otimes A$.

Lemma 4.1 *Any tensor $\theta \in A \otimes A$ such that $\theta(x \otimes 1) = (1 \otimes x)\theta$ for all $x \in A$ is a scalar multiple of P. Moreover $P(1 \otimes x) = (x \otimes 1)P$ for any $x \in A$.*

Lemma 4.2 [3, Lemma 1.3] *Let $r(u; y_1, y_2)$ be a non-degenerate unitary solution of (1). Assume that $r(u; y_1, y_2)$ has a pole along $y_1 = y_2$. Then this pole is simple and $\lim_{y_2 \to y_1} (y_1 - y_2)\, r(u; y_1, y_2) = c \cdot P$ for some $c \in \mathbb{C}$.*

Proof. Write $r(u; y_1, y_2) = \alpha(u; y_1 - y_2) + \beta(u; y_1, y_2)$ and assume that no summand of $\beta(u; y_1, y_2)$ depends only on u and $y = y_1 - y_2$. Let $\alpha(u; y) = \frac{\theta(u)}{y^k} + \frac{\eta(u)}{y^{k-1}} + \dots$ be the Laurent expansion near $y = 0$. In order to see that $k \le 1$, we consider the polar parts in (1) as $y_3 \to y_1$, which yields

$$\theta^{13}(u+v)\, r^{12}(-v; y_1, y_2) + r^{23}(v; y_2, y_1)\, \theta^{13}(u) = 0. \tag{10}$$

Analogously, for $y_2 \to y_1$

$$\theta^{12}(u)\, r^{23}(u+v; y_1, y_3) - r^{13}(u+v; y_1, y_3)\, \theta^{12}(-v) = 0. \tag{11}$$

Let $V \subseteq A$ be the minimal subspace such that $\theta(u) \in V \otimes A$ for all u where $\theta(u)$ is defined. Obviously $r^{23}(u; y_1, y_2)\, \theta^{13}(u) \in V \otimes A \otimes A$ hence by (10) $\theta^{13}(u+v)\, r^{12}(-v; y_1, y_2) \in V \otimes A \otimes A$ as well. Thus, $r^{12}(u; y_1, y_2) \in A_1 \otimes A$, where

$$A_1 = \{a \in A |\, \theta(u)\, (a \otimes 1) \in V \otimes A \,\text{for all}\; u\}.$$

By non-degeneracy $A_1 = A$, thus $VA \subseteq V$. Similarly, using (11), we get $AV \subseteq V$, so that V is a two-sided non-zero ideal in A. Hence $V = A$.

Let us come back to (1). We want to have a look at the coefficient of $(y_1 - y_2)^{1-k}$ in the expansion of (1) near $y_2 - y_1 = h$ equal to zero. The terms contributing to this only depend on $r^{12}(u; y_1, y_1 + h)\, r^{23}(u+v; y_1 + h, y_3)$ and $r^{13}(u+v; y_1, y_3)\, r^{12}(-v; y_1, y_1 + h)$. Thus the coefficient of h^{1-k} consists of two summands, the first one being $\eta^{12}(u)\, r^{23}(u + v; y_1, y_3) - r^{13}(u+v; y_1, y_3)\, \eta^{12}(-v)$ and second one being $\theta^{12}(u)$ times the coefficient of h in $r^{23}(u + v; y_1 + h, y_3)$. Note that for this last summand to be non-zero we must assume $k > 1$. Now $r^{23}(u+v; y_1+h, y_3) - r^{23}(u+v; y_1, y_3)$ equals the summand of $r^{23}(u+v; y_1+h, y_3)$ divisible by h, hence the coefficient of h^{1-k} is exactly

$$\eta^{12}(u)\, r^{23}(u+v; y_1, y_3) - r^{13}(u+v; y_1, y_3)\, \eta^{12}(-v) + \theta^{12}(u)\, \frac{\partial r^{23}}{\partial y_1}(u+v; y_1, y_3).$$

Examining the polar parts in the above expression for $y_1 - y_3$ in a neighborhood of zero, we deduce that $\theta^{12}(u)\, \theta^{23}(u+v) = 0$. Setting $v = 0$ this amounts to saying that $\theta(u) = \{a \otimes b |\, ab = 0\}$. Since $V = A$ this is a contradiction. Therefore $k = 1$.

Next, we have a look at the polar parts in (7) near $y_3 = y_2$. We deduce $\theta^{23}(u+v)\, r^{12}(u; y_1, y_2) = r^{13}(u; y_1, y_2)\, \theta^{23}(v)$. Hence $r(u; y_1, y_2) \in A \otimes A(u)$, where

$$A(u) = \{a \in A \,|\theta(u+v)(x \otimes 1) = (1 \otimes x)\theta(v) \,\text{for all}\, v\}.$$

Since A is non-degenerate this implies $A(u) = A$ for generic u, in which case $1 \in A(u)$ and thus $\theta(u + v) = \theta(u)$. Hence $\theta = \theta(0)$ is constant. Recalling lemma 4.1 finishes the proof. □

Corollary 4.3 [3, Lemma 1.5] *Let $r(u; y_1, y_2)$ be a non-degenerate unitary solution of (1) of the form (2). Then $r(u; y_1, y_2)$ has a simple pole along $y_1 = y_2$ with residue a scalar multiple of P.*

Proof. This is essentially the same proof as that of lemma 1.5 in [3], using lemma 4.2 where Polishchuk refers to lemma 1.3 of his paper. □

5. Quantization of solutions of CYBE coming from solutions of AYBE

In this section we prove part i) of theorem 1.1:

Theorem 5.1 [3, Theorem 1.4] *Let $r(u; y_1, y_2)$ be a non-degenerate unitary solution of (1) of the form (2) and let $\overline{r}_0(y_1, y_2) = (\mathrm{pr} \otimes \mathrm{pr})\,(r_0(y_1, y_2))$.*

(i) *$\overline{r}_0(y_1, y_2)$ is a non-degenerate unitary solution of the CYBE (3).*

(ii) *The following conditions are equivalent:*

  (a) *for fixed $u \in \mathbb{C}^\times$, $r(u; y_1, y_2)$ satisfies the QYBE (4).*

  (b) *there exits a scalar function $\varphi(u; y_1, y_2)$ such that*

$$r(u; y_1, y_2)\, r(-u; y_1, y_2) = \varphi(u; y_1, y_2)\,(1 \otimes 1)\,.$$

  (c) *for $i \in \{1, 2\}$ there exists a scalar function $\psi_i(y_1, y_2)$ such that*

$$\frac{\partial}{\partial y_i}\,(r_0(y_1, y_2) - \overline{r}_0(y_1, y_2)) = \psi_i(y_1, y_2)\,(1 \otimes 1)\,.$$

  (d) *we have*

$$(\mathrm{pr}\otimes\mathrm{pr}\otimes\mathrm{pr})\,\left[\overline{r}_0^{12}(y_1, y_2)\,\overline{r}_0^{13}(y_1, y_3) - \overline{r}_0^{23}(y_2, y_3)\,\overline{r}_0^{12}(y_1, y_2) + \overline{r}_0^{13}(y_1, y_3)\,\overline{r}_0^{23}(y_2, y_3)\right] = 0.$$

(iii) *These conditions are satisfied if $\overline{r}_0(y_1, y_2)$ has no infinitesimal symmetries.*

Before proving this statement, we first need to establish some auxiliary results. The reader might wish to postpone checking them and to go to the proof of theorem 5.1 at the end of this section immediately.

Lemma 5.2 [3, Lemma 1.6] *For any triple of variables u_1, u_2, u_3 set $u_{ij} = u_i - u_j$. Let $r(u; y_1, y_2)$ be any unitary solution of (1) and $s(u; y_1, y_2) = r(u; y_1, y_2)\, r(-u; y_1, y_2)$. Then*

$$\begin{aligned} r^{12}(u_{12}; y_1, y_2)\, r^{13}(u_{23}; y_1, y_3)\, r^{23}(u_{12}; y_2, y_3) &- r^{23}(u_{23}; y_2, y_3)\, r^{13}(u_{12}; y_1, y_3)\, r^{12}(u_{23}; y_1, y_2) \\ &= s^{23}(u_{23}; y_2, y_3)\, r^{13}(u_{13}; y_1, y_3) - r^{13}(u_{13}; y_1, y_3)\, s^{23}(u_{21}; y_2, y_3) \\ &= r^{13}(u_{13}; y_1, y_3)\, s^{12}(u_{32}; y_1, y_2) - s^{12}(u_{12}; y_1, y_2)\, r^{13}(u_{13}; y_1, y_3). \end{aligned}$$

Proof. Let us write $r^{ij}(u)$ as short-hand for $r^{ij}(u; y_i, y_j)$. Since we may assume $u = u_{12}$, $v = u_{23}$ and $u + v = u_{13}$, (1) may be written as

$$r^{12}(u_{12})\, r^{23}(u_{13}) = r^{13}(u_{13})\, r^{12}(u_{32}) + r^{23}(u_{23})\, r^{13}(u_{12}). \tag{12}$$

Analogously, putting $u = u_{13}$ and $v = u_{21}$, (7) reads

$$r^{23}(u_{23})\, r^{12}(u_{13}) = r^{12}(u_{12})\, r^{13}(u_{23}) + r^{13}(u_{13})\, r^{23}(u_{21}). \tag{13}$$

Multiplying (13) with $r^{23}(u_{12})$ from the right yields

$$r^{23}(u_{23})\,r^{12}(u_{13})\,r^{23}(u_{12}) = r^{12}(u_{12})\,r^{13}(u_{23})\,r^{23}(u_{12}) + r^{13}(u_{13})\,s^{23}(u_{21})$$

while switching u_2 and u_3 in (12) followed by multiplication with $r^{23}(u_{23})$ from the left yields

$$r^{23}(u_{23})\,r^{12}(u_{13})\,r^{23}(u_{12}) = r^{23}(u_{23})\,r^{13}(u_{12})\,r^{12}(u_{23}) + s^{23}(u_{23})\,r^{13}(u_{13}).$$

Subtracting these equations, we end up with

$$r^{12}(u_{12})\,r^{13}(u_{23})\,r^{23}(u_{12}) - r^{23}(u_{23})\,r^{13}(u_{12})\,r^{12}(u_{23}) = s^{23}(u_{23})\,r^{13}(u_{13}) - r^{13}(u_{13})\,s^{23}(u_{21}).$$

Switching indices 1 and 3 and using unitarity of r yields the other identity. □

For the next statement we need the notion of an infinitesimal symmetry of a solution r of (1), which is simply that of an element $a \in \mathfrak{sl}_n(\mathbb{C})$ such that $\left[r(u; y_1, y_2), a^1 + a^2\right] = 0$, where $a^1 = a \otimes 1$ and $a^2 = 1 \otimes a$.

Lemma 5.3 [3, Lemma 1.7] *Let $r(u; y_1, y_2)$ be a unitary solution of (1) of the form (2) and $s(u; y_1, y_2) = r(u; y_1, y_2)\,r(-u; y_1, y_2)$. Assuming that $r(u; y_1, y_2)$ has a simple pole along $y_1 = y_2$ with residue cP for some $c \in \mathbb{C}$, we have*

$$s(u; y_1, y_2) = a \otimes 1 + 1 \otimes a + (f(u) + g(y_1, y_2))\, 1 \otimes 1$$

where $f(u) = f(-u)$, $g(y_1, y_2) = g(y_2, y_1)$ and $a \in \mathfrak{sl}_n(\mathbb{C})$ is an infinitesimal symmetry of $r(u; y_1, y_2)$. Moreover, we may write

$$r_0(y_1, y_2) = \overline{r}_0(y_1, y_2) + \alpha(y_2) \otimes 1 - 1 \otimes \alpha(y_1) + h(y_1, y_2) 1 \otimes 1$$

with $\overline{r}_0(y_1, y_2)$ mapping to $\mathfrak{sl}_n(\mathbb{C}) \otimes \mathfrak{sl}_n(\mathbb{C})$, $\alpha(y)$ to $\mathfrak{sl}_n(\mathbb{C})$, $h(y_1, y_2)$ a scalar function and

$$\alpha(y) = \alpha(0) + \frac{y}{cn} a.$$

Proof. By assumption $r(u; y_1, y_2) = \frac{c}{y_1 - y_2} P + \tilde{r}(u; y_1, y_2)$ where $\tilde{r}(u; y_1, y_2)$ does not have a pole along $y_1 = y_2$. Let us write $r(u; y_{ij})$ and $\tilde{r}(u; y_{ij})$ as short-hand for $r(u; y_i, y_j)$ and $\tilde{r}(u; y_i, y_j)$ respectively. Then starting from (7) we derive that for $v = -u + h$:

$$\begin{aligned} r^{13}(u; y_{13})\, r^{23}(-u + h; y_{23}) &= r^{23}(h; y_{23})\, r^{12}(u; y_{12}) - r^{12}(u - h; y_{12})\, r^{13}(h; y_{13}) \qquad (14)\\ &= \left[r^{23}(h; y_{23})\, r^{12}(u; y_{12}) - r^{12}(u; y_{12})\, r^{13}(h; y_{13})\right] \\ &\quad + \left[r^{12}(u; y_{12}) - r^{12}(u - h; y_{12})\right] r^{13}(h; y_{13}). \end{aligned}$$

Let us rewrite the expression in the first bracket on the right-most side as

$$\left(r^{23}(h; y_{23}) \frac{c}{y_1 - y_2} P^{12} - \frac{c}{y_1 - y_2} P^{12}\, r^{13}(h; y_{13})\right) + r^{23}(h; y_{23})\, \tilde{r}^{12}(u; y_{12}) - \tilde{r}^{12}(u; y_{12})\, r^{13}(h; y_{13}).$$

Using lemma 4.1, we know that $P^{12}\, r^{13}(h; y_{13}) = r^{23}(h; y_{13})\, P^{12}$, hence the right-most side of (14) equals

$$\begin{aligned} &\frac{r^{23}(h; y_{23}) - r^{23}(h; y_{13})}{y_1 - y_2}\, cP^{12} + r^{23}(h; y_{23})\, \tilde{r}^{12}(u; y_{12}) - \tilde{r}^{12}(u; y_{12})\, r^{13}(h; y_{13}) \\ &\qquad + \left[\tilde{r}^{12}(u; y_{12}) - \tilde{r}^{12}(u - h; y_{12})\right] r^{13}(h; y_{13}). \end{aligned}$$

Passing to the limit $y_2 \to y_1$, we see that

$$\begin{aligned} r^{13}(u;y_{13})\,r^{23}(-u+h;y_{13}) = &-\frac{\partial r^{23}}{\partial y_1}(h;y_{13})\,cP^{12} + r^{23}(h;y_{13})\,\tilde{r}^{12}(u;y_{11}) \\ &+ \left[\tilde{r}^{12}(u;y_{11}) - \tilde{r}^{12}(u-h;y_{11})\right]\,r^{13}(h;y_{13}) \\ &- \tilde{r}^{12}(u;y_{11})\,r^{13}(h;y_{13}) \end{aligned} \quad (15)$$

We want to apply the operator $\mu \otimes \mathrm{id} : A \otimes A \otimes A \to A \otimes A$ to this equation, where μ is the product in A. Observe that

$$(\mu \otimes \mathrm{id})\left(a^{13}b^{23}\right) = ab,\ (\mu \otimes \mathrm{id})\left(a^{23}b^{12} - b^{12}a^{13}\right) = 0$$

where $a, b \in A \otimes A$ and the notation is best explained by the example $a^{13} = a_1 \otimes 1 \otimes a_2$ for $a = a_1 \otimes a_2$. Moreover, using that $\sum_{i,j} e_{ij} a e_{ji} = \mathrm{tr}(a)1$ for any $a \in A$ clearly, we derive that for $\mathrm{tr}_1 = \mathrm{tr} \otimes \mathrm{id} : A \otimes A \to A$ we have

$$(\mu \otimes \mathrm{id})\left(a^{23}P^{12}\right) = 1 \otimes \mathrm{tr}_1(a).$$

Hence applying $\mu \otimes \mathrm{id}$ to (15) yields

$$\begin{aligned} r(u;y_{13})\,r(-u+h;y_{13}) = &(\mu \otimes \mathrm{id})\left(\left[\tilde{r}^{12}(u;y_{11}) - \tilde{r}^{12}(u-h;y_{11})\right]\,r^{13}(h;y_{13})\right) \\ &- c \cdot 1 \otimes \mathrm{tr}_1\left(\frac{\partial r}{\partial y_1}(h;y_{13})\right). \end{aligned}$$

Now, take the limit $h \to 0$. The left-hand side of yields $s(u;y_1,y_3)$. As for the right-hand side, we invoke our assumption on the existence of a certain Laurent expansion (2) to derive that

$$\lim_{h\to 0}\frac{\partial r}{\partial y_1}(h;y_{13}) = \frac{\partial r_0}{\partial y_1}(y_1,y_3)\,.$$

Moreover

$$\lim_{h\to 0}\left(\left[\tilde{r}^{12}(u;y_{11}) - \tilde{r}^{12}(u-h;y_{11})\right]\,r^{13}(h;y_{13})\right) = \frac{\partial \tilde{r}^{12}}{\partial u}(u;y_{11})\left(\lim_{h\to 0} r^{13}(h;y_{13})\cdot h\right).$$

Again using (2), we see that the second factor of this last term is simply $1 \otimes 1 \otimes 1$. Putting all this together, we end up with

$$s(u;y_1,y_3) = -c \cdot 1 \otimes \mathrm{tr}_1\left(\frac{\partial r_0}{\partial y_1}(y_1,y_3)\right) + \mu\left(\frac{\partial \tilde{r}}{\partial u}(u;y_1,y_1)\right) \otimes 1.$$

Hence we may write $s(u;y_1,y_2) = 1 \otimes \beta(y_1,y_2) + \gamma(u,y_1) \otimes 1$. Note that $\beta(y_1,y_2) = \mathrm{pr}\,(\beta(y_1,y_2)) + \frac{\mathrm{tr}\,(\beta(y_1,y_2))}{n}\,1$. Using the same trick for $\gamma(u,y_1)$, we may actually write

$$s(u;y_1,y_2) = a(u,y_1) \otimes 1 + 1 \otimes b(y_1,y_2) + (f(u,y_1) + g(y_1,y_2))\,1 \otimes 1$$

where now both

$$a(u,y_1) = \mathrm{pr}\,\mu\left(\frac{\partial \tilde{r}}{\partial u}(u;y_1,y_1)\right),\ b(y_1,y_2) = -c \cdot \mathrm{pr}\ \mathrm{tr}_1\left(\frac{\partial r_0}{\partial y_1}(y_1,y_2)\right)$$

map to $\mathfrak{sl}_n(\mathbb{C})$. Note that unitarity of $r(u; y_1, y_2)$ implies that $s^{21}(-u; y_2, y_1) = s^{12}(u; y_1, y_2)$. Applying $\mathrm{pr} \otimes 1$ to this equation yields $a(u, y_1) = b(y_2, y_1)$. It follows that both a and b depend on the second variable only and actually coincide, hence

$$s(u; y_1, y_2) = a(y_1) \otimes 1 + 1 \otimes a(y_2) + (f(u, y_1) + g(y_1, y_2))\, 1 \otimes 1.$$

In order to show the statement concerning the form of s, we have to prove that $a(y_1)$ is constant. To this end, we substitute the form of s just calculated into the second equation of the equality stated in lemma 5.2. We derive that

$$\begin{aligned} r^{13}(u_{13}; y_1, y_3) \cdot (f(u_{32}, y_1) + f(u_{21}, y_2) - f(u_{12}, y_1) - f(u_{23}, y_2)) \\ = \left[a^1(y_1) + a^3(y_3), r^{13}(u_{13}; y_1, y_3)\right]. \end{aligned} \tag{16}$$

Let us focus on the left-hand side. This equals

$$\left[a^1(y_1) + a^3(y_3), \frac{cP^{13}}{y_1 - y_3}\right] + \left[a^1(y_1) + a^3(y_3), \tilde{r}^{13}(u_{13}; y_1, y_3)\right].$$

By lemma 4.1, we may rewrite the first summand as

$$\frac{a^1(y_1) - a^1(y_3)}{y_1 - y_3} cP^{13} + \frac{a^3(y_3) - a^3(y_1)}{y_1 - y_3} cP^{13}$$

thus the limit $y_3 \to y_1$ of the left-hand side of (16) is given by

$$\frac{d}{dy_1}\left(a^1(y_1) - a^3(y_1)\right) cP^{13} + \left[a^1(y_1) + a^3(y_1), \tilde{r}^{13}(u_{13}; y_1, y_1)\right].$$

In particular, the limit $y_3 \to y_1$ of the right-hand side of (16) exists as well. But $r^{13}(u_{13}; y_1, y_3)$ has a pole along $y_1 = y_3$ and the other factor on the right-hand side is independent of y_3. Hence we conclude that

$$f(u_{32}, y_1) + f(u_{21}, y_2) - f(u_{12}, y_1) - f(u_{23}, y_2) = 0.$$

In particular, the left-hand side of (16) equals zero. Focusing on the polar part yields

$$\left[a^1(y_1) + a^3(y_3), \frac{cP^{13}}{y_1 - y_3}\right] = 0$$

which, by the above, implies

$$\frac{d}{dy_1}\left(a^1(y_1) - a^3(y_1)\right) cP^{13} = 0.$$

But then $\frac{da}{dy}(y)$ must be zero, so that $a(y) = a \in \mathfrak{sl}_n(\mathbb{C})$ is constant. Therefore $s(u; y_1, y_2) = a \otimes 1 + 1 \otimes a + (f(u, y_1) + g(y_1, y_2))\, 1 \otimes 1$. By (16) we also see that a is an infinitesimal symmetry of $r(u; y_1, y_2)$.
Finally, we want to prove the statement concerning $r_0(y_1, y_2)$. Clearly, we may write

$$r_0(y_1, y_2) = \overline{r}_0(y_1, y_2) + \alpha(y_2, y_1) \otimes 1 - 1 \otimes \alpha(y_1, y_2) + h(y_1, y_2) 1 \otimes 1$$

with α mapping to $\mathfrak{sl}_n(\mathbb{C})$. Note that in the discussion above we derived that

$$a = b(y_1, y_2) = -c \cdot \mathrm{pr}\ \mathrm{tr}_1\left(\frac{\partial r_0}{\partial y_1}(y_1, y_2)\right).$$

Journal of Physics: Conference Series **346** (2012) 012007 doi:10.1088/1742-6596/346/1/012007

Since both $\overline{r}_0$ and α map to $\mathfrak{sl}_n(\mathbb{C})$, so will their partial derivatives. This implies

$$a = -cn \cdot \mathrm{pr}\left(- \left(\frac{\partial}{\partial y_1} \right) \alpha(y_1, y_2) + \frac{\partial h}{\partial y_1}(y_1, y_2) \cdot 1 \right).$$

Hence $\frac{\partial}{\partial y_1}\alpha(y_1, y_2) = \frac{a}{cn}$, which gives the formula for α. In particular, $\alpha(y_1, y_2)$ does not depend on the second argument. This completes the proof of the formula for r_0. □

Before we finally prove theorem 5.1, we need to state one more easy lemma:

Lemma 5.4 *The follwing hold:*

(i) For any $x, y, \phi \in A, i \in \{1,2\}$ and $\phi^1 = \phi \otimes 1$ respectively $\phi^2 = 1 \otimes \phi$, we have

$$(\mathrm{pr} \otimes \mathrm{pr}) \left[x \otimes y, \phi^i \right] = \left[(\mathrm{pr} \otimes \mathrm{pr}) (x \otimes y), \phi^i \right].$$

(ii) Let r be a non-degenerate solution of (1) and $[r, 1 \otimes a] = 0$ for some $a \in \mathfrak{sl}_n(\mathbb{C})$. Then $a = 0$.

Proof. *(i)* is straightforward. As to *(ii)*, write $r = \sum_{i \in I} r_i' \otimes r_i''$ for some index set I and let $\varphi : A \otimes A \to \mathrm{End}(A)$ denote the isomorphism given by $X \otimes Y \mapsto (Z \mapsto X \,\mathrm{tr}\,(YZ))$. Then $0 = \varphi([r, 1 \otimes a])(b) = \sum_{i \in I} r_i' \,\mathrm{tr}\,([r_i'', a]\, b) = \sum_{i \in I} r_i' \,\mathrm{tr}\,(r_i''\, [a, b]) = \varphi(r)\,([a, b])$ for all $b \in A$. Now r is non-degenerate, hence $\varphi(r)$ is an isomorphism. This yields $[a, b] = 0$ for all $b \in A$, especially all $b \in \mathfrak{sl}_n(\mathbb{C})$. But the Lie bracket is non-degenerate on $\mathfrak{sl}_n(\mathbb{C})$, hence $a = 0$. □

Proof of Theorem 5.1.
(1) By lemma 2.5 $\overline{r}_0$ is a unitary solution of the CYBE (3). The rest is immediate by lemma 4.3 and the fact that $(\mathrm{pr} \otimes \mathrm{pr})\,(P)$ is the Casimir element of $\mathfrak{sl}_n(\mathbb{C}) \otimes \mathfrak{sl}_n(\mathbb{C})$ with respect to the trace form $(x, y) \mapsto \mathrm{tr}(x \cdot y)$.

(2) Setting $u_1 = u$, $u_2 = 0$ and $u_3 = -u$ in lemma 5.2, we derive that $r(u; y_1, y_2)$ satisfies the QYBE (4) for u fixed if and only if

$$s^{23}(u; y_2, y_3)\, r^{13}(2u; y_1, y_3) = r^{13}(2u; y_1, y_3)\, s^{23}(-u; y_2, y_3).$$

Applying lemma 5.3 this is equivalent to

$$[r(u; y_1, y_2), 1 \otimes a] = 0 \tag{17}$$

which, by lemma 5.4 (ii) is equivalent to $a = 0$. By lemma 5.3 this last condition holds if and only if either of conditions (b) or (c) of the theorem are satisfied. It remains to show equivalence with condition (d). To this end, recall (8). Denote the right-hand side of this equation by $AYBE[r_0](y_1, y_2, y_3)$. Then (d) simply reads $(\mathrm{pr} \otimes \mathrm{pr} \otimes \mathrm{pr})\, AYBE[\overline{r}_0](y_1, y_2, y_3) = 0$. To show the equivalence of this with $a = 0$, express r_0 in terms of $\overline{r}_0$ as stated in lemma 5.3. Then

$$\begin{aligned} &- cn \cdot (\mathrm{pr} \otimes \mathrm{pr} \otimes \mathrm{pr})\,(AYBE\,[r_0]\,(y_1, y_2, y_3) - AYBE\,[\overline{r}_0]\,(y_1, y_2, y_3)) \\ &= (y_1 - y_3)\, \overline{r}_0^{13}(y_1, y_3)\, a^2 + (y_1 - y_2)\, \overline{r}_0^{12}(y_1, y_2)\, a^3 + (y_3 - y_2)\, \overline{r}_0^{23}(y_2, y_3)\, a^1. \end{aligned}$$

It is immediate by (8) that $(\mathrm{pr} \otimes \mathrm{pr} \otimes \mathrm{pr})\, AYBE\,[r_0]\,(y_1, y_2, y_3) = 0$. Hence if a is zero, this implies $(\mathrm{pr} \otimes \mathrm{pr} \otimes \mathrm{pr})\, AYBE[\overline{r}_0] = 0$. On the other hand, assuming $(\mathrm{pr} \otimes \mathrm{pr} \otimes \mathrm{pr})\, AYBE[\overline{r}_0] = 0$ we deduce

$$(y_1 - y_3)\, \overline{r}_0^{13}(y_1, y_3)\, a^2 + (y_1 - y_2)\, \overline{r}_0^{12}(y_1, y_2)\, a^3 + (y_3 - y_2)\, \overline{r}_0^{23}(y_2, y_3)\, a^1 = 0.$$

We will show that this implies $a = 0$. Indeed, by lemma 4.3 we know that $r_0(y_1, y_2) = \frac{cP}{y_1-y_2} + \tilde{r}_0(y_1, y_2)$ with $\tilde{r}_0$ being defined along $y_1 = y_2$ and similarly for $\bar{r}_0$. Hence passing to the limit $y_1, y_2, y_3 \to y$ yields

$$(\mathrm{pr} \otimes \mathrm{pr} \otimes \mathrm{pr}) \left[P^{13}a^2 + P^{12}a^3 + P^{23}a^1\right] = 0.$$

Let us write $a = \sum a_{ij}e_{ij}$. Looking at the coefficient of $e_{ij} \otimes e_{ji} \otimes e_{ij}$ in the above equation for $i \neq j$, we derive $a_{ij} = 0$. But then projecting the above equation to $e_{12} \otimes e_{21} \otimes \mathfrak{sl}_n(\mathbb{C})$, we may deduce that $a = 0$.

(3) As we just saw, the conditions of (2) are satisfied if $a = 0$. But a is an infinitesimal symmetry of r by lemma 5.3, hence one of r_0. Invoking lemma 5.4 a), we deduce that a is an infinitesimal symmetry of $\bar{r}_0$ and so $a = 0$. □

6. Uniqueness of lifts from CYBE to AYBE

In this section we will prove part ii) of theorem 1.1:

Theorem 6.1 [2, Theorem 6] *Let $r(u; y_1, y_2)$ and $s(u; y_1, y_2)$ be a unitary solutions of (1) of the form (2). Assume that the corresponding solution $\bar{r}_0(y_1, y_2) = (\mathrm{pr} \otimes \mathrm{pr})(r_0(y_1, y_2))$ of the CYBE (3) is non-degenerate, has no infinitesimal symmetries and that $\bar{s}_0(y_1, y_2) = \bar{r}_0(y_1, y_2)$. Then there exists a meromorphic function $g : \mathbb{C} \to \mathbb{C}$ such that $s(u; y_1, y_2) = \exp\big(u(g(y_2) - g(y_1))\big)r(u; y_1, y_2)$.*

Proof. First, we show that r is uniquely determined by r_0. By lemma 3.5 r is uniquely determined by r_0 and r_1 and moreover r_1 is a solution of a certain equation in r_0 which is given by (8). If $r_1' \neq r_1$ was a solution of (8) with the same properties as r_1, then taking the difference we would obtain a meromorphic function $\alpha : (\mathbb{C}^2, 0) \to A \otimes A$ with $\alpha^{21}(y_2, y_1) = \alpha(y_1, y_2)$ and

$$\alpha^{12}(y_1, y_2) + \alpha^{13}(y_1, y_3) + \alpha^{23}(y_2, y_3) = 0. \tag{18}$$

Using lemma 4.2, we also know that the residue of $r(u; y_1, y_2)$ near $y_1 = y_2$ is independent of u. Comparing this to the Laurent expansion (2), we derive that $r_1(y_1, y_2)$ has no poles along $y_1 = y_2$, hence the same is true for $\alpha(y_1, y_2)$. To prove that r_1 is determined by r_0, we need only show that α is already zero. Choosing $y_3 = y_2$ and then applying $\mathrm{pr}\otimes\mathrm{id}\otimes\mathrm{id}$ to (18) we derive that $(\mathrm{pr}\otimes\mathrm{id})(\alpha(y_1, y_2)) = 0$. Similarly, $(\mathrm{id}\otimes\mathrm{pr})(\alpha(y_1, y_2)) = 0$, hence $\alpha(y_1, y_2) = f(y_1, y_2)\, 1\otimes 1$ where f is a meromorphic function such that $f(y_1, y_2) + f(y_1, y_3) + f(y_2, y_3) = 0$. Since $r_1(y_1, y_2)$ has no pole along $y_1 = y_2$ by lemma 4.2, $\alpha(y_1, y_1)$ exists. We may deduce that $2f(y_1, y_2) = -f(y_2, y_2)$, so f depends only on the second variable. But then choosing $y_2 = y_1 = y_3$ we read $3f(y_1, y_1) = 0$, thus $f = 0$.We have proved that r is uniquely determined by r_0.

It remains to prove that, provided $\bar{r}_0$ has no infinitesimal symmetries, r can be uniquely recovered from $\bar{r}_0$ up to the factor $\exp\left(u\left(g(y_2) - g(y_1)\right)\right)$ for some meromorphic function $g : \mathbb{C} \to \mathbb{C}$. Note that this is equivalent to showing that $r_0(y_1, y_2)$ is uniquely determined by $\bar{r}_0(y_1, y_2) = (\mathrm{pr} \otimes \mathrm{pr})(r_0(y_1, y_2))$ up to a summand of the form $(g(y_2) - g(y_1))\, 1 \otimes 1$. By assumption $(s_0(y_1, y_2), s_1(y_1, y_2))$ is another tuple satisfying (8) such that

$$s_0^{21}(y_2, y_1) = -s_0(y_1, y_2),\ s_1^{21}(y_2, y_1) = s_1(y_1, y_2).$$

We claim that

$$s_0(y_1, y_2) = r_0(y_1, y_2) + (g(y_2) - g(y_1))\, 1 \otimes 1.$$

Since $\overline{s}_0(y_1, y_2) = \overline{r}_0(y_1, y_2)$, we may write

$$s_0(y_1, y_2) = r_0(y_1, y_2) + \phi^1(y_1, y_2) - \phi^2(y_2, y_1) + \psi(y_1, y_2)\, 1 \otimes 1$$

for a $\mathfrak{sl}_n(\mathbb{C})$ valued function ϕ and a scalar function ψ. Denoting the left-hand side of (8) by $LHS(r)$, we have

$$\begin{aligned} 0 &= (\mathrm{pr} \otimes \mathrm{pr} \otimes \mathrm{pr})\,(LHS(s) - LHS(r)) \\ &= \overline{r}_0^{12}(y_1, y_2)\, \left[\phi^3(y_3, y_2) - \phi^3(y_3, y_1)\right] + \overline{r}_0^{23}(y_2, y_3)\, \left[\phi^1(y_1, y_3) - \phi^1(y_1, y_2)\right] \\ &\quad + \overline{r}_0^{13}(y_1, y_3)\, \left[\phi^2(y_2, y_3) - \phi^2(y_2, y_1)\right]. \end{aligned}$$

If the function ϕ is not constant then contracting this equation with a generic functional in the third component we derive that $\overline{r}_0$ is a sum of two decomposable tensors, that is $\overline{r}_0 = a_1 \otimes b_1 + a_2 \otimes b_2$ where all terms depend on y_1, y_2. But $\overline{r}_0$ is non-degenerate by assumption, so $\mathrm{span}_{\mathbb{C}}(\{a_1, a_2\}) \cong \mathfrak{sl}_n(\mathbb{C})$, which is impossible for any $n \geq 2$. Thus $\phi \in \mathfrak{sl}_n(\mathbb{C})$ is constant. Applying $(\mathrm{pr} \otimes \mathrm{pr} \otimes \mathrm{id})$ to $LHS(s) - LHS(r)$ yields

$$\begin{aligned} (\mathrm{pr} \otimes \mathrm{pr} \otimes \mathrm{id})\left(s_1^{12}(y_1, y_2) - r_1^{12}(y_1, y_2)\right) &= -\phi^1\phi^2 + (\psi(y_1, y_3) - \psi(y_2, y_3))\, \overline{r}^{12}(y_1, y_2) \\ &\quad + (\mathrm{pr} \otimes \mathrm{pr} \otimes \mathrm{id})\left(r_0^{12}(y_1, y_2)\phi^1 - \phi^2 r_0^{12}(y_1, y_2)\right) \end{aligned} \tag{19}$$

This implies that $\psi(y_1, y_3) - \psi(y_2, y_3)$ is actually independent of y_3, hence equal to some function $\beta(y_1, y_2)$. Also, we know by unitarity of r that $\psi(y_1, y_2) = -\psi(y_2, y_1)$, thus $\beta(y_1, y_2) = \psi(y_1, y_3) + \psi(y_3, y_2)$. It follows from lemma 4.2 that r_0 and s_0 have the same pole along $y_1 = y_2$, hence $\psi(y_1, y_1)$ exists and we may deduce that $\beta(y_1, y_2) = \psi(y_1, y_1) + \psi(y_1, y_2) = \psi(y_1, y_2)$. Thus the definition of β reads $\psi(y_1, y_2) = \psi(y_1, y_3) - \psi(y_2, y_3)$. Therefore, defining $g(y) = \psi(y, a)$ for some fixed $a \in \mathbb{C}$, we have $\psi(y_1, y_2) = g(y_1) - g(y_2)$. Altogether

$$s_0(y_1, y_2) = r_0(y_1, y_2) + \phi^1 - \phi^2 + (g(y_1) - g(y_2))\; 1 \otimes 1 \tag{20}$$

Since s_0 and r_0 are both meromorphic, so is ψ and thus also g.

Next, we replace $r(u; y_1, y_2)$ by $\exp\left(u\,(g(y_2) - g(y_1))\right) r(u; y_1, y_2)$ and hence may assume that $g = 0$ in the above formula for s_0. Thus (19) yields

$$(\mathrm{pr} \otimes \mathrm{pr})\,(s_1(y_1, y_2) - r_1(y_1, y_2)) = (\mathrm{pr} \otimes \mathrm{pr})\left(r_0(y_1, y_2)\phi^1 - \phi^2 r_0(y_1, y_2)\right) - \phi^1\phi^2.$$

We exchange the first two components, make the substitutions $y_1 \leftrightarrow y_2, y_2 \leftrightarrow y_1$ and use unitarity of r for both sides of the resulting equation. Comparing the result with the above equation, we derive

$$(\mathrm{pr} \otimes \mathrm{pr})\left(r_0(y_1, y_2)\phi^1 - \phi^2 r_0(y_1, y_2)\right) = (\mathrm{pr} \otimes \mathrm{pr})\left(-r_0(y_1, y_2)\phi^2 + \phi^1 r_0(y_1, y_2)\right).$$

By lemma 5.4 (i) we deduce that $\left[\overline{r}_0(y_1, y_1), \phi^1 + \phi^2\right] = 0$. But then ϕ is an infinitesimal symmetry of $\overline{r}_0$, so $\phi = 0$. Thus $s_0 = r_0$. □

Remark. By theorem 5.1 (1), the assumption of theorem 6.1 on the non-degeneracy of $\overline{r}_0$ is automatically satisfied if r itself is non-degenerate. Moreover, we deduce from the proof of theorem 6.1 that in that case r is already uniquely determined by r_0.

Corollary 6.2 *In the notations of theorem 6.1, assume that $r_0(y_1, y_2)$ and $s_0(y_1, y_2)$ have the same poles on $\left(\mathbb{C}^2 \setminus V((y_1 - y_2))\right)$. Then g is a holomorphic function. Thus, r and s are gauge equivalent.*

Proof. It follows from the assumption and lemma 4.2 that the poles of r_0 and s_0 coincide. By (20) this implies that g is holomorphic. The remaining statement follows from the discussion in example 3.2. □

Combining the above corollary with theorem 2.3 yields the final result of this section, finishing the proof of theorem 1.1 ii).

Corollary 6.3 *In the notations of theorem 6.1, assume that both r and s can be obtained by the procedure described in section 2. Then r and s are gauge equivalent.*

Acknowledgments
This work was supported by the DFG project Bu–1866/2–1.

References

[1] Burban I and Kreußler B 2007 vector bundles on degenerations of elliptic curves and Yang–Baxter equations Preprint 0708.1685
[2] Polishchuk A 2002 classical Yang–Baxter equation and the A_∞–constraint *Adv. Math.* **168** 56–95
[3] Polishchuk A 2009 Massey products on cycles of projective lines and trigonometric solutions of the Yang–Baxter equations *Algebra, arithmetic, and geometry: in honor of Yu. I. Manin* (Progr. Math. **270** vol 2) ed Y Tschinkel and Y Zarhin (Boston: Birkhäuser Boston) pp 573–617
[4] Stolin A 1991 on rational solutions of Yang–Baxter equation for $\mathfrak{sl}(n)$ *Math. Scand.* **69** 57–80

Algebra, Geometry, and Mathematical Physics 2010 IOP Publishing
Journal of Physics: Conference Series **346** (2012) 012008 doi:10.1088/1742-6596/346/1/012008

Automorphic products, generalized Kac-Moody algebras and string amplitudes

Stefan Hohenegger

Max–Planck–Institut für Physik, Werner–Heisenberg–Institut, Föhringer Ring 6, 80805 München, Germany

E-mail: shoheneg@mppmu.mpg.de

Abstract. We review automorphic products and generalized Kac-Moody algebras from a physics point of view. We discuss the appearance of automorphic products in BPS-saturated one-loop quantities in heterotic string theory. At particular points in moduli space, with enhanced gauge symmetry $\mathfrak{h}$, these products can be used to define a generalized Kac-Moody algebra $\mathcal{G}(\mathfrak{g}^{++})$ as an automorphic correction of the Lorentzian Kac-Moody algebra $\mathfrak{g}^{++}$, which is obtained through double extension of the complement $\mathfrak{g} = (\mathfrak{e}_8 \oplus \mathfrak{e}_8)/\mathfrak{h}$. The root multiplicities of $\mathcal{G}(\mathfrak{g}^{++})$ are then encoded in the Fourier coefficients of certain modular forms, which appear directly in the integrand of the one-loop quantities. We review particular examples of this extension for compactifications with $\mathcal{N} = 2$ and $\mathcal{N} = 4$ space-time supersymmetry.

MSC 2010: 81R10, 17B65, 17B67, 17B81

1. Introduction

The BPS spectrum is a particular set of states in the Hilbert space of theories with extended supersymmetry. Especially in the context of superstring theory it has received a lot of attention during the recent years. One reason for this is that observables which receive contributions only from BPS states (*BPS-saturation*) are protected when extrapolated from weak to strong string coupling. The study of such quantities therefore allows invaluable insights into non-perturbative aspects of string theory and in the past has for example been the foremost arena for studying string dualities [2, 43]. Besides that, such BPS-saturated quantities are in many cases strongly related to mathematically interesting structures. Examples which have been studied extensively in the past include helicity supertraces [40, 41, 18] and topological amplitudes [1, 10, 3, 5, 4, 6, 8]. Physically, degeneracies of BPS states can be related to the entropy of particular supersymmetric black holes (see [49] for a review). In this way, string theory is capable of a microscopic explanation of thermodynamic (macroscopic) properties of (certain) black holes. This connection is not only qualitative in nature, but in many examples precise agreement is found even to subleading order (in the black-hole charges) (see e.g. [18, 17]). Degeneracies of BPS-states (or more precisely BPS-indices) are locally constant functions on the moduli space of string compactifications. However, they may jump at (real) co-dimension one hypersurfaces where bound states of BPS-states may decay or recombine. Such *walls of marginal stability* have been extensively studied in the recent past (see *e.g.* [48, 19, 14]).

In [34, 35] Harvey and Moore have undertaken a closer investigation of mathematical properties of the BPS-spectrum in string theory. In particular they argued that the space of BPS-states should form an algebra. Indeed they discovered that part of the threshold corrections to the gauge couplings in four-dimensional $\mathcal{N} = 2$ heterotic string theory can be written as infinite product representations of automorphic forms on the Grassmannian $SO(2, 2+n)/(SO(2) \times SO(2+n))$. The latter, through work of Borcherds [12], are in turn related to denominator formulae for *Generalized Kac-Moody algebras* (GKM), which was seen as evidence that the algebra of BPS-states is indeed a GKM. However, although these findings are very intriguing and suggestive, a direct connection between these infinite product formulae and the algebra of BPS-states was not established.

A different — physics inspired — approach was initiated by the work of Dijkgraaf, Verlinde and Verlinde [20], who conjectured that the degeneracies of (non-perturbative) 1/4 BPS dyon-states in four-dimensional $\mathcal{N} = 4$ heterotic compactifications are encoded in the Fourier coefficients of a certain Siegel modular form, known as the *Igusa cusp form* Φ_{10}. As it had been shown in the mathematics literature [30], Φ_{10} has an infinite product representation which relates it to the denominator formula of a certain rank 3 (super) GKM-algebra, which was called $\mathfrak{g}_{1,\text{II}}$ in [30]. In this way, the degeneracies of dyons also become directly related to root multiplicities of a GKM-algebra. Further hints for the role of the algebra $\mathfrak{g}_{1,\text{II}}$ in physics were found in [14] (see also [15, 16] and [27, 28, 29] for related work), where it was shown that the wall-crossing behaviour of the dyon spectrum is controlled by the hyperbolic Weyl group $\mathcal{W}(\mathfrak{g}_{1,\text{II}})$ of this GKM-algebra.

Recently, further efforts [23, 36] have been made to understand the appearance of GKM algebras in BPS-saturated couplings in string theory. In [23] a microscopic analysis of the algebraic properties of the BPS spectrum has been undertaken. Explicitly, within the framework of the $E_8 \times E_8$ heterotic string theory compactified on T^2, it was shown that the space of BPS states forms a representation of a certain GKM algebra $\mathcal{G}$, which was constructed explicitly. Moreover, within the same heterotic model a particular one-loop BPS-saturated amplitude $\mathcal{F}_1$ (first discussed in [3]) was analyzed. By explicit computation it was proven that (at a generic point in moduli space) part of this amplitude encodes the root-multiplicities of a slightly larger algebra $\mathcal{G}_{\text{ext}}$, of which $\mathcal{G}$ is a subalgebra $\mathcal{G} \subset \mathcal{G}_{\text{ext}}$. In [36] $\mathcal{F}_1$ has been systematically studied at particular points in moduli space where the gauge group is enhanced to $\mathfrak{h} \subset \mathfrak{e}_8 \oplus \mathfrak{e}_8$. It was shown that after some proper regularization, part of $\mathcal{F}_1$ can be interpreted as the denominator formula of a GKM which is the Borcherds extension of $\mathfrak{g}^{++}$, where the latter is the hyperbolic extension of the commutant of $\mathfrak{h}$ in $\mathfrak{e}_8 \oplus \mathfrak{e}_8$.

The organization of this article is as follows: In section 2 we will review basic properties of Generalized Kac-Moody algebras. In particular, we will explain how they can be obtained through an automorphic correction of a hyperbolic Kac-Moody algebra. The key ingredients in this construction are so-called *automorphic products*, which in turn can be understood as Borcherds-lifts of modular forms of $\mathrm{SL}(2,\mathbb{Z})$. Within the framework of the (singular) theta correspondence, such automorphic products appear naturally as one-loop BPS saturated quantities in string theory. As an example, we will review in section 3 the work of Harvey and Moore [34, 35] and study threshold corrections to the gauge couplings in $\mathcal{N} = 2$ heterotic string compactifications. Finally, in section 4 we will review aspects of [36] and analyze a particular one-loop topological amplitude in heterotic string theory on T^2, at a point in moduli space where the gauge algebra is enhanced to a Lie algebra $\mathfrak{h}$. We will extract from this amplitude the denominator formula for a GKM which can be constructed as an automorphic correction of the hyperbolic extension of the commutant of $\mathfrak{h}$ in $\mathfrak{e}_8 \oplus \mathfrak{e}_8$. For simplicity we will focus on $\mathfrak{g}$ being simple — the treatment of the semisimple case can be found in [23, 36].

2. Generalized Kac-Moody algebras

2.1. General properties

Before considering the appearance of generalized Kac-Moody algebras in the context of physical quantities, let us first review some of their properties. For any simple Lie algebra $\mathfrak{g}$ of rank r the *Cartan matrix* $\mathcal{C}_{ij}$ (with $i = 1, \ldots, r$) is a matrix with the following properties

$$\mathcal{C}_{ii} = 2, \quad \mathcal{C}_{ij} = 0 \Leftrightarrow \mathcal{C}_{ji} = 0 \quad \mathcal{C}_{ij} \in \mathbb{Z}_- \; \forall i \neq j\,. \tag{1}$$

We can distinguish three different cases, depending on whether $\mathcal{C}$ is

- positive definite: $\mathfrak{g}$ is a finite dimensional Lie algebra falling into the Cartan-Killing classification of finite simple Lie algebras (see e.g. [39] for an overview)
- positive semi definite: $\mathfrak{g}$ is an *affine Kac-Moody* algebra; affine KM can be constructed as extensions of finite dimensional Lie algebras $\mathfrak{h}$ (to indicate this we will mostly write $\mathfrak{g} = \mathfrak{h}^+$)
- indefinite: in this work we focus on the case that $\mathcal{C}$ is of Lorentzian signature such that $\mathfrak{g}$ is a *Lorentzian Kac-Moody* algebra; a full classification exists only for particular subclasses e.g. hyperbolic Kac-Moody algebras. In these notes we will denote with $\mathfrak{g} = \mathfrak{h}^{++}$ the hyperbolic extension of a finite dimensional Lie algebra $\mathfrak{h}$ (as an example we have depicted the Dynkin diagram of $\mathfrak{a}_k^{++}$ in figure 1)

The algebras that we will be concerned with in this review are of a more general type (i.e. their Cartan matrices generalize (1)). A *Generalized Kac-Moody* algebra (GKM)[1] $\mathcal{G}$ has finite rank and relaxes the first condition of (1). In particular, $\mathcal{G}$ is characterized by the fact that some of the diagonal entries $\mathcal{C}_{ii} < 0$. If we view the Cartan matrix as the inner product matrix of the simple roots α_i of $\mathcal{G}$, this entails that $\mathcal{G}$ has two types of simple roots

$$\begin{aligned} &(\alpha_i|\alpha_i) > 0 \quad \ldots \quad \text{real simple roots} \\ &(\alpha_i|\alpha_i) \leq 0 \quad \ldots \quad \text{imaginary simple roots} \end{aligned}$$

For further convenience, we will denote the set of all roots of $\mathcal{G}$ as $\Delta(\mathcal{G})$. GKMs share in common with finite dimensional Lie algebras the existence of a *Weyl group* $\mathcal{W}(\mathcal{G})$ defined as the group of reflections in the simple roots. *I.e.* if $\{\alpha_I\}$ (with $I = 1, \ldots, n$) denotes the set of simple real roots of $\mathcal{G}$, $\mathcal{W}(\mathcal{G})$ is generated by n fundamental reflections

$$w_I \; : x \longmapsto x - 2\frac{(x|\alpha_I)}{(\alpha_I|\alpha_I)}\alpha_I \quad \text{with} \quad \in \Lambda_{\mathcal{G}} \otimes \mathbb{C}\,.$$

GKMs also posses a Weyl vector ρ which satisfies

$$(\rho|\alpha) \leq -\frac{1}{2}(\alpha|\alpha) \quad \text{for} \quad \alpha \in \Delta(\mathcal{G})\,,$$

and equality holds if and only if α is a simple root of $\mathcal{G}$. The existence of $\mathcal{W}(\mathcal{G})$ and ρ also allows to formulate a generalization of the Weyl-Kac-character formula. For any (integrable) lowest weight representation $R(\lambda)$ of $\mathcal{G}$ we have [39, 11]

$$\text{ch}\, R(\lambda) = \frac{\sum_{w\in\mathcal{W}} \epsilon(w) w(S) e^{-\rho}}{\prod_{\alpha\in\Delta_+}(1 - e^{\alpha})^{\text{mult}\,\alpha}}\,. \tag{2}$$

Here $\epsilon(w) = (-1)^{\ell(w)}$, where $\ell(w)$ denotes the length of the Weyl element w (see e.g. [38]). Notice that the product runs over the positive roots Δ^+ of $\mathcal{G}$ and $\text{mult}\,\alpha$ denotes the multiplicity of

[1] Some authors also use the name *Borcherds-Kac-Moody* algebra (BKM) or just *Borcherds* algebra.

Journal of Physics: Conference Series **346** (2012) 012008 doi:10.1088/1742-6596/346/1/012008

$\alpha \in \Delta^+$. The main difference of (2) to the 'standard' Weyl-Kac character formula (for e.g. finite Lie algebras) is the correction factor S which is modified due to the presence of imaginary simple roots in $\mathcal{G}$ [11]:

$$S = e^{\lambda+\rho} \sum_{\alpha \in \Lambda^+_{\mathcal{G}}} \xi(\alpha) e^{\alpha} .$$

where $\Lambda^+_{\mathcal{G}}$ is the positive part of the root lattice of $\mathcal{G}$ and $\xi(\alpha) = (-1)^m$ if α is a sum of m distinct pairwise orthogonal imaginary simple roots which are orthogonal to λ, and $\xi(\alpha) = 0$ otherwise. Thus, if $\mathcal{G}$ contains no imaginary roots, we recover the familiar expression [39]. In this review we will mainly be interested in (2) for the trivial representation with $\lambda = 0$ and $R(0) = 1$. For this particular case (2) reduces to the so-called *denominator formula*

$$\sum_{w \in \mathcal{W}} \epsilon(w) w(S) = e^{\rho} \prod_{\alpha \in \Delta_+} (1 - e^{\alpha})^{\mathrm{mult}\,\alpha} . \tag{3}$$

This formula relates a sum over the Weyl group $\mathcal{W}(\mathcal{G})$ to an infinite product over all positive roots of $\mathcal{G}$.

2.2. Automorphic correction

An important question is how to construct GKMs. A method which will be very important for the further sections of this review is the *automorphic correction* of a hyperbolic Kac-Moody algebra. Morally speaking, this method extends the root system of a hyperbolic Kac-Moody algebra $\mathfrak{g}^{++}$ by an infinite set of imaginary simple roots. To guarantee consistency, a further ingredient is needed [12, 30], namely a weak Jacobi form $f(\tau, z)$ of $SL(2, \mathbb{Z})$ with Fourier expansion

$$f(\tau, z) = \sum_{\lambda \in \Lambda_{\mathfrak{g}^{++}}} c(-\lambda^2/2)\, q^{-\frac{1}{2}(\lambda|\lambda)} e^{2\pi i (z|\lambda)} , \tag{4}$$

where $\Lambda_{\mathfrak{g}^{++}}$ is the root lattice of $\mathfrak{g}^{++}$. If the dimension $\dim(\Lambda_{\mathfrak{g}^{++}}) = r+2$, then $f(\tau, z)$ has weight $-r/2$. Notice that these modular properties guarantee that all Fourier coefficients $c(-\lambda^2/2)$ of (4) are integers, which allows us to define the *modular product* (with respect to $O(2, 2+r)$) [12]

$$e^{-2\pi i(\rho|y)} \prod_{\lambda \in \Lambda^+_{\mathfrak{g}^{++}}} \left(1 - e^{-2\pi i(\lambda|y)}\right)^{c(-\lambda^2/2)} , \quad y \in \Lambda_{\mathfrak{g}^{++}} \otimes \mathbb{C} , \tag{5}$$

where ρ is the lattice Weyl-vector of $\mathfrak{g}^{++}$. The idea of constructing a GKM is to identify (5) with the right hand side of (3) and to interpret the additional terms in (5) as new roots on top of the ones already present in $\Delta^+_{\mathfrak{g}^{++}}$. Because of the crucial minus sign in the exponent of q in (4), these additional roots are generically imaginary. Moreover, since the product in (5) is infinite, we generically add an infinite set of additional simple roots. It was shown in [11, 12] that there exists indeed a GKM $\mathcal{G}$ with these roots. In order to emphasize that the latter was constructed from $\mathfrak{g}^{++}$ we will in many cases write $\mathcal{G} \equiv \mathcal{G}(\mathfrak{g}^{++})$. It is, however, important to realize that the extension of $\mathfrak{g}^{++}$ is not unique, since different modular products (i.e. different $f(\tau, z)$ in (4)) will lead to different algebras $\mathcal{G}$.

Notice that identifying the modular product (5) with the right hand side of (3) suggests that it has an alternative presentation as an infinite sum (corresponding to the left hand side of (3)). While the presentation (5) is usually called the *multiplicative* (or Borcherds) *lift* [30, 32, 33, 11] of the seed $f(\tau, z)$, the inifinite sum representation can be thought of as an *additive* (or arithmetic) *lift* [44, 31] of some different automorphic form.

2.3. Automorphic products as one-loop amplitudes in string theory

All ingredients for the construction of a GKM that we have mentioned in the previous paragraph— in particular the automorphic products (5)— arise naturally in one-loop expressions in string theory. In toroidally compactified heterotic string theory, the latter are typically of the following form

$$\mathcal{I} = \int_{\mathbb{F}} \frac{d^2\tau}{\tau_2^2}\, \tau_2^{m/2}\, f(\tau,\bar\tau) \sum_{p\in\Gamma^{m,n}} q^{\frac{1}{2}|p^L|^2} \bar q^{\frac{1}{2}|p^R|^2}, \quad \text{with} \quad \begin{array}{l} m,n\in\mathbb{N} \\ q=e^{2\pi i\tau} \end{array} . \tag{6}$$

Here the integral is over the modular parameter of the world-sheet torus $\tau = \tau_1 + i\tau_2 \in \mathbb{H}^+$ [here $\mathbb{H}^+$ is the upper half-plane] with the integration region $\mathbb{F}$ the fundamental domain of $SL(2,\mathbb{Z})$

$$\mathbb{F} = \{\tau\in\mathbb{H}^+ | \tau_1^2+\tau_2^2 \geq 1 \text{ and } -\tfrac{1}{2}\leq\tau_1<\tfrac{1}{2}\}, \tag{7}$$

$f(\tau,\bar\tau)$ is an (almost) modular form of weight $\frac{m-n}{2}$ while $\Gamma^{m,n}$ denotes the Narain lattice of signature (m,n) with the left- and right-moving momenta (p^L,p^R). The latter also encode the full (target space) moduli dependence of $\mathcal{I}$. Recalling that the modular invariant integral measure of $SL(2,\mathbb{Z})$ is $d^2\tau/\tau_2^2$ one can readily check that the integrand in (6) is modular invariant, which is necessary for the physical consistency of the amplitude $\mathcal{I}$. In the following we will mainly be interested in the case $m=2$ but n generic. In this case, f has weight $1-n/2$ and we will see in many cases that it has exactly the right properties to define an automorphic product as in (5).

From a more mathematical point of view, integrals of the type (6) fall into the category of so-called *(singular) theta correspondences.* To understand this, we recall that the summation over $\Gamma^{m,n}$ in (6) is a particular class of theta-function[2] $\Theta_n(\tau,v)$ with

$$v\in\frac{O(2,n)}{O(2)\times O(n)},$$

representing the moduli of the toroidal heterotic compactification (we recall that $m=2$). $O(2,n)$ and $SL(2,\mathbb{R})$ form a dual reductive pair[3] inside the metaplectic group (the double cover of the symplectic group $Sp(2,n)$). The decomposition of representations of the metaplectic group into tensor representations of the dual reductive pair induces a correspondence between representations of the two subgroups. In the case of $SL(2,\mathbb{R})$ and $O(2,n)$ this correspondence can also be extended to automorphic forms associated with these representations [37, 24, 47], which concretely takes the form

$$\Pi(v,f) = \int_{\mathbb{F}} \frac{d^2\tau}{\tau_2}\, \Theta_n(\tau,v) f(\tau),$$

with Π a modular form of $O(2,n,\mathbb{Z})$. Since in general the integral in (6) will have singularities at particular points in moduli space (see e.g. section 4.4), we are usually dealing in string theory with a *singular* theta correspondence [12, 13].

3. Automorphic products with $\mathcal{N}=2$ supersymmetry and the BPS algebra

3.1. $\mathcal{N}=2$ threshold corrections

As a first appearance of automorphic products in string theory we will briefly review the results of [34, 35]. We consider the $E_8\times E_8$ heterotic string compactified on $K3\times T^2$ which preserves

[2] In the literature it is indeed often called *Siegel-Narain theta-function.*

[3] Two subgroups of a group G form a dual reductive pair if each of them is the commutant of the other in G. Moreover, at the level of representation theory, we are thinking in terms of the continuous group $SL(2,\mathbb{R})$.

$\mathcal{N}=2$ space-time supersymmetry. Furthermore, we choose to work at a point in the moduli space where the gauge group is given by[4]

$$G=\left[U(1)^2\right]_{\text{left}}\times\left[E_7\times U(1)^{10}\right]_{\text{right}},$$

such that the original $\Gamma^{6,22}$ Narain lattice decomposes as

$$\Gamma^{6,22}=\Gamma^{4,4}\oplus\Gamma^{0,8}\oplus\Gamma^{2,10}.$$

In this setting it was found in [34, 35] that the one-loop threshold correction to the gauge coupling contains a term which can be written as

$$\frac{1}{g^2}=\int_{\mathbb{F}}\frac{d^2\tau}{\tau_2}\frac{\bar{E}_4^2(\bar{\tau})}{\bar{\eta}^{24}(\bar{\tau})}\sum_{p\in\Gamma^{2,10}}q^{\frac{1}{2}|p^L|^2}\bar{q}^{\frac{1}{2}|p^R|^2}+\dots. \tag{8}$$

Comparing this with (6) we have $m=2$ and $n=10$ and we extract the modular form

$$f(\bar{\tau})=\frac{\bar{E}_4^2(\bar{\tau})}{\bar{\eta}^{24}(\bar{\tau})}=\sum_{n=-1}^{\infty}c(n)\bar{q}^n=\bar{q}^{-1}+504+73764\bar{q}+2695040\bar{q}^2+\dots, \tag{9}$$

which has weight -4. Here $E_4(\tau)$ is an Eisenstein series of weight 4 and $\eta(\tau)$ is the Dedekind eta-function. The integration over τ in (8) is rather non-trivial, however, can be performed explicitly using e.g. methods first developed in [21] and the explicit result can be written as [34]

$$\frac{1}{g^2}\sim-2\log||\Phi||^2-c(0)\left(\log(-(\Re y)^2)+\text{const.}\right)+\dots, \tag{10}$$

with the following modular form of $O(2,10)$

$$\Phi(y)=e^{-2\pi\rho\cdot y}\prod_{r>0}\left(1-e^{-2\pi r\cdot y}\right)^{c(-r^2/2)}\quad\text{with}\quad\in\mathbb{R}^{1,11}\otimes\mathbb{C}.$$

It was shown in [34] that ρ can be identified with the lattice Weyl vector of $\mathfrak{e}_{10}=\mathfrak{e}_8^{++}$ (which is the hyperbolic extension of $\mathfrak{e}_8$) and the range of the infinite product can be written as a product over $\Lambda^+_{\mathfrak{e}_{10}}$

$$\Phi(y)=e^{-2\pi i(\rho|y)}\prod_{\lambda\in\Lambda^+_{\mathfrak{e}_{10}}}\left(1-e^{-2\pi i(\lambda|y)}\right)^{c(-\lambda^2/2)}. \tag{11}$$

The $||\cdot||^2$ in (10) takes into account a similar contribution with $r<0$ which can be rewritten as a product over $\Lambda^-_{\mathfrak{e}_{10}}$. Comparing (11) with (3) we see that $\Phi(y)$ can be identified with the infinite product part of the denominator formula. Therefore, Φ can be used to extend $\mathfrak{e}_{10}$ to the GKM $\mathcal{G}(\mathfrak{e}_{10})$ whose root multiplicities are encoded in the Fourier coefficients $c(-\lambda^2/2)$: The simple real roots have $\lambda^2=2$, such that their multiplicity is $c(-1)=1$ (see (9)). However, in addition to the latter there is an infinite set of further simple imaginary roots, which are the hallmark of a GKM as discussed in section 2.1.

[4] For simplicity, we are focusing here on the special case $s=8$ of the slightly more general class of examples discussed in [34]. Note that for later convenience we have exchanged left- and right moving side compared to [34].

3.2. Algebra of BPS states

As already alluded to in the introduction, the result presented in the previous section has been interpreted in [34, 35] as an indication for particular algebraic properties of the spectrum of BPS states in string theory. Indeed, it has been argued that BPS states [which form a very particular class of states within the spectrum of string theory] should form an algebra. Moreover, threshold corrections of the type considered in (8) are *BPS-saturated* in the sense that they only receive contributions from such BPS states. The fact that part of these corrections can be written as the denominator formula of a GKM (10) was interpreted as a sign that this BPS algebra is a GKM. This line of argumentation was extended in [35] where it was attempted to construct this algebra microscopically and to identify a particular inner product structure which represents the corresponding GKM. An interesting question, however, is in how far these results depend on the particular model we have chosen. It is interesting to study different BPS-saturated objects than $\mathcal{N} = 2$ threshold corrections in theories with more supersymmetry and to explore the particular moduli dependence of such objects. In the following section we will therefore consider a particular class of so-called topological amplitudes and show their relation to particular GKM algebras.

4. Automorphic products with $\mathcal{N} = 4$ supersymmetry

4.1. Topological amplitudes

The first step in extending the results of [34, 35] to theories with more supersymmetry is to identify suitable BPS-saturated one-loop quantities. We will consider toroidally compactified heterotic string theory where in the recent years a number of so-called topological amplitudes has been found and studied extensively [9, 3, 5, 7] (see also [4] for a review). These are string theory amplitudes which can also be written as correlation functions in suitably twisted two-dimensional topological theories (for further examples in the context of $\mathcal{N} = 2$ supersymmetry see also [1, 6, 8]). From a field-theoretic perspective, they only receive contributions from BPS states and are therefore BPS-saturated which makes them the ideal candidate to play a similar role as the $\mathcal{N} = 2$ threshold corrections in the study of GKM algebras within string theory. To be precise, we will focus on a particular class of $\mathcal{N} = 4$ topological amplitudes which has first been discovered and extensively described in [3]. These quantities compute BPS couplings in the string effective action, which in harmonic superspace take the form (see [5] for further details)

$$S = \int d^4x \int du \int d^4\theta^+ \int d^4\bar{\theta}_- (K^{++}_{\mu\nu} K^{++,\mu\nu})^{g+1} \mathcal{F}_g(Y^{++}_A, u) \,. \tag{12}$$

Here $u^{\pm a}_I$ are the (bosonic) harmonic coordinates which parameterize the coset

$$u^{\pm a}_I \in \frac{SU(4)}{S(U(2) \times U(2))} \,.$$

We will use the notation that $A = 1, \ldots, 22$ is an index of $SO(22)$, $I = 1, \ldots, 4$ is an index of $SU(4) \cong SO(6)$, while $a = 1, 2$ and $\dot{a} = 1, 2$ are indices of either of the two $SU(2)$'s, and the $\pm$ signs denote the charge with respect to the diagonal $U(1)$. Moreover $K^{IJ}_{\mu\nu}$ is a particular super-descendant of the (linearized) $\mathcal{N} = 4$ supergravity multiplet, while Y^{IJ}_A is a linearized $\mathcal{N} = 4$ vector-multiplet, whose lowest components y^{IJ}_A form the moduli space

$$\mathcal{M}_{(6,22)} = \frac{SO(6, 22)}{SO(6) \times SO(22)} \,,$$

of the $\mathcal{N} = 4$ string compactification. $K^{++}_{\mu\nu}$ and Y^{++}_A are respective projections with the harmonic coordinates.

The class of amplitudes $\mathcal{F}_g$ in (12) corresponds to a series of g-loop amplitudes in type II string theory compactified on $K3 \times T^2$, while their dual counterparts in heterotic string theory compactified on T^6 start receiving contributions at the one-loop level. However, the latter will generically receive perturbative and non-perturbative corrections. Focusing on the leading contribution in the (heterotic) perturbative regime it was found in [3, 5] to be given by

$$\mathcal{F}_g(y) = \int_{\mathbb{F}} \frac{d^2\tau}{\bar{\eta}^{24}} \tau_2^{2g-1} G_{g+1}(\tau,\bar{\tau}) \Theta_g^{(6,22)}(\tau,\bar{\tau},y)\,, \tag{13}$$

where the integral is over the fundamental domain $\mathbb{F}$ of $SL(2,\mathbb{Z})$ as defined in (7). Moreover, the expression

$$\Theta_g^{(6,22)}(\tau,\bar{\tau},y) = \begin{cases} \sum_{p\in\Gamma^{6,22}} (p^L_{++})^{2g-2}\, q^{\frac{1}{2}|p^L|^2}\, \bar{q}^{\frac{1}{2}|p^R|^2} & g>1 \\ \sum_{\substack{p\in\Gamma^{6,22}\\ p\neq 0}} q^{\frac{1}{2}|p^L|^2}\, \bar{q}^{\frac{1}{2}|p^R|^2} & g=1 \end{cases} \tag{14}$$

is again a Siegel-Narain theta-function of the even unimodular lattice $\Gamma^{6,22}$ with harmonically projected momentum insertions

$$p^L_{++}(y) = \frac{1}{2}\epsilon^{ab}\bar{u}^I_{+a}\bar{u}^J_{+b}\,p^L_{IJ}(y), \quad \text{with} \quad y\in\mathcal{M}_{(6,22)}\,.$$

Finally, as above, y^{IJ}_A are the moduli of the heterotic compactification and span $\mathcal{M}_{(6,22)}$ of the $\mathcal{N}=4$ string compactification (for details see [5]). For $g=1$ we do not sum over $p=0$ in the definition (14): since the amplitude $\mathcal{F}_1$ has no explicit p^L-insertions it would receive contributions from $p=0$. However, integrating this contribution over τ will diverge for any generic value of y^{IJ}_A, thereby rendering $\mathcal{F}_1$ infinite. In order to avoid this singularity, we have chosen to regularize $\mathcal{F}_1$ by performing the summation in (14) only over $p\neq 0$.

The function $G_g(\tau,\bar{\tau})$ in (13) is a weight $2g$ non-antiholomorphic modular form and can be obtained as the coefficient of λ^{2g} of the generating functional [2] (see also [8])

$$G(\lambda,\tau,\bar{\tau}) = \left(\frac{2\pi i\lambda\bar{\eta}^3}{\bar{\theta}(\lambda,\bar{\tau})}\right)^2 e^{-\frac{\pi\lambda^2}{\tau_2}} = \sum_{g=0}^{\infty}\lambda^{2g}G_g(\tau,\bar{\tau})\,. \tag{15}$$

G_g can be written in terms of Eisenstein series in the following manner [45]

$$G_g(\tau,\bar{\tau}) = -\mathcal{S}_g\left(\hat{\bar{\mathcal{E}}}_2, \frac{1}{2}\bar{\mathcal{E}}_4,\ldots,\frac{1}{2g}\bar{\mathcal{E}}_{2g}\right), \quad \text{with} \quad \begin{aligned} &\mathcal{S}_k(x_1,\ldots,x_k) = x_k+\cdots+x_1^k/(k!)\,,\\ &\hat{\bar{\mathcal{E}}}_2(\tau,\bar{\tau}) = \tfrac{\pi^2}{3}\left(\bar{E}_2(\bar{\tau})-\tfrac{3}{\pi\tau_2}\right),\\ &\mathcal{E}_{2k}(\tau) = 2\zeta(2k)E_{2k}(\tau)\,, \end{aligned} \tag{16}$$

where $\mathcal{S}$ are the Schur-polynomials and E_{2k} the Eisenstein series of weight $2k$. Notice that $\bar{E}_2$ does not only transform with a weight under modular transformations but receives an additional anomalous shift-term. We have therefore introduced the quantity $\hat{\bar{\mathcal{E}}}_2$ which only transforms with weight 2 but is non-antiholomorphic in τ.

4.2. Harmonicity relation and analytic amplitude

Upon inspecting the superspace coupling (12) we see that it only depends on one particular harmonic projection of the vector multiplets, which is compatible with the Grassmann

Journal of Physics: Conference Series **346** (2012) 012008 doi:10.1088/1742-6596/346/1/012008

component of the integral measure. As was discussed in [5], for $g > 1$ this particular analytic dependence ('G-analyticity') manifests itself as two second-order differential equations for $\mathcal{F}_g$.

$$\epsilon_{ab}\epsilon^{IJKL}\frac{\partial}{\partial \bar{u}^J_{+b}}D_{KL,A}\mathcal{F}_g = (2g-2)\bar{u}^I_{+a}D_{++,A}\mathcal{F}_{g-1}\,,$$
$$\left(\epsilon^{IJKL}D_{IJ,A}D_{KL,B} + 4(g+1)\delta_{AB}\right)\mathcal{F}_g = 4D_{++,A}D_{++,B}\mathcal{F}_{g-1}\,,$$

where $D_{++,A}$ are harmonic projections of the covariant derivatives $D_{IJ,A}$ in the moduli space $\mathcal{M}_{(6,22)}$. We will refer to these equations as the harmonicity and second order relation, respectively. In both equations the amplitude $\mathcal{F}_{g-1}$ appears on the right hand side which from a field-theoretic perspective is an anomalous contribution in string theory, which generalizes the holomorphic anomaly [10] in $\mathcal{N} = 2$. It can be understood as an anomalous violation of the above mentioned G-analyticity of $\mathcal{F}_g$ at the quantum level. Moreover, as has been pointed out in [23], analyzing the explicit computation of the anomaly in [5], it follows that its origin can be traced back to the holomorphic τ-dependence of $G_g(\tau,\bar{\tau})$ in (15). We can use this circumstance to split G_g into an analytic (anomaly-free) and a non-analytic (anomalous) part

$$G_g(\tau,\bar{\tau}) = G_g^{\text{analy}}(\bar{\tau}) + G_g^{\text{non-analy}}(\tau,\bar{\tau})$$

with the explicit expressions (using (16))

$$G_g^{\text{analy}}(\bar{\tau}) = -\mathcal{S}_g\left(0, \frac{1}{2}\bar{\mathcal{E}}_4, \ldots, \frac{1}{2g}\bar{\mathcal{E}}_{2g}\right)\,,$$
$$G_g^{\text{non-analy}}(\tau,\bar{\tau}) = -\mathcal{S}_g\left(\hat{\bar{\mathcal{E}}}_2, \frac{1}{2}\bar{\mathcal{E}}_4, \ldots, \frac{1}{2g}\bar{\mathcal{E}}_{2g}\right) + \mathcal{S}_g\left(0, \frac{1}{2}\bar{\mathcal{E}}_4, \ldots, \frac{1}{2g}\bar{\mathcal{E}}_{2g}\right)\,,$$

which both have weight $2g$ under modular transformations. It is therefore consistent to define the purely analytic contribution to the amplitude as

$$\mathcal{F}_g^{\text{analy}}(y) = \int_{\mathbb{F}}\frac{d^2\tau}{\bar{\eta}^{24}}\tau_2^{2g-1}G_{g+1}^{\text{analy}}(\bar{\tau})\Theta_g^{(6,22)}(\tau,\bar{\tau},y)\,, \tag{17}$$

which yields a vanishing anomaly when inserted into the harmonicity and second order relation. In the following we will study more carefully $\mathcal{F}_{g=1}^{\text{analy}}(y)$ to highlight its connection to the denominator formula of a GKM.

4.3. Gauge group and Wilson line

In this section we will follow the discussion of [36]. In order to simplify the expression for $\mathcal{F}_1$, we will first split the internal T^6 of the heterotic $\mathcal{N} = 4$ compactification into $T^4 \times T^2$ and take the large volume limit of T^4. The moduli space of such a compactification is described by the Kähler (T) and complex structure (U) moduli of T^2, as well as by two real Wilson lines $\vec{v}_{1,2} \in \mathbb{R}^{16}$ which we combine into a single complex vector $\vec{V} = \vec{v}_1 + i\vec{v}_2$. For the one-loop integrand of (17) this assumption implies that the Siegel-Narain theta function of the original $\Gamma^{6,22}$ Narain-lattice will be decomposed as

$$\frac{G_2^{\text{analy}}(\bar{\tau})\tau_2^2}{\bar{\eta}^{24}}\Theta_{g=1}^{(6,22)} \sim \text{Vol}\,\frac{G_2^{\text{analy}}(\bar{\tau})}{\bar{\eta}^{24}}\Theta_{g=1}^{(2,18)}\,,$$

where Vol is the volume of T^4 and $\Theta_{g=1}^{(2,18)}$ is the corresponding Siegel-Narain theta function for the lattice $\Gamma^{2,18}$. For later use we will denote the inner product on $\Gamma^{2,18}$ as $\langle\cdot|\cdot\rangle$. Moreover, in the following we will consider the subspace of the moduli space where $\vec{V}$ breaks the $\mathfrak{e}_8 \oplus \mathfrak{e}_8$ gauge symmetry to

$$\mathfrak{e}_8 \oplus \mathfrak{e}_8 \quad \longrightarrow \quad \mathfrak{h}, \quad \text{with} \quad \mathfrak{h} \oplus \mathfrak{g} \subset \mathfrak{e}_8 \oplus \mathfrak{e}_8 \ , \tag{18}$$

where $\mathfrak{h}$ and $\mathfrak{g}$ are fixed subalgebras[5] of $\mathfrak{e}_8 \oplus \mathfrak{e}_8$. This breaking of the gauge groups can best be understood through the decomposition of the Narain lattice $\Gamma^{2,18}$ (for more details see [23]): given $\vec{V}$ we denote by $\Lambda_\mathfrak{h}$ the sublattice of $\Lambda_{\mathfrak{e}_8} \oplus \Lambda_{\mathfrak{e}_8}$ consisting of all vectors that are orthogonal to the complex Wilson line $\vec{V}$

$$\Lambda_\mathfrak{h} = \left\{ \vec{d} \in \Lambda_{\mathfrak{e}_8} \oplus \Lambda_{\mathfrak{e}_8} \ : \ \vec{d} \cdot \vec{V} = 0 \right\} \ .$$

The vectors of length squared two in $\Lambda_\mathfrak{h}$ are the roots of what we would like to call the *unbroken Lie algebra* $\mathfrak{h}$. The commutant of $\mathfrak{h}$ in $\mathfrak{e}_8 \oplus \mathfrak{e}_8$ defines the Lie algebra $\mathfrak{g}$, whose root lattice is spanned by the roots of $\mathfrak{e}_8 \oplus \mathfrak{e}_8$ that are orthogonal to $\Lambda_\mathfrak{h}$. Adding to the corresponding root lattice the T^2 torus directions in $\Gamma^{2,2}$ leads to the sublattice $\Gamma_\mathfrak{g} \subseteq \Gamma^{2,18}$. Since $\Lambda_\mathfrak{h}$ is naturally a sublattice of $\Gamma^{2,18}$, we have

$$\Gamma_\mathfrak{g} \oplus \Lambda_\mathfrak{h} \subseteq \Gamma^{2,18} \ ,$$

and the sublattice on the left hand side is of maximal rank. Generically, $\Gamma_\mathfrak{g} \oplus \Lambda_\mathfrak{h}$ is a proper sublattice of $\Gamma^{2,18}$ with index s, thus we can write the decomposition

$$\Gamma^{2,18} = (\Gamma_\mathfrak{g} \oplus \Lambda_\mathfrak{h}) \ \oplus \ \bigoplus_{\mu=1}^{s-1} (\lambda_\mu + \Gamma_\mathfrak{g} \oplus \Lambda_\mathfrak{h}) \ , \tag{19}$$

where λ_μ denotes the different cosets.

4.4. *Singularities of the one-loop integral*

Before explicitly performing the τ integration for the choice of gauge group (18), it is instructive to first consider possible singularities of $\mathcal{F}^{\text{analy}}_{g=1}(y)$. As we will see, this will already provide us some hints at interesting algebraic structures in the integral. To begin, we will write the BPS-integral in the following manner

$$\mathcal{F}^{\text{analy}}_{g=1} = \int_\mathbb{F} \frac{d^2\tau}{\bar{\eta}^{24}} \, \tau_2 G^{\text{analy}}_2(\bar{\tau}) \sum_{\substack{p \in \Gamma^{2,18} \\ p \neq 0}} \bar{q}^D e^{-\pi\tau_2 |p^L|^2} \ , \tag{20}$$

where $D = \frac{1}{2}\left(|p^R|^2 - |p^L|^2\right)$. To understand the mechanism by which a singularity might occur in (20) (see also [12, 34, 13] for related discussions), we observe that the upper boundary of the τ_2 integration is at infinity. At a generic point in moduli space, the integrand is rendered finite for $\tau_2 \to \infty$ due to the exponential suppression of the Narain momenta. However, at particular points in moduli space this damping might fail, thus leading to a logarithmic divergence. A necessary condition for such a divergence to appear is

$$|p^L| = 0 \ , \tag{21}$$

[5] For simplicity we will assume in this review that $\mathfrak{g}$ is simple. For a discussion of the semi-simple case see [23].

since then the suppression induced by the factor $e^{-\pi\tau_2|p^L|^2}$ in (20) is absent. However, condition (21) alone is not sufficient. Indeed, for $|p^L| = 0$ the τ_1-integration will pick the coefficient of $\bar{q}^0$ in the expansion

$$\frac{G_2^{\text{analy}}(\bar{\tau})}{\bar{\eta}^{24}(\bar{\tau})} \sum_{\substack{p\in\Gamma^{2,18}\\ p\neq 0,\, p^L=0}} \bar{q}^{\frac{1}{2}|p^R|^2} = \sum_{\substack{p\in\Gamma^{2,18}\\ p\neq 0,\, p^L=0}} \sum_{n=-1}^{\infty} d(n)\bar{q}^{n+D}\,.$$

Therefore, if $d(-D) \neq 0$ for some vector in the sum over $\Gamma^{2,18}$, we indeed find a logarithmic divergence. Following [36], we will focus on the singularities with $D = 1$, which have an interesting algebraic interpretation. To see this, we fix a vector $p \in \Gamma^{2,18}$ with $\langle p|p\rangle = 2$, and determine those values of the moduli $y = (U, T, \vec{V})$ such that $p^L = 0$. As it was explained in [23], due to $SO(2,18,\mathbb{Z})$ invariance of the integral [this is just the familiar T-duality action] it is in fact sufficient to focus on vectors $\hat{p} \in \Gamma^{1,17} \subset \Gamma^{2,18}$. Recalling moreover the choice of gauge group (18), out of these only the vectors $\hat{p} \in \Gamma^{1,17} \cap \Gamma_{\mathfrak{g}} = \Lambda_{\mathfrak{g}^{++}}$ will be of interest. Here $\Lambda_{\mathfrak{g}^{++}}$ is the root lattice of the hyperbolic extension of $\mathfrak{g}$ (see section 2.1; as an example, the Dynkin diagram of $\mathfrak{a}_k^{++}$ is depicted in figure 1). The norm condition

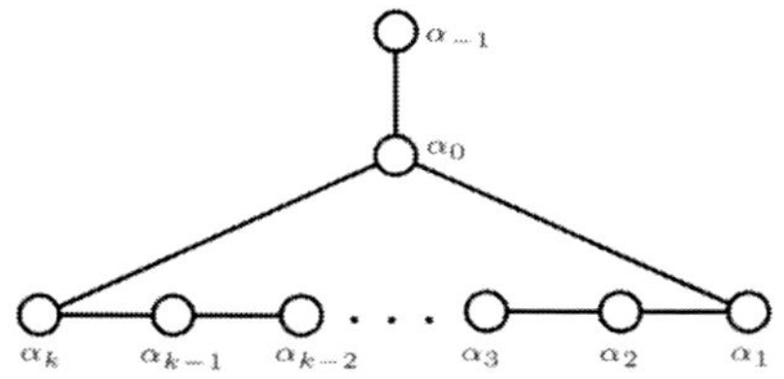

Figure 1. Dynkin diagram of the double extensions $\mathfrak{a}_k^{++}$ of $\mathfrak{a}_k$. The the nodes corresponding to the roots α_0 and α_{-1} are called affine- and hyperbolic node respectively and are the key-features of the hyperbolic extension.

$$D = 1 \quad\Longleftrightarrow\quad \langle\hat{p}|\hat{p}\rangle = 2\,,$$

then implies that only the real roots of $\mathfrak{g}^{++}$ are of importance, which means that we can identify the singular loci of $\mathcal{F}_{g=1}^{\text{analy}}(y)$ with fixed points of the Weyl reflections $w_\alpha \in \mathcal{W}(\mathfrak{g}^{++})$. The latter act as

$$w_\alpha \;:\; y \longmapsto y - (y|\alpha)\,\alpha\;,$$

on the moduli, where $(\cdot|\cdot)$ is the inner product on the lattice $\Lambda_{\mathfrak{g}^{++}}$, which is inherited from $\Gamma^{2,18}$. The group $\mathcal{W}(\mathfrak{g}^{++})$ is generated by the $k+2$ reflections w_I, where $I = -1, 0, \ldots, k = \text{rk}(\mathfrak{g})$ and w_I is the reflection associated to the simple root α_I of $\mathfrak{g}^{++}$. The latter may be taken to be

$$\alpha_{-1} = (1,-1;\vec{0}), \quad \alpha_0 = (-1,0;-\vec{\theta}), \quad \alpha_i = (0,0;\vec{e}_i)\;, \quad i = 1,\ldots,k\;,$$

where $\vec{\theta}$ is the highest root of $\mathfrak{g}$, and $\vec{e}_i$ are the simple roots of $\mathfrak{g}$. The Weyl reflection associated to any real root in $\mathfrak{g}^{++}$ can then be written as a finite product of the Weyl reflections associated to the simple roots. With the basis given above, the singular divisors can then be explicitly given to be

$$\begin{aligned}
\mathcal{D}_{-1} &= \{y \in \mathcal{M}_{2,2+k} \mid (y|\alpha_{-1}) = U - T = 0\}\\
\mathcal{D}_0 &= \left\{y \in \mathcal{M}_{2,2+k} \mid (y|\alpha_0) = T - \vec{\theta}\cdot\vec{V} = 0\right\}\\
\mathcal{D}_i &= \left\{y \in \mathcal{M}_{2,2+k} \mid (y|\alpha_i) = \vec{e}_i\cdot\vec{V} = 0\right\}\,.
\end{aligned}$$

Note that the divisor $\mathcal{D}_i$ describes a boundary component of $\mathcal{M}_{(2,2+k)}$ since for $y \in \mathcal{D}_i$ a root of $\mathfrak{g}$ becomes orthogonal to $\vec{V}$, and hence should have been part of $\mathfrak{h}$. Put differently, for a given point y in the moduli space $\mathcal{M}_{(2,2+k)}$ the divisors $\mathcal{D}_I$ represent the 'dominant' walls of the complexified Weyl chamber of $\mathfrak{g}^{++}$

$$\mathcal{C}_{\mathbb{C}} = \{y \in \mathcal{M}_{2,2+k} \mid \mathcal{D}_I \geq 0,\ I = -1, 0, 1, \ldots k\} ,$$

and the only singularities appear at the boundary of $\mathcal{C}_{\mathbb{C}}$.

4.5. Torus integral and denominator formula

After having discussed the singularities of the integral (17) which has given us first hints to a connection with a GKM algebra, we will now demonstrate the appearance of a denominator formula. To see the latter, one has to explicitly perform the world-sheet torus integration. This has been done in [23, 36] using the method of orbits first introduced in [21] which have been further developed in [34].[6] The first step is to implement the choice of gauge group in (18) as the decomposition of the Narain lattice (19) in the integrand (20). Explicitly we can write

$$\frac{G_2^{\text{analy}}(\bar{\tau})}{\bar{\eta}^{24}}\,\Theta^{(2,18)}_{g=1} = \sum_{\mu=0}^{r-1} \mathcal{P}_\mu^{(k)}(\bar{\tau})\,\Theta_\mu^{(2,2+k)}(\tau,\bar{\tau},y) , \tag{22}$$

where $\Theta_\mu^{(2,2+k)}(\tau,\bar{\tau},y)$ is the theta series associated to the $\Gamma_{\mathfrak{g}}$ coset in the decomposition of (19)

$$\Theta_\mu^{(2,2+k)}(y) = \sum_{x\in\Gamma_{\mathfrak{g}}+\lambda_\mu^{\mathfrak{g}}} \bar{q}^{\frac{1}{2}\langle x|x\rangle}\, e^{2\pi\tau_2 \frac{|\langle x|u(y)\rangle|^2}{(\Im y|\Im y)^2}} ,$$

while $\mathcal{P}_\mu^{(k)}(\bar{\tau})$ captures the contributions from $\Lambda_{\mathfrak{h}}$ (as well as the parts coming from G_2^{analy} and the powers of the $\bar{\eta}$-function).

$$\mathcal{P}_\mu^{(k)}(\bar{\tau}) = \frac{G_2^{\text{analy}}}{\bar{\eta}^{24}}\,\Theta_\mu^{\mathfrak{h}}(\bar{\tau}) = \frac{G_2^{\text{analy}}}{\bar{\eta}^{24}} \sum_{\vec{\ell}\in\Lambda_{\mathfrak{h}}+\lambda_\mu^{\mathfrak{h}}} \bar{q}^{\frac{1}{2}\vec{\ell}\cdot\vec{\ell}}, \tag{23}$$

where $\lambda_\mu^{\mathfrak{g}}$ and $\lambda_\mu^{\mathfrak{h}}$ are the projections of the glue vector λ_μ in (19) onto $\mathfrak{g}$ and $\mathfrak{h}$ respectively. Since $\Lambda_{\mathfrak{h}}$ is a sublattice of $\mathfrak{e}_8 \oplus \mathfrak{e}_8$ that is orthogonal to $\vec{V}$, $\mathcal{P}_\mu^{(k)}(\bar{\tau})$ does not depend on the moduli $y = (U, T; \vec{V})$. Using (22), the full integral (20) was given explicitly in [36] (see also [23])

$$\begin{aligned}
\mathcal{F}_{g=1}^{\text{analy}} = \sum_{\mu=1}^{s}\Bigg\{ \sum_{\vec{\ell}\in\Lambda_{\mathfrak{g}}+\lambda_\mu^{\mathfrak{g}}} \Bigg[&\frac{2\pi Y}{3U_2}\left(c_\mu(0,\vec{\ell}) - 24 c_\mu(-1,\vec{\ell})\right) + 2\log\left|1 - e^{2\pi i \vec{\ell}\odot\vec{V}}\right|^{c_\mu(0,\vec{\ell})} \\
&+ 2\log \prod_{\substack{n',r\in\mathbb{Z}\\ r>0}} \left|1 - e^{2\pi i(rT + n'U + \vec{\ell}\odot\vec{V})}\right|^{c_\mu(n'r,\vec{\ell})} + 2\log\prod_{n=1}^{\infty}\left|1 - e^{2\pi i(nU + \vec{\ell}\odot\cdot\vec{V})}\right|^{c_\mu(0,\vec{\ell})}\Bigg] \\
&+ c_\mu(0,\vec{0})\left(\frac{\pi U_2}{3} - \ln Y + K\right) + 2\log\prod_{n=1}^{\infty}|1 - e^{2\pi i n U}|^{c_\mu(0,0)}\Bigg\} \\
+ \frac{2U_2}{3\pi} + \frac{2\pi}{U_2}&(\vec{\ell}\odot\Im\vec{V})\left((\vec{\ell}\odot\Im\vec{V}) + U_2\right) ,
\end{aligned} \tag{24}$$

[6] In the mathematics literature it is usually known as the Rankin-Selberg method .

where $K = \gamma_E - 1 - \ln \frac{8\pi}{3\sqrt{3}}$, with γ_E being the Euler-Mascheroni constant and $Y = (\Im y|\Im y)$. Furthermore, we have introduced the shorthand notation for the modified scalar-product

$$\vec{\ell} \odot \vec{V} = \ell \cdot \Re \vec{V} + i \left| \vec{\ell} \cdot \Im \vec{V} \right| .$$

The coefficients $c_\mu(n'r, \vec{\ell})$ arise from the Fourier expansion

$$\sum_{\mu=0}^{s-1} \mathcal{P}_\mu^{(k)}(\bar{\tau}) \sum_{\vec{\ell} \in \Lambda_\mathfrak{g} + \lambda_\mu^\mathfrak{g}} \bar{q}^{\frac{1}{2}\vec{\ell}\cdot\vec{\ell}} e^{2\pi i \vec{\ell}\cdot\vec{z}} = \sum_{\mu=0}^{s-1} \sum_{n=-1}^{\infty} \sum_{\vec{\ell} \in \Lambda_\mathfrak{g} + \lambda_\mu^\mathfrak{g}} c_\mu(n, \vec{\ell}\,)\, \bar{q}^n\, e^{2\pi i \vec{\ell}\cdot\vec{z}} . \tag{25}$$

By construction the left hand side of (25) transforms as a weak Jacobi form under $SL(2, \mathbb{Z})$, and in fact every summand in the sum over μ transforms as a weak Jacobi form under some congruence subgroup. Thus the coefficients $c_\mu(n, \vec{\ell}\,)$ only depend on $(n, \vec{\ell}\,)$ through the combination [22]

$$c_\mu(n, \vec{\ell}\,) \equiv c_\mu \left(n - \tfrac{1}{2} \vec{\ell} \cdot \vec{\ell} \right) .$$

Moreover, by inspection of (23) it is clear that $\mathcal{P}_\mu^{(k)}(\bar{\tau})$ has a simple pole at $\bar{\tau} \to i\infty$, and hence

$$c_\mu \left(n - \tfrac{1}{2} \vec{\ell} \cdot \vec{\ell} \right) = 0 \qquad \forall\, n - \tfrac{1}{2} \vec{\ell} \cdot \vec{\ell} < -1 .$$

In the following we shall mainly be interested in the contribution of the trivial conjugacy class labeled by $\mu = 0$. We will show that this contribution can be identified with the infinite product side of the denominator formula for the Borcherds extension $\mathcal{G}(\mathfrak{g}^{++})$, where $\mathfrak{g}^{++}$ is the double extension of the unbroken gauge algebra $\mathfrak{g}$. Moreover, within the $\mu = 0$ contribution we will concentrate on the logarithmic terms. Most of the non-logarithmic terms contribute to the Weyl vector ρ, appearing in the exponential prefactor of the denominator formula, and ensure that the whole denominator formula will be invariant under a congruence subgroup of $SO(2, 2+r, \mathbb{Z})$. Thus, the relevant part of (24) can be written as

$$\log ||\Phi_\mathfrak{g}(y)||^2 , \quad \text{where} \quad \Phi_\mathfrak{g}(y) = \prod_{(r,n';\vec{\ell})>0} \left(1 - e^{2\pi i (rT + n'U + \vec{\ell}\cdot\vec{V})} \right)^{c_0(n'r, \vec{\ell}\,)} , \tag{26}$$

where the range of the product $(r, n'; \vec{\ell}\,) > 0$ is defined by

$$n'r - \frac{1}{2} \vec{\ell} \cdot \vec{\ell} \geq -1, \quad \text{and} \quad \begin{cases} r > 0,\ n' \in \mathbb{Z},\ \vec{\ell} \in \Lambda_\mathfrak{g} & \text{or} \\ r = 0,\ n' > 0,\ \vec{\ell} \in \Lambda_\mathfrak{g} & \text{or} \\ r = n' = 0,\ \vec{\ell} \in \Lambda_\mathfrak{g}^+ . \end{cases} \tag{27}$$

The norm $|| \cdot ||^2$ in (26) takes into account that there are contributions with $(r, n'; \vec{\ell}\,) > 0$ and contributions with $(r, n'; \vec{\ell}\,) < 0$. It was shown in [36] that (27) are just the conditions that characterize the elements α of $\Lambda_{\mathfrak{g}^{++}}^+$ with norm $\alpha^2 \leq 2$. (Note that the correspondence only holds provided that we require this norm condition.) Thus we can then write $\Phi_\mathfrak{g}(y)$ as a product over the positive roots $\Delta_{\mathcal{G}_\mathfrak{g}}^+$ of the GKM-algebra $\mathcal{G}_\mathfrak{g}$

$$\Phi_\mathfrak{g}(y) = \prod_{\alpha \in \Delta_{\mathcal{G}_\mathfrak{g}}^+} \left(1 - e^{2\pi i (\alpha|y)} \right)^{c_0(-\alpha^2/2)} , \tag{28}$$

where we have used that $c_0(n) = 0$ for $n < -1$. Following the reasoning of [11, 12], the automorphic correction $\mathcal{G}(\mathfrak{g}^{++})$ is by construction a generalized Kac-Moody algebra. This implies in particular that the product includes imaginary simple roots. While the multiplicity of all real positive roots is given by $c_0(-1) = 1$, the multiplicities of the imaginary roots are encoded via (28) in the remaining Fourier coefficients $c_0\left(n'r - \frac{\vec{\ell}\cdot\vec{\ell}}{2}\right)$. Finally, the norm $||\cdot||^2$ in (26) is now interpreted as splitting the infinite product into contributions from positive and negative roots.

5. Conclusions

In this work we have reviewed the appearance of automorphic products in BPS-saturated one-loop quantities in heterotic string theory with $\mathcal{N} = 2$ and $\mathcal{N} = 4$ supersymmetry. Following the work of Borcherds [11, 12], these products can be used to construct generalized Kac-Moody algebras as automorphic corrections of hyperbolic Kac-Moody algebras, which are related in a natural manner to the gauge group of the heterotic model. At a particular point in moduli space, the Weyl group of this Borcherds algebra appears directly in the physical quantities, since it governs the singular loci in moduli space.

Here we have focused on two particular examples, namely the extension of $\mathfrak{e}_{10}$ in the case of $\mathcal{N} = 2$ supersymmetry and the extension of $\mathfrak{g}^{++}$ (with $\mathfrak{g} \subset \mathfrak{e}_8$) in the case of $\mathcal{N} = 4$. Additional examples can be found in the original works, i.e. [34, 35] and [23, 36] respectively. In the latter case, the analysis was performed for an arbitrary choice of the gauge group (determined by the choice of Wilson-line moduli) and thereby a full family of Borcherds algebras was constructed.

For the future it would be interesting to further extend this analysis: Particularly in the $\mathcal{N} = 4$ example, the limit of large T^4-volume in the four-dimensional heterotic compactification has been assumed. As was explained in [36] this is the reason why the Borcherds algebra which we have constructed is an extension of a hyperbolic Kac-Moody algebra. It would be interesting to repeat the analysis with finite T^4 volume and evaluate the integral $\mathcal{F}_1$ for the full Narain lattice of the six-torus T^6. In this case one might expect that the $\mathcal{F}_1$ integral encodes the denominator formula of a GKM which corresponds to the automorphic correction of a more general indefinite Kac-Moody algebra. However, since the Narain moduli space $SO(6,22)/(SO(6) \times SO(22))$ is then no longer a hermitian symmetric domain and it is therefore unclear whether the theta correspondence affords an infinite product representation which can be related to a denominator formula [13] (see also [42, 46] for related discussions).

Moreover, it would be highly desirable to have a further interpretation of the Fourier coefficients which enter into the automorphic product. From the algebraic point of view they encode the root-multiplicities of the imaginary roots of the GKM algebras. However, since the coupling $\mathcal{F}_1$ is topological, one would expect that they can be understood as some generalization of Gopakumar-Vafa invariants from the $\mathcal{N} = 2$ context (see [25, 26]). In particular, since heterotic string theory is compactified on T^6 is dual to type II string theory on $K3 \times T^2$, one might wonder whether these coefficients encode (basic) topological data of $K3$.

Finally, in this review we have focused on extracting denominator formulae from one-loop objects which appear in the heterotic effective action. Our approach was therefore in a sense macroscopical and in particular did not involve an understanding of the GKM algebra at the level of the BPS-(micro)states, as originally envisaged in the work of Harvey and Moore [34, 35]. A closer study of this aspect has been attempted in [23] (see also [36], where explicitly a Borcherds algebra $\mathcal{G}$ was constructed which acts naturally on BPS states. Put differently, BPS states fall into particular representations of $\mathcal{G}$. Notice that this is a slightly modification of the original idea of Harvey and Moore [34, 35]: There it was proposed that the full space of BPS-states would be made into a Lie algebra. In connection with the physical amplitudes it was moreover shown in [23, 36] that (at a generic point in moduli space) $\mathcal{F}_1$ encodes the root multiplicities of some slightly bigger GKM-algebra, of which $\mathcal{G}$ is a subalgebra. It will be clearly interesting to obtain a better understanding of the connection between the microscopic construction of the algebra and further macroscopic quantities. For example it might be interesting to consider the appearance of $\mathcal{G}$ in different BPS-saturated objects than $\mathcal{F}_1$ (see e.g. [3]).

Acknowledgments

I am grateful to the organizers of the AGMP-6 workshop in Tjärnö for giving me the opportunity to present this work there. Moreover I want to thank M. Gaberdiel and D. Persson for collaboration on the subject presented in this proceedings and for enlightening and very helpful discussions. I am particularly indebted to D. Persson for a careful reading of the draft and many valuable comments and improvements.

References

[1] Antoniadis I, Gava E, Narain KS and Taylor TR 1994 Topological amplitudes in string theory *Nuclear Phys. B* **413** 162 (Preprint hep-th/9307158)
[2] Antoniadis I, Gava E, Narain KS and Taylor TR 1995 N=2 type II heterotic duality and higher derivative F terms *Nuclear Phys. B* **455** 109 (Preprint hep-th/9507115)
[3] Antoniadis I, Hohenegger S and Narain KS 2007 N=4 topological amplitudes and string effective action *Nuclear Phys. B* **771** 40 (Preprint hep-th/0610258)
[4] Antoniadis I and Hohenegger S 2007 Topological amplitudes and physical couplings in string theory, *Nuclear Phys. Proc. Suppl.* **171** 176-195 (Preprint hep-th/0701290)
[5] Antoniadis I, Hohenegger S, Narain KS and Sokatchev E 2008 Harmonicity in N=4 supersymmetry and its quantum anomaly *Nuclear Phys. B* **794** 348 (Preprint 0708.0482 [hep-th])
[6] Antoniadis I, Hohenegger S, Narain KS and E. Sokatchev 2009 A New Class of N=2 Topological Amplitudes *Nuclear Phys. B* **823** 448-508 (Preprint 0905.3629 [hep-th])
[7] Antoniadis I and Hohenegger S, 2010 N=4 Topological amplitudes and black hole entropy *Nuclear Phys. B* **837** 61 (Preprint 0910.5596 [hep-th])
[8] Antoniadis I, Hohenegger S, Narain KS and Taylor TR 2010 Deformed Topological Partition Function and Nekrasov Backgrounds *Nuclear Phys. B* **838** 253 (Preprint 1003.2832 [hep-th])
[9] Berkovits N and Vafa C 1995 N=4 topological strings *Nuclear Phys. B* **433** 123 (Preprint hep-th/9407190)
[10] Bershadsky M, Cecotti S, Ooguri H and Vafa C 1994 Kodaira-Spencer theory of gravity and exact results for quantum string amplitudes *Commun. Math. Phys.* **165** 311 (Preprint hep-th/9309140)
[11] Borcherds RE 1988 Generalized Kac-Moody algebras *J. Algebra* **115** 501.
[12] Borcherds RE 1995 Automorphic forms on O(2,n) and infinite products *Invent. Math.* **120** 161
[13] Borcherds RE 1998 Automorphic forms with singularities on Grassmannians *Invent. Math.* **132** 491
[14] Cheng MCN and E.P. Verlinde 2008 Wall crossing, discrete attractor flow, and Borcherds algebra *SIGMA* **4** 068 (Preprint 0806.2337 [hep-th])
[15] Cheng MCN 2008 The spectra of supersymmetric states in string theory Preprint 0807.3099 [hep-th]
[16] Cheng MCN and Dabholkar A 2009 Borcherds-Kac-Moody symmetry of N=4 dyons *Commun. Num. Theor. Phys.* **3** 59 (Preprint 0809.4258 [hep-th])
[17] Dabholkar A 2005 Exact counting of black hole microstates *Phys. Rev. Lett.* **94** 241301 (Preprint hep-th/0409148)
[18] Dabholkar A, Denef F, Moore GW and Pioline B 2005 Precision counting of small black holes, *J. High Energy Phys.* **0510** 096 (Preprint hep-th/0507014)
[19] Denef F and Moore GW 2007 Split states, entropy enigmas, holes and halos Preprint hep-th/0702146
[20] Dijkgraaf R, Verlinde EP and Verlinde HL 1997 Counting dyons in N=4 string theory *Nuclear Phys. B* **484** 543 (Preprint hep-th/9607026)
[21] Dixon LJ, Kaplunovsky V and Louis J 1991 Moduli dependence of string loop corrections to gauge coupling constants *Nuclear Phys. B* **355** 649
[22] Eichler M and Zagier D 1985 The Theory of Jacobi Forms (*Birkhäuser*)
[23] Gaberdiel MR, Hohenegger S and Persson D 2011 Borcherds Algebras and N=4 Topological Amplitudes *J. High Energy Phys.* **1106** 125 (Preprint 1102.1821 [hep-th])
[24] Gelbart S 1979 Examples of dual reductive pairs *Proceedings of Symposia in Pure Mathematics* **33**, part 1, p 287
[25] Gopakumar R and Vafa C 1998 M theory and topological strings. Part 1 (Preprint hep-th/9809187)
[26] Gopakumar R and Vafa C 1998 M theory and topological strings. Part 2 (Preprint hep-th/9812127)

[27] Govindarajan S and Gopala Krishna K 2009 Generalized Kac-Moody algebras from CHL dyons *J. High Energy Phys.* **0904** 032 (Preprint 0807.4451 [hep-th])
[28] Govindarajan S and Gopala Krishna K 2010 BKM Lie superalgebras from dyon spectra in $\mathbb{Z}_N$ CHL orbifolds for composite N *J. High Energy Phys.* **1005** 014 (Preprint 0907.1410 [hep-th])
[29] Govindarajan S 2010 BKM Lie superalgebras from counting twisted CHL dyons *J. High Energy Phys.* 1105 089 (Preprint 1006.3472 [hep-th])
[30] Gritsenko VA and Nikulin VV 1991 Siegel automorphic form corrections of some Lorentzian Kac-Moody algebras *Amer. J. Math.* **119** 181
[31] Gritsenko VA 1995 Modular forms and moduli spaces of abelian and k3 surfaces *St. Petersburg Math. Jour.* **6** 1179
[32] Gritsenko VA and Nikulin VV Automorphic forms and Lorentzian Kac-Moody algebras. Part 1 (Preprint alg-geom/9610028).
[33] Gritsenko VA and Nikulin VV Automorphic forms and Lorentzian Kac-Moody algebras. Part 2 (Preprint alg-geom/9610022)
[34] Harvey JA and Moore GW 1996 Algebras, BPS states, and strings *Nuclear Phys. B* **463** 315 (Preprint hep-th/9510182)
[35] Harvey JA and Moore GW 1998 On the algebras of BPS states *Commun. Math. Phys.* **197** 489 (Preprint hep-th/9609017)
[36] Hohenegger S and Persson D 2011 Enhanced Gauge Groups in N=4 Topological Amplitudes and Lorentzian Borcherds Algebras Preprint 1107.2301 [hep-th]
[37] Howe R 1979 θ-seris and invariant theory *Proceedings of Symposia in Pure Mathematics* vol. 33, part 1, p. 275
[38] Humphreys JE 1990 Reflection Groups and Coxeter Groups *Cambridge Studies in Advanced Mathematics* vol 29, Cambridge University Press
[39] Kac VG 1990 Infinite-dimensional Lie algebras (*Cambridge University Press*)
[40] Kiritsis E 1997 Introduction to nonperturbative string theory (Preprint hep-th/9708130)
[41] Kiritsis E 1998 Introduction to superstring theory *Leuven Univ. Pr.* 315 pp (Preprint hep-th/9709062)
[42] Kiritsis E and Obers NA 1997 Heterotic/type-I duality in $D < 10$ dimensions, threshold corrections and D-instantons *J. High Energy Phys.* **9710** 004 (Preprint hep-th/9709058).
[43] Kiritsis E, Obers NA and Pioline B 2000 Heterotic/type II triality and instantons on K3 *J. High Energy Phys.* **0001** 029 (Preprint hep-th/0001083)
[44] Maaß H 1979 Über eine Spezialschar von Modulformen zweiten Grades i *Invent. math.* **52** 95; Maaß H 1979 Über eine Spezialschar von Modulformen zweiten Grades ii *Invent. math.* **53** 249; Maaß H 1979 Über eine Spezialschar von Modulformen zweiten Grades iii *Invent. math.* **53** 255
[45] Marino M and Moore GW 1999 Counting higher genus curves in a Calabi-Yau manifold *Nuclear Phys. B* **543** 592 (Preprint hep-th/9808131)
[46] Obers NA and Pioline B 2000 Eisenstein series and string thresholds *Commun. Math. Phys.* **209** 275 (Preprint hep-th/9903113)
[47] Rallis S 1979 On a relation between SL_2 cusp forms and automorphic forms on orthogonal groups *Proceedings of Symposia in Pure Mathematics,* **33** part 1, p 297
[48] Sen A 2007 Walls of marginal stability and dyon spectrum in N=4 supersymmetric string theories *J. High Energy Phys.* **0705** 039 (Preprint hep-th/0702141)
[49] Sen A 2008 Black Hole Entropy Function, Attractors and Precision Counting of Microstates *Gen. Rel. Grav.* **40** 2249 (Preprint 0708.1270)

Geometric properties of Frobenius algebras

Jaroslav Hrdina[1]

[1]Institute of Mathematics, Faculty of Mechanical Engineering, Brno University of Technology, Czech Republic.

E-mail: `hrdina@fme.vutbr.cz`

Abstract. We discuss the properties of k–algebras and their modules. A useful observation in this direction suggests that one may choose Frobenius algebras to avoid problems with generic rank. In particular, we prove that the study of generic rank reduces to the study of Frobenius algebras.

MSC 2010: 16D10, 53C15

1. Algebras and modules

In this chapter we fix a ground field $k := \mathbb{R}$ and vector spaces V and W over k. A pairing of V and W is by definition $\beta := \langle | \rangle : V \otimes W \to k$ and we can write $v \otimes w \mapsto \langle v|w \rangle$. A pairing β is called nondegenerate in the variable V if there is copairing $\gamma : k \to W \otimes V$, such that following composition is equal to the identity map of V.

$$\begin{array}{ccccc} V & & (V \otimes W) \otimes W & \xrightarrow{\beta \otimes id_V} & k \otimes V \\ \| & & \| & & \| \\ V \otimes k & \xrightarrow{id_V \otimes \gamma} & (V \otimes W) \otimes W & & k \otimes V \end{array}$$

Similarly, β is called nondegenerate in the variable W and the paring β is called nondegenerate if it is simultaneously nondegenerate in V and W. Let us remark that in that case the coparing automatically holds. In particular, if we fix second argument $w \in W$ in pairing $\langle | \rangle$ we get a linear functional

$$|w\rangle : V \to k, \;\; v \mapsto \langle v|w \rangle$$

and this defines a linear map

$$|\rangle : W \to V^*, \;\; W \mapsto |w\rangle.$$

Similarly we could fix $v \in V$ to get a linear functional $\langle v| : W \to k$ and $\langle | : V \to W^*$.

Lemma 1. *Given a pairing $\langle | \rangle : V \otimes W \to k$ between two finite–dimensional vector spaces V and W, the following are equivalent.*

(i) $\langle | \rangle$ *is nondegenerate*

(ii) $\langle |$ *is an isomorphism*

(iii) $|\rangle$ *is an isomorphism*

A k–algebra is a k–vector space leave A together with two k–linear maps $\mu : A \otimes A \to A$, $\nu : k \to A$. In the terms of elements, if we have $\mu(x \otimes y) = xy$ and $\nu(1) = 1_A$, then we can write the axioms of associativity and unity: $(xy)z = x(yz)$ a $1_A x = x 1_A = x$. Finally, left A–module is a vector space M together with a k–linear map

$$\alpha : A \otimes M \to M$$

In the terms of elements, have write $\alpha(a \otimes x) = a \cdot x$ and we can write the axioms:

$$a \cdot (b \cdot x) = (ab) \cdot x \text{ and } 1_A \cdot x = x.$$

Let V be a left A–module and W be a right A–module, a pairing $\beta : V \otimes W \to k$ is said to be associative if

$$\langle x \cdot a | y \rangle = \langle x | a \cdot y \rangle,$$

for every $x \in V$, $y \in W$, $a \in A$.

From now, we assume that A if of finite dimension. A linear functional $\epsilon : A \to k$ defines a hyperplane

$$Null(\epsilon) := \{x \in A | \epsilon(x) = 0\},$$

this hyperplane is called a nullspace.

Definition 1.1. A *Frobenius algebra* is a k–algebra A of a finite dimension, equipped with a linear functional $\epsilon : A \to k$ whose nullspace contains no nontrivial left ideals. The functional $\epsilon \in A^*$ is called *Frobenius form.*

Let us remind that a linear functional $\epsilon : A \to k$ canonically determines a pairing $A \otimes A \to k$, namely $x \otimes y \mapsto \epsilon(xy)$ and conversely, given an associative pairing $\langle | \rangle : A \otimes A \to k$ a linear functional $\langle 1_A |$ is canonically determined. In particular, a Frobenius algebra is a k–algebra A of a finite dimension, equipped with an associative pairing $\langle | \rangle$ and we call this pairing *Frobenius pairing.* See [3] for more details in this direction.

Definition 1.2. A Frobenius algebra A is called a *symmetric Frobenius algebra* if one (and hence all) of the following equivalent conditions hold.

(i) The Frobenius form $\epsilon : A \to k$ is *central*; this means $\epsilon(ab) = \epsilon(ba)$ for all $a, b \in A$.

(ii) The Frobenius pairing $\langle | \rangle$ is symmetric, i.e. $\langle a|b \rangle = \langle b|a \rangle$ for all $a, b \in A$.

Example 1. *The ring $Mat_n(k)$ is a Frobenius algebra with the usual trace map. To see that bilinear pairing resulting from trace is nondegenerate, take the linear basis of $Mat_n(k)$ consisting of matrices E_{ij} with only one nonzero entry $e_{ij} = 1$. Clearly E_{ij} is the dual basis element to E_{ji} under this pairing and this pairing is symmetric.*

2. Structural constants and structural polynomial matrices

In this chapter we fix a ground field $k := \mathbb{R}$, a k–algebra A with basis $\{F_i\}$, where $i = 1, \ldots, n$, $F_1 := E$, with structure constants $C_{ij}^k \in k$ (i.e. $F_i F_j = C_{ij}^s F_s$) and finite dimensional A–module V. In particular, one can easily see that $F_i = F_1 F_i = C_{1i}^s F_s = \delta_i^s F_s$, i.e. $C_{1i}^s = \delta_i^s$.

We introduce matrices

$$\hat{C}_i = (C_{ji}^k), \quad \hat{C}_i^* = (C_{ik}^j),$$

where j is a number of rows. Then the associativity condition can be written as

$$\hat{C}_j \hat{C}_k = C_{jk}^s \hat{C}_s \text{ or } \hat{C}_j^* \hat{C}_k^* = C_{jk}^s \hat{C}_s^*$$

and unity can be written as $\hat{C}_1 = E$. Now, consider an element $F = a_1F_1 + \cdots + a_nF_n \in A$ and compute

$$F_1F = b_1^1F_1 + \cdots + b_n^1F_n$$
$$\vdots$$
$$F_nF = b_1^\ell F_1 + \cdots + b_n^nF_n.$$

Then the *structure polynomial matrix of* A is a matrix

$$B := \begin{pmatrix} b_1^1 & \cdots & b_n^1 \\ \vdots & \cdots & \vdots \\ b_1^n & \cdots & b_n^n \end{pmatrix} \in Mat_n(k[a_1, \cdots, a_n]).$$

Clearly,

$$B = \sum_{i=1}^{n} a_i\hat{C}_i.$$

See [4, 2] for more details in this direction.

Lemma 2. *Let A be a finite dimensional k–algebra and B a structure polynomial matrix of A. The determinant $det(B) \in k[a_1, \cdots, a_n]$ is not identically zero.*

Proof. In this case, a straightforward computations quickly prove our lemma:

$$B\mu = 0 \Rightarrow \sum_{i=1}^{n} a_i\hat{C}_i\mu = 0 \Rightarrow \hat{C}_i\mu = 0 \text{ for any } i = 1, \ldots n$$

$$\Rightarrow \hat{C}_1\mu = 0 \Rightarrow E\mu = 0 \Rightarrow \mu = 0$$

□

Definition 2.1. Let A be a n–dimensional k–algebra, where $k := \mathbb{R}$ and let V be a finite dimensional A–module. We say that A–module V has weak generic rank n if the subset of vectors $X \in V$, such that the A–hull $A(X) := \{FX|F \in A\}$ generating a vector subspace of dimension n is open and dense in V.

Hence, if there exists $X \in V$ such that $FX = 0$, then $F_1FX = \cdots = F_nFX = 0$ and the determinant

$$\begin{vmatrix} b_1^1 & \cdots & b_n^1 \\ \vdots & \cdots & \vdots \\ b_1^n & \cdots & b_n^n \end{vmatrix} = \sigma(a_1, \ldots, a_n)$$

has to be zero for some $a_i \in k$.

Remark. In differential geometry, in the theory of A–structures, A is assumed to be an algebra and M a smooth manifold, such that $dim(M) = m$ and we suppose that any $F \in A$ acts on TM as an affinor. If A–module TM has weak generic rank ℓ then the A–structure has generic rank $2\ell < m$. For more details see [2, 1].

Theorem 1. *Let $X_1, \ldots, X_m$ be a basis of A–module V, A be a n–dimensional k–algebra, where $n < m$. If there exists $X \in V$ such that $dim(A(X)) = n$ then the A–module has weak generic rank n.*

Journal of Physics: Conference Series **346** (2012) 012009 doi:10.1088/1742-6596/346/1/012009

Proof. We prove equivalent statement, A–module V does not have a generic rank ℓ if and only if there is a vector $X \in V$, such that for any vector $Y \in V$ there is an affinor G_Y, such that $G_Y(X - \epsilon Y) = 0$, for small ϵ. Therefore, for any vector $Y \in V$ there is an affinor G_Y such that

$$G_Y(X) = \epsilon G_Y(Y),$$

for small ϵ. Hence, the affinor $\frac{1}{\sqrt{\epsilon}} G_Y$ maps $\frac{1}{\sqrt{\epsilon}} X$ to a vector $G_Y(Y)$ and therefore, for any vector $Y \in V$ there is an affinor H_Y, such that $H_Y(\frac{1}{\sqrt{\epsilon}} X) = H_Y(Y)$. In particular, there is an affinor S such that $S(\frac{1}{\sqrt{\epsilon}} X) = S(Y + \frac{1}{\sqrt{\epsilon}} X)$ and therefore, for any $Y \in V$ there is an affinor S_Y such that $S_Y(Y) = 0$. □

Example 2. *Let $A = \langle E, P\rangle$ be a para–complex structure, i.e. $P^2 = E$, then the matrices of structure constants are the following*

$$\hat{C}_1 = \begin{pmatrix} 1 & 0 \\ 0 & 1 \end{pmatrix}, \hat{C}_2 = \begin{pmatrix} 0 & 1 \\ 1 & 0 \end{pmatrix}$$

and the structure polynomial matrix is

$$B = \begin{pmatrix} a_1 & a_2 \\ a_2 & a_1 \end{pmatrix}$$

therefore if $x_1 \neq x_2 \neq 0$ then $dim(A(X)) = 2$, where $X = x_1, x_2)$.

3. Frobenius algebras

A linear functional $\epsilon : A \to k$ is determined by the choice of a n–dimensional vector $\lambda = (\lambda_1, \dots, \lambda_n)$. Now, for $\epsilon(F_i) = \lambda_i$ and for $F = \sum_{i=1}^n a_i F_i \in A$ we can see immediately that $\epsilon(F) = \sum_{i=1}^n a_i \lambda_i$ and finally $F_i F_j = C_{ij}^s F_s$.

Theorem 2. *Let A be a k–algebra and $\lambda \in k^n$ be a vector such that the matrix $G = (g_{ij})$ is regular, where $g_{ij} := C_{ij}^s \lambda_s$. The functional $\epsilon : A \to k$, such that $\epsilon : \sum_{i=1}^n a_i F_i \mapsto \sum_{i=1}^n a_i \lambda_i$ is a Frobenius form.*

Proof. Consider $F \in Null(\epsilon)$ such that F is an element of a notrivial left ideal $I \subset A$. Therefore $\epsilon(a_1 F_1 + \cdots a_n F_n) = 0 \Rightarrow F_i \epsilon(a_1 F_1 + \cdots a_n F_n) = 0$ for every $i = 1, \dots n$ and $\epsilon(a_1 C_{i1}^s F_s + \cdots + a_n C_{in}^s F_s) = a_1 C_{i1}^s \lambda_s + \cdots + a_n C_{in}^s \lambda_s = 0$ for any $i = 1, \dots n$. This means

$$a_1 C_{11}^s \lambda_s + \cdots + a_n C_{1n}^s \lambda_s = 0, \cdots, a_1 C_{n1}^s \lambda_s + \cdots + a_n C_{nn}^s \lambda_s = 0,$$

and therefore $a_1 g_{11} + \cdots + a_n g_{1n} = 0, \cdots, a_1 g_{n1} + \cdots + a_n g_{nn} = 0$ and

$$G \begin{pmatrix} a_1 & \cdots & a_n \end{pmatrix} = 0 \Rightarrow \begin{pmatrix} a_1 & \cdots & a_n \end{pmatrix} = 0 \Rightarrow F = 0.$$

□

Example 3. *Let $A = \langle E, P\rangle$ be a para–complex structure, the matrices of structure constant are*

$$\hat{C}_1 = \begin{pmatrix} 1 & 0 \\ 0 & 1 \end{pmatrix}, \hat{C}_2 = \begin{pmatrix} 0 & 1 \\ 1 & 0 \end{pmatrix}$$

and for $\lambda = (\lambda_1, \lambda_2)$, such that $\lambda_1 \neq \lambda_2 \neq 0$ we have

$$G = (\hat{C}_1 \lambda | \hat{C}_2 \lambda) = \begin{pmatrix} \lambda_1 & \lambda_2 \\ \lambda_2 & \lambda_1 \end{pmatrix}$$

and $det(G) = \lambda_1^2 - \lambda_2^2 \neq 0$.

Journal of Physics: Conference Series **346** (2012) 012009 doi:10.1088/1742-6596/346/1/012009

In fact, we need the property having a of generic rank for the computation of morphisms with respect to a connection [1, 2]. The formula for generic rank n from the Theorem 1. reads:

"$\exists X \in V$ such that $\{F_i X\}$ are linearly independent $\Rightarrow V$ has a generic rank "

On the other hand, we need the property of Frobenius algebra for the computations of connection [4]. The formula for Frobenius algebra from the Theorem 2 reads:

"$\exists \lambda \in k^n$ such that $\{\hat{C}_i \lambda\}$ are linearly independent $\Rightarrow A$ is a Frobenius algebra"

This indicates that these properties lead to similar conditions. In other words, if A is a k–algebra and the matrices $\hat{C}_i$ are structural matrices of A, then there is A–module k^n, where $B = \langle \hat{C}_1, \dots, \hat{C}_n \rangle$ such that $a \cdot x$ is matrix multiplication. Therefore, the k–algebra A is a Frobenius algebra if and only if the A–module k^n has a generic rank n.

Acknowledgement
Published results were acquired using the subsidization of the GA ČR, grant no. 201/09/0981.

[1] Hrdina J and Slovák J 2006 Generalized planar curves and quaternionic geometry *Ann. Glob. Anal. Geom* **29** 349
[2] Hrdina J 2011 Notes on connections attached to A–structures *Differ. Geom. Appl.* **29 S1** 91
[3] Kock J 2003 *Frobenius Algebras and 2D Topological Quantum Field Theories*. London Mathematical Society Student Texts Series (Cambridge University Press)
[4] Shirokov A P 2002 Spaces over algebras and their applications *J. Math. Sci.* **108** 232

Algebra, Geometry, and Mathematical Physics 2010 IOP Publishing
Journal of Physics: Conference Series **346** (2012) 012010 doi:10.1088/1742-6596/346/1/012010

Subring depth below an ideal

Lars Kadison
Departamento de Matematica, Faculdade de Ciências da Universidade do Porto, Rua Campo Alegre 687, 4169-007 Porto, Portugal
E-mail: lkadison@fc.up.pt

In memory of Gerhard Hochschild

Abstract. A minimum depth is assigned to a ring homomorphism and a bimodule over its codomain. When the homomorphism is an inclusion and the bimodule is the codomain, the recent notion of depth of a subring in a paper by Boltje-Danz-Külshammer is recovered . Subring depth below an ideal gives a lower bound for BDK's subring depth of a group algebra pair or a semisimple complex algebra pair.

1. Introduction and Preliminaries

Algebras, coalgebras and Hopf algebras are some of the interesting objects with structure in representation categories of commutative rings. In the representation category of a noncommutative ring, these objects become ring extensions, corings and Hopf algebroids. Some basic algebras of interest are the cohomological dimension 0 and ∞ cases of separable algebra and Frobenius algebra; which become separable extensions and Frobenius extensions in noncommutative representation theory. Also, QF rings, semisimple rings, and Azumaya algebras generalize to ring extensions; however depth is not such a notion, originating as a tool of induced representation theory. Depth is essentially constant on (especially projective) algebras over a commutative ring, but gives different and interesting outcomes for ring extensions.

The depth of many subgroups are recently computed, both as induced complex representations [6] and as induced representations over general commutative rings of group algebras [1]. For example, the depth of the permutation groups $S_n \subset S_{n+1}$ is $2n-1$ over any ground ring and depends only on a combinatorial depth of subgroups defined in terms of bisets in [1]. The authors of [1] show that combinatorial depth $d_c(H,G)$ of a subgroup H in a finite group G satisfies $d_c(H,G) \leq 2n$ for $n \geq 1$ (respectively, $d_c(H,G) \leq 2n-1$ for $n > 1$) $\Leftrightarrow$ for any $x_1,\ldots,x_n \in G$, there is $y_1,\ldots,y_{n-1} \in G$ such that $H \cap_{i=1}^{n} x_i H x_i^{-1} = H \cap_{i=1}^{n-1} y_i H y_i^{-1}$ (respectively, the latter condition and additionally $x_1 h x_1^{-1} = y_1 h y_1^{-1}$, all $h \in H \cap_{i=1}^{n} x_i H x_i^{-1}$). All notions of depth ≤ 2 are the same and occur precisely if H is a normal subgroup. However, depth of subalgebras over base rings (of varying characteristic denoted by a subscript) for $R = k[G]$ and $S = k[H]$ and combinatorial depth diverge in a string of inequalities given in [1] as follows:

$$d_0(H,G) \leq d_p(H,G) \leq d_{\mathbb{Z}}(H,G) \leq d_c(H,G) \leq 2[G : N_G(H)]. \tag{1}$$

Also $d_k(H,G) \leq d_c(H,G)$ showing that all extensions of finite dimensional group algebras have finite depth.

The authors begin in [1] with a new notion of subring depth $d(S, R)$, given below in (4). They show in an appendix how it is based on and equal to a previous notion where S and R are semisimple complex algebras given below in (5). Such a pair $R \supseteq S$ is a special case of (split separable) Frobenius extensions; in [13, Theorem 5.3] I show that subring depth is equal to tower depth of Frobenius extensions [11] satisfying a generator module condition. The authors of [1] define a left and right even depth and show these are the same on group algebra extensions; [13, Theorem 3.4] shows this equality holds for all QF extensions.

In this paper a change is made to the definition of subring depth; we define an I-depth $d^I(S \to R)$ of a ring homomorphism $S \to R$ with R-bimodule I, which we use in place of R in the n-fold tensor products over S in the definition (4) of $d(S, R)$ (as well as a converse, automatic in the presence of units). When I is an ideal of a semisimple complex algebra R with semisimple subalgebra S the I-depth $d^I(S \to R)$ gives a lower bound, $d^I(S \to R) \leq d(S, R)$ discussed in Section 2 in terms of the part of the bipartite graph of the inclusion which is directly below the ideal I.

There are intriguing similarities and relations between relative homological algebra and subring depth theory. For example, the depth two condition on a subring $S \subseteq R$ leads in [10] to an isomorphism of differential graded algebras between the relative Hochschild R-valued cochains with cup product and the Amitsur complex of a coring with grouplike element (on the endomorphism ring $\mathrm{End}\,{}_SR_S$ over the centralizer subring R^S). Also the paper [12] contains some relations between depth 2 and notions of relative homological algebra carried over to corings in [4]. The tower of iterated endomorphism rings above a ring extension becomes in the case of Frobenius extensions a tower of rings on the bar resolution groups $C_n(R, S)$ ($n = 0, 1, 2, \ldots$) with Frobenius and Temperley-Lieb structures explicitly calculated from their more usual iterative definition in the paper [13]. At the same time Frobenius extensions of depth more than 2 are known to have depth 2 further out in the tower: I extend this observation in [11] with new proofs to include other ring extensions satisfying the hypotheses of [13, Prop. 4.3].

1.1. H-equivalent modules

Let R be a ring. Two left R-modules, ${}_RN$ and ${}_RM$, are said to be *h-equivalent*, denoted by ${}_RM \overset{h}{\sim} {}_RN$ if two conditions are met. First, for some positive integer r, N is isomorphic to a direct summand in the direct sum of r copies of M, denoted by

$$ {}_RN \oplus * \cong {}_RM^r \quad \Leftrightarrow \quad N \mid M^r \;\Leftrightarrow \tag{2}$$

$$\exists f_i \in \mathrm{Hom}\,({}_RM, {}_RN),\; g_i \in \mathrm{Hom}\,({}_RN, {}_RM), i = 1, \ldots, r \;:\; \sum_{i=1}^{r} f_i \circ g_i = \mathrm{id}_N$$

Second, symmetrically there is $s \in \mathbb{Z}_+$ such that $M \mid N^s$. It is easy to extend this definition of h-equivalence (sometimes referred to as similarity) to h-equivalence of two objects in an abelian category, and to show that it is an equivalence relation.

If two modules are h-equivalent, ${}_RN \overset{h}{\sim} {}_RM$, then they have Morita equivalent endomorphism rings, $\mathcal{E}_N := \mathrm{End}\,{}_RN$ and $\mathcal{E}_M := \mathrm{End}\,{}_RM$. This is quite easy to see since a Morita context of bimodules is given by $H(M, N) := \mathrm{Hom}\,({}_RM, {}_RN)$, which is an $\mathcal{E}_N$-$\mathcal{E}_M$-bimodule via composition, and the bimodule ${}_{\mathcal{E}_M}H(N, M)_{\mathcal{E}_N}$; these are progenerator modules, by applying to (2) or its reverse, $M \mid N^s$, any of the four Hom-functors such as $\mathrm{Hom}\,({}_R-, {}_RM)$ from the category of left R-modules into the category of left E_M-modules showing that ${}_{\mathcal{E}_M}H(N, M)$ is finite projective; similarly, generator. Then the explicit conditions on mappings for h-equivalence show that $H(M, N) \otimes_{\mathcal{E}_M} H(N, M) \to \mathcal{E}_N$ and the reverse mapping given by composition are both bimodule isomorphisms as required. Since $\mathcal{E}_M$ and $\mathcal{E}_N$ are Morita equivalent rings, their centers are isomorphic:

$$\mathrm{End}\,{}_RM_{\mathcal{E}_M} \cong \mathrm{End}\,{}_RN_{\mathcal{E}_N}.$$

Journal of Physics: Conference Series **346** (2012) 012010 doi:10.1088/1742-6596/346/1/012010

The theory of h-equivalent modules applies to bimodules ${}_TM_S \overset{h}{\sim} {}_TN_S$ by letting $R = T \otimes_{\mathbb{Z}} S^{\mathrm{op}}$ which sets up an equivalence of abelian categories between T-S-bimodules and left R-modules. Two additive functors $F, G : \mathcal{C} \hookrightarrow \mathcal{D}$ are h-equivalent if there are natural split epis $F(X)^n \to G(X)$ and $G(X)^m \to F(X)$ for all X in $\mathcal{C}$. We leave the proof of the lemma below as an elementary exercise.

Lemma 1.1 *Suppose two R-modules are h-equivalent, $M \overset{h}{\sim} N$ and two additive functors from R-modules to an abelian category are h-equivalent, $F \overset{h}{\sim} G$. Then $F(M) \overset{h}{\sim} G(N)$.*

For example, the following substitution in equations involving the $\overset{h}{\sim}$-equivalence relation follows from the lemma:

$$ {}_RP_T \overset{h}{\sim} {}_RQ_T \quad {}_TU_S \overset{h}{\sim} {}_TV_S \Rightarrow {}_RP \otimes_T U_S \overset{h}{\sim} {}_RQ \otimes_T V_S \tag{3}$$

Example 1.2 If R is a semisimple artinian ring with simples $\{P_1, \ldots, P_t\}$ (representatives from each isomorphism class), all finitely generated modules M_R and N_R have a unique factorization into simple components. Denote the simple constituents of M_R by $\mathrm{Simples}\,(M) = \{P_i \mid [P_i, M] \neq 0\}$ where $[P_i, M]$ is the number of factors in M isomorphic to P_i. Then $M \mid N^q$ for some positive q if $\mathrm{Simples}\,(M) \subseteq \mathrm{Simples}\,(N)$; and $M \overset{h}{\sim} N$ iff $\mathrm{Simples}\,(M) = \mathrm{Simples}\,(N)$.

Suppose R has central primitive idempotents $e_1, \ldots, e_t$ such that each $[P_i, e_iR] = n_i$, so that R decomposes into the product of matrix rings over each of the division rings $D_i := \mathrm{End}\,(P_i)_R$: $R \cong M_{n_1}(D_1) \times \cdots \times M_{n_t}(D_t)$. If M and N are h-equivalent f.g. R-modules, then the endomorphism rings E_M and E_N are explicitly Morita equivalent as they are both products of matrix rings over the same subset of division rings $D_1, \ldots, D_t$.

Example 1.3 Via some more category theory, we may see that positive integers n and m are h-equivalent if $n \mid m^r$ and $m \mid n^s$ for some positive integers r, s; whence there are primes $p_1, \ldots, p_k$ such that n and m lie in the same h-equivalence class $\{p_1^{r_1} \cdots p_k^{r_k} \mid r_1, \ldots, r_k \geq 1\}$. This explains the notation in (2).

1.2. Depth two

A subring pair $S \subseteq R$ is said to have left depth 2 (or be a left depth two extension [9]) if $R \otimes_S R \overset{h}{\sim} R$ as natural S-R-bimodules. Right depth 2 is defined similarly in terms of h-equivalence of natural R-S-bimodules. In [9] it was noted that the left condition implies the right and conversely if R is a Frobenius extension of S. Also in [9] a Galois theory of Hopf algebroids was defined on the endomorphism ring $H := \mathrm{End}\,{}_SR_S$ as total ring and the centralizer $C := R^S$ as base ring. The antipode is the restriction of the natural anti-isomorphism stemming from following the arrows,

$$\mathrm{End}\,R_S \xrightarrow{\cong} R \otimes_S R \xrightarrow{\cong} (\mathrm{End}\,{}_SR)^{\mathrm{op}}.$$

The Galois properties may then be summarized by the invariants under the obvious action of H, $R^H = S$ if R_S is faithfully flat, and $\mathrm{End}\,R_S \cong R\#H$ a smash product product ring structure on $R \otimes_C H$: the details are in [9]. There is also a duality structure by going a step further along in the tower above $S \subseteq R \hookrightarrow \mathrm{End}\,R_S \hookrightarrow \mathrm{End}\,R \otimes_S R_R$, where the dual Hopf algebroid $H' := (R \otimes_S R)^S$ plays a role [9].

Conversely, Galois extensions have depth 2, which is most easily seen from the Galois map of an H-comodule algebra A with invariant subalgebra B and finite dimensional Hopf algebra H over a base field k, which is given by $A \otimes_B A \xrightarrow{\cong} A \otimes_k H$, $a' \otimes a \mapsto a'a_{(0)} \otimes a_{(1)}$, whence $A \otimes_B A \cong A^{\dim H}$ as A-B-bimodules. The Hopf subalgebras within a finite dimensional Hopf algebra which have depth 2 are precisely the normal Hopf subalgebras; if normal, it has depth

2 by applying the Hopf-Galois observation just made. The converse follows from an argument discovered by [2, Boltje-Külshammer] which divides the normality notion into right and left just like depth 2, where left normal is invariance under the left adjoint action. Note their argument given in the context of any augmented algebra A (such as a quasi-Hopf algebra) next. Let $\varepsilon : A \to k$ be the algebra homomorphism into a base ring k. Let A^+ denote $\ker \varepsilon$, and for a subalgebra $B \subseteq A$, let B^+ denote $\ker \varepsilon \cap B$. For example, it may be shown that if a (quasi-)Hopf algebra H has (quasi-)Hopf subalgebra that is invariant under the left adjoint action of H, then $HK^+ \subseteq K^+H$.

Proposition 1.4 *Suppose $B \subseteq A$ is a subalgebra of an augmented algebra. If $B \subseteq A$ has right depth 2, then $AB^+ \subseteq B^+A$.*

Proof. To $A \otimes_B A \,|\, A^q$ as A-B-bimodules, apply the additive functor $k_\varepsilon \otimes_A -$, which results in $A/B^+A \,|\, k^q$ as right B-modules. The annihilator of k^q restricted to B is of course B^+, which then also annihilates A/B^+A, so $AB^+ \subseteq B^+A$.

The opposite inclusion is of course satisfied by a left depth 2 extension of augmented algebras.

Also subalgebra pairs of semisimple complex algebras have depth 2 exactly when they are normal in a classical sense of Rieffel. We note the theorem in [6] below and give a new proof in one direction along the lines of the previous proposition.

Theorem 1.5 *[6, Theorem 4.6] Suppose $B \subseteq A$ is a subalgebra pair of semisimple complex algebras. Then $B \subseteq A$ has depth 2 if and only if for every maximal ideal I in A, one has $A(I \cap B) = (I \cap B)A$.*

Proof. ($\Leftarrow$) See [6, Section 4]. ($\Rightarrow$) Given maximal ideal I in A, there is an ideal J with identity element 1_J such that $A = I \oplus J$, and algebra homomorphism $\varepsilon : A \to A/I \cong J$. Denote $I = A^+$, $I \cap B = B^+$, and note that the A-module $J_A = J_\varepsilon$. Given the right depth 2 condition ${}_AA \otimes_B A_B \,|\, A^q$, tensor from the left by J_A obtaining $J \otimes_B A_B \,|\, J^q$.

Note the B-module homomorphism $A/B^+A \to J \otimes_B A$ given by $a + B^+A \mapsto 1_J \otimes_B a$ (well-defined since $1_J \cdot B^+ = 0$) which we claim is monic. For suppose that $1_J \otimes a = 0$ for a in the projective module ${}_BA$, so $a = \sum_i f_i(a)e_i$ in some free module B^n. Then

$$0 = 1_J \otimes a = \sum_i \varepsilon(f_i(a)) \otimes e_i \Rightarrow \varepsilon(f_i(a)) = 0 \Rightarrow f_i(a) \in B^+, \forall i = 1, \ldots, n$$

hence $a = \sum_i f_i(a)e_i \in B^+A$ so $a + B^+A = 0$ which proves the claim.

Since B^+ annihilates J^q_B, it annihilates $J \otimes_B A_B$ and therefore A/B^+A via the monomorphism. Thus $AB^+ \subseteq B^+A$. The opposite inclusion follows from a similar argument applied to the left depth 2 condition.

2. Ideal depth of a ring homomorphism

Let S and R be unital associative rings and $S \to R$ a ring homomorphism where $1_S \mapsto 1_R$. Suppose ${}_RI_R$ is a bimodule. With no further ado, we will restrict I to bimodules ${}_SI_R$, ${}_RI_S$ or ${}_SI_S$ via the homomorphism $S \to R$. Note that the kernel of $S \to R$ is contained in the annihilator ideal in S of the left (or right) S-module I denoted by $\mathrm{ann}\,{}_SI$.

We let $C^I_0(S \to R) = S$, and for $n \geq 1$,

$$C^I_n(S \to R) = I \otimes_S \cdots \otimes_S I \quad (n \text{ times } I)$$

For $n \geq 1$, the $C^I_n(S \to R)$ has a natural R-R-bimodule (briefly R-bimodule) structure which restricts to S-R-, R-S- and S-bimodule structures occuring in the next definition.

Definition 2.1 *The ring homomorphism $S \to R$ has I-depth $2n+1 \geq 1$ if as S-bimodules $C_n^I(S \to R) \overset{h}{\sim} C_{n+1}^I(S \to R)$. The ring homomorphism $S \to R$ has left (right) I-depth $2n \geq 2$ if $C_n^I(S \to R) \overset{h}{\sim} C_{n+1}^I(S \to R)$ as S-R-bimodules (respectively, R-S-bimodules).*

It is clear that if $S \to R$ has either I-depth $2n$, it has I-depth $2n+1$ by restricting the h-equivalence condition to S-bimodules. If it has I-depth $2n+1$, it has I-depth $2n+2$ by tensoring the h-equivalence by $- \otimes_S I$ or $I \otimes_S -$. The *minimum I-depth* is denoted by $d^I(S \to R)$.

Note that the minimum left and right minimum *even* I-depths may differ by 2 (in which case $d^I(S \to R)$ is the least of the two). In the next section we provide a general condition, which includes a Hopf subalgebra pair $S \subseteq R$ of symmetric Frobenius algebras with I an ideal in R, where the left and right minimum even I-depths coincide.

We also remark that once $S \to R$ has I-depth $2n+1$ the $C_{n+m}^I(S \to R)$'s stop growing as $m \to \infty$ in terms of adding new indecomposables in a category of modules with unique factorization, since $C_n^I(S \to R) \overset{h}{\sim} C_{n+m}^I(S \to R)$ for all $m \geq 0$ (see the example in the previous section). This corresponds well with the classical notion of finite depth in subfactor theory.

Lemma 2.2 *Let $S \to R$ have kernel K, $\overline{S} := S/K$ and $\overline{S} \hookrightarrow R$ be the induced ring monomorphism. Then the left or right minimum depth $d^I(S \to R) = d^I(\overline{S} \hookrightarrow R)$ unless $d^I(\overline{S} \hookrightarrow R) = 1$, in which case equality holds if the quotient homomorphism $p : S \to \overline{S}$ has a section.*

Proof. Note that if $M_R \,|\, N_R^q$, then $\operatorname{ann} N_R \subseteq \operatorname{ann} M_R$. Since K is in $\operatorname{ann} C_n^I(R,S)$ for all $n \geq 1$ and $C_n^I(R,S) \cong C_n^I(R,\overline{S})$ as $\overline{S}$-modules, it follows that $C_n^I(R,S) \overset{h}{\sim} C_{n+1}^I(R,S)$ implies $C_n^I(R,\overline{S}) \overset{h}{\sim} C_{n+1}^I(R,\overline{S})$ for the bimodules at issue. The converse is easy by pullback along p.

$S \to R$ has I-depth 1 iff there are central elements $w_j, z_i \in I^S$ and mappings $f_j, g_i \in \operatorname{Hom}({}_SI_S, {}_SS_S)$ such that $x = \sum_i z_i g_i(x)$ for all $x \in I$ and $\sum_j f_j(w_j) = 1_S$. By composing with the quotient homomorphism $S \to \overline{S}$, we obtain $\tilde{f}_j, \tilde{g}_i \in \operatorname{Hom}({}_{\overline{S}}I_{\overline{S}}, {}_{\overline{S}}\overline{S}_{\overline{S}})$ and $z_i \in I^{\overline{S}}$ such that $x = \sum_i \tilde{g}_i(x) z_i$ and $\sum_j \tilde{f}_j(w_j) = 1_{\overline{S}}$. The converse may be proven with the extra hypothesis in the lemma, since all mappings in $\operatorname{Hom}({}_{\overline{S}}I_{\overline{S}}, {}_{\overline{S}}\overline{S}_{\overline{S}})$ have a lifting to $\operatorname{Hom}({}_SI_S, {}_SS_S)$ along p via a section $\sigma : \overline{S} \hookrightarrow S$ satisfying $p \circ \sigma = \mathrm{id}_{\overline{S}}$.

Example 2.3 Suppose S is a subring of R (where $1_S = 1_R$). Let $S \to R$ be the inclusion monomorphism and $I = R$, the natural R-bimodule. The minimum depth of the subring $S \subseteq R$ as defined in [1, Boltje-Danz-Külshammer] is denoted by $d(S,R)$. We note that $d(S,R) = d^R(S \hookrightarrow R)$. In fact, $C_n^R(S \to R) = R \otimes_S \cdots \otimes_S R := C_n(R,S)$ (n times R) for $n > 0$, and the depth $2n+1$ condition in [1] is that

$$C_{n+1}(R,S) \,|\, C_n(R,S)^q \tag{4}$$

as S-bimodules (some $q \in \mathbb{Z}_+$). The left depth $2n$ condition in [1] is (4) more strongly as natural S-R-bimodules (and as R-S-bimodules for the right depth $2n$ condition). But (using a pair of classical face and degeneracy maps of homological algebra) we always have $C_n(R,S) \,|\, C_{n+1}(R,S)$ as R-S-, S-R- or S-bimodules, so that the depth $2n$ as well as $2n+1$ conditions coincide in the case of subring with the I-depth $2n$ and $2n+1$ conditions above where $I = R$. (Note though that R-depth 1 is slightly stronger than subring depth 1 since R is not just centrally projective over S (i.e., $R \,|\, S^q$ as S-bimodules) but also R is a split extension of S as S-bimodules since $S \,|\, R^q$ implies $S \,|\, R$; the split extension condition is satisfied by all group algebra extensions and subfactor examples of finite depth.)

Example 2.4 Let $S \subseteq R$ be a subring pair of semisimple complex algebras. Then the minimum depth $d(S,R)$ may be computed from the inclusion matrix M, alternatively an n by m induction-restriction table of n S-simples induced to non-negative integer linear combination of m R-simples along rows, and by Frobenius reciprocity, columns show restriction of R-simples in terms of S-simples). The procedure to obtain $d(S,R)$ given in the paper [6] is the following: let $M^{[2n]} = (MM^t)^n$ and $M^{[2n+1]} = M^{[2n]}M$ (and $M^{[0]} = I_n$), then the matrix M has depth $n \geq 1$ if for some $q \in \mathbb{Z}_+$

$$M^{[n+1]} \leq qM^{[n-1]} \tag{5}$$

The minimum depth of M is equal to $d(S,R)$ by [1, appendix] (or Theorem 5.3 in [13] combined with [5, 6]).

In terms of the bipartite graph of the inclusion $S \subseteq R$, $d(S,R)$ is the lesser of the minimum odd depth and the minimum even depth [6]. The matrix M is an incidence matrix of this bipartite graph if all entries greater than 1 are changed to 1, while zero entries are retained as 0: let the S-simples be represented by n white dots in a bottow row of the graph, and R-simples by m black dots in a top row, connected by edges joining black and white dots (or not) according to the 0-1-matrix entries obtained from M. The minimum odd depth of the bipartite graph is 1 plus the diameter in edges of the row of white dots (indeed an odd number), while the minimum even depth is 2 plus the largest of the diameters of the bottom row where a subset of black dots under one white dot is identified together.

Now suppose I is an ideal in R. Let the primitive central idempotents of R be given by $e_1, \ldots, e_m$ and those of S by $f_1, \ldots, f_n$. Then I is itself a semisimple complex algebra with unit $e = e_1 + \cdots + e_r$ (assumed with no loss of generality). Now suppose $f_i e_j = 0$ for $i > s$ and all $j \leq r$, while $f_i e_j \neq 0$ for $i \leq s$ and some $j \leq r$. Let $J = f_1 S \oplus \cdots \oplus f_s S$, a semisimple subalgebra of S: this ideal satisfies $J \oplus \operatorname{ann}({}_S I) = S$. Then it is not hard to see that I-depth of $S \subseteq R$ is computed as the depth of the subring pair of semisimple algebras $J \hookrightarrow I$ via $s \mapsto es$:

$$d^I(S,R) = d(J,I), \tag{6}$$

the minimum depth of the $s \times r$ submatrix M_1 in the upper lefthand corner of M. This follows from the lemma where $S/\operatorname{ann}({}_S I) = J$ and the realization that $I \otimes_J \cdots \otimes_J I$ is induction and restriction n times of I-simples as explained in the appendix of [1].

Example 2.5 As a sub-example of the previous example, let $R = \mathbb{C}\,S_4$, the complex group algebra of the permutation group on four letters, and $S = \mathbb{C}\,S_3$. The inclusion diagram pictured below with the degrees of the irreducible representations, is determined from the character tables of S_3 and S_4 or the branching rule (for the Young diagrams labelled by the partitions of n and representing the irreducibles of S_n).

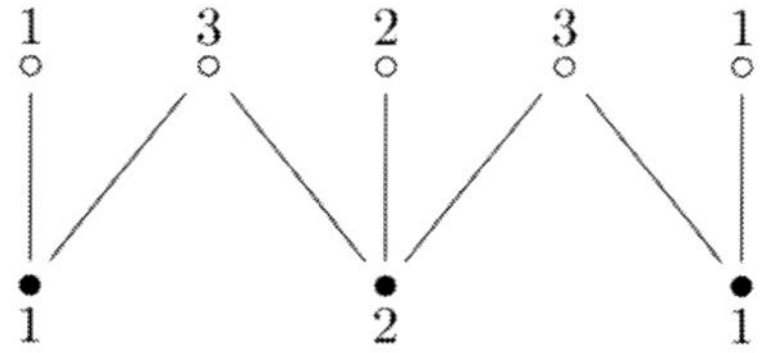

This graph has minimum odd depth 5 and minimum even depth 6, whence $d(S,R) = 5$. Alternatively, the inclusion matrix M is given by

$$M = \begin{pmatrix} 1 & 1 & 0 & 0 & 0 \\ 0 & 1 & 1 & 1 & 0 \\ 0 & 0 & 0 & 1 & 1 \end{pmatrix}$$

whose bracketed powers defined above satisfy a depth 5 inequality (5).

Now let I be the ideal in R associated with the two-dimensional representation, the white dot labelled 2. Then $d(J, I)$ is the depth of the matrix (1), so $d^I(S, R) = 1$. If I is the ideal of R associated with the first three white dots in the diagram above, then J is the ideal in S associated to the first two black dots, and $d(J, I)$ is the minimum depth of the (upper-left hand corner) matrix

$$M' = \begin{pmatrix} 1 & 1 & 0 \\ 0 & 1 & 1 \end{pmatrix}$$

which has minimum depth 3. If I is the ideal associated to the three white dots labelled 3,2, and 3, we similarly compute $d^I(S, R) = 4$. Finally, if I is ideal associated to the first four white dots in the diagram above, the $d^I(S, R) = 5$.

Proposition 2.6 *Suppose $S \subseteq R$ is a subring pair of semisimple complex algebras and $I \subseteq R$ is an ideal. Then $d^I(S, R) \leq d(S, R)$.*

Proof. This follows from the observation above that $d^I(S, R) = d(J, I)$ where $J \subseteq S$ and $I \subseteq R$ are both subring pairs of semisimple algebras. But $d(J, I)$ is the depth of a subgraph of the inclusion graph of $S \subseteq R$. By the description of depth of a bipartite graph as the minimum of the odd and even depths in terms of diameter of the row of black dots, it is clear that $d(J, I) \leq d(S, R)$.

The next theorem states a general case when left and right even depth of a subring below an ideal are the same; its proof is a straightforward generalization of [13, Theorem 3.4], therefore omitted. An R-bimodule I is said to be *QF relative to a subring $S \subseteq R$* below if I_S and ${}_SI$ are f.g. projectives, ${}_SI_R \overset{h}{\sim} {}_S\mathrm{Hom}\,(I_S, S_S)_R$ and ${}_RI_S \overset{h}{\sim} {}_R\mathrm{Hom}\,({}_SI, {}_SS)_S$. We also suppose of the R-bimodule I that it is a ring with multiplication, associative in all respects with the bimodule structure, such as $(x_1 \cdot r)x_2 = x_1(r \cdot x_2)$ for all $x_1, x_2 \in I, r \in R$ (called a *multiplicative bimodule* by R.S. Pierce). For example, an ideal I in a semisimple complex algebra R with semisimple subalgebra S satisfies this hypothesis.

Theorem 2.7 *Suppose I is a multiplicative R-bimodule with unit e and is QF relative to a subring $S \subseteq R$. Then $S \subseteq R$ has left I-depth $2n$ if and only if $S \subseteq R$ has right I-depth $2n$.*

Acknowledgments

The author thanks Sebastian Burciu, Mio Iovanov, Christian and Paula Lomp for interesting conversations. Research in this paper was funded by the European Regional Development Fund through the programme COMPETE and by the Portuguese Government through the FCT under the project PE-C/MAT/UI0144/2011.

References

[1] Boltje R, Danz S and Külshammer B 2011 On the depth of subgroups and group algebra extensions, *J. Algebra* **335** 258
[2] Boltje R and Külshammer B 2010 On the depth 2 condition for group algebra and Hopf algebra extensions, *J. Algebra* **323** 1783
[3] Boltje R and Külshammer B 2011 Group algebra extensions of depth one, *Algebra Number Theory* **5** 63
[4] Brzezinski T and Wisbauer R 2001 *Corings and Comodules* (*Cambridge Univ. Press*)
[5] Burciu S and Kadison L 2010 Subgroups of depth three (Perspectives in Mathematics and Physics, Proc. in honor of I.M. Singer, Cambridge, Mass. 5/2009, Surveys in Diff. Geom. vol XV) eds T Mrowka T and S-T Yau pp 17–36.
[6] Burciu S, Kadison L and Külshammer B 2011 On subgroup depth *I.E.J.A.* **9** 133
[7] Farnsteiner R 1994 On Frobenius extensions defined by Hopf algebras *J. Algebra* **166** 130
[8] Hochschild G 1056 Relative homological algebra, *Trans. Amer. Math. Soc.* **82** 246
[9] Kadison L and Szlachanyi K 2003 Bialgebroid actions on depth two extensions and duality, *Adv. in Math.* **179** 75

[10] Kadison L 2007 Simplicial Hochschild cochains as an Amitsur complex, *J. Gen. Lie Th. Appl.* **2** 180
[11] Kadison L 2008 Finite depth and Jacobson-Bourbaki correspondence *J. Pure Appl. Alg.* **212** 1822
[12] Kadison L 2009 Skew Hopf algebras, irreducible extensions and the Π-method, *Münster J. Math.* **2** 183
[13] Kadison L 2010 Ideal depth of QF extensions (*Preprint arXiv.1012.1754*)
[14] Morita K 1965 Adjoint pairs of functors and Frobenius extensions, *Sc. Rep. T.K.D. Sect. A* **9** 40

Algebra, Geometry, and Mathematical Physics 2010 IOP Publishing
Journal of Physics: Conference Series **346** (2012) 012011 doi:10.1088/1742-6596/346/1/012011

The Belavin-Drinfeld theorem on non-degenerate solutions of the classical Yang-Baxter equation

Lisa Kierans, Bernd Kreussler
Mary Immaculate College, South Circular Road, Limerick, Ireland
E-mail: bernd.kreussler@mic.ul.ie

Abstract. We give a coordinate free proof of Belavin and Drinfeld's Theorem about the classification of non-degenerate solutions of the classical Yang-Baxter equation. The equivalence of different characterisations of non-degeneracy is also shown in such a way.

1. Introduction

What is now called the (Quantum) Yang-Baxter Equation was introduced around 1970 independently by C. N. Yang and R. J. Baxter. The Classical Yang-Baxter Equation (CYBE), see Section 3 equation (3.1), can be derived from it by considering the linear part of a solution which is a power series in an auxiliary parameter. It now plays an important role in the study of Lie bialgebras and quantum groups. More recently, A. Polishchuk [10] discovered an interesting connection between solutions of CYBE and structural properties of derived categories of coherent sheaves on projective curves of arithmetic genus one. Further work on this connection was done by I. Burban and B. Kreussler [6] which includes some explicit examples and a relative version of Polishchuk's construction which can be used to explain degeneration of solutions.

Despite the developments in the past three decades, the results obtained by A. Belavin and V. Drinfeld in [2] are still fundamental for the study of solutions of CYBE. In this note we focus on non-degenerate solutions of CYBE and the trichotomy of solutions: rational, trigonometric and elliptic. Our contribution is a reformulation of the technical details of the proofs in [2] in a coordinate free way. We think that the key steps of the proofs become more transparent this way. We also filled in some details which we felt were a bit sketchy originally.

2. Preliminaries

All Lie algebras considered in this article will be finite dimensional complex Lie algebras. Let $\mathfrak{g}$ be a semi-simple finite dimensional complex Lie algebra. If $\beta : \mathfrak{g} \otimes \mathfrak{g} \longrightarrow \mathbb{C}$ is a non-degenerate (symmetric invariant) bilinear form, we denote by $\varphi^\beta : \mathfrak{g} \longrightarrow \mathfrak{g}^*$ the corresponding isomorphism, given by $\varphi^\beta(a) = \beta(a, \cdot)$. A bilinear form β is called *invariant* or *associative* if for all $a, b, c \in \mathfrak{g}$ we have $\beta(a, [b, c]) = \beta([a, b], c)$. If $\beta = \kappa$ is the Killing form, which is non-degenerate by Cartan's Theorem and symmetric and invariant, we simply write $\varphi = \varphi^\kappa$.

The isomorphisms $\mathbb{1}_\mathfrak{g} \otimes \varphi : \mathfrak{g} \otimes \mathfrak{g} \longrightarrow \mathfrak{g} \otimes \mathfrak{g}^*$ and $\mathbb{1}_\mathfrak{g} \otimes \varphi \otimes \varphi : \mathfrak{g} \otimes \mathfrak{g} \otimes \mathfrak{g} \longrightarrow \mathfrak{g} \otimes \mathfrak{g}^* \otimes \mathfrak{g}^*$ give us isomorphisms of vector spaces

$$\varphi_1 : \mathfrak{g} \otimes \mathfrak{g} \longrightarrow \mathsf{Hom}(\mathfrak{g}, \mathfrak{g}) \quad \varphi_2 : \mathfrak{g} \otimes \mathfrak{g} \otimes \mathfrak{g} \longrightarrow \mathsf{Hom}(\mathfrak{g} \otimes \mathfrak{g}, \mathfrak{g})$$

Journal of Physics: Conference Series **346** (2012) 012011 doi:10.1088/1742-6596/346/1/012011

which are explicitly given by

$$\varphi_1(a\otimes b)(x)=\kappa(b,x)\cdot a \quad \varphi_2(a\otimes b\otimes c)(x\otimes y)=\kappa(b,x)\kappa(c,y)\cdot a.$$

We call the tensor $\Omega\in\mathfrak{g}\otimes\mathfrak{g}$, which is characterised by $\varphi_1(\Omega)=\mathbb{1}_{\mathfrak{g}}$, the *Casimir element*. If $I_1,\dots,I_n$ is an orthonormal basis of $\mathfrak{g}$ with respect to the Killing form, then $\Omega=\sum_{\mu=1}^n I_\mu\otimes I_\mu$.

These definitions work for any non-degenerate symmetric invariant bilinear form β. If $\beta\neq\kappa$, we write $\varphi_1^\beta,\varphi_2^\beta$ and Ω^β instead of φ_1,φ_2 and Ω respectively. For $\lambda\in\mathbb{C}^*$ we then have $\varphi_1^{\lambda\beta}=\lambda\varphi_1^\beta,\varphi_2^{\lambda\beta}=\lambda^2\varphi_2^\beta$ and $\lambda\Omega^{\lambda\beta}=\Omega^\beta$. As we will later focus on simple Lie algebras, this shows that it is sufficient to work with κ, because on a simple Lie algebra over $\mathbb{C}$ any non-degenerate symmetric invariant bilinear form is a multiple of κ.

Example 2.1. The Lie algebra $\mathfrak{sl}(2)$, whose elements are the 2×2 matrices of trace zero, is simple. The standard basis elements

$$e=\begin{pmatrix}0&1\\0&0\end{pmatrix},\quad f=\begin{pmatrix}0&0\\1&0\end{pmatrix},\quad h=\begin{pmatrix}1&0\\0&-1\end{pmatrix}$$

enjoy the identities $[e,f]=h,[h,e]=2e,[h,f]=-2f$. The trace form with respect to the natural representation is given by $\beta(a,b)=\mathrm{tr}(ab)$ for $a,b\in\mathfrak{sl}(2)$. It is related to the Killing form $\kappa(a,b)=\mathrm{tr}\big(\mathrm{ad}(a)\,\mathrm{ad}(b)\big)$ by $\kappa=4\beta$. Hence, $\Omega=\frac{1}{4}\Omega^\beta$ and a straightforward calculation shows $4\Omega=\Omega^\beta=\frac{1}{2}h\otimes h+e\otimes f+f\otimes e$.

Lemma 2.2. *If $\mathfrak{g}$ is simple, $f\in\mathsf{Hom}(\mathfrak{g},\mathfrak{g})$ and $[\mathrm{ad}(x),f]=0$ for all $x\in\mathfrak{g}$, then $f\in\mathbb{C}\cdot\mathbb{1}_{\mathfrak{g}}$.*

Proof. This follows from Schur's Lemma, because the assumption that $[\mathrm{ad}(x),f]=0$ for all $x\in\mathfrak{g}$ is equivalent to $f:\mathfrak{g}\longrightarrow\mathfrak{g}$ being a homomorphism of $\mathfrak{g}$-modules. □

Lemma 2.3. *If $\mathfrak{g}$ is a simple Lie algebra and $\psi:\mathfrak{g}\longrightarrow\mathfrak{g}$ a homomorphism of Lie algebras, then* $\det(\psi)\in\{-1,0,1\}$.

Proof. Because $\mathfrak{g}$ is simple, $\ker(\psi)$ can only be 0 or $\mathfrak{g}$. Hence, $\det(\psi)=0$ iff $\psi=0$. Assume now $\psi\neq 0$, then ψ is an automorphism of $\mathfrak{g}$. We show that then $\kappa(\psi(x),\psi(y))=\kappa(x,y)$ for all $x,y\in\mathfrak{g}$. To do so, we rewrite $\psi[x,y]=[\psi(x),\psi(y)]$ as $\psi\circ\mathrm{ad}(x)=\mathrm{ad}\big(\psi(x)\big)\circ\psi$. This implies $\psi\circ\mathrm{ad}(x)\circ\psi^{-1}=\mathrm{ad}\big(\psi(x)\big)$ from which we obtain $\psi\circ\mathrm{ad}(x)\circ\mathrm{ad}(y)\circ\psi^{-1}=\mathrm{ad}\big(\psi(x)\big)\circ\mathrm{ad}\big(\psi(y)\big)$ for all $x,y\in\mathfrak{g}$. Because $\mathrm{tr}(MCM^{-1})=\mathrm{tr}(C)$, this implies $\kappa(\psi(x),\psi(y))=\kappa(x,y)$ for all $x,y\in\mathfrak{g}$. This means that $\psi^*=\psi^{-1}$, hence $\det(\psi)^2=1$ and the claim follows. □

Lemma 2.4. *Let $f,g\in\mathsf{Hom}(\mathfrak{g},\mathfrak{g})$ and denote by g^* the adjoint of g with respect to the Killing form. Then, for all $a,b\in\mathfrak{g}$,*

$$\varphi_1\big(f(a)\otimes g(b)\big)=f\circ\varphi_1(a\otimes b)\circ g^*.$$

This means that the following diagram is commutative

$$\begin{array}{ccc}\mathfrak{g}\otimes\mathfrak{g} & \xrightarrow{\varphi_1} & \mathsf{Hom}(\mathfrak{g},\mathfrak{g})\\ {\scriptstyle f\otimes g}\big\downarrow & & \big\downarrow{\scriptstyle f\circ\_\circ g^*}\\ \mathfrak{g}\otimes\mathfrak{g} & \xrightarrow{\varphi_1} & \mathsf{Hom}(\mathfrak{g},\mathfrak{g}).\end{array}$$

Proof. For $a,b,x\in\mathfrak{g}$ we have $\varphi_1\big(f(a)\otimes g(b)\big)(x)=\kappa(g(b),x)f(a)=f\big(\kappa(g(b),x)a\big)=f\big(\kappa(b,g^*(x))a\big)=f\big(\varphi_1(a\otimes b)(g^*(x))\big)$ where we have used the symmetry of κ. □

Corollary 2.5. *For each $A \in \mathfrak{g} \otimes \mathfrak{g}$ we have $(\varphi_1(A) \otimes \mathbb{1}_{\mathfrak{g}})(\Omega) = A$.*

Proof. Because $\varphi_1(\Omega) = \mathbb{1}_{\mathfrak{g}}$, we have $\varphi_1(\varphi_1(A) \otimes \mathbb{1}_{\mathfrak{g}})(\Omega) = \varphi_1(A) \circ \varphi_1(\Omega) = \varphi_1(A)$. The claim follows as φ_1 is an isomorphism. □

Lemma 2.6. *Let $\tau : \mathfrak{g} \otimes \mathfrak{g} \longrightarrow \mathfrak{g} \otimes \mathfrak{g}$ denote the swapping map $\tau(a \otimes b) = b \otimes a$ and $A \in \mathfrak{g} \otimes \mathfrak{g}$. Then $\varphi_1(\tau A) = \varphi_1(A)^*$ (the adjoint with respect to the Killing form).*

Proof. It is sufficient to consider $A = a \otimes b$. Using the symmetry of the Killing form, we obtain $\kappa(\varphi_1(a \otimes b)(x), y) = \kappa(\kappa(b, x)a, y) = \kappa(b, x)\kappa(a, y) = \kappa(x, \kappa(a, y)b) = \kappa(x, \varphi_1(b \otimes a)(y))$. □

A tensor $A \in \mathfrak{g} \otimes \mathfrak{g}$ is called *unitary* if $\tau A = -A$ as this is equivalent to $\varphi_1(A)^* = -\varphi_1(A)$.

Definition 2.7. A meromorphic function $r : U \to \mathfrak{g} \otimes \mathfrak{g}$, defined on an open disc $U \subset \mathbb{C}$ centred at the origin, is called *unitary* iff for all $u \in U$ at which r is holomorphic, $\tau r(u) = -r(-u)$.

A calculation similar to the one in the proof of Lemma 2.4 gives the following.

Lemma 2.8. *Let $f, g, h \in \mathsf{Hom}(\mathfrak{g}, \mathfrak{g})$ and denote by g^*, h^* the adjoint of g, h with respect to the Killing form. Then, for all $a, b, c \in \mathfrak{g}$,*

$$\varphi_2(f(a) \otimes g(b) \otimes h(c)) = f \circ \varphi_2(a \otimes b \otimes c) \circ (g^* \otimes h^*).$$

This means that the following diagram is commutative

$$\begin{array}{ccc} \mathfrak{g} \otimes \mathfrak{g} \otimes \mathfrak{g} & \xrightarrow{\varphi_2} & \mathsf{Hom}(\mathfrak{g} \otimes \mathfrak{g}, \mathfrak{g}) \\ \downarrow{\scriptstyle f \otimes g \otimes h} & & \downarrow{\scriptstyle f \circ \_ \circ (g^* \otimes h^*)} \\ \mathfrak{g} \otimes \mathfrak{g} \otimes \mathfrak{g} & \xrightarrow{\varphi_2} & \mathsf{Hom}(\mathfrak{g} \otimes \mathfrak{g}, \mathfrak{g}). \end{array}$$

Corollary 2.9. *For each $A \in \mathfrak{g} \otimes \mathfrak{g} \otimes \mathfrak{g}$ we have $(\varphi_2(A) \otimes \mathbb{1}_{\mathfrak{g}} \otimes \mathbb{1}_{\mathfrak{g}})(\Omega') = A$, where $\Omega' = \tau_{23}(\Omega \otimes \Omega)$ and $\tau_{23} : \mathfrak{g}^{\otimes 4} \longrightarrow \mathfrak{g}^{\otimes 4}$ is the swapping map $\tau_{23}(a_1 \otimes b_1 \otimes a_2 \otimes b_2) = a_1 \otimes a_2 \otimes b_1 \otimes b_2$.*

Proof. If we set $\varphi_2' = (\varphi_1 \otimes \varphi_1) \circ \tau_{23}$, then $\varphi_2'(\Omega') = \mathbb{1}_{\mathfrak{g}} \otimes \mathbb{1}_{\mathfrak{g}} = \mathbb{1}_{\mathfrak{g} \otimes \mathfrak{g}}$. A straightforward calculation shows $\varphi_2 \circ (\varphi_2(A) \otimes \mathbb{1}_{\mathfrak{g}} \otimes \mathbb{1}_{\mathfrak{g}}) = \varphi_2(A) \circ \varphi_2'$, from which we obtain the claim, because φ_2 is an isomorphism. □

Definition 2.10. We call $A \in \mathfrak{g} \otimes \mathfrak{g}$ *non-degenerate* iff $\det(\varphi_1(A)) \neq 0$.

For any $A \in \mathfrak{g} \otimes \mathfrak{g}$ we define $V_A = \mathrm{im}\,(\varphi_1(A)) \subset \mathfrak{g}$. In general, this is a subspace only, and we have

$$A \text{ is non-degenerate} \iff V_A = \mathfrak{g}.$$

If $V \subset \mathfrak{g}$ is a subspace and $A \in V \otimes \mathfrak{g}$, then it is clear from the definition of φ_1 that $V_A \subset V$. On the other hand, Corollary 2.5 implies $A \in V_A \otimes \mathfrak{g}$. Therefore, we have

$$A \in V \otimes \mathfrak{g} \iff V_A \subset V. \tag{2.1}$$

Similarly, using Corollary 2.9, we obtain for each subspace $V \subset \mathfrak{g}$ and $A \in \mathfrak{g} \otimes \mathfrak{g} \otimes \mathfrak{g}$

$$A \in V \otimes \mathfrak{g} \otimes \mathfrak{g} \iff \mathrm{im}(\varphi_2(A)) \subset V. \tag{2.2}$$

Definition 2.11. A meromorphic function $r : U \longrightarrow \mathfrak{g} \otimes \mathfrak{g}$, defined on an open subset $U \subset \mathbb{C}^n$, is called *non-degenerate* iff there exists $u_0 \in U$ such that the tensor $r(u_0)$ is non-degenerate, i.e. $V_{r(u_0)} = \mathfrak{g}$.

Remark 2.12. If $r : U \longrightarrow \mathfrak{g} \otimes \mathfrak{g}$ is non-degenerate, then the meromorphic function $u \mapsto \det(\varphi_1(r(u)))$ is not identically zero, hence vanishes only on a subset of codimension one in U. This means that a non-degenerate function r will have $r(u)$ non-degenerate for all u in an open dense subset of U, provided that U is connected.

Definition 2.13. If $\mathcal{A}$ is an associative unital $\mathbb{C}$-algebra, we define linear maps $\phi_{ij} : \mathcal{A} \otimes \mathcal{A} \longrightarrow \mathcal{A} \otimes \mathcal{A} \otimes \mathcal{A}$ for all $i, j \in \{1, 2, 3\}$ with $i > j$ by $\phi_{12}(a \otimes b) = a \otimes b \otimes 1$, $\phi_{13}(a \otimes b) = a \otimes 1 \otimes b$ and $\phi_{23}(a \otimes b) = 1 \otimes a \otimes b$. We shall often write $(a \otimes b)^{ij}$ instead of $\phi_{ij}(a \otimes b)$.

Any associative algebra $\mathcal{A}$ becomes a Lie algebra if we define a Lie-bracket by

$$[A, B] = A \cdot B - B \cdot A.$$

Any Lie algebra $\mathfrak{g}$ is a subalgebra of its universal enveloping algebra $U(\mathfrak{g})$, which is an associative unital algebra. Using the Lie bracket on $U(\mathfrak{g}) \otimes U(\mathfrak{g}) \otimes U(\mathfrak{g})$, for $a, b, c, d \in \mathfrak{g}$, we obtain the following expressions in $\mathfrak{g} \otimes \mathfrak{g} \otimes \mathfrak{g}$

$$\begin{aligned}
[(a \otimes b)^{12}, (c \otimes d)^{13}] &= [a, c] \otimes b \otimes d, \\
[(a \otimes b)^{12}, (c \otimes d)^{23}] &= a \otimes [b, c] \otimes d, \\
[(a \otimes b)^{13}, (c \otimes d)^{23}] &= a \otimes c \otimes [b, d].
\end{aligned}$$

The Classical Yang-Baxter Equation involves terms of this type only, see Section 3.

Proposition 2.14. *For $A, B \in \mathfrak{g} \otimes \mathfrak{g}$, $x, y \in \mathfrak{g}$ we have*

(a) $\varphi_2([A^{12}, B^{13}])(x \otimes y) = [\varphi_1(A)(x), \varphi_1(B)(y)]$

(b) $\varphi_2([A^{12}, B^{23}])(x \otimes y) = -\varphi_1(A)([x, \varphi_1(B)(y)])$

(c) $\varphi_2([A^{13}, B^{23}])(x \otimes y) = \varphi_1(A)([\varphi_1(\tau B)(x), y])$

(d) $\varphi_1([A, y \otimes 1])(x) = [\varphi_1(A)(x), y]$

(e) $\varphi_1([A, 1 \otimes y])(x) = \varphi_1(A)([y, x])$

Proof. It is sufficient to consider $A = a \otimes b$ and $B = c \otimes d$ with $a, b, c, d \in \mathfrak{g}$. Using the definitions, we obtain $\varphi_2([A^{12}, B^{13}])(x \otimes y) = \varphi_2([a, c] \otimes b \otimes d)(x \otimes y) = \kappa(b, x)\kappa(d, y) \cdot [a, c] = [\kappa(b, x)a, \kappa(d, y)c] = [\varphi_1(A)(x), \varphi_1(B)(y)]$.
To show (b), we use the invariance of the Killing form:

$$\begin{aligned}
&\varphi_2([A^{12}, B^{23}])(x \otimes y) = \varphi_2(a \otimes [b, c] \otimes d)(x \otimes y) = \kappa([b, c], x)\kappa(d, y) \cdot a \\
&= -\kappa(b, [x, c])\kappa(d, y) \cdot a = -\kappa(b, [x, \kappa(d, y)c]) \cdot a = -\varphi_1(A)([x, \varphi_1(B)(y)]).
\end{aligned}$$

For equation (c) we use the invariance again

$$\begin{aligned}
&\varphi_2([A^{13}, B^{23}])(x \otimes y) = \varphi_2(a \otimes c \otimes [b, d])(x \otimes y) = \kappa(c, x)\kappa([b, d], y) \cdot a \\
&= \kappa(c, x)\kappa(b, [d, y]) \cdot a = \kappa(b, [\kappa(c, x)d, y]) \cdot a = \varphi_1(A)([\varphi_1(\tau B)(x), y]).
\end{aligned}$$

Finally, (d) follows directly from the definition and (e) with the aid of the invariance:

$$\begin{aligned}
\varphi_1([a \otimes b, y \otimes 1])(x) &= \varphi_1([a, y] \otimes b)(x) = \kappa(b, x)[a, y] = [\kappa(b, x)a, y] = [\varphi_1(a \otimes b)(x), y] \\
\varphi_1([a \otimes b, 1 \otimes y])(x) &= \varphi_1(a \otimes [b, y])(x) = \kappa([b, y], x)a = \kappa(b, [y, x])a = \varphi_1(a \otimes b)([y, x]).
\end{aligned}$$

□

Corollary 2.15. *Let $\mathfrak{g}$ be a simple Lie algebra and $A \in \mathfrak{g} \otimes \mathfrak{g}$.*

(a) $[\Omega^{12}, A^{13} + A^{23}] = 0$ *and* $[A^{12} + A^{13}, \Omega^{23}] = 0$
(b) $[\Omega^{12}, \Omega^{13}] = -[\Omega^{12}, \Omega^{23}] = [\Omega^{13}, \Omega^{23}]$
(c) $\varphi_2([\Omega^{12}, \Omega^{13}])(x \otimes y) = [x, y]$ *for all* $x, y \in \mathfrak{g}$
(d) $[A^{12}, \Omega^{23}] = 0$ *implies* $A = 0$
(e) $[\Omega^{12} - \Omega^{23}, A^{13}] = 0$ *implies* $A = 0$

Proof. For $x, y \in \mathfrak{g}$ we obtain from Proposition 2.14

$$\begin{aligned}\varphi_2([\Omega^{12}, A^{13} + A^{23}])(x \otimes y) &= [\varphi_1(\Omega)(x), \varphi_1(A)(y)] - \varphi_1(\Omega)([x, \varphi_1(A)(y)]) \\ &= [x, \varphi_1(A)(y)] - [x, \varphi_1(A)(y)] = 0,\end{aligned}$$

hence $[\Omega^{12}, A^{13} + A^{23}] = 0$. The proof of the second equality in (a) is similar, but uses $\tau\Omega = \Omega$, which follows from Lemma 2.6. With $A = \Omega$ in (a) we obtain (b). Part (c) is a direct consequence of Proposition 2.14 (a) and $\varphi_1(\Omega) = \mathbb{1}_{\mathfrak{g}}$.

If $[A^{12}, \Omega^{23}] = 0$, we obtain from Proposition 2.14 (b) that $\varphi_1(A)[x, y] = 0$ for all $x, y \in \mathfrak{g}$. As $\mathfrak{g}$ is simple, $[\mathfrak{g}, \mathfrak{g}] = \mathfrak{g}$, and so we conclude $\varphi_1(A) = 0$, hence $A = 0$ and (d) is shown.

Finally, using Proposition 2.14, we translate $[\Omega^{12} - \Omega^{23}, A^{13}] = 0$ into

$$[x, \varphi_1(A)(y)] + \varphi_1(A)[x, y] = 0 \tag{2.3}$$

for all $x, y \in \mathfrak{g}$, which can be rewritten as $\mathrm{ad}(x) \circ \varphi_1(A) + \varphi_1(A) \circ \mathrm{ad}(x) = 0$ for all $x \in \mathfrak{g}$. If we apply $\mathrm{ad}(y) \circ \_ + \_ \circ \mathrm{ad}(y)$ to this identity, we obtain for all $x, y \in \mathfrak{g}$

$$\begin{aligned}\mathrm{ad}(y) \circ \mathrm{ad}(x) \circ \varphi_1(A) + \varphi_1(A) \circ \mathrm{ad}(x) \circ \mathrm{ad}(y) + \mathrm{ad}(x) \circ \varphi_1(A) \circ \mathrm{ad}(y) \\ + \mathrm{ad}(y) \circ \varphi_1(A) \circ \mathrm{ad}(x) = 0.\end{aligned}$$

Taking the difference between this expression and the one obtained from it by swapping x and y gives $\mathrm{ad}([x, y]) \circ \varphi_1(A) - \varphi_1(A) \circ \mathrm{ad}([x, y]) = 0$. As $\mathfrak{g}$ is simple, we have $[\mathfrak{g}, \mathfrak{g}] = \mathfrak{g}$ and so $[\mathrm{ad}(x), \varphi_1(A)] = 0$ for all $x \in \mathfrak{g}$. Lemma 2.2 implies now that $\varphi_1(A) = \lambda \mathbb{1}_{\mathfrak{g}}$ for some $\lambda \in \mathbb{C}$. Using this in (2.3), we obtain $2\lambda[x, y] = 0$ for all $x, y \in \mathfrak{g}$, hence $\lambda = 0$ and so also $A = 0$. □

Lemma 2.16. *If* $A \in \mathfrak{g} \otimes \mathfrak{g}$ *is non-degenerate, then* $[A^{12}, A^{13}] \neq 0$.

Proof. If $[A^{12}, A^{13}] = 0$, Proposition 2.14 (a) implies $[\varphi_1(A)(x), \varphi_1(A)(y)] = 0$ for all $x, y \in \mathfrak{g}$. This means $[X, Y] = 0$ for all $X, Y \in V_A$. As A is non-degenerate, we have $V_A = \mathfrak{g}$. But $\mathfrak{g}$ is not abelian, hence $[A^{12}, A^{13}]$ cannot vanish. □

Lemma 2.17. *Let* $A, B \in \mathfrak{g} \otimes \mathfrak{g}$ *and* $\mathfrak{a} = \{x \in \mathfrak{g} \mid [x, v] \in V_B$ *for all* $v \in V_B\}$*, then*

(a) $\mathfrak{a} \subset \mathfrak{g}$ *is a Lie subalgebra and* $\mathfrak{a} = \mathfrak{g}$ *if and only if* V_B *is an ideal in* $\mathfrak{g}$*;*
(b) If $[B^{12}, A^{13}] \in V_B \otimes \mathfrak{g} \otimes \mathfrak{g}$*, then* $A \in \mathfrak{a} \otimes \mathfrak{g}$*;*
(c) If $[A^{12}, B^{23}] \in \mathfrak{g} \otimes V_B \otimes \mathfrak{g}$*, then* $A \in \mathfrak{g} \otimes \mathfrak{a}$.

Proof. The proof of (a) is a consequence of the Jacobi identity and the definitions. If $[B^{12}, A^{13}] \in V_B \otimes \mathfrak{g} \otimes \mathfrak{g}$, the image of $\varphi_2([B^{12}, A^{13}])$ is contained in V_B, see (2.2). Proposition 2.14 (a) implies now that $[\varphi_1(B)(x), \varphi_1(A)(y)] \in V_B$ for all $x, y \in \mathfrak{g}$. As $V_B = \mathrm{im}(\varphi_1(B))$ this means that $\varphi_1(A)(y) \in \mathfrak{a}$ for all $y \in \mathfrak{g}$, i.e. $V_A \subset \mathfrak{a}$. This implies $A \in \mathfrak{a} \otimes \mathfrak{g}$, see (2.1), hence we have shown (b). To prove (c) we use the swapping operator $\tau_{12} : \mathfrak{g} \otimes \mathfrak{g} \longrightarrow \mathfrak{g} \otimes \mathfrak{g}$, given by $\tau_{12}(a \otimes b \otimes c) = b \otimes a \otimes c$. Because $\tau_{12}[A^{12}, B^{23}] = [(\tau A)^{12}, B^{13}]$, the assumption implies $[(\tau A)^{12}, B^{13}] \in V_B \otimes \mathfrak{g} \otimes \mathfrak{g}$. Like before, this implies $[\varphi_1(\tau A)(x), \varphi_1(B)(y)] \in V_B$ for all $x, y \in \mathfrak{g}$, from which we get $\varphi_1(\tau A)(x) \in \mathfrak{a}$ for all $x \in \mathfrak{g}$. Therefore, $\tau A \in \mathfrak{a} \otimes \mathfrak{g}$, i.e. $A \in \mathfrak{g} \otimes \mathfrak{a}$. □

Lemma 2.18. *Let $A, B \in \mathfrak{g}\otimes\mathfrak{g}$ such that $[B^{12}, A^{13}+A^{23}] = 0$, then $A \in \mathfrak{a}\otimes\mathfrak{g}$, where $\mathfrak{a} = \{x \in \mathfrak{g} \mid [\mathrm{ad}(x), \varphi_1(B)] = 0\} \subset \mathfrak{g}$, which is a Lie subalgebra.*

Proof. Because $\mathrm{ad} : \mathfrak{g} \longrightarrow \mathsf{End}(\mathfrak{g})$ is a homomorphism of Lie algebras and centralisers are subalgebras, $\mathfrak{a} \subset \mathfrak{g}$ is a Lie subalgebra. If $[B^{12}, A^{13}+A^{23}] = 0$, we obtain from Proposition 2.14 for all $x, y \in \mathfrak{g}$

$$\begin{aligned}0 &= \varphi_2\big([B^{12}, A^{13}+A^{23}]\big)(x\otimes y) = [\varphi_1(B)(x), \varphi_1(A)(y)] - \varphi_1(B)\big[x, \varphi_1(A)(y)\big]\\ &= \varphi_1(B)\big[\varphi_1(A)(y), x\big] - \big[\varphi_1(A)(y), \varphi_1(B)(x)\big]\\ &= \Big(\varphi_1(B)\circ\mathrm{ad}\big(\varphi_1(A)(y)\big) - \mathrm{ad}\big(\varphi_1(A)(y)\big)\circ\varphi_1(B)\Big)(x),\end{aligned}$$

hence $\big[\varphi_1(B), \mathrm{ad}\big(\varphi_1(A)(y)\big)\big] = 0$ and so $\varphi_1(A)(y) \in \mathfrak{a}$ for all $y \in \mathfrak{g}$. Because of (2.1) this implies $A \in \mathfrak{a}\otimes\mathfrak{g}$. □

3. The Classical Yang Baxter Equation

The Classical Yang Baxter Equation, referred to as (CYBE), is the equation

$$[r^{12}(u), r^{13}(u+v)] + [r^{12}(u), r^{23}(v)] + [r^{13}(u+v), r^{23}(v)] = 0 \tag{3.1}$$

for $r : U \longrightarrow \mathfrak{g}\otimes\mathfrak{g}$ being a meromorphic function defined on an open neighbourhood $U \subset \mathbb{C}$ of the origin. In this equation, it is assumed that $u, v, u+v \in U$. This equation makes also sense if U is a neighbourhood of the origin in $\mathbb{C}^n$, but should not be confused with the classical Yang Baxter equation with two spectral parameters if $n = 2$.

Throughout this section, $\mathfrak{g}$ denotes a finite-dimensional simple complex Lie algebra.

Lemma 3.1. *If $r : U \longrightarrow \mathfrak{g}\otimes\mathfrak{g}$ is a non-degenerate holomorphic solution of (CYBE), defined on an open disc $U \subset \mathbb{C}$ centred at the origin, then $r(0)$ is a non-degenerate tensor.*

Proof. There exists $u_0 \in U$ such that $r(u_0)$ is non-degenerate. Let us denote $T_u = \varphi_1(r(u))$ and $\psi_u = T_u \circ T_{u_0}^{-1}$ for all $u \in U$. We show now that ψ_u is a homomorphism of Lie algebras.

Let $v = 0$ in (3.1) and apply φ_2. Using Proposition 2.14, this produces

$$[T_u(x), T_u(y)] = T_u\big([x, T_0(y)] - [T_0^*(x), y]\big).$$

Setting $u = u_0$ and applying ψ_u makes this into

$$\psi_u[T_{u_0}(x), T_{u_0}(y)] = T_u\big([x, T_0(y)] - [T_0^*(x), y]\big).$$

Both equations have the same right hand side. Therefore, with $X = T_{u_0}(x)$ and $Y = T_{u_0}(y)$ we obtain $\psi_u[X, Y] = [\psi_u(X), \psi_u(Y)]$. Because T_{u_0} is an isomorphism, this equation shows that ψ is a homomorphism of Lie algebras. Because $\det(\psi_u)$ is a continuous function on U which can, by Lemma 2.3, only take the three values $-1, 0, 1$, it must be constant. Therefore $\det(\psi_0) = \det(\psi_{u_0}) = 1$ and T_0 is invertible. This shows that $r(0)$ is non-degenerate. □

3.1. Constant Solutions

Let $\mathfrak{g}$ be a simple Lie algebra. The aim of this subsection is to show that the Classical Yang-Baxter Equation (3.1) does not have non-degenerate constant solutions. This result will be needed in the proof of Theorem 3.6.

A constant solution of (3.1) is given by a tensor $r_0 \in \mathfrak{g}\otimes\mathfrak{g}$ which satisfies

$$[r_0^{12}, r_0^{13}] + [r_0^{12}, r_0^{23}] + [r_0^{13}, r_0^{23}] = 0. \tag{3.2}$$

Within this subsection we shall frequently use the abbreviation $T = \varphi_1(r_0)$. A tensor r_0 is unitary iff $\tau r_0 = -r_0$, or equivalently, if the adjoint with respect to the Killing form on $\mathfrak{g}$ satisfies $T^* = -T$, see Lemma 2.6.

Lemma 3.2. *Let $r_0 \in \mathfrak{g} \otimes \mathfrak{g}$ and $T = \varphi_1(r_0)$.*
(a) Equation (3.2) is equivalent to

$$[T(x), T(y)] - T\Big([x, T(y)] - [T^*(x), y]\Big) = 0 \qquad \forall\, x, y \in \mathfrak{g}. \tag{3.3}$$

(b) For $\lambda \in \mathbb{C}$, the equation $r_0 + \tau r_0 = \lambda\Omega$ is equivalent to $T + T^* = \lambda \mathbb{1}_{\mathfrak{g}}$.
(c) If $\lambda \in \mathbb{C}$ and $r_0 + \tau r_0 = \lambda\Omega$, then equation (3.2) is equivalent to

$$[T(x), T(y)] = T\Big([x, T(y)] + [T(x), y] - \lambda[x, y]\Big) \qquad \forall\, x, y \in \mathfrak{g}. \tag{3.4}$$

Proof. From Proposition 2.14 we obtain

$$\varphi_2\big([r_0^{12}, r_0^{13}]\big)(x \otimes y) = [T(x), T(y)], \quad \varphi_2\big([r_0^{12}, r_0^{23}]\big)(x \otimes y) = -T\,[x, T(y)]$$
$$\text{and} \quad \varphi_2\big([r_0^{13}, r_0^{23}]\big)(x \otimes y) = T\,[T^*(x), y] \quad \text{which gives (a).}$$

Part (b) follows from Lemma 2.6, and (c) is clear from (a) and (b). □

We need the following result, a proof of which can be found in [4, Theorem 9.2].

Theorem 3.3. *If $\psi : \mathfrak{g} \longrightarrow \mathfrak{g}$ is an automorphism of the simple Lie algebra $\mathfrak{g}$, then there exists a non-zero element $x \in \mathfrak{g}$ such that $\psi(x) = x$, i.e. $\det(\psi - \mathbb{1}_{\mathfrak{g}}) = 0$.*

Lemma 3.4. *If $r_0 \in \mathfrak{g} \otimes \mathfrak{g}$ is a non-degenerate solution of (3.2), then it is unitary.*

Proof. With notation as above, we have equation (3.3) for all $x, y \in \mathfrak{g}$. Interchanging x and y and using the skew-symmetry of the Lie-bracket, we obtain from it

$$-[T(x), T(y)] - T\Big(-[T(x), y] + [x, T^*(y)]\Big) = 0. \tag{3.5}$$

Adding (3.3) and (3.5) and using that T is an isomorphism by assumption, we obtain $\big[(T+T^*)(x), y\big] = \big[x, (T+T^*)(y)\big]$ for all $x, y \in \mathfrak{g}$. This equation implies, using the associativity of the Killing form and that $(T + T^*)$ is self-adjoint,

$$\begin{aligned}\kappa\big(x, (T+T^*)[y,z]\big) &= \kappa\big((T+T^*)(x), [y,z]\big) = \kappa\big(\big[(T+T^*)(x), y\big], z\big)\\ &= \kappa\big(\big[x, (T+T^*)(y)\big], z\big) = \kappa\big(x, \big[(T+T^*)(y), z\big]\big)\end{aligned}$$

for each $z \in \mathfrak{g}$. As κ is non-degenerate, we obtain $(T+T^*)[y,z] = \big[(T+T^*)(y), z\big]$ for all $y, z \in \mathfrak{g}$, i.e. $(T + T^*) \circ \mathrm{ad}(z) = \mathrm{ad}(z) \circ (T + T^*)$ for all $z \in \mathfrak{g}$. As $\mathfrak{g}$ is simple, Lemma 2.2 implies now that $(T + T^*) = \lambda \cdot \mathbb{1}_{\mathfrak{g}}$ for some $\lambda \in \mathbb{C}$.

We have to show that $\lambda = 0$. For a proof by contradiction, we assume $\lambda \neq 0$. Because we could then replace r_0 by r_0/λ, we may even assume $\lambda = 1$. In this case, subtracting $T[T(x), T(y)]$ from both sides of (3.4) gives

$$(T - \mathbb{1}_{\mathfrak{g}})\big[T(x), T(y)\big] = T\big[(T - \mathbb{1}_{\mathfrak{g}})(x), (T - \mathbb{1}_{\mathfrak{g}})(y)\big]. \tag{3.6}$$

As T was assumed to be invertible, $T^* = \mathbb{1}_{\mathfrak{g}} - T$ is invertible as well. Let $\psi = T \circ (T - \mathbb{1}_{\mathfrak{g}})^{-1} = (T - \mathbb{1}_{\mathfrak{g}})^{-1} \circ T$ and apply $(T - \mathbb{1}_{\mathfrak{g}})^{-1}$ to (3.6). This gives

$$\big[T(x), T(y)\big] = \psi\big[(T - \mathbb{1}_{\mathfrak{g}})(x), (T - \mathbb{1}_{\mathfrak{g}})(y)\big].$$

If we replace x, y with $(T - \mathbb{1}_{\mathfrak{g}})^{-1}(x)$ and $(T - \mathbb{1}_{\mathfrak{g}})^{-1}(y)$ respectively, we arrive at $[\psi(x), \psi(y)] = \psi[x, y]$, i.e. ψ is an automorphism of Lie algebras. We have $\psi - \mathbb{1}_{\mathfrak{g}} = T \circ (T - \mathbb{1}_{\mathfrak{g}})^{-1} - (T - \mathbb{1}_{\mathfrak{g}}) \circ (T - \mathbb{1}_{\mathfrak{g}})^{-1} = (T - \mathbb{1}_{\mathfrak{g}})^{-1}$, hence $\det(\psi - \mathbb{1}_{\mathfrak{g}}) \neq 0$ in contradiction to Theorem 3.3. This proves that $\lambda = 0$ and so T is unitary. □

Proposition 3.5. *There are no non-degenerate solutions of* (3.2).

Proof. Assume r_0 is a non-degenerate solutions of (3.2) and let $T = \varphi_1(r_0)$ as before and $S = T^{-1}$. By Lemma 3.4, T is unitary. Therefore, from Lemma 3.2 (c) with $\lambda = 0$ we have $[T(x), T(y)] = T[x, T(y)] + T[T(x), y]$. Applying S and replacing x, y with $S(x)$ and $S(y)$ respectively, gives

$$S[x, y] = [S(x), y] + [x, S(y)],$$

that is S is a derivation of $\mathfrak{g}$. For semi-simple $\mathfrak{g}$ each derivation is equal to $\mathrm{ad}(a)$ for some $a \in \mathfrak{g}$. But $\mathrm{ad}(a)$ is never an isomorphism, so it cannot be equal to S. This contradiction shows that there cannot be a non-degenerate solutions r_0 of (3.2). □

3.2. Characterisation of non-degeneracy

In the proof of the theorem below we use the following identity, which holds for arbitrary holomorphic functions $f, g : U \longrightarrow \mathfrak{g}$ defined on an open set $U \subset \mathbb{C}$ and which follows from the product rule for derivatives:

$$\frac{d}{du}[f(u), g(u)] = [f'(u), g(u)] + [f(u), g'(u)]. \tag{3.7}$$

Theorem 3.6 (Belavin-Drinfeld). *Let $\mathfrak{g}$ be a finite-dimensional simple complex Lie algebra, $U \subset \mathbb{C}$ an open disc with centre 0 and $r : U \longrightarrow \mathfrak{g} \otimes \mathfrak{g}$ a meromorphic solution of CYBE. Then the following are equivalent.*

(A) The solution r is non-degenerate.

(B) The function r has at least one pole in U and there does not exist a proper Lie subalgebra $\mathfrak{a} \subset \mathfrak{g}$ such that $r(u) \in \mathfrak{a} \otimes \mathfrak{a}$ for all u in a small neighbourhood of $0 \in U$.

(C) All poles of r are simple and the residue of r at 0 is equal to $\lambda\Omega$ for some $\lambda \in \mathbb{C}^*$.

(D) The function r has a simple pole at 0 with residue equal to $\lambda\Omega$ for some $\lambda \in \mathbb{C}^*$.

Proof. (A)$\Longrightarrow$(B) If r is a non-degenerate holomorphic solution of (3.1), Lemma 3.1 implies that $r(0)$ is a non-degenerate constant solution of (3.1). This contradicts Proposition 3.5, hence r must have at least one pole in U.

Assume $\mathfrak{a} \subset \mathfrak{g}$ is a Lie subalgebra such that there exists a small neighbourhood U' of 0 such that $r(u) \in \mathfrak{a} \otimes \mathfrak{a}$ for all $u \in U'$ at which r is holomorphic. For such u, $V_{r(u)} \subset \mathfrak{a}$ and so $r(u)$ can only be non-degenerate if $\mathfrak{a} = \mathfrak{g}$. As r was assumed to be non-degenerate, $r(u)$ must be non-degenerate for $u \in U'$ except at finitely many points, see Remark 2.12. Hence, $\mathfrak{a} = \mathfrak{g}$.

(B)$\Longrightarrow$(C) Let $\gamma \in U$ be a pole of r and $k \geq 1$ the order of this pole. Let $\theta_\gamma = \lim_{u \to \gamma}(u-\gamma)^k r(u)$ be the leading coefficient of the Laurent series of r at γ. If we multiply (3.1) by $(v - \gamma)^k$ and let v tend to γ, we obtain

$$[r^{12}(u) + r^{13}(u + \gamma), \theta_\gamma^{23}] = 0 \tag{3.8}$$

for all $u \in U$ with $u + \gamma \in U$. Similarly, after multiplication by $(u - \gamma)^k$ we obtain

$$[\theta_\gamma^{12}, r^{13}(v + \gamma) + r^{23}(v)] = 0 \tag{3.9}$$

for all $v \in U$ for which $v + \gamma \in U$. Recall that $V_{\theta_\gamma} = \mathrm{im}\,(\varphi_1(\theta_\gamma)) \subset \mathfrak{g}$ is equal to $\mathfrak{g}$ if and only if θ_γ is a non-degenerate element of $\mathfrak{g} \otimes \mathfrak{g}$.

Note that $[\theta_\gamma^{12}, r^{23}(v)] \in V_{\theta_\gamma} \otimes \mathfrak{g} \otimes \mathfrak{g}$ and $[r^{13}(u + \gamma), \theta_\gamma^{23}] \in \mathfrak{g} \otimes V_{\theta_\gamma} \otimes \mathfrak{g}$, because $\theta_\gamma \in V_{\theta_\gamma} \otimes \mathfrak{g}$. From (3.8) and (3.9) we therefore get $[\theta_\gamma^{12}, r^{13}(v+\gamma)] \in V_{\theta_\gamma} \otimes \mathfrak{g} \otimes \mathfrak{g}$ and $[r^{12}(u), \theta_\gamma^{23}] \in \mathfrak{g} \otimes V_{\theta_\gamma} \otimes \mathfrak{g}$. If $\mathfrak{a} = \{x \in \mathfrak{g} \mid [x, v] \in V \text{ for all } v \in V\}$, Lemma 2.17 implies that $r(u) \in \mathfrak{g} \otimes \mathfrak{a}$ and $r(v+\gamma) \in \mathfrak{a} \otimes \mathfrak{g}$ for all $u \in U$ for which $u+\gamma \in U$ and for all $v \in U$ with $v+\gamma \in U$. Hence, $r(u) \in \mathfrak{a} \otimes \mathfrak{g} \cap \mathfrak{g} \otimes \mathfrak{a} = \mathfrak{a} \otimes \mathfrak{a}$

whenever $u-\gamma, u, u+\gamma \in U$. As U is an open disc with centre 0 and $\gamma \in U$, this condition is satisfied for u in a small neighbourhood of 0. According to our assumption this implies $\mathfrak{a} = \mathfrak{g}$, hence $V_{\theta_\gamma} \subset \mathfrak{g}$ is an ideal. As $\mathfrak{g}$ is simple and $\theta_\gamma \neq 0$ by definition, we get $V_{\theta_\gamma} = \mathfrak{g}$, i.e. θ_γ is non-degenerate.

By Lemma 2.16 we obtain $[\theta_\gamma^{12}, \theta_\gamma^{13}] \neq 0$. If r does not have a pole of order at least k at the origin, multiplying (3.9) by v^k and letting v tend to 0 gives us $[\theta_\gamma^{12}, \theta_\gamma^{13}] = 0$. Hence, r has a pole of order l at the origin and $l \geq k \geq 1$. Therefore, we may set $\gamma = 0$ above and see that $\theta = \theta_0 = \lim_{u\to 0} u^l r(u)$ is non-degenerate and $[\theta^{12}, \theta^{13}] \neq 0$.

Assume $l \geq 2$, then multiplying (3.1) by u^l gives the equation

$$[u^l r^{12}(u), r^{13}(u+v) + r^{23}(v)] + u^l[r^{13}(u+v), r^{23}(v)] = 0 \tag{3.10}$$

in which all ingredients, including $u^l r^{12}(u)$, are holomorphic functions of u in a small neighbourhood of 0, provided that $v \in U$ is fixed and close to 0. Using (3.7) and $l \geq 2$, the derivative of (3.10) with respect to u at $u = 0$ evaluates as

$$[\theta^{12}, \left(r'(v)\right)^{13}] + [A^{12}, r^{13}(v) + r^{23}(v)] = 0$$

where A denotes the derivative of $u^l r(u)$ at 0. This equation holds for all $v \neq 0$ in a small neighbourhood of the origin. Note that the leading term of $r'(v)$ in its Laurent expansion is $-l\theta/v^{l+1}$. If we multiply the above equation by v^{l+1} and let v tend to 0 we obtain $[\theta^{12}, -l\theta^{13}] = 0$, in contradiction to $[\theta^{12}, \theta^{13}] \neq 0$. Therefore, $l = 1$ and so also $k = 1$. This proves the first part of (B).

To find the residue of the simple pole at the origin, we consider (3.9) with $\gamma = 0$ and the equation obtained from (3.8) with $\gamma = 0$ by applying the cyclic permutation operator $\tau_{231}(a \otimes b \otimes c) = b \otimes c \otimes a$, i.e.

$$[\theta^{12}, r^{13}(u) + r^{23}(u)] = 0 \qquad \text{and} \qquad [\theta^{12}, (\tau r)^{13}(u) + (\tau r)^{23}(u)] = 0. \tag{3.11}$$

If we define $\mathfrak{a} = \{x \in \mathfrak{g} \mid [\mathrm{ad}(x), \varphi_1(\theta)] = 0\}$, by Lemma 2.18, the equations (3.11) imply $r(u), \tau r(u) \in \mathfrak{a} \otimes \mathfrak{g}$, hence $r(u) \in \mathfrak{a} \otimes \mathfrak{g} \cap \mathfrak{g} \otimes \mathfrak{a} = \mathfrak{a} \otimes \mathfrak{a}$ for all $u \in U$ where r is holomorphic. By assumption this can only happen if $\mathfrak{a} = \mathfrak{g}$ and so $\varphi_1(\theta) = \lambda \mathbb{1}_{\mathfrak{g}}$ by Lemma 2.2 for some $\lambda \in \mathbb{C}$. Hence, $\theta = \lambda\Omega$ with $\lambda \in \mathbb{C}$ and $\lambda \neq 0$ as $\theta \neq 0$.

(C)$\Longrightarrow$(D) This is obvious.

(D)$\Longrightarrow$(A) By assumption, $\lim_{u\to 0} u \cdot r(u) = \lambda\Omega$, $\lambda \in \mathbb{C}^*$, hence $\lim_{u\to 0} u \cdot \varphi_1(r(u)) = \lambda \mathbb{1}_{\mathfrak{g}}$, and so $\lim_{u\to 0} \det(u \cdot \varphi_1(r(u))) \neq 0$. For $u \neq 0$ in a small neighbourhood of the origin we obtain $\det(\varphi_1(r(u))) \neq 0$, i.e. $r(u)$ is non-degenerate for such u. This means that r is non-degenerate. □

Proposition 3.7. *Let $\mathfrak{g}$ be simple and $U \subset \mathbb{C}^n$ an open neighbourhood of the origin. Then, all non-degenerate meromorphic solutions $r : U \longrightarrow \mathfrak{g} \otimes \mathfrak{g}$ of* (3.1) *are unitary, i.e. satisfy $\tau r(u) = -r(-u)$.*

Proof. If we replace u, v by $-u$ and $u+v$ respectively, we obtain from (3.1), after applying the swapping operator $\tau_{12}(a \otimes b \otimes c) = b \otimes a \otimes c$, the equation

$$\left[(\tau r)^{12}(-u), r^{23}(v)\right] + \left[(\tau r)^{12}(-u), r^{13}(u+v)\right] + \left[r^{23}(v), r^{13}(u+v)\right].$$

We choose v such that $r(v)$ is non-degenerate. If we add this to (3.1), we obtain

$$\left[\left(r(u) + \tau r(-u)\right)^{12}, r^{23}(v) + r^{13}(u+v)\right] = 0.$$

We replace v by tv with $t \in \mathbb{C}$, multiply the above expression by t and take the limit if t approaches 0. Because r, restricted to the one-dimensional subspace $\langle v \rangle \subset \mathbb{C}^n$, has (by Theorem 3.6) a simple pole at the origin with residue $\lambda\Omega \neq 0$, we obtain

$$\left[\big(r(u) + \tau r(-u)\big)^{12}, \lambda\Omega^{23}\right] = 0.$$

The claim follows now from Corollary 2.15 (d). □

4. The Classification Theorem of Belavin and Drinfeld

The purpose of this section is to prove the main classification result of Belavin and Drinfeld, Theorem 4.8. The proof presented here is a reformulation of the original proof in [2], with some details filled in. We have benefited from other versions of this proof in [4], [7] and [9]. Throughout this section, $\mathfrak{g}$ denotes a finite-dimensional simple complex Lie algebra.

Definition 4.1. Two meromorphic functions $r, s : \mathbb{C} \longrightarrow \mathfrak{g} \otimes \mathfrak{g}$ are called *equivalent* if there exists a holomorphic function $\psi : \mathbb{C} \longrightarrow \mathsf{Aut}(\mathfrak{g})$ such that, for all $u_1, u_2 \in \mathbb{C}$

$$s(u_1 - u_2) = \big(\psi(u_1) \otimes \psi(u_2)\big) r(u_1 - u_2).$$

It is easy to see that s is a solution of (3.1) if and only if r is so and that s is non-degenerate if and only if r is so. Note that, for general ψ, the right hand side may depend on (u_1, u_2) and not only on the difference $u_1 - u_2$.

For the proof of the main theorem, we need to introduce birational group laws and their relation to algebraic groups. This theory goes back to A. Weil [11]. Generalisations of his results can be found in [1] and [5] which are written in a language more familiar nowadays.

Definition 4.2 ([5], §5.1, Def. 1). A *birational group law* on a scheme X over $\mathbb{C}$ is a rational map $P : X \times X \longrightarrow X$, written as $P(x, y) = x * y$, such that

(i) the rational maps $\Phi : X \times X \longrightarrow X \times X$, given by $\Phi(x, y) = (x, x * y)$ and $\Psi : X \times X \longrightarrow X \times X$, given by $\Psi(x, y) = (x * y, y)$ are birational;

(ii) P is associative, i.e. $(x * y) * z = x * (y * z)$ whenever both sides are defined.

The birational group law P is called *commutative* if $x * y = y * x$ whenever both sides are defined.

A typical example is a dense open subset of a group scheme. The relation to group schemes is clarified by Weil's Theorem which is given below. Because the neutral element and some inverses could be missing in X, the usual requirement that such elements exist is replaced by condition (i).

Theorem 4.3 (A. Weil [11], see also [5], §5.1, Thm. 5). *Let (X, P) be a birational group law on a smooth separated scheme X of finite type over $\mathbb{C}$. Then there exists an algebraic group G, an open dense subscheme $X' \subset X$ and an open dense immersion $X' \subset G$ such that the restriction of the group law of G coincides with P on X'.*

If P was commutative then G will also be a commutative algebraic group, because commutativity on an open dense subset implies commutativity everywhere.

Proposition 4.4. *Let $\mathfrak{g}$ be a finite-dimensional simple complex Lie algebra, $U \subset \mathbb{C}^n$ an open neighbourhood of the origin which is convex and invariant under the involution $u \mapsto -u$. Let $\widetilde{U} = \{(u, v) \mid u + v \in U\} \subset U \times U$ and $r : U \longrightarrow \mathfrak{g} \otimes \mathfrak{g}$ be a non-degenerate meromorphic solution of CYBE.*

Then, there exist rational functions $P, Q : (\mathfrak{g} \otimes \mathfrak{g}) \times (\mathfrak{g} \otimes \mathfrak{g}) \longrightarrow (\mathfrak{g} \otimes \mathfrak{g})$ such that

$$r(u + v) = P\big(r(u), r(v)\big) \qquad \textit{for all } (u, v) \in \widetilde{U} \textit{ and}$$
$$r(u - v) = Q\big(r(u), r(v)\big) \qquad \textit{for all } (u, -v) \in \widetilde{U}.$$

Proof. Let $Y_{u,v} : \mathfrak{g} \otimes \mathfrak{g} \longrightarrow \mathfrak{g} \otimes \mathfrak{g} \otimes \mathfrak{g}$ denote the linear map which is given by $Y_{u,v}(A) = [r^{12}(u) - r^{23}(v), A^{13}]$. We first show that $Y_{u,v}$ is injective for (u, v) in an open subset of $\widetilde{U}$. To do so, we fix $u_0 \in U$ with $(u_0, u_0) \in \widetilde{U}$ such that $r(u_0)$ is non-degenerate and consider the family of linear maps tY_{tu_0,tu_0}, which depend holomorphically on $t \in \mathbb{C}$ in a small neighbourhood of 0. Without loss of generality, we may assume that the restriction of r to the one-dimensional subspace $\langle u_0 \rangle \subset \mathbb{C}^n$ has residue Ω at the origin. The limit of tY_{tu_0,tu_0} if t approaches 0 exists and is then equal to the linear map which sends $A \in \mathfrak{g} \otimes \mathfrak{g}$ to $\left[\Omega^{12} - \Omega^{23}, A^{13}\right]$. This map is injective by Corollary 2.15 (e). As injectivity is an open property, there exists $t_0 \neq 0$ with $(t_0u_0, t_0u_0) \in \widetilde{U}$ such that $t_0Y_{t_0u_0,t_0u_0}$ and so also $Y_{t_0u_0,t_0u_0}$ is injective. By the same reason, there exists an open neighbourhood $\widetilde{U}' \subset \widetilde{U}$ of (t_0u_0, t_0u_0) such that $Y_{u,v}$ is injective for all $(u, v) \in \widetilde{U}'$.

The assumption that r satisfies CYBE is equivalent to the equation

$$Y_{u,v}\big(r(u + v)\big) = [r^{23}(v), r^{12}(u)].$$

The injectivity of $Y_{u,v}$ implies that $r(u + v)$ can be obtained by applying to the right hand side of this equation the inverse of a certain square minor of a matrix representation of $Y_{u,v}$. Because the entries of the inverse of a square matrix are rational expressions of the entries of the original matrix, this shows that $r(u+v)$ depends rationally on $r(u)$ and $r(v)$, provided that $(u, v) \in \widetilde{U}'$. This shows that, there exists a rational function $P : (\mathfrak{g} \otimes \mathfrak{g}) \times (\mathfrak{g} \otimes \mathfrak{g}) \longrightarrow (\mathfrak{g} \otimes \mathfrak{g})$ such that $r(u + v) = P\big(r(u), r(v)\big)$ for all $(u, v) \in \widetilde{U}'$. Tracing this argument backwards, we see that the function r satisfies CYBE for $(u, v) \in \widetilde{U}'$ if and only if $r(u + v) = P\big(r(u), r(v)\big)$ for $(u, v) \in \widetilde{U}'$.

The meromorphic function $f(u, v) = r(u + v) - P\big(r(u), r(v)\big)$ is defined on $\widetilde{U}$ and identically zero on the open subset $\widetilde{U}'$. As U was assumed to be convex, $\widetilde{U}$ is connected and so f vanishes identically on $\widetilde{U}$. This proves $r(u + v) = P\big(r(u), r(v)\big)$ for all $(u, v) \in \widetilde{U}$.

From Proposition 3.7 we know that r is unitary, i.e. $r(-v) = -\tau r(v)$. As τ is linear, $Q(A, B) = P(A, -\tau B)$ is rational again. If $(u, -v) \in \widetilde{U}$, we have $u, v \in U$ and $r(u - v) = P\big(r(u), r(-v)\big) = P\big(r(u), -\tau r(v)\big) = Q\big(r(u), r(v)\big)$ as required. □

Corollary 4.5. *If $\mathfrak{g}$ is a finite-dimensional simple complex Lie algebra, $U \subset \mathbb{C}$ an open disc with centre 0 and $r : U \longrightarrow \mathfrak{g} \otimes \mathfrak{g}$ a non-degenerate meromorphic solution of CYBE, then r extends to a meromorphic function $r : \mathbb{C} \longrightarrow \mathfrak{g} \otimes \mathfrak{g}$.*

Proof. For any disc $D(0, \rho) \subset \mathbb{C}$ we say that r satisfies CYBE *on* $D(0, \rho)$, if (3.1) is satisfied for all $(u, v) \in \widetilde{D}(0, \rho) := \{(u, v) \mid u, v, u + v \in D(0, \rho)\} \subset \mathbb{C}^2$. Fix $\varepsilon > 0$ such that $D(0, \varepsilon) \subset U$, for example $D(0, \varepsilon) = U$. If $\rho > \varepsilon$ and r extends to a meromorphic function $r : D(0, \rho) \longrightarrow \mathfrak{g}\otimes\mathfrak{g}$, then r satisfies CYBE on $D(0, \rho)$. This follows from the Identity Theorem because the meromorphic function, which is the left hand side of CYBE, vanishes identically on the open subset $\widetilde{D}(0, \varepsilon)$ of $\widetilde{D}(0, \rho)$, hence vanishes on $\widetilde{D}(0, \rho)$ as well.

Assume that r does not extend to a meromorphic function on $\mathbb{C}$. Then there exists a real number $\rho \geq \varepsilon$ such that r extends to $D(0, \rho)$ but not to $D(0, \rho + \varepsilon)$. For each fixed $v \in D(0, \varepsilon)$ we define a meromorphic function $s : D(v, \rho) \longrightarrow \mathfrak{g}\otimes\mathfrak{g}$ on the disc with centre v and radius ρ by $s(u) = P\big(r(u-v), r(v)\big)$, where P is the rational function from Proposition 4.4 for which we have $r(u) = P\big(r(u-v), r(v)\big)$ whenever $u, v, u-v \in D(0, \rho)$. Hence the two functions r and s coincide on $D(0, \rho)\cap D(v, \rho)$. Therefore, s defines a meromorphic extension of r to $D(0, \rho)\cup D(v, \rho)$. If we vary $v \in D(0, \varepsilon)$ we obtain an extension of r to $D(0, \rho + \varepsilon) = \bigcup_{v\in D(0,\varepsilon)} D(v, \rho)$ in contradiction to our assumption. Therefore r extends to the whole complex plane. □

Proposition 4.6. *The set of poles $\Gamma \subset \mathbb{C}$ of a meromorphic non-degenerate solution $r : \mathbb{C} \longrightarrow \mathfrak{g} \otimes \mathfrak{g}$ of (3.1) is a discrete subgroup and there exists a homomorphism $\Gamma \longrightarrow \mathsf{Aut}(\mathfrak{g})$ sending γ to A_γ such that $r(u + \gamma) = (A_\gamma \otimes \mathbb{1}_\mathfrak{g})r(u)$ for all $u \in \mathbb{C}$ and $\gamma \in \Gamma$.*

Proof. As the set of poles of a meromorphic function is always discrete, we only need to show that Γ is a subgroup. By Theorem 3.6, $0 \in \Gamma$ and all poles of r are simple. After rescaling, we are able to assume that the residue at the origin is equal to the Casimir element Ω. Let θ_γ denote the residue of r at $\gamma \in \Gamma$. Then $A_\gamma := \varphi_1(\theta_\gamma) : \mathfrak{g} \longrightarrow \mathfrak{g}$ is a non-zero linear map. As before, we use the abbreviation $T_u = \varphi_1(r(u))$. Because r is unitary by Proposition 3.7, we have $T_\gamma = -T^*_{-\gamma}$, hence $\gamma \in \Gamma$ implies $-\gamma \in \Gamma$. If we multiply (3.1) by $(u-\gamma)$ and take the limit if u approaches γ, we obtain the equation

$$\left[\theta_\gamma^{12}, r^{13}(v+\gamma) + r^{23}(v)\right] = 0. \tag{4.1}$$

After multiplying it by v and letting v tend to 0, we get $\left[\theta_\gamma^{12}, \theta_\gamma^{13} + \Omega^{23}\right] = 0$. Applying the map φ_2 and using Proposition 2.14, this equation is seen to be equivalent to $[A_\gamma(x), A_\gamma(y)] = A_\gamma[x,y]$ for all $x, y \in \mathfrak{g}$. This means that A_γ is a homomorphism of Lie algebras. It is even an isomorphism, as $\mathfrak{g}$ is simple and $A_\gamma \neq 0$. Now we apply φ_2 to (4.1) and use Proposition 2.14 to obtain

$$0 = [A_\gamma(x), T_{v+\gamma}(y)] - A_\gamma[x, T_v(y)] = [A_\gamma(x), T_{v+\gamma}(y)] - [A_\gamma(x), A_\gamma T_v(y)]$$

for all $x, y \in \mathfrak{g}$. Because A_γ is surjective, this implies $T_{v+\gamma} = A_\gamma \circ T_v$ for all $v \in \mathbb{C}$, which is, by Lemma 2.4, equivalent to the desired equality $r(v+\gamma) = (A_\gamma \otimes \mathbb{1}_\mathfrak{g})r(v)$. Hence, if $v \in \Gamma$ then $v + \gamma \in \Gamma$ and Γ is closed under addition.

Using $v \in \mathbb{C}$ with invertible T_v, we see that $A_{\gamma_1+\gamma_2} = A_{\gamma_1} \circ A_{\gamma_2}$ which concludes the proof. □

Lemma 4.7. *Let $R : \mathbb{C}^n \longrightarrow \mathfrak{g} \otimes \mathfrak{g}$ be a non-degenerate meromorphic function which satisfies (3.1) with $u, v \in \mathbb{C}^n$. Then there exists a vector space $V \subset \mathbb{C}^n$ of codimension 1 such that for all $\mathbf{e}, \mathbf{e}' \in \mathbb{C}^n \setminus \{0\}$ with $\mathbf{e}' - \mathbf{e} \in V$, the functions $r(u) = R(u\mathbf{e})$ and $r'(u) = R(u\mathbf{e}')$ are equivalent.*

Proof. We write $R(z) = Y(z)/f(z)$ with $Y : \mathbb{C}^n \longrightarrow \mathfrak{g} \otimes \mathfrak{g}$ and $f : \mathbb{C}^n \longrightarrow \mathbb{C}$ holomorphic such that $S = \{z \in \mathbb{C}^n \mid f(z) = 0, Y(z) \neq 0\}$ is not empty. The set $\widetilde{S}$ of poles of R contains S, but might not coincide with it. We shall first show that there exist maps $\Phi : S \longrightarrow \mathsf{Aut}(\mathfrak{g})$ and $\lambda : S \longrightarrow \mathbb{C}^*$ such that for all $z \in \mathbb{C}^n, h \in S$

$$Y(h) = \lambda(h) \cdot (\Phi(h) \otimes \mathbb{1}_\mathfrak{g})(\Omega) \tag{4.2}$$

$$R(z+h) = \big(\Phi(h) \otimes \mathbb{1}_\mathfrak{g}\big)R(z). \tag{4.3}$$

As before, we use the abbreviation $T_z = \varphi_1(R(z))$. Fix $h \in S$, multiply CYBE (3.1) for R with variables u and $v = z$ by $f(u)$ and take the limit if u approaches h, to obtain $\big[Y^{12}(h), R^{13}(h+z) + R^{23}(z)\big] = 0$. By Proposition 2.14 this is equivalent to

$$\big[\varphi_1(Y(h))(x), T_{h+z}(y)\big] - \varphi_1(Y(h))\big[x, T_{z(y)}\big] = 0 \quad \text{for all } x, y \in \mathfrak{g}.$$

Using Proposition 2.14 (d), (e) this is seen to be equivalent to

$$\big[Y(h), T_{h+z}(y) \otimes 1 + 1 \otimes T_z(y)\big] = 0. \tag{4.4}$$

Consider now $\mathfrak{g} \oplus \mathfrak{g}$ as a subalgebra of $U(\mathfrak{g}) \otimes U(\mathfrak{g})$ via $(a,b) \mapsto a \otimes 1 + 1 \otimes b$. This embedding is indeed an algebra homomorphism since $[a \otimes 1, 1 \otimes b] = 0$. For fixed

$$z \in W = \left\{z \;\middle|\; z, h+z \notin \widetilde{S} \text{ and } T_z, T_{h+z} \text{ are invertible}\right\} \subset \mathbb{C}^n$$

we let $\mathfrak{g}_0 \subset \mathfrak{g} \oplus \mathfrak{g}$ be the Lie subalgebra generated by $T_{h+z}(y) \otimes 1 + 1 \otimes T_z(y)$ for all $y \in \mathfrak{g}$. The two projections $\mathfrak{g} \oplus \mathfrak{g} \longrightarrow \mathfrak{g}$ induce Lie-algebra homomorphism $p_i : \mathfrak{g}_0 \longrightarrow \mathfrak{g}$, $i = 1, 2$ which are surjective because $z \in W$. As $\mathfrak{g}$ is simple, the kernel $\ker(p_2) = \mathfrak{g}_0 \cap (\mathfrak{g} \oplus 0) \subset \mathfrak{g} \oplus 0 \cong \mathfrak{g}$

is either $\mathfrak{g} \oplus 0$ or 0. If $\ker(p_2) = \mathfrak{g} \oplus 0$, the dimension formula implies $\mathfrak{g}_0 = \mathfrak{g} \oplus \mathfrak{g}$ and so, in particular, $y \otimes 1 \in \mathfrak{g}_0$. The definition of $\mathfrak{g}_0$ and equation (4.4) imply that $[Y(h), A] = 0$ for all $A \in \mathfrak{g}_0$, hence $[Y(h), y \otimes 1] = 0$ for all $y \in \mathfrak{g}$. But then, by Proposition 2.14 (d), $0 = \varphi_1\big([Y(h), y \otimes 1]\big)(x) = \big[\varphi_1(Y(h))(x), y\big]$ for all $x, y \in \mathfrak{g}$. As $\mathfrak{g}$ in not abelian, this is only possible if $Y(h) = 0$ in contradiction to $h \in S$. This shows that $\ker(p_2) = 0$ and $p_2 : \mathfrak{g}_0 \longrightarrow \mathfrak{g}$ is an isomorphism of Lie algebras. Let $\Phi(h, z) = p_1 \circ p_2^{-1} \in \mathsf{Aut}(\mathfrak{g})$, then each element of $\mathfrak{g}_0$ can be written as $(\Phi(h, z)(x), x) \in \mathfrak{g} \oplus \mathfrak{g}$. If we apply this to the generators of $\mathfrak{g}_0$ we obtain $T_{h+z} = \Phi(h, z) \circ T_z$, which is equivalent to

$$R(h + z) = \big(\Phi(h, z) \otimes 1\!\!1_{\mathfrak{g}}\big) R(z). \tag{4.5}$$

If we plug this in (4.4), we get $\big[Y(h), (\Phi(h, z) \circ T_z)(y) \otimes 1 + 1 \otimes T_z(y)\big] = 0$. Because $\Phi(h, z)$ is a Lie-algebra isomorphism, this is equivalent to

$$\big[\big(\Phi(h, z)^{-1} \otimes 1\!\!1_{\mathfrak{g}}\big) Y(h), T_z(y) \otimes 1 + 1 \otimes T_z(y)\big] = 0.$$

With the aid of Proposition 2.14 (d), (e) this translates into $[\varphi_1(B)(x), T_z(y)] - \varphi_1(B)\,[x, T_z(y)] = 0$ for all $x, y \in \mathfrak{g}$, where we have set $B = \big(\Phi(h, z)^{-1} \otimes 1\!\!1_{\mathfrak{g}}\big) Y(h)$. As T_z is surjective, this is equivalent to $[\mathrm{ad}(y), \varphi_1(B)] = 0$ for all $y \in \mathfrak{g}$. Lemma 2.2 implies now $\varphi_1(B) \in \mathbb{C} \cdot 1\!\!1_{\mathfrak{g}}$. As $Y(h) \neq 0$, this shows that there exists a non-zero number $\lambda(h, z)$ such that $\big(\Phi(h, z)^{-1} \otimes 1\!\!1_{\mathfrak{g}}\big) Y(h) = \lambda(h, z)\Omega$, i.e.

$$Y(h) = \lambda(h, z) \cdot \big(\Phi(h, z) \otimes 1\!\!1_{\mathfrak{g}}\big)(\Omega). \tag{4.6}$$

This is equivalent to $\varphi_1(Y(h)) = \lambda(h, z) \cdot \Phi(h, z)$. As the left hand side does not depend on z and there is only one multiple of the linear map $\varphi_1(Y(h))$ which is a Lie-algebra homomorphism, both $\lambda(h, z)$ and $\Phi(h, z)$ are determined by $Y(h)$ and do not depend on z. From (4.5) and (4.6) we see that the functions $\lambda(h) = \lambda(h, z)$ and $\Phi(h) = \Phi(h, z)$ satisfy (4.2) and (4.3) for all $z \in W$ and $h \in S$. As W is dense in $\mathbb{C}^n$ these two equations will be satisfied for all $z \in \mathbb{C}^n$.

Next, we show that there exists a vector space $V \subset \mathbb{C}^n$ of codimension 1 and a holomorphic homomorphism of groups $\Phi : V \longrightarrow \mathsf{Aut}(\mathfrak{g})$ such that for all $z \in \mathbb{C}^n, h \in V$

$$R(z + h) = \big(\Phi(h) \otimes 1\!\!1_{\mathfrak{g}}\big) R(z) \tag{4.7}$$

$$R(z) = \big(\Phi(h) \otimes \Phi(h)\big) R(z). \tag{4.8}$$

To find the subspace V we denote by H the subgroup of $\mathbb{C}^n$ which is generated by S. Because of (4.3), which is equivalent to $T_{h+z} = \Phi(h) \circ T_z$, we can extend Φ to a homomorphism $\Phi : H \longrightarrow \mathsf{Aut}(\mathfrak{g})$ which satisfies (4.3) for all $z \in \mathbb{C}^n$ and $h \in H$. From $T_{h+z} = \Phi(h) \circ T_z$ we see that the set $\widetilde{S}$ of poles of R is invariant under H, hence $H \neq \mathbb{C}^n$. Because H contains the analytic subset S of codimension one in $\mathbb{C}^n$, this group contains a codimension one linear subspace $V \subset \mathbb{C}^n$.

To see that $\Phi : V \longrightarrow \mathsf{Aut}(\mathfrak{g})$ is holomorphic, we recall that R was assumed to be non-degenerate. This implies that for each $h_0 \in V$ the set which consist of those $z \in \mathbb{C}^n$ for which R is holomorphic and non-degenerate at $h_0 + z$, is open and dense in $\mathbb{C}^n$. Hence, for each $h_0 \in V$ there exists $z_0 \in \mathbb{C}^n$ such that R is holomorphic at z_0 and at $h_0 + z_0$ and T_{z_0} is an isomorphism. Because $\Phi(h) = T_{h+z_0} \circ T_{z_0}^{-1}$, we see now that Φ is holomorphic in a neighbourhood of $h_0 \in V$.

To show equation (4.8), we recall that R is unitary by Proposition 3.7. This means that $T_z^* = -T_z$. Equation (4.7) implies $T_{-z} = \Phi(h) \circ T_{-z-h}$ and $T_{z+h} = \Phi(h) \circ T_z$, hence

$$T_z = -T_{-z}^* = -T_{-z-h}^* \circ \Phi(h)^* = T_{z+h} \circ \Phi(h)^* = \Phi(h) \circ T_z \circ \Phi(h)^*,$$

which is equivalent to (4.8) by Lemma 2.4.

Finally, let $h = (u_1 - u_2)(\mathbf{e}' - \mathbf{e}) \in V$ for given $u_1, u_2 \in \mathbb{C}$. With $z = (u_1 - u_2)\mathbf{e}$, equation (4.7) implies $r'(u_1 - u_2) = R(h + z) = \big(\Phi(h) \otimes \mathbb{1}_\mathfrak{g}\big)R(z) = \big(\Phi\big((u_1 - u_2)(\mathbf{e}' - \mathbf{e})\big) \otimes \mathbb{1}_\mathfrak{g}\big)r(u_1 - u_2)$. From equation (4.8) we obtain $r(u_1 - u_2) = \big(\Phi\big(u_2(\mathbf{e}' - \mathbf{e})\big) \otimes \Phi\big(u_2(\mathbf{e}' - \mathbf{e})\big)\big)r(u_1 - u_2)$ and as Φ is a homomorphism, we have $\Phi\big((u_1 - u_2)(\mathbf{e}' - \mathbf{e})\big) = \Phi\big(u_1(\mathbf{e}' - \mathbf{e})\big) \circ \Phi\big(u_2(\mathbf{e}' - \mathbf{e})\big)^{-1}$. Therefore, $r'(u_1 - u_2) = \big(\Phi\big(u_1(\mathbf{e}' - \mathbf{e})\big) \otimes \Phi\big(u_2(\mathbf{e}' - \mathbf{e})\big)\big)r(u_1 - u_2)$, i.e. r and r' are equivalent. □

Theorem 4.8 (Belavin-Drinfeld). *Let $r : \mathbb{C} \longrightarrow \mathfrak{g} \otimes \mathfrak{g}$ be a non-degenerate meromorphic solution of (3.1) and $\Gamma \subset \mathbb{C}$ its lattice of poles. Then exactly one of the following three cases occurs.*

(a) If $\mathrm{rk}(\Gamma) = 2$, *r is* elliptic, *i.e. it has two periods that are independent over $\mathbb{R}$ and are contained in Γ.*

(b) If $\mathrm{rk}(\Gamma) = 1$, *there exists a rational function $f : \mathbb{C} \longrightarrow \mathfrak{g} \otimes \mathfrak{g}$ and a constant $\lambda \in \mathbb{C}$ such that the function $u \mapsto f(e^{\lambda u})$ is equivalent to r. Such solutions are called* trigonometric.

(c) If $\mathrm{rk}(\Gamma) = 0$, *there exists a rational function $f : \mathbb{C} \longrightarrow \mathfrak{g} \otimes \mathfrak{g}$ which is equivalent to r. Such solutions f are called* rational.

Proof. Let $X_0 = \{r(u) \mid u \in \mathbb{C} \setminus \Gamma\} \subset \mathfrak{g} \otimes \mathfrak{g}$ and denote by $X \subset \mathfrak{g} \otimes \mathfrak{g}$ the closure of X_0 in the Zariski topology. This means that X is the smallest subset of the vector space $\mathfrak{g} \otimes \mathfrak{g}$ which contains X_0 and which is the zero set of finitely many polynomials. As r is meromorphic and not necessarily rational, the dimension of X could be larger than one. Let now P, Q be rational functions as in Proposition 4.4. Because $P(x, y) \in X_0$ for $x, y \in X_0$, the restriction of P is a rational map $X \times X \longrightarrow X$. Because $r(u+v) = P\big(r(u), r(v)\big)$ for all $u, v \in \mathbb{C}$ and addition on $\mathbb{C}$ is associative and commutative, we obtain $P(x, y) = P(y, x)$ and $P(x, P(y, z)) = P(P(x, y), z)$ for all $x, y, z \in X_0$. As the set where two rational functions coincide is Zariski-closed, these two identities hold for all $x, y, z \in X$ where they are defined.

The only thing left to verify that (X, P) is a birational group law is the birationality of the maps Φ and Ψ from Definition 4.2. We use the rational map Q from Proposition 4.4 to define an inverse of Φ on $X_0 \times X_0$ by $(x, z) \mapsto (x, Q(z, x))$. Similarly, an inverse of Ψ on $X_0 \times X_0$ is given by $(z, y) \mapsto (Q(z, y), y)$. Using the same argument as before, these maps are rational inverses of Φ and Ψ respectively on $X \times X$, hence these two maps are birational.

By Weil's Theorem 4.3, there exists an algebraic group G and a birational map $\psi : X \to G$ under which the group law of G corresponds to P. Denote by $f : \mathbb{C} \to G$ the meromorphic map $\psi \circ r$ and by $\widetilde{R} : G \longrightarrow \mathfrak{g} \otimes \mathfrak{g}$ the rational map ψ^{-1} followed by the embedding of X into $\mathfrak{g} \otimes \mathfrak{g}$.

As (X, P) was commutative, G is a commutative algebraic group. By Chevalley's Theorem, the group G is an extension of an abelian variety by a finite product of additive and multiplicative groups, see for example [5, §9.2, Thms. 1, 2] and [8, Ch. IV, §3, no. 6]. This implies that the variety G is isomorphic to $\mathbb{C}^a \times (\mathbb{C}^*)^b \times \mathbb{A}$, where $\mathbb{A}$ is an abelian variety. Therefore, the universal covering of G is a vector space $\mathbb{C}^n$, with $n = a + b + \dim(\mathbb{A})$. If $\pi : \mathbb{C}^n \longrightarrow G$ denotes a covering map which is a homomorphism of groups, we obtain the following commutative diagram

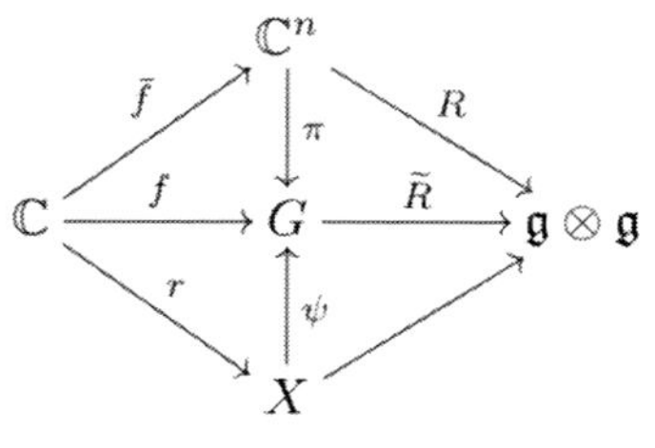

in which $\bar{f}$ is the unique lift of f. By definition, f is a homomorphism of groups, and therefore $\bar{f}$ as well. This implies that there exists $\mathbf{e} \in \mathbb{C}^n$ such that $\bar{f}(u) = u\mathbf{e}$. The commutativity of the diagram implies now $r(u) = R(u\mathbf{e})$. Because addition on an open dense subset of $\mathbb{C}^n$

corresponds to P under $\psi^{-1} \circ \pi$, we have $R(\mathbf{e} + \mathbf{e}') = P(R(\mathbf{e}), R(\mathbf{e}'))$ for all $\mathbf{e}, \mathbf{e}' \in \mathbb{C}^n$. From the proof of Proposition 4.4 we see that this implies that R satisfies (3.1) with $u, v \in \mathbb{C}^n$ and so we can apply Lemma 4.7. By $V \subset \mathbb{C}^n$ we denote a hyperplane which satisfies the conditions of Lemma 4.7.

We choose a basis $\mathbf{e}_1, \mathbf{e}_2, \dots, \mathbf{e}_n$ of $\mathbb{C}^n$ such that π maps $W_{\text{rat}} = \langle \mathbf{e}_1, \dots, \mathbf{e}_a \rangle$ isomorphically onto $\mathbb{C}^a$ and $W_{\text{trig}} = \langle \mathbf{e}_{a+1}, \dots, \mathbf{e}_{a+b} \rangle$ onto $(\mathbb{C}^*)^b$ in such a way that $\pi\left(\sum_{k=a+1}^{a+b} x_k \mathbf{e}_k\right) = (\exp(x_{a+1}), \dots, \exp(x_{a+b})) \in (\mathbb{C}^*)^b$. Let $W = W_{\text{rat}} \oplus W_{\text{trig}}$ and $\widetilde{\Gamma} = \pi^{-1}(0)$ be the full period lattice of the quotient map. Then $\widetilde{\Gamma}_k = \widetilde{\Gamma} \cap \langle \mathbf{e}_k \rangle$ is equal to zero if $k \le a$ and equal to the free abelian group $2\pi i\mathbb{Z}\mathbf{e}_k$ of rank 1 if $a < k \le a + b$. This is the full lattice of periods of the function $r_k(u) = R(u\mathbf{e}_k)$. Moreover, if $k \le a + b$, $G_k = \langle \mathbf{e}_k \rangle / \widetilde{\Gamma}_k \subset G$ is one of the additive or multiplicative subgroups, in particular it is a closed subvariety of G.

Let $\gamma_1, \dots, \gamma_{2m} \in \widetilde{\Gamma}$, $m = n - a - b = \dim(\mathbb{A})$, be such that $\mathbb{A}$ is the quotient of $\mathbb{C}^n/W$ under the lattice of maximal rank generated by the images $\overline{\gamma}_1, \dots, \overline{\gamma}_{2m}$ of $\gamma_1, \dots, \gamma_{2m}$ in $\mathbb{C}^n/W$. Then $\widetilde{\Gamma} = (2\pi i\mathbb{Z})^b \oplus \langle \gamma_1, \dots, \gamma_{2m} \rangle$. Let $\lambda_j \in \mathbb{C}$ be determined by $\gamma_j - \lambda_j \mathbf{e} \in V$. As the set of poles of R is invariant under V (see Lemma 4.7), we have $\lambda_j \in \Gamma$.

If $W \subset V$, then the images of $\gamma_1, \dots, \gamma_{2m}$ under the surjection $\mathbb{C}^n \to \mathbb{C}^n/W \to \mathbb{C}^n/V$ generate a lattice of maximal rank. The one-dimensional subspace $\langle \mathbf{e} \rangle$ maps isomorphically onto $\mathbb{C}^n/V$ and the image of $\lambda_j \mathbf{e}$ coincides with the image of γ_j. Hence, the assumption $W \subset V$ implies $\text{rk}(\Gamma) = 2$.

If $\text{rk}(\Gamma) < 2$ we, therefore, find $\mathbf{e}_k \notin V$ with $1 \le k \le a+b$ and there exists a non-zero constant λ such that $\mathbf{e} - \lambda \mathbf{e}_k \in V$, hence $r_k(\lambda u)$ and $r(u)$ are equivalent by Lemma 4.7. This implies that both functions have the same set of poles, which is a discrete subgroup by Proposition 4.6. As the rational function $\widetilde{r}_k = \widetilde{R}_{|G_k}$ has only finitely many poles, the poles and the periods of $r_k(\lambda u)$ have to be lattices of the same rank, that is $\text{rk}\, \widetilde{\Gamma}_k = \text{rk}\, \Gamma$. Hence, if $\text{rk}\, \Gamma = 1$, $\mathbf{e}_k \in W_{\text{trig}}$ and $r_k(\lambda u)$ is a rational function of $\exp(\lambda u)$ and in the case $\text{rk}\, \Gamma = 0$ we have $\mathbf{e}_k \in W_{\text{rat}}$ and $r_k(\lambda u)$ is a rational function.

If $\text{rk}\, \Gamma = 2$ we cannot argue in the same way, because it is not clear that we can find a one-dimensional subspace, not contained in W, which contains two independent elements of the lattice $\langle \gamma_1, \dots, \gamma_{2m} \rangle$. Instead we use the homomorphism $\Gamma \longrightarrow \mathsf{Aut}(\mathfrak{g})$ from Proposition 4.6 sending γ to A_γ and prove directly that r is an elliptic function.

If the image $\Gamma' \subset \mathsf{Aut}(\mathfrak{g})$ of this homomorphism was infinite, the smallest algebraic subgroup $\overline{\Gamma}'$ of $\mathsf{Aut}(\mathfrak{g})$ which contains Γ' would be commutative and of positive dimension. Therefore, its Lie algebra, which is a commutative subalgebra of $\mathfrak{g}$, contains a non-zero element a. For this element we have $A_\gamma(a) = a$ for all $\gamma \in \Gamma$. Let $f : \mathbb{C} \longrightarrow \mathfrak{g}$ be the meromorphic map given by $f(u) = T_u^*(a)$, where $T_u = \varphi_1(r(u))$. We have $f(u + \gamma) = T_{u+\gamma}^*(a) = T_u^* A_\gamma^*(a) = T_u^*(a) = f(u)$, because $T_{u+\gamma} = A_\gamma \circ T_u$ and $A_\gamma^* = A_\gamma^{-1} = A_{-\gamma}$ (Proposition 4.6). This means that f is Γ-periodic. But f has a simple pole at each point of Γ and no other poles, which is impossible for a doubly periodic function. This contradiction shows that Γ' is a finite group.

Let γ_1, γ_2 be generators of Γ. Then $A_{\gamma_1}, A_{\gamma_2}$ are both of finite order $n_i = \text{ord}\,(A_{\gamma_i})$. From $T_{u+\gamma} = A_\gamma \circ T_u$ we obtain that $T_{u+n_1\gamma_1} = T_u = T_{u+n_2\gamma_2}$ and this means that T_u and so also r is an elliptic (doubly periodic with periods $n_1\gamma_1, n_2\gamma_2$) function. □

References

[1] Artin M 1970 Théorème de Weil sur la construction d'un groupe à partir d'une loi rationelle *Schémas en Groupes* SGA 3 Exp. XVIII (Berlin: Springer) pp 632–53

[2] Belavin A and Drinfeld V 1982 Solutions of the classical Yang-Baxter equation for simple Lie algebras *Funktsional. Anal. i ego Prilozhen.* **16** 1

[3] Belavin A and Drinfeld V 1983 The classical Yang-Baxter equation for simple Lie algebras *Funktsional. Anal. i ego Prilozhen.* **17** 69

[4] Belavin A and Drinfeld V 1998 *Triangle Equations and Simple Lie Algebras* (Amsterdam: Harwood Academic Publishers)
[5] Bosch S, Lütkebohmert W and Raynaud M 1990 *Néron Models* (Berlin: Springer)
[6] Burban I and Kreussler B 2007 Vector bundles on degenerations of elliptic curves and Yang–Baxter equations *Preprint* math.AG/0708.1685, to appear in *Memoirs of the AMS*
[7] Chari V and Pressley A 1994 *A Guide to Quantum Groups* (Cambridge: Cambridge University Press)
[8] Demazure M and Gabriel P 1970 *Groupes Algébriques I* (Paris, Amsterdam: Masson, North-Holland)
[9] Etingof P and Schiffmann O 2002 *Lectures on Quantum Groups* (Somerville: International Press)
[10] Polishchuk A 2002 Classical Yang-Baxter equation and the A_∞-constraint *Adv. Math.* **168** 56
[11] Weil A 1955 On Algebraic Groups of Transformations *Amer. J. Math.* **77** 355

Algebra, Geometry, and Mathematical Physics 2010 IOP Publishing
Journal of Physics: Conference Series **346** (2012) 012012 doi:10.1088/1742-6596/346/1/012012

Tensor powers for non-simply laced Lie algebras B_2-case

P P Kulish[1], **V D Lyakhovsky**[2] **and O V Postnova**[3]

[1] Sankt-Petersburg Branch of V A Steklov Mathematical Institute RAS, Sankt-Petersburg, Russia

[2 3] High Energy Physics and Elementary Particles Department, Sankt-Petersburg State University, Sankt-Petersburg, Russia

E-mail: [1] kulish@pdmi.ras.ru

E-mail: [2] lyakvladimir@yandex.ru

E-mail: [3] olgapostnova@mail.ru

Abstract. We study the decomposition problem for tensor powers of B_2-fundamental modules. To solve this problem singular weight technique and injection fan algorithms are applied. Properties of multiplicity coefficients are formulated in terms of multiplicity functions. These functions are constructed showing explicitly the dependence of multiplicity coefficients on the highest weight coordinates and the tensor power parameter. It is thus possible to study general properties of multiplicity coefficients for powers of the fundamental B_2-modules.

1. Introduction

Consider the Brauer centralizer algebras for the spinor groups acting on the tensor space $(\otimes^p \mathbf{V}^n)$. Here $\mathbf{V}^n$ is the fundamental representation of $\mathrm{Spin}(n)$. This centralizer algebra of the diagonal action of orthogonal algebra on $(\otimes^p \mathbf{V}^n)$ is generated by the symmetric group S_p and the contractions and immersions of the invariant form. To proceed further one needs the list of $\mathrm{Spin}(n)$-irreducible submodules in the decomposition $(\otimes^p \mathbf{V}^n) = \sum_{\mu \in P} m_\mu^p \mathbf{V}^{(\mu)}$ (P is the $\mathrm{Spin}(n)$ weight space) and their multiplicities m_μ^p. Since we are interested in an arbitrary power p of the fundamental module $\mathbf{V}^n_{\mathrm{Spin}(n)}$ our main problem is to find multiplicities of submodules in a form of *multiplicity functions* $M(\mu, p)$ explicitly depending on the corresponding highest weight μ and the power parameter p.

There are numerous combinatorial studies of the problem [1, 2, 3, 4] and also series of works dealing with fermionic formulas, some of them based on crystal basis approach [5, 6, 7, 8]. On this way important general results were obtained [9, 10, 11]. On the other hand practical computations with the corresponding formulas are scarcely possible for all but the simplest examples. In most of these studies the simply laced algebras are considered and as a rule the multiplicities formulas are connected with complicated path countings.

We should also mention an algorithm for tensor product decompositions proposed in [12] and improved in [13]. It is used in our investigations.

Summing up, we are to find multiplicities m_μ^p in the decomposition $(\otimes^p \mathbf{V}^n) = \sum_{\mu \in P} m_\mu^p \mathbf{V}^{(\mu)}$ as a function of μ and p . To solve this problem we propose an algorithm based on singular

Published under licence by IOP Publishing Ltd

weights properties [16] and the injection fan technique [17, 18]. We study the multiplicities m_μ^p formulated in terms of multiplicity functions $M_{\mathfrak{g}}(\mu, p)$. The latter have the weight space $\mathcal{L}P$ for the domain of definition. On the sublattice P^{++} of dominant weights the multiplicity function gives us the desired multiplicities, $M(\mu, p)|_{\mu \in P^{++}} = m_\mu^p$. In this paper we shall show how to adopt these tools to non-simply laced algebras and shall demonstrate how they work by studying the tensor powers $\left(L_{B_n}^{\omega_i}\right)^{\otimes p}$ of the fundamental module $L_{B_2}^{\omega_i}$ of B_2.

2. Basic definitions and relations

Throughout the paper the following notation will be used.

$\mathfrak{g}$ – simple Lie algebra of the series B_n,

L^μ – the integrable module of $\mathfrak{g}$ with the highest weight μ ;

r – the rank of the algebra $\mathfrak{g}$;

Δ – the root system; Δ^+ – the positive root system for $\mathfrak{g}$;

$\mathcal{N}^\mu$ – the weight diagram of L^μ;

W – the Weyl group;

$C^{(0)}$ – the fundamental Weyl chamber, $\overline{C^{(0)}}$ – its closure;

ρ – the Weyl vector;

$\epsilon(w) := \det(w)$, $w \in W$;

α_i – the i-th simple root for $\mathfrak{g}$; $i = 0, \ldots, r$;

ω_i – the i-th fundamental weight for $\mathfrak{g}$; $i = 0, \ldots, r$;

$L_{\mathfrak{g}}^{\omega_i}$ – the i-th fundamental module;

$\left\{e^i\right\}_{|i=1,\ldots,r}$ – the natural Euclidean basis of the weight space (the e-basis), $\{v_i\}$ – the coordinates of a weight in the e-basis;

P – the weight lattice $\mathcal{L}P$ – the weight space;

Q – the root lattice;

$\mathcal{E}$ – the group algebra of the group P;

$\widehat{\Psi^{(\mu)}}$ – the set of singular weights $\psi \in P$ for the Verma module V^μ [14, 15], we supply singular weights with coordinates $(\psi, \epsilon(w(\psi)))\,|_{\psi = w(\psi)\circ(\mu+\rho)-\rho}$;

$\Psi^{(\mu)} := \sum_{w \in W} \epsilon(w) e^{w\circ(\mu+\rho)-\rho}$ – the singular element for the $\mathfrak{g}$-module L^μ, i.e. the element in $\mathcal{E}$ corresponding to the set of singular weights of Verma module V^μ with coefficients $\epsilon(w)$;

In what follows we attribute singular weights and singular elements to the highest weight module L^μ without mentioning the corresponding Verma module V^μ.

$\mathrm{ch}(L^\mu)$ – the formal character of L^μ;

$\mathrm{ch}(L^\mu) = \frac{\sum_{w\in W} \epsilon(w) e^{w\circ(\mu+\rho)-\rho}}{\prod_{\alpha\in\Delta^+}(1-e^{-\alpha})} = \frac{\Psi^{(\mu)}}{\Psi^{(0)}}$ – the Weyl formula;

$R := \prod_{\alpha\in\Delta^+}(1 - e^{-\alpha}) = \Psi^{(0)}$ – the denominator;

$M_{\mathfrak{g}}^{\omega_i}(\mu, p)$ – the multiplicity function corresponding to the decomposition $(L_{\mathfrak{g}}^{\omega_i})^{\otimes p} = \sum m_\mu^{(i)p} L_{\mathfrak{g}}^\mu$, $M_{\mathfrak{g}}^{\omega_i}(\mu, p)\,|_{\mu\in P^{++}} = m_\mu^{(i)p}$.

3. Some useful properties.

Lemma 1 *The projection $\Psi_{\downarrow\mathfrak{g}}^{(\nu,\xi)}$ of the singular element $\Psi_{\mathfrak{g}\oplus\mathfrak{g}}^{(\nu,\xi)}$ for the irreducible representation $L^{(\nu)} \otimes L^{(\xi)}$ of the direct sum $\mathfrak{g} \oplus \mathfrak{g}$ on the weight space of the diagonal subalgebra $\mathfrak{g} \to \mathfrak{g} \oplus \mathfrak{g}$ is equal to the product*

$$\Psi_{\downarrow\mathfrak{g}}^{(\nu,\xi)} = \Psi_{\mathfrak{g}}^{\nu}\Psi_{\mathfrak{g}}^{\xi}.$$

Let $\left\{\psi_k^{(\nu)} \mid \psi_k^{(\nu)} \in \Psi^{(\nu)}, k = 1, \ldots, |W|\right\}$ and $\left\{\psi_p^{(\xi)} \mid \psi_p^{(\xi)} \in \Psi^{(\xi)}, p = 1, \ldots, |W|\right\}$ be the sets of singular weights for the modules $L^{(\nu)}$ and $L^{(\xi)}$ correspondingly then the set $\widehat{\Psi_{\downarrow\mathfrak{g}}^{(\nu,\xi)}}$ consists of the

Journal of Physics: Conference Series **346** (2012) 012012 doi:10.1088/1742-6596/346/1/012012

weights $\left\{\psi_k^{(\nu)}+\psi_p^{(\xi)},\epsilon\left(w\left(\psi_k^{(\nu)}\right)\right)\epsilon\left(w\left(\psi_p^{(\xi)}\right)\right)\right\}$.

Proof. Let $\left\{e^1,e^2,\dots,e^r,e^{r+1},\dots e^{2r}\right\}$ be the weight space basis for $\mathcal{L}P(\mathfrak{g}\oplus\mathfrak{g})$, $(\nu_1,\dots,\nu_r,\xi_1,\dots,\xi_r)$ – the coordinates for the highest weight (ν,ξ) naturally belonging to the space $\mathcal{L}P(\mathfrak{g}\oplus\mathfrak{g})$. The weights $v^{(\nu)}\in\mathcal{N}^\nu$, $u^{(\xi)}\in\mathcal{N}^\xi$ and the singular vectors $\psi_k^{(\nu)}$ and $\psi_p^{(\xi)}$ also are lifted to the space $\mathcal{L}P(\mathfrak{g}\oplus\mathfrak{g})$

$$\begin{aligned} lv_a^{(\nu)} &\Rightarrow \left(\nu_{a1}^{(\nu)},\dots,\nu_{ar}^{(\nu)},0,\dots,0\right);lu_b^{(\xi)}\Rightarrow\left(0,\dots,0,u_{b1}^{(\eta)},\dots,u_{br}^{(\eta)}\right)\\ l\psi_k^{(\nu)} &\Rightarrow \left(\psi_{k1}^{(\nu)},\dots,\psi_{kr}^{(\nu)},0,\dots,0\right);l\psi_p^{(\xi)}\Rightarrow\left(0,\dots,0,\psi_{p1}^{(\xi)},\dots,\psi_{pr}^{(\xi)}\right)\end{aligned}$$

The set $\left\{lv_a+lu_b\mid a=1,\dots,\dim\left(L^\nu\right),b=1,\dots,\dim\left(L^\xi\right)\right\}$ forms the weight diagram $\mathcal{N}_{\mathfrak{g}\oplus\mathfrak{g}}^{(\nu,\xi)}$ of $L_{\mathfrak{g}\oplus\mathfrak{g}}^{(\nu,\xi)}$. As far as for the Weyl group $W_{\mathfrak{g}\oplus\mathfrak{g}}$ we have $W_{\mathfrak{g}\oplus\mathfrak{g}}=W\times W$ and the Weyl vector is $\rho_{\mathfrak{g}\oplus\mathfrak{g}}=(\rho,\rho)$, the set of singular weights $\widehat{\Psi_{\mathfrak{g}\oplus\mathfrak{g}}^{(\nu,\xi)}}$ is formed by the vectors whose first $2r$ coordinates are $\left\{l\psi_k^{(\nu)}+l\psi_p^{(\xi)}\mid k,p=1,\dots,|W|\right\}$ and the last one is equal to the product $\epsilon\left(w\left(\psi_k^{(\nu)}\right)\right)\epsilon\left(w\left(\psi_p^{(\xi)}\right)\right)$:

$$\widehat{\Psi_{\mathfrak{g}\oplus\mathfrak{g}}^{(\nu,\xi)}}=\left\{l\psi_k^{(\nu)}+l\psi_p^{(\xi)},\epsilon\left(w\left(\psi_k^{(\nu)}\right)\right)\epsilon\left(w\left(\psi_p^{(\xi)}\right)\right)\mid k,p=1,\dots,|W|\right\}.$$

The vector $(c_1,\dots,c_r,c_{r+1},\dots,c_{2r})\in\mathcal{L}P(\mathfrak{g}\oplus\mathfrak{g})$ being projected to the diagonal subalgebra weight space $\mathcal{L}P$ in the basis $\left\{\left(e^1+e^{r+1}\right)/2,\dots,\left(e^r+e^{2r}\right)/2\right\}$ has the coordinates $((c_1+c_{r+1}),\dots,(c_r+c_{2r}))$. The latter means that

$$\begin{aligned}\widehat{\Psi_{\downarrow\mathfrak{g}}^{(\nu,\xi)}} &= \left\{\psi_k^{(\nu)}+\psi_p^{(\xi)},\epsilon\left(w\left(\psi_k^{(\nu)}\right)\right)\epsilon\left(w\left(\psi_p^{(\xi)}\right)\right)\right\},\\ k,p &= 1,\dots,|W|.\end{aligned}$$

Q.E.D. ■

One of the main tools to study the decomposition properties is the *injection fan* $\Gamma_{\mathfrak{g}_{\rm diag}\to\mathfrak{g}\oplus\mathfrak{g}}$. [17, 18] . To use this instrument we consider the decomposition of tensor products as a special case of branching. The latter corresponds to the injection of the diagonal subalgebra into the direct sum: $\mathfrak{g}_{\rm diag}\to\mathfrak{g}\oplus\mathfrak{g}$.

Lemma 2 *The vectors of the injection fan* Γ *for* $\mathfrak{g}_{\rm diag}\to\mathfrak{g}\oplus\mathfrak{g}$ *consists of the opposites to the singular weights of the trivial module* $L_{\mathfrak{g}}^{(0)}$

$$\Gamma_{\mathfrak{g}_{\rm diag}\to\mathfrak{g}\oplus\mathfrak{g}}=-S\circ\widehat{\Psi^{(0)}}\setminus(0,\dots,0),$$

(here S *is the full reflection).*

Proof. According to the definition [17] the vectors γ of the fan $\Gamma_{\mathfrak{g}_{\rm diag}\to\mathfrak{g}\oplus\mathfrak{g}}$ are fixed by the relation $1-\prod_{\alpha\in\left(\Delta^+_{\mathfrak{g}\oplus\mathfrak{g}\downarrow\mathfrak{g}_{\rm diag}}\right)}\left(1-e^{-\alpha}\right)^{\mathrm{mult}_{\mathfrak{g}\oplus\mathfrak{g}}(\alpha)-\mathrm{mult}_{\mathfrak{g}_{\rm diag}}(\alpha)}=\sum_{\gamma\in\Gamma_{\mathfrak{g}_{\rm diag}\to\mathfrak{g}\oplus\mathfrak{g}}}s(\gamma)\,e^{-\gamma}$. The projections of the $\mathfrak{g}\oplus\mathfrak{g}$ -roots to the diagonal subalgebra obviously reproduce the set $\{\alpha_i\}_{i=0,\dots,r}$ in $\mathcal{L}P(\mathfrak{g})$:

$$\left.\begin{array}{l}(\alpha_i,0)_{\downarrow\mathfrak{g}}\\(0,\alpha_i)_{\downarrow\mathfrak{g}}\end{array}\right\}=\alpha_i.$$

Thus $\mathrm{mult}_{\mathfrak{g}}(\alpha) = 2$ while $\mathrm{mult}_{\mathfrak{g}\oplus\mathfrak{g}}(\alpha) = 1$ and we have

$$\begin{aligned}\sum_{\gamma\in\Gamma_{\mathfrak{g}_{\mathrm{diag}}\to\mathfrak{g}\oplus\mathfrak{g}}} s(\gamma)\, e^{-\gamma} &= 1 - \prod_{\alpha\in(\Delta^+_{\mathfrak{g}\oplus\mathfrak{g}\downarrow\mathfrak{g}})} \left(1-e^{-\alpha}\right)^{\mathrm{mult}_{\mathfrak{g}\oplus\mathfrak{g}}(\alpha)-\mathrm{mult}_{\mathfrak{g}}(\alpha)} = \\ &= 1 - \prod_{\alpha\in(\Delta^+)} \left(1-e^{-\alpha}\right).\end{aligned}$$

Q.E.D. ■

Lemma 3 *The singular element* $\Psi_{\mathfrak{g}}^{(\xi)}$ *for the module* $L^{\mu}\otimes L^{\nu}$ *can be presented in two equivalent forms:*

$$\mathrm{ch}\left(L^{\mu}_{\mathfrak{g}}\right)\Psi^{\nu}_{\mathfrak{g}} = \Psi^{\mu}_{\mathfrak{g}}\mathrm{ch}\left(L^{\nu}_{\mathfrak{g}}\right). \tag{1}$$

Proof. In the Weyl formula for $L^{\mu}\otimes L^{\nu}$,

$$\mathrm{ch}\left(L^{\mu}\otimes L^{\nu}\right)_{\downarrow P^+_{\mathfrak{g}}} = \sum_{\xi\in P^{++}_{\mathfrak{g}}} m^{\mu\nu}_{\xi}\mathrm{ch}\left(L^{\xi}\right),$$

apply the result of Lemma 1:

$$\left(\frac{\Psi^{(\nu,\xi)}_{\mathfrak{g}\oplus\mathfrak{g}}}{\Psi^{0}_{\mathfrak{g}\oplus\mathfrak{g}}}\right)_{\downarrow P_{\mathfrak{g}}} = \frac{\Psi^{\mu}_{\mathfrak{g}}\Psi^{\nu}_{\mathfrak{g}}}{\Psi^{0}_{\mathfrak{g}}\Psi^{0}_{\mathfrak{g}}} = \sum_{\xi\in P^{++}_{\mathfrak{g}}} m^{\mu\nu}_{\xi}\frac{\Psi^{\xi}_{\mathfrak{g}}}{\Psi^{0}_{\mathfrak{g}}}.$$

$$\left(\left(\Psi^{0}_{\mathfrak{g}}\right)^{-1}\Psi^{\mu}_{\mathfrak{g}}\right)\Psi^{\nu}_{\mathfrak{g}} = \Psi^{\mu}_{\mathfrak{g}}\left(\left(\Psi^{0}_{\mathfrak{g}}\right)^{-1}\Psi^{\nu}_{\mathfrak{g}}\right) = \sum_{\xi\in P^{++}_{\mathfrak{g}}} m^{\mu\nu}_{\xi}\Psi^{\xi}_{\mathfrak{g}}.$$

Thus we have

$$\sum_{\xi\in P^{++}_{\mathfrak{g}}} m^{\mu\nu}_{\xi}\Psi^{\xi}_{\mathfrak{g}} = \mathrm{ch}\left(L^{\mu}_{\mathfrak{g}}\right)\Psi^{\nu}_{\mathfrak{g}} = \Psi^{\mu}_{\mathfrak{g}}\mathrm{ch}\left(L^{\nu}_{\mathfrak{g}}\right). \tag{2}$$

Q.E.D. ■

Now put $\mu=\omega$, $\nu=(p-1)\,\omega$

$$\mathrm{ch}\left(L^{(\omega)}\right)\Psi^{\left(\otimes^{(p-1)}\omega\right)} = \sum_{\xi\in P} M^{\omega_1}_{\mathfrak{g}}\left(\xi,p\right)\Psi^{(\xi)}, \tag{3}$$

$M^{\omega_1}_{\mathfrak{g}}(\xi,p)$ defines the singular element $\Psi^{(\otimes^{p}\omega)}$. On the other hand these equations can be considered as a system of recurrent relations for the multiplicity function $M^{\omega_1}_{\mathfrak{g}}(\xi,p)$.

Proposition 4 *Let* $\mathfrak{g} = B_2$, L^{μ} *and* $L^{\omega_1} = L^{vect}$ *be the highest weight modules with the highest weights* μ *and* ω_1 *(the first fundamental weight). Then the tensor product decomposition* $\left(L^{\mu}\otimes L^{vect}\right)\downarrow_{\mathfrak{g}_{\mathrm{diag}}} = \oplus_{\gamma}L^{\gamma}$ *is multiplicity free.*

Proof. According to Lemma 1 the projected singular element for a $\mathfrak{g}\oplus\mathfrak{g}$ -module $L^{\mu}\otimes L^{vect}$ is $\Psi^{(\mu,vect)}_{\mathfrak{g}\oplus\mathfrak{g}\downarrow\mathfrak{g}_{\mathrm{diag}}} = \Psi^{\mu}_{\mathfrak{g}}\Psi^{vect}_{\mathfrak{g}}$ and the set of singular weights is

$$\begin{aligned}\widehat{\Psi^{(\mu,vect)}_{\downarrow\mathfrak{g}_{\mathrm{diag}}}} &= \left\{\psi^{(\mu)}_{k} + \psi^{(vect)}_{p}, \epsilon\left(w\left(\psi^{(\mu)}_{k}\right)\right)\epsilon\left(w\left(\psi^{(vect)}_{p}\right)\right)\right\}, \\ k,p &= 1,\ldots,|W|.\end{aligned}$$

Suppose $\mu = (\mu_1, \mu_2)$ is greater than $\omega_1 = (1, 0)$ and $\mu_1 > \mu_2 \geq 1$. The singular weights of L^μ in the fundamental chamber and its nearest neighbours are

$$\begin{aligned}\left\{\psi_s^{(\mu)}\right\} &= (\mu_1, \mu_2, (+1)), (\mu_1, -\mu_2 - 1, (-1)), (\mu_2 - 1, \mu_1 + 1, (-1)) \\ s &= 1, 2, 3\end{aligned}$$

They will give rise to the 24 weights of the type

$$\begin{aligned}\left\{\psi_s^{(\mu)} + \psi_p^{(vect)}\right\} &= (\mu_1, \mu_2, (+1)) + \psi_p^{(vect)}, \\ &\quad (\mu_1, -\mu_2 - 1, (-1)) + \psi_p^{(vect)}, \\ &\quad (\mu_2 - 1, \mu_1 + 1, (-1)) + \psi_p^{(vect)} \\ s &= 1, 2, 3; p = 1, \ldots, \#W.\end{aligned}$$

Applying successively the fan $\Gamma_{\mathfrak{g}_{\text{diag}} \to \mathfrak{g} \oplus \mathfrak{g}}$ to the set $\widehat{\Psi^{(\mu, vect)}_{\downarrow \mathfrak{g}_{\text{diag}}}}$ in the three selected chambers (starting with the highest weight $(\mu_1 + 1, \mu_2, (+1)))$ we find the weights:

$$\begin{aligned}&(\mu_1, \mu_2, (+1)), (\mu_1 \pm 1, \mu_2, (+1)), (\mu_1, \mu_2 \pm 1, (+1)), \\ &(\mu_1, -\mu_2 - 1, (-1)), (\mu_1 \pm 1, -\mu_2 - 1, (-1)), (\mu_1, -\mu_2, (-1)), (\mu_1, -\mu_2 - 2, (-1)), \\ &(\mu_2 - 1, \mu_1 + 1, (-1)), (\mu_2, \mu_1 + 1, (-1)), (\mu_2 - 2, \mu_1 + 1, (-1)), \\ &(\mu_2 - 1, \mu_1 + 2, (-1)), (\mu_2 - 1, \mu_1, (-1)).\end{aligned}$$

Only the first 5 weights are in $\overline{C^{(0)}}$. They are the highest singular weights for $\left(L^\mu \otimes L^{vect}\right)_{\downarrow \mathfrak{g}_{\text{diag}}}$ and we have the decomposition:

$$\left(L^\mu \otimes L^{vect}\right)_{\downarrow \mathfrak{g}_{\text{diag}}} = L^\mu \oplus L^{(\mu_1+1, \mu_2)} \oplus L^{(\mu_1-1, \mu_2)} \oplus L^{(\mu_1, \mu_2+1)} \oplus L^{(\mu_1, \mu_2-1)}.$$

There are two special cases where the highest weights are on the borders of $\overline{C^{(0)}}$. For $\mu = n\omega_1 = (n, 0)$ and for $\mu = n\omega_2 = (n/2, n/2)$ the same algorithm gives:

$$\left(L^\mu \otimes L^{vect}\right)_{\downarrow \mathfrak{g}_{\text{diag}}} = L^{(\mu_1+1, 0)} \oplus L^{(\mu_1-1, 0)} \oplus L^{(\mu_1, \mu_2+1)},$$

$$\left(L^\mu \otimes L^{vect}\right)_{\downarrow \mathfrak{g}_{\text{diag}}} = L^\mu \oplus L^{(\mu_1+1, \mu_2)} \oplus L^{(\mu_1, \mu_2-1)}.$$

correspondingly. Q.E.D. ■

4. Singular elements and fans. B_2-case

For $\mathfrak{g} = B_2$, $r = 2$ the simple roots in e-basis are $\alpha_1 = e_1 - e_2, \alpha_2 = e_2$, the fundamental weights are $\omega_2 = \frac{1}{2}(e_1 + e_2), \omega_1 = e_1$ and the fundamental modules – L^{ω_2}(spinor) and $\dim L^{\omega_2} = 4$, L^{ω_1}(vector), $\dim L^{\omega_1} = 5$.

Consider the modules $(L^{\omega_i})^{\otimes p}|_{p \in Z_+, i=1,2}$ and the decompositions $(L^{\omega_i})^{\otimes p} = \sum_\nu m_\nu^{(i)p} L^\nu$.

Our aim is to find multiplicities $m_\nu^{(i)p}$ as functions of ν and p. To solve the problem we propose to use the *singular elements formalism* [16], the polynomial dependence property that is a consequence of the relation (3) and the injection fan technique [17] [18].

4.1. Constructing the fan $\Gamma_{B_2^{\mathrm{diag}} \to \oplus^p B_2}$

Consider the injection $B_2^{\mathrm{diag}} \to \oplus^p B_2$. The fan $\Gamma_{B_2^{\mathrm{diag}} \to \oplus^p B_2} \equiv \Gamma_p$ is the $(p-1)$-th tensor power of the trivial module singular element that is of $\Psi_{B_2}^{(0)}$.

Proposition 5 *Place the origin of the space $\mathcal{L}P$ at the end of the lowest weight vector of the fan. The structure of the fan Γ_p is as follows:*

(i) Along the line $p\alpha_1$ in the k-th root lattice point the multiplicity is $(-1)^k C_p^{k-1}$; $k = 1, \ldots, p+1$.

(ii) Each weight with the coordinates $(k-1, 1-k)$ is an origin of the set $S_{(k-1,1-k)}$ of singular weights described below.

(iii) The set $S_{(k-1,1-k)}$ is composed of the tensor product of the singular elements $\Psi_x^{[0]<}$ and $\Psi_y^{[v]<}$ of the trivial and vector lowest weight modules of the algebras A_1 with the roots x and y correspondingly. The list of modules for the set $S_{(k-1,1-k)}$ is completely defined by its coordinates $(k-1, 1-k)$:

$$S_{(k-1,1-k)} = \left(\Psi_{e_2}^{[v]<}\right)^{\otimes(k-1)} \otimes \left(\Psi_{e_1}^{[0]<}\right)^{\otimes(k-1)} \otimes \left(\Psi_{e_1}^{[v]<}\right)^{\otimes(p-k+1)} \otimes \left(\Psi_{e_2}^{[0]<}\right)^{\otimes(p-k+1)}.$$

Proof. Let Γ_p be the fan with the properties described above. Remember that the set $S_{(k-1,1-k)}$ has itself the multiplicity $(-1)^k C_p^{k-1}$. Multiply the fan Γ_p by the element $\Psi^{(0)}$, i.e. pass to the power $(p+1)$. This means that the set $S_{(k-1,1-k)}$ will be transformed to

$$(-1)^k C_p^{k-1} \left(\Psi_{e_2}^{[v]<}\right)^{\otimes(k-1)} \otimes \left(\Psi_{e_1}^{[0]<}\right)^{\otimes(k-1)} \otimes \left(\Psi_{e_1}^{[v]<}\right)^{\otimes(p-k+2)} \otimes \left(\Psi_{e_2}^{[0]<}\right)^{\otimes(p-k+2)}.$$

The set $S_{(k-2,2-k)}$ will become $S_{(k-1,1-k)}$ with the multiplicity $(-1)^{k+1} C_p^{k-2}$. According to the Pascal triangle property,

$$(-1)^k C_p^{k-1} + (-1)^{k+1} C_p^{k-2} = (-1)^k C_{p+1}^{k-1},$$

the first set $S_{(0,0)}$ takes the form

$$\left(\Psi_{e_1}^{[v]<}\right)^{\otimes(p+1)} \otimes \left(\Psi_{e_2}^{[0]<}\right)^{\otimes(p+1)},$$

while the last one becomes $S_{(p,p)}$,

$$(-1)^{p+2} \left(\Psi_{e_2}^{[v]<}\right)^{\otimes(p+1)} \otimes \left(\Psi_{e_1}^{[0]<}\right)^{\otimes(p+1)}.$$

Thus the structure of this product coinsides with the previously defined fan Γ_q with $q = p+1$. Q.E.D. ∎

An explicit expression for the fan Γ_p is obtained by substituting $\Psi_{e_i}^{[0]<}$ and $\Psi_{e_i}^{[v]<}$ (as formal algebra elements) by their expressions in terms of a function

$$\widehat{C}_j^i = \begin{cases} C_j^i & \text{for } 0 < i \leq j > 0 \\ 0 & \text{otherwise.} \end{cases}$$

Finally we get

$$\Gamma_p = \sum_{a,b} \gamma_p(a,b)\, e^{(a,b)}; \quad \begin{cases} a = k-1,\dots,3p-k-2, \\ b = 1-k,\dots,p+k-2, \end{cases}$$

$$\gamma_p(a,b) = \sum_{k=1}^{p}\sum_{l_k=1}^{k}\sum_{m_k=1}^{p-k+1} e^{(a,b)}(-1)^{k+a+b-2(l_k+m_k)} \times \times \widehat{C}_{p-1}^{k-1}\widehat{C}_{k-1}^{l_k-1}\widehat{C}_{p-k}^{m_k-1}\widehat{C}_{p-k}^{b+k-3l_k+2}\widehat{C}_{k-1}^{a-k-3m_k+4}, \quad (4)$$

Here the fan is a function of the parameter p and the coordinates (a,b) of the highest weight. The zero point has the multiplicity (-1). The result is that for any $\mu \in P$ the singular weights diagram $\Psi\left(\sum_\nu m_\nu^{(i)p} L^\nu\right) = \sum e^{(\mu)}\epsilon\left(w\left(\psi\left(\mu\right)\right)\right)$ has the following fundamental property:

$$\sum_{a,b}\gamma_p(a,b)\,\epsilon\left(w\left(\psi\left(\mu+(a,b)\right)\right)\right) = 0,$$

described by the fan Γ_p.

4.2. Singular element for $(L^{\omega_2})^{\otimes p}$ – the spinor case

Let us construct the singular element for the p-th power of the second (spinor) fundamental module L^{ω_2} (coordinates of singular weights here are half-integer and the W-invariant vector is $(0,0)$).

(i) Define the system S_k consisting of blocks enumerated by a pair of indices (l_k, m_k) where $l_k = 1,\dots,k+1$ and $m_k = 1,\dots,p-k+2$ and attached to the points $\left(\frac{p}{2}-k+1-4(m_k-1), \frac{p}{2}+k-1-4(l_k-1)\right)$. The multiplicities of these blocks are $(-1)^{l_k+m_k-2}\,\widehat{C}_k^{l_k-1}\widehat{C}_{p-k+1}^{m_k-1}$

(ii) Localize the systems S_k along the line $\left(\frac{p}{2},\frac{p}{2}\right)-p\alpha_1$: the first system S_1 has the origin at the point $\left(\frac{p}{2},\frac{p}{2}\right)$, the k-th – at the point $\left(\frac{p}{2}-k+1, \frac{p}{2}+k-1\right)$, the last one, S_{p+1}, – at $\left(-\frac{1}{2}p, \frac{3}{2}p\right)$. These systems have the multiplicities $(-1)^{k-1}\,C_p^{k-1}$; $k = 1,\dots,p+1$.

(iii) The numbers $\left(\frac{p}{2}-k+1-4(m_k-1), \frac{p}{2}+k-1-4(l_k-1)\right)$ are the coordinates of the upper right corner of the (l_k,m_k)-block. The blocks have the structure dual to the structure of the system S_k but the intervals in the blocks are doubled. The weights in the block are enumerated by the indices (i_k,j_k) where $j_k = 1,\dots,k+1$ and $i_k = 1,\dots,p-k+2$.

(iv) Thus the (l_k,m_k)-block in S_k has the form:

$$\sum_{i_k=1}^{p-k+2}\sum_{j_k=1}^{k+1}(-1)^{i_k+j_k-2}\,\widehat{C}_k^{j_k-1}\widehat{C}_{p-k+1}^{i_k-1}e^{\left(\frac{p}{2}-k+1-4m_k-2j_k+7, \frac{p}{2}+k-1-4l_k-2i_k+5\right)}.$$

(v) Now the system S_k can be composed:

$$\sum_{i_k,m_k=1}^{p-k+2}\sum_{j_k,l_k=1}^{k+1}(-1)^{l_k+m_k+i_k+j_k-4} \times \times \widehat{C}_k^{l_k-1}\widehat{C}_{p-k+1}^{m_k-1}\widehat{C}_k^{j_k-1}\widehat{C}_{p-k+1}^{i_k-1}e^{\left(\frac{p}{2}-k+1-4m_k-2j_k+7, \frac{p}{2}+k-1-4l_k-2i_k+5\right)}.$$

(vi) Finally, the singular element $\Psi^{((\omega_2)^{\otimes p})} \equiv \Psi^{((s)^{\otimes p})}$ is fixed as

$$\begin{aligned}\Psi^{((s)^{\otimes p})} &= \sum_{k=1}^{p+1}\sum_{i_k,m_k=1}^{p-k+2}\sum_{j_k,l_k=1}^{k+1}(-1)^{l_k+m_k+i_k+j_k-4}\times\\ &\times\widehat{C}_k^{l_k-1}\widehat{C}_{p-k+1}^{m_k-1}\widehat{C}_k^{j_k-1}\widehat{C}_{p-k+1}^{i_k-1}\times\\ &\times e^{\left(\frac{p}{2}-k+1-4m_k-2j_k+7,\frac{p}{2}+k-1-4l_k-2i_k+5\right)}\end{aligned}$$

The singular multiplicities are functions of p and c,d:

$$\begin{aligned}\psi^{((s)^{\otimes p})}(c,d) &= e^{(c,d)}\sum_{k=1}^{p+1}\sum_{l_k=1}^{k}\sum_{m_k=1}^{p-k+2}(-1)^{k+\frac{1}{2}(c-d)-(l_k+m_k)+1}\times\\ &\times\widehat{C}_p^{k-1}\widehat{C}_k^{l_k-1}\widehat{C}_k^{\frac{1}{2}\left(4(1-m_k)-k+c-\frac{1}{2}p+1\right)}\times\\ &\times\widehat{C}_{p-k+1}^{m_k-1}\widehat{C}_k^{\frac{1}{2}\left(2-4m_k+k-d+\frac{1}{2}p+1\right)}.\end{aligned}$$

4.3. Singular element for $(L^{\omega_1})^{\otimes p}$ – the vector case

The construction procedure in the vector case is analogous to that of the spinor. It results in obtaining the expression:

$$\begin{aligned}\Psi^{((\omega_1)^{\otimes p})} &\equiv \Psi^{((v)^{\otimes p})} = \sum_{k=1}^{p+1}\sum_{j,n=0}^{k-1}\sum_{i,m=0}^{p-k+1}e^{(2k+j+5m-4p-2,-2p+i-k+5n+1)}\times\\ &\times(-1)^{i+j+k+m+n-1}\widehat{C}_{p-k+1}^{m}\widehat{C}_{k-1}^{n}\widehat{C}_{p-k}^{j}\widehat{C}_p^{k-1}\widehat{C}_{p-k+1}^{i}.\end{aligned}$$

The corresponding singular multiplicities function depending on p and c,d are

$$\begin{aligned}\psi^{((v)^{\otimes p})}(c,d) &= \sum_{k=1}^{p+1}\sum_{l_k=1}^{k}\sum_{m_k=1}^{p-k+2}e^{(c,d)}\times\\ &\times(-1)^{k-d-c+p-4(l_k+m_k)+7}\times\\ &\times\widehat{C}_p^{k-1}\widehat{C}_{k-1}^{l_k-1}\widehat{C}_{p-k+1}^{m_k-1}\widehat{C}_{p-k+1}^{-d+2k-5(l_k-1)-2}\widehat{C}_{k-1}^{p-c-2k-5(m_k-1)+2}.\end{aligned} \tag{5}$$

4.4. Recursive procedure for the vector case

To illustrate the recursive algorithm we perform calculations that must give us the multiplicities $m_\nu^{(1)p}$. The starting value is always known – this is the multiplicity of the highest weight $\nu = (p,0)$ of $(L^{\omega_1})^{\otimes p}$ that equals 1. Suppose we have found the values of the multiplicity function $m_\nu^{(1)p}$ for the 14 first weights:

$c\setminus d$	-2	-1	0	$+1$	$+2$	$+3$
p	0	-1	$+1$	0	0	0
$p-1$	$1-p$	0	0	$p-1$	0	0
$p-2$					$\frac{1}{2}p(p-3)$	0

Applying the fan Γ_p (4) we find the multiplicity for the next weight in the third line, it has the coordinates $(p-2,1)$. The first line of the fan weights contributes the value

$$(p-1)\times\frac{1}{2}p(p-3),$$

the second line –

$$-(p-1)(p-2)\times(p-1)$$

and the third line –

$$\left(+\frac{1}{2}(p-1)(p-2)(p-3)\right)\times(+1)+$$
$$+\left((p-1)-\frac{1}{2}(p-1)(p-2)\right)\times(-1)$$

Now we must calculate the singular element contribution – the value of the singular weights function $\psi^{\left((v)^{\otimes p}\right)}(p-2,1)$ for the weight $(p-2,1)$. The "singular contribution" is

$$\psi^{\left((v)^{\otimes p}\right)}(p-2,1)=p(p-1).$$

The multiplicity $m_{(p-2,1)}^{(1)p}$ is the sum:

$$\begin{aligned} m_{(p-2,1)}^{(1)p} &= (p-1)\left(\begin{array}{c} \frac{1}{2}p(p-3)-(p-1)(p-2)+p \\ +\frac{1}{2}(p-2)+\frac{1}{2}(p-2)(p-3)-1 \end{array}\right) \\ &= \frac{1}{2}(p-1)(p-2). \end{aligned}$$

Notice that here we consider the line $\nu=(p-2,1)$ in the space $P\times\mathbf{R}^1$. As a result we obtain polynomials characterizing the p-dependence of the multiplicity for a fixed distance between the highest weight and the weight ν.

This example shows that the tools elaborated above (the injection fan and singular elements) are effective in solving the reduction problem for tensor power modules $(L^{\omega_i})^{\otimes p}$.

5. Alternative approach.

According to Lemma 3 the multiplicity coefficients have additional recurrence properties generated by the fundamental module weights system $N\left(L^{(\omega_i)}\right)$:

$$\sum_{\mu\in P^{++}} m_\mu^{(i)p}\Psi^{(\mu)}=\operatorname{ch}\left(L^{(\omega_i)}\right)\Psi^{\left(\otimes^{(p-1)}\omega_i\right)}. \tag{6}$$

This relation can be decomposed using the multiplicity functions (defined on P),

$$\sum_{\mu\in P} M^{\omega_i}(\mu,p)\,e^{\mu}=\operatorname{ch}\left(L^{(\omega_i)}\right)\Psi^{\left(\otimes^{(p-1)}\omega_i\right)}, \tag{7}$$

remember that $M^{\omega_i}(\mu,p)\mid_{\mu\in\overline{C^{(0)}}}=m_\mu^{(i)p}$. Thus instead of the highest weight search for each singular element $\Psi^{(\mu)}$ we use their anti-symmetry properties. This leads to the recurrent relation:

$$M^{\omega_i}(\mu,p)=\sum_{\zeta\in N\left(L^{(\omega_i)}\right)} n_\zeta\left(L^{(\omega_i)}\right)M^{\omega_i}(\mu-\zeta,p-1), \tag{8}$$

where $n_\zeta\left(L^{(\omega_i)}\right)=\operatorname{mult}_{L^{(\omega_i)}}(\zeta)$. Obviously such relations are especially useful when $\dim\left(L^{(\omega_i)}\right)$ is small, thus for our needs (when the module $L^{(\omega_i)}$ is fundamental with the trivial weights multiplicities $n_\zeta\left(L^{(\omega_i)}\right)=1$) the obtained recurrence must be effective.

Formula (8) tells us what happens when we pass from the $(p-1)$-th power to the p-th. In particular for the spinor module L^{ω_2} to find the value of $M^{\omega_2}(\mu, p)$ the coordinates must be shifted by the vectors of $\mathcal{N}\left(L^{(\omega_2)}\right)$ diagram:

$$M^{\omega_2}(\mu, p) = \sum_{\zeta = N\left(L^{(\omega_2)}\right)} M^{\omega_2}(\mu - \zeta, p-1). \tag{9}$$

In the recurrence starting point the value of the multiplicity function is known $M^{\omega_2}(p\omega_2, p) = 1$ and for all $\nu > p\omega_2$ it has zero values. In the natural coordinates this means:

$$M^{\omega_2}((a,b), p) = \sum_{\lambda = (a,b) - \left\{\left(\frac{1}{2},\frac{1}{2}\right)\left(-\frac{1}{2},\frac{1}{2}\right)\left(\frac{1}{2},-\frac{1}{2}\right)\left(-\frac{1}{2},-\frac{1}{2}\right)\right\}} M^{\omega_2}(\lambda, p-1).$$

For the vector module L^{ω_1} and its tensor powers we can construct similar relations:

$$\begin{aligned} M^{\omega_1}((a,b), p) &= \sum_{\zeta = N\left(L^{(\omega_1)}\right)} M^{\omega_1}((a,b) - \zeta, p-1) = \\ & \sum_{\lambda = (a,b) - \{(1,0)(0,1)(-1,0)(0,-1),(0,0)\}} M^{\omega_1}(\lambda, p-1), \end{aligned} \tag{10}$$

with the similar boundary condition

$$M^{\omega_1}((p,0), p) = 1. \tag{11}$$

The obtained recurrence relations indicate an important property of $M^{\omega_i}(\mu, p)$:

Statement 6 *The multiplicity function $M^{\omega_i}(\mu, p)$ is a polynomial on p over* **Q** *(rational numbers).*

Notice that when μ belongs to the correlated (different!) boundaries of the area where the function is nontrivial the values $M^{\omega_i}(\mu, p)$ for $i = 1$ and $i = 2$ coincide.

Statement 7 *The multiplicities $M^{\omega_1}(\mu, p)$ of the "upper diagonal" highest weights ($\mu = (p,0) - n\alpha_1$) for $(L^{\omega_1})^{\otimes p}$ coinside with the multiplicities $M^{\omega_2}(\mu, p)$ of the "upper line" highest weights ($\mu = \left(\frac{p}{2}, \frac{p}{2}\right) - n\alpha_2$) for $(L^{\omega_2})^{\otimes p}$.*

On the left boundary of the Weyl chamber $\overline{C^{(0)}}$ the multiplicities $M^{\omega_1}(\mu, p)$ are subject to the presence of the "reflected" singular weights in the left adjacent Weyl chamber. This observation is important because for an analogous boundary in the spinor case the situation is different: the adjacent Weyl chamber had no influence on the values of $M^{\omega_2}(\mu, p)$ with $\mu \in \overline{C^{(0)}}$ (on the corresponding subdiagonal the function has zero values). In particular this results in the following property.

Statement 8 *The "second diagonal" of the highest weights for $(L^{\omega_1})^{\otimes p}$ starts with zero: $M^{\omega_1}((p-1)\omega_1, p) = 0$.*

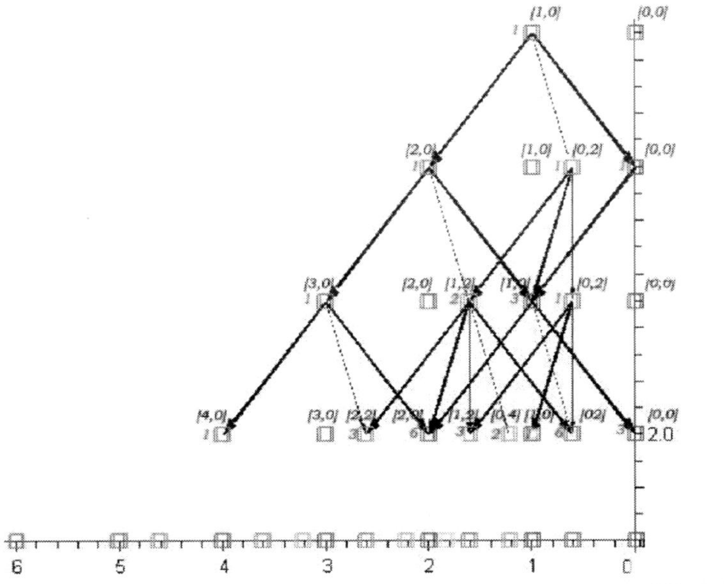

Figure 1. Bratteli-like diagram for B_2 vector module L^{ω_1} for $p = 1, \ldots, 4$. The maximal number of paths that connect the $p-1$ slice with the p-th one is five.

6. Solutions for recurrence relations

We have found out that the multiplicity functions $M^{\omega_i}(\mu, p)$ are subject to an infinite system of coupled algebraic equations with simple and obvious boundary conditions. They can be solved step by step.

For example consider the Bratteli-like diagram for B_2 vector module L^{ω_1} and let $p = 1, \ldots, 6$. The maximal number of paths that connect a point in the $p-1$ slice with the points in the p-th one is five.

Notice that the path counting procedure here is very complicated because of the boundary effects. Such complexities grow up considerably if we try to apply that counting procedures to algebras with higher rank. This fact stimulates special interest to direct studies of the recurrence relations systems. Moreover if the corresponding equations could be solved this will give an explicit p-dependence of the multiplicity function – the result that scaresly could be achieved by combinatorial methods.

We can construct the solution for the recurrence equations successively and the answer is limited only by the number of equations in the system solved . This gives the explicit multiplicity dependence on p but for a finite number of successive weights. Thus having solved the first five equations for the spinor case we get the following table of functions $M^{\omega_2}((a, b), p)$ (here the coordinates (a, b) are fixed by the relation $\mu = p\omega_2 - a\alpha_2 - (b-1)\alpha_1$):

$b\setminus a$	$5/2$	2	$3/2$	1	$1/2$	0
1		$\frac{1}{2}p\left(p-3\right)$		$p-1$		1
	0		0		0	
2		$\frac{1}{3}\left(p-1\right)\cdot\left(p+1\right)\left(p-3\right)$		$\frac{1}{2}p\left(p-1\right)$		0
	0		0		0	
3		$\frac{1}{12}\left(p-1\right)\left(p-2\right)\cdot\left(p-3\right)\left(p+2\right)$		0		

Correspondingly having solved the first fifteen equations for the vector case we get the following table of functions $M^{\omega_1}\left(\left(a,b\right),p\right)$,

$b\setminus a$	0	1	2	3
p	1	0	0	0
$p-1$	0	$p-1$	0	0
$p-2$	$\frac{1}{2}p\left(p-1\right)$	$\frac{1}{2}\left(p-1\right)\left(p-2\right)$	$\frac{1}{2}p\left(p-3\right)$	0
$p-3$	$\frac{1}{6}\left(p-1\right)\left(p-2\right)\left(p-3\right)$	$\frac{1}{2}p\left(p-1\right)\left(p-3\right)$	$\frac{1}{3}p\left(p-2\right)\left(p-4\right)$	$\frac{1}{6}p\left(p-1\right)\left(p-5\right)$

7. Weyl symmetry and solutions for recurrence equations

When the algebra is simply laced, for example $\mathfrak{g}=A_n$, the Weyl symmetry was proven to be a highly effective tool to solve the set of recurrences equations for the powers of the first fundamental module $L_{A_n}^{\omega_1}$ [19]. In the simplest case $\mathfrak{g}=A_1$ the complete set of multiplicity functions for powers of an arbitrary irreducible module were thus constructed.

In the case of B_2 the difficulties start when the vector fundamental module is tensored. The recurrence equation can be solved successively, as was shown in the previous section, but the complete solution for the function $M^{\omega_1}\left(\left(a,b\right),p\right)$ was not found.

Nevertheless the recurrence property (9) permits to describe the general dependence of the multiplicities on one of the coordinates. To see this consider the coordinates (s,t) defined by the relation $\mu=p\omega_2-t\alpha_1-\left(s-1\right)e_1$. The t-dependence will be explicitly described, but only for limited values of $s=1,2,\ldots$. This description is based on the fact that the proposition 7 gives us an explicit answer to the "first diagonal" of multiplicities $\left\{m_{(1,t)}^{(1)p}|t=0,1,\ldots\right\}$. Starting with this expression and using the relation (9) reformulated for the "diagonal lines" of functions we can find explicit expressions for any such line provided the previous lines are known:

$$M^{\omega_1}\left(\left(1,t\right),p\right)=M^{\omega_2}\left(\left(\frac{p}{2},\frac{p}{2}-t\right),p\right)=\frac{\Gamma\left(p+1\right)\left(p+1-2t\right)}{\Gamma\left(p+2-t\right)\Gamma\left(t+1\right)},$$

$$M^{\omega_1}\left(\left(2,t\right),p\right)=\frac{\Gamma\left(p+1\right)\left(p-t\right)\left(p-2t\right)}{\Gamma\left(p+2-t\right)\Gamma\left(t\right)\left(t+1\right)},$$

$$M^{\omega_1}\left(\left(3,t\right),p\right)=\frac{\Gamma\left(p+1\right)\left(p-2t-1\right)}{2\Gamma\left(p-t\right)\Gamma\left(t+1\right)},$$

$$\begin{aligned}M^{\omega_1}\left(\left(4,t\right),p\right) &= \frac{\Gamma\left(p+1\right)\left(p-2t-2\right)}{6\Gamma\left(p+1-t\right)\Gamma\left(t+3\right)}\cdot\\ &\cdot\left(\begin{array}{c}\left(t^2+6t+2\right)p^2-2\left(t+2\right)^2\left(t+1\right)p+\\ +t^4+4t^3+8t^2+8t+6\end{array}\right),\end{aligned}$$

$$M^{\omega_1}((5,t),p) = \frac{\Gamma(p+1)(p-2t-3)}{24\Gamma(p-t)\Gamma(t+3)} \cdot \cdot \begin{pmatrix} (t^2+11t+6)p^2-(2t+4)(t+1)(t+2)p+ \\ +(t^4+4t^3+8t^2+8t+6) \end{pmatrix},$$

$$M^{\omega_1}((6,t),p) = \frac{\Gamma(p+1)(p-2t-4)}{120\Gamma(p-t)\Gamma(t+4)} \cdot \cdot \begin{pmatrix} (t^3+21t^2+86t+36)p^3- \\ -(3t^4+54t^3+309t^2+654t+276)p^2+ \\ +(3t^5+45t^4+326t^3+1086t^2+1408t+516)p- \\ -(t+1)(t+3)(t^4+8t^3+68t^2+208t+12) \end{pmatrix},$$

and so on. We see that beginning from $M^{\omega_1}_{B_2}((4,t),p)$ only some factor of the multiplicity function can be presented as a product of simple binomials like $(p-x)$. In the forthcoming publications we shall discuss this property in details.$M^{\omega_1}((4,b),p)$

8. Conclusions

The tensor powers decomposition algorithm based on singular weights and injection fan technique was proven to be an effective tool in multiplicity property studies. Its abilities were demonstrated on tensor powers decompositions of B_2-fundamental modules. This algorithm is universal and can be applied to investigate decomposition properties in case of an arbitrary simple Lie algebra and its arbitrary module.

As it was predicted in [19] in non-simply laced case the Weyl symmetry properties are insufficient to provide the final solution for the corresponding set of recurrence relations for multiplicity functions (at least this appeared to be true for the vector fundamental modules). Nevertheless (this was shown above in our studies of fundamental B_2-modules) important properties of multiplicity coefficients for any highest weight ν can be found by constructing the functions $M^{\omega_i}((\nu),p)$ successively i.e. by constructing the solution for a final part of the full set of recurrence relations.

Acknowledgments

Supported by the Russian Foundation for Fundamental Research grant N 09-01-00504 and the "Dynasty" Foundation.

References

[1] Kirillov A N and Reshetikhin N Yu 1996 *Journ. of Math. Sci.* **80** 3
[2] Kleber M 1996 Combinatorial Structure of Finite Dimensional Representations of Yangians: the Simply-Laced Case Preprint q-alg/9611032v2
[3] Kleber M 1998 Finite Dimensional Representations of Quantum Affine Algebras Preprint math.QA/9809087v1
[4] Chari V On the fermionic formula and the Kirrilov-Reshetikhin proposition Preprint math.QA/0006090v2
[5] Hatayama G, Kuniba A, Okado M, Takagi T and Yamada Y 1998 Remarks on Fermionic Formula Preprint math.QA/9812022v3
[6] Hatayama G, Kuniba A, Okado M and Takagi T 2000 Combinatorial R-matrices for a family of crystals: $B_n(1)$, $D_n(1)$, $A_{2n}(2)$ and $D_{n+1}(2)$ cases Preprint math.QA/0012247v1
[7] Hatayama G, Kuniba A, Okado M, Takagi T and Tsuboi Z 2001 Paths, Crystals and Fermionic Formulae Preprint math.QA/0102113v1
[8] Naito S and Sagaki D 2006 *Commun. Math. Phys.* **263** 749
[9] Leduc R and Ram A 1997 *Adv. Math.* **125** 1
[10] Kulish P P, Manojlovic N and Nagy Z 2010 *J. Math. Phys.* **51** 043516

Journal of Physics: Conference Series **346** (2012) 012012 doi:10.1088/1742-6596/346/1/012012

[11] Kumar Shrawan 2010 *Proc. Int. Congr. of Mathematicians 19-27 August 2010 (Hyderabad)* vol 2 (World Scientific) pp 1226-1261

[12] Klimyk A U 1968 *Amer. Math. Soc. Transl.* **2(76)** 63

[13] Klimyk A and Schmudgen K 1997 *Quantum groups and their representations* (Berlin: Springer)

[14] Verma D N 1968 *Bull. Amer. Math. Soc.* **74** 160

[15] Bernstein I N, Gel'fand I M and Gel'fand S I 1971 *Functional. Analiz i Prilozhen.* **8** 1

[16] Feigin B L, Fuchs D B and Malikov F G 1986 *Funct. Anal. Appl.* **20** 25

[17] Ilyin M, Kulish P and Lyakhovsky V 2009 *Algebra i Analiz* **21** 52

[18] Lyakhovsky V and Nazarov A 2011 *J. Phys. A* **44** 075205

[19] Kulish P P, Lyakhovsky V D and Postnova O V Multiplicity function for tensor powers of A_n-modules, accepted for publication in *Teoreticheskaya i matematicheskaya fisika*

Algebra, Geometry, and Mathematical Physics 2010 IOP Publishing
Journal of Physics: Conference Series **346** (2012) 012013 doi:10.1088/1742-6596/346/1/012013

Initial-boundary value problems for linear equations of electrodynamics with nonlinear boundary conditions

Konstantin Lukin

Institute for Radiophysics and Electronics, National Academy of Sciences of Ukraine, Kharkov, 61085 Ukraine

E-mail: Lukin.konstantin@gmail.com

Abstract. A new class of electromagnetic (EM) problems is introduced, initial-boundary value problem for linear wave equation with nonlinear boundary conditions, and methods for their solution are elaborated. Theory of nonlinear 1D maps and difference-differential equations have been applied to solve the problem. It has been shown that spatiotemporal dynamics of EM field in a resonator with nonlinear reflecting surfaces may be either regular or chaotic in spite of the major part of the system is formed by linear media. The approach may be used for analysis of chaotic processes in many electromagnetic and electronic devices.

1. Introduction

In many physical systems and electronic devices the volume of the active nonlinear elements is essentially smaller than that occupied by electromagnetic (EM) field (see in [1]) and therefore interaction between the field and such a nonlinear medium may be treated as a multiple scattering process. In general case the analysis of such systems requires formulation of initial-boundary value problems for 3D wave equation with nonlinear boundary conditions for the EM fields which is really hard task even for computer integration. That's why we restrict ourselves with considerations of simplest problem statements, enabling description of initial-boundary value problem for 1D wave equation. This will allow us to understand physical properties of the investigated processes, to choose analysis technique and to consider ways of solving more complicated problems of that class.

In the paper we have considered the problem of multiple scattering of nonstationary plane EM waves by a nonlinear reflecting surface, such as a thin layer containing currents of free and/or bound charges with their nonlinear dependences on the electric field. Appropriate boundary conditions are formulated and the expression for energy conservation law is derived. The simplest case of the multiple scattering of plane linear polarized wave at the normal reflection from an infinite plane nonlinear reflecting structure is considered as an example. It is shown that at rather strong nonlinearity oscillations with complicated dynamics and wideband frequency spectrum could be generated in the systems under consideration. Such oscillations can be interpreted as a turbulent state of EM field. Actually, we have formulated a new class of EM problems: initial-boundary value problem for linear wave equation with nonlinear boundary conditions and elaborated methods for their solving [2].

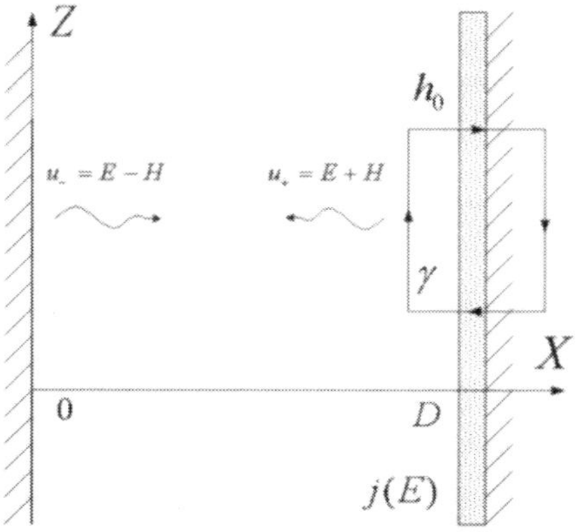

Figure 1. Geometry of 1D resonator with nonlinear reflecting surface

2. Boundary conditions and energy conservation law for electromagnetic fields in 1D resonator with nonlinear reflecting surface

Consider multiple scattering of a plane EM wave, propagating along OX axis, by nonlinearly reflecting structure, located at the plane $x = D$. Figure 1 shows geometry of the problem. Assume that a possible smallest space scale $\lambda_{\min}$ of the scattered field is much greater than the width d of the nonlinear layer: $\lambda_{\min} \gg d$. In this case interaction between the field and the layer can be described locally along the coordinate x and the scattering surface may be treated as a boundary with nonlinear reflection. For derivation of the above boundary condition for the EM field at this surface we use a well known approach. Magnetic field $\vec{H}$ and electric field $\vec{E}$ inside the layer obey the Maxwell equation

$$rot\vec{E} = -\frac{1}{c}\frac{\partial \vec{H}}{\partial t}; \; rot\vec{H} = \frac{1}{c}\frac{\partial \vec{E}}{\partial t} + \frac{4\pi}{c}\left\{\frac{\partial \vec{P}}{\partial t} + \vec{j}\right\} \tag{1}$$

where $\vec{j}$ is the current density of free charges; $\vec{P}$ is the polarization vector of bound charges; c is the light velocity in vacuum.

In order to derive boundary condition for EM fields at the nonlinear reflecting surface we have to integrate Eq. (1) over the area γ that contains all three media (Figure 1) and after that turn the height h_0 of the area γ to zero. Applying standard reasoning and taking into account condition $\lambda_{\min} \gg d$ and perfect conductivity of the metal, we may simplify both sides of Eq.(1) to obtain

$$[\vec{n}, \vec{H}]_{|x\in(D,D-d)|} \cong \frac{4\pi}{c}\left\{\frac{\partial \vec{P}^{surf}(E,y,z,t)}{\partial t} + \vec{j}^{surf}(E,z,y,t)\right\} \tag{2}$$

where volume current density and the polarization vector in Eq. (1) have been replaced with corresponding surface values in Eq. (2).

Boundary condition (2) is valid for an arbitrary EM field. It determines nonlinear relation between fields $\vec{E}$ and $\vec{H}$ at the considered boundary ($x \cong D$) that contains the layer with nonlinear currents.

Taking scalar product of Eq. (2) with vector $c\vec{E}/4\pi$, making cyclic permutation of the vectors in the left hand side and integrating over a surface S_0, we obtain the following expression:

$$S_x^{NL} \equiv \frac{c}{4\pi}\int_{S_0}[\vec{E}, \vec{H}]\,\vec{n}\,dS = -\int_{S_0}\vec{E}\frac{\partial \vec{P}^{surf}}{\partial t}dS - \int_{S_0}\vec{j}^{surf}\vec{E}dS. \tag{3}$$

Eq. (3) may be treated as an energy conservation law for the system under consideration. Left hand term in Eq. (3) is the Pointing vector of EM wave describing the EM energy flow through the surface S_0 perpendicular to it, and the right one is the work of the electric field over the surface currents of bounded and free charges of the layer within the area S_0. The work of the electric field E over the currents is positive if the layer absorbs its energy. If the layer has an active media, then its energy can be transferred to the EM field as it happens in self-oscillatory systems. Following Eq. (3), in the first case the energy flow is directed against the normal vector $\vec{n}$ to the surface, and we have an EM energy sink, while in the second case the EM energy flows along the opposite direction, and we have an EM energy source. Power reflection coefficient is determined by the ratio of X-components of Poynting vectors for the incident and reflected waves:

$$\Gamma = 1 + \frac{S_x^{NL}}{|S_x^L|},$$

where S_x^{NL} is determined by Eq. (3); S_x^L is X-component of the Poynting vector for the incident wave. It is clear from Eq. (3) that the reflection coefficient can exceed unity (active layer), be less than unity (partial absorbance) and can be equal to 0 (absolute absorbance) depending on the sign and the value of the E-field work over the layer current.

3. Multiple scattering of nonstationary waves

Consider linear polarized plane wave with components H_z and E_y at normal incidence angel on the nonlinear reflecting structure. Assume that there is no dependence of the current density and polarization vector in the layer on the coordinates z and y. Then the scattered field will not depend on these coordinates either. If we'll put at the plane $x = 0$ a perfect conductive surface, where general boundary condition is valid

$$E_y\,|_{x=0} = 0\,, \tag{4}$$

then the excited wave will scatter repeatedly on the nonlinearly reflecting structure. In this way a 1D resonator with nonlinear reflection of EM waves has been formed. Spatiotemporal dynamics of nonstationary plane waves in this resonator is described by equations, following from homogeneous Maxwell equations

$$\begin{aligned} \frac{\partial E_y}{\partial x} &= -\frac{1}{c}\frac{\partial H_z}{\partial t} \\ \frac{\partial E_z}{\partial x} &= -\frac{1}{c}\frac{\partial E_y}{\partial t} \end{aligned} \tag{5}$$

being supplied with the proper initial conditions for electric and magnetic fields and also boundary condition Eq. (2) and Eq. (4)

Linear equations of that type Eq. (5) with nonlinear boundary conditions have been considered, for example, in [3], [4], [5] for analysis of elastic oscillations in a bar with nonlinear elastic fastening of its ends; waves in periodical structures with local insertions of nonlinear elements, and nonlinear oscillations in transmission lines, respectively. However those problems have been solved approximately using asymptotic methods.

Introducing Riemann invariants $u_\pm \equiv H_z \pm E_y$ we may transform Eq. (5) to the following form

$$\frac{\partial u_\pm}{\partial \tau} \pm \frac{\partial u_\pm}{\partial \xi} = 0, \tag{6}$$

where $\xi = x/D$, $\tau = ct/D$.

Initial-boundary value problem for the functions $u_{\pm}(x,t)$ can be formulated as follows: it's necessary to find functions $u_{\pm}(x,t)$ obeying equations (6) and satisfying both the initial conditions

$$u_{\pm}(\xi,0)=U_{\pm}\equiv H_0(\xi)\pm E_0(\xi)$$

and boundary conditions (2) and (4) rewritten for the functions $u_{\pm}(x,t)$:

$$(u_+(1,\tau)+u_-(1,\tau))/2=-\frac{4\pi}{c}\left\{\frac{\partial P_y^{surf}[(u_+-u_-)/2]}{\partial t}+j_y^{surf}[(u_+-u_-)/2]\right\}_{|\xi=1}$$

$$u_+(0,\tau)=u_-(0,\tau). \tag{7}$$

For simplicity we neglect the dipole current of bound charges $\partial P_y^{surf}/\partial t=0$ in the second Eq. (7). Then initial boundary problem (6) – (7) is reduced to initial problem for nonlinear difference equation with continuous time using the method of characteristics (see for instance [5]).

Second condition (7) may be rewrite as follows

$$u_-(1,\tau)=-\nu(1,\tau)-\frac{4\pi}{c}j_y^{surf}[\nu(1,\tau)]\equiv\Psi(\nu) \tag{8}$$

where $\nu(1,\tau)\equiv\frac{1}{2}[u_+(1,\tau)-u_-(1,\tau)]$.

It follows from Eq. (8)

$$u_+(1,\tau)=u_-(1,\tau)+2\varphi[u_-(1,\tau)]\doteq\tilde{\varphi}[u_-(1,\tau)] \tag{9}$$

where $\varphi(u_-)$ is the inverse function of $\Psi(\bullet)$.

Solving equation (9) with respect to $u_-(1,\tau)$, we obtain

$$u_-(1,\tau)=g[u_+(1,\tau)], \tag{10}$$

where $g(u_+)$ is the inverse function of $\tilde{\varphi}(u\_)$.

Solutions for Eq. (6) are constant at their characteristics $\frac{d\xi}{d\tau}=\pm1$. Using this property of the solutions to Eq. (6) and also boundary conditions (7) and (10) we derive the following nonlinear difference equation with the continuous argument for the function $u_+(1,\tau)$:

$$u_+(1,\tau+2)=g[u_+(1,\tau)]. \tag{11}$$

This equation relates two values of the function $u_+(\xi,\tau)$ at the nonlinear structure ($\xi=1$) for two different instants of time separated by the delay interval $T=2D/c$, which is equal to round-trip propagation of the wave between the resonator planes. Solution $f(\tau)$ to the difference equation (11) with initial conditions

$$f(\tau)\Big|_{\tau\in[-1,1)}=U(\tau)\equiv\begin{cases}U_+ : \tau\in[-1,0)\\ U_- : \tau\in[0,1)\end{cases} \tag{12}$$

gives solutions for the initial-boundary value problem (6) – (7) in the following form

$$u_{\pm}(\xi,\tau)=f(\tau\mp\xi).$$

Electric and magnetic fields can be calculated from the formulae for Reimann invariants introduced before Eq. (6):

$$E_y(x;t)=\frac{1}{2}\left[f(\omega t-hx)-f(\omega t+hx)\right]$$

Journal of Physics: Conference Series **346** (2012) 012013 doi:10.1088/1742-6596/346/1/012013

$$H_z(x;t) = \frac{1}{2}\left[f(\omega t - hx) + f(\omega t + hx)\right], \tag{13}$$

where $h = D^{-1}$; $\omega = c/D$.

Thus, spatiotemporal dynamics of EM plane wave in 1D resonator with nonlinear reflecting boundary is fully determined by the solutions of difference equation (11). The solutions to Eq. (11) are determined by trajectories of 1D map $g(u)$. In the theory of dynamical systems defined by 1D maps, fundamental results have been obtained which allowed to define the character of the system motion and predicting its behavior with varying of map parameter. This also enabled to specify criteria for identification of system motion modes: regular, pre-turbulent and turbulent, etc. One of the main results of the theory is the famous Sharkovsky's theorem [6], [5]. According to this theorem any continuous map of an interval $g(u) : J \to J$ along with a cycle of period m includes also cycles of all previous periods m' in the following order (Sharkovsky's order):

$$1 \prec 2 \prec 2^2 \prec 2^3 \prec ... \prec 5 \cdot 2^2 \prec 3 \cdot 2^2 \prec ...5 \cdot 2 \prec 3 \cdot 2 \prec ...7 \prec 5 \prec 3 \tag{14}$$

In particular, it means that if for certain value of the parameter, the map $g(u)$ has a stable cycle of period 3, then it also has cycles with all periods defined in Sharkovsky's order (14), which are unstable ones. So we obtain the mechanism for transition (with map parameter change) to chaotic solution of deterministic system via consequence of period-doubling bifurcations of cyclic motions that was obtained independently and properly studied much later by Fejhenbaum (1980).

Earlier considered problems of this type are usually reduced (approximately in most cases) to 1D maps with *discrete* time. Unlike these approaches the initial-boundary value problem under consideration has been reduced (mathematically rigorous) to the equivalent 1D map with *continuous* time, In particular this means that:

(i) Phase space of the system under consideration is infinite;

(ii) For an analysis of the system behavior one has to know the structure of the trajectories set for the map $g(u)$ and its intersection with "initial set", initial conditions (12);

(iii) Having solutions for difference equation (11) one can calculate exactly the system behavior at any instant of time at any point of the interval $(0, D)$.

Numerical analysis of Eq. (11) for the map function $g(u) = A\ u\ (1 - u^2)$ has been done in [8] when investigating generators with delayed feedback. Dependence of Eq. (11) solutions and their asymptotic behavior (at $t \to \infty$) on a set of map $g(u)$ trajectories and its intersection with "initial set" are specified in [5].

According to these papers the behavior of solution $u(t)$ depends on what kinds of the attractive subintervals $J_i \subset J$ of the set of map attractive cycles $g(u)$ intersect the initial set. These integrals are specified by the combining of finite or countable number of opened nonintersecting intervals $J \backslash W$, where W is the set of points dividing the interval J into the subintervals. This set has a property of repulsion trajectories. If map $g(u)$ has cycles only with period 1 or 2, then W is either finite or countable. If $g(u)$ has cycles, period of which differs from 1, 2 and T_n, then the set W is uncountable. In the last case W is a mixing repeller, which is homeomorphic to the Cantor's set.

From the derivation of the map $g(u)$ (see Eq. (8) – Eq. (11)) it is obvious that it preserves the topology of the map $\Psi(\nu)$, that is defined by nonlinear dependence of surface current density on the electric field of the wave. This allows to make qualitative conclusions about the behavior of the EM fields in real systems with a known law of nonlinearity $j(E)$. In particular, a cubic nonlinearity of the current dependence $j(E)$ is common for dielectrics with center of inversion,

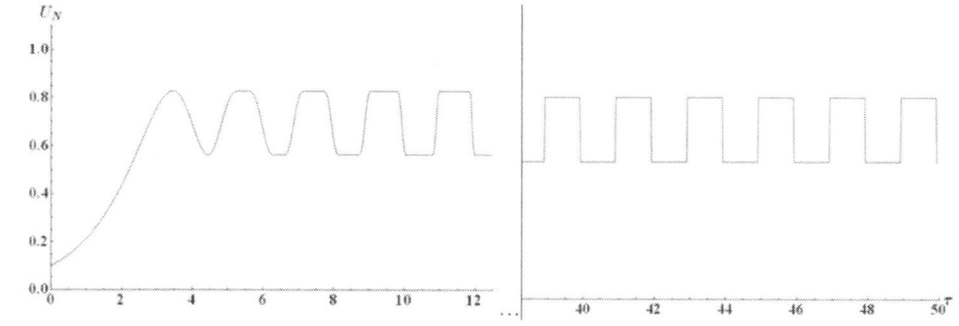

a

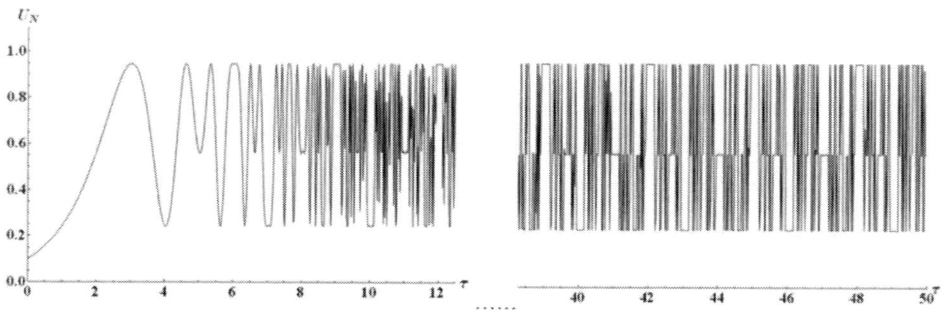

b

Figure 2. Solutions for Eq. (11) with the map function (15):
a – regular periodic solution for $A = 2.15$; b – "dry turbulence" solution for $A = 2.46$

electron beams synchronous with a wave, semiconductor devices, plasmas, etc. So it will be of high interest to consider the following map

$$g[u(\dot{t})] = A\, u(t)[1 - u^2(t)], \tag{15}$$

where $u(t) \equiv u_+(1, t)$.

If parameter A lays within the interval $0 < A < 2.6$ then Eq. (15) defines the map of the interval [0,1] into itself. When $2 < A < 2.286$ the map Eq. (15) has a cycle of period 2 [8]. With further increase in A, bifurcations of period doubling and further bifurcations, defined by Sharkovskiy's order take place. For $A = 2.46$ this mapping has the unique attracting cycle with the period 3 ($u_1 = 0.2439...$, $u_2 = 0.5643...$, $u_3 = 0.9461...$) and mixing repeller according to [5]. In this case exponential increasing of oscillation frequencies characterizes the solutions for this equation, and so called "dry turbulence" takes place as has been shown in [5].

Figure 2 shows examples of two qualitatively different solutions of Eq. (11) with the map (15) obtained for two different values of parameter A. Periodic solution was obtained for A= 2.15 and is shown in the Figure 2a. For A= 2.58 the solution has chaotic structure and depicted in Figure 2b.

It follows from Eq. (13) that behavior of the electric and magnetic fields of the scattered waves might be similar. Indeed using the solutions for Eq. (11) we may readily construct spatiotemporal dependences for electric and magnetic fields in the resonator applying Eq. (13). Figures 3 and 4 shows initial stage of spatiotemporal evolution of electric field, E, in the resonator

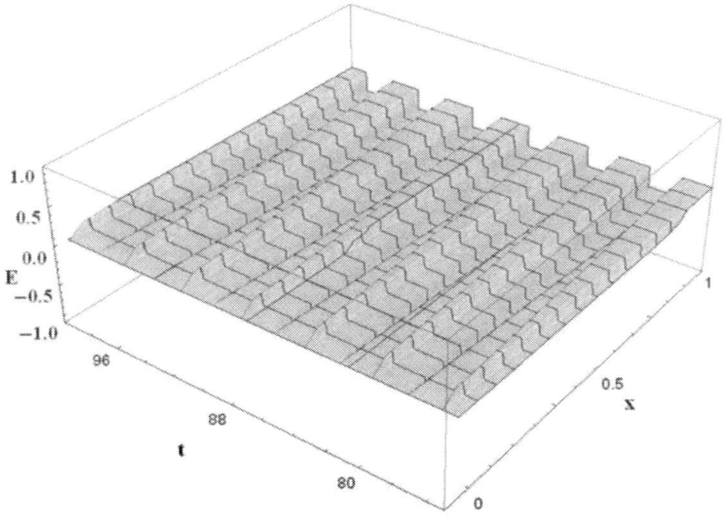

Figure 3. Spatiotemporal evolution of electric field, E, in 1D resonator with nonlinear reflecting wall ($x = 1$) for the case of regular periodic solution for Eq. (11) with cubic parabola map (15; cycle of period 2; A= 2.15. time interval {75 – 100}

under consideration with nonlinear reflecting wall is placed at $x = 1$ in the case of regular period 2 solution for Eq. (11) with cubic parabola map (15): $A = 2.15$. It is clearly seen formation of squared surfaces in the areas where the E and H- fields have maximum values and that those areas are in opposite walls of the resonator as it should be according to boundary conditions (7).

Figure 5 and 6 shows initial and developed stages, respectively, of spatiotemporal dynamics of magnetic field, for the case of "dry turbulent" solution of Eq. (11) with map (15), A = 2.46. It is seen as smooth spatiotemporal pattern of the resonator magnetic field is transforming into rather complicated irregular structure, which may be called as turbulent state of the electromagnetic field in the resonator with nonlinear reflections.

So, multiple/repeated scattering of plane EM waves by nonlinear reflecting structure has a complicated spatiotemporal dynamics and is accompanied with an exponential growth of the frequency spectrum.

4. Finite conductivity of metal and reflection inertia effects

The above results have been obtained for an idealized situation. At first, perfect conductivity of metal surfaces was assumed, that is equivalent to neglecting of EM energy dissipation and inertia of EM field reflections taking place in the system. EM energy losses and reflections inertia in the systems under consideration emerge because of finite conductivity of the metal surfaces, energy dissipation in nonlinear layer, and also as a result of wave radiation out of the resonator through holes in metal surfaces which also may depart from plane geometry. Due to Ohmic losses inside metal surfaces, the boundary condition (4) is to be changed to the impedance one. It has been shown in [9] that for nonstationary fields with relatively narrow Fourier spectrum

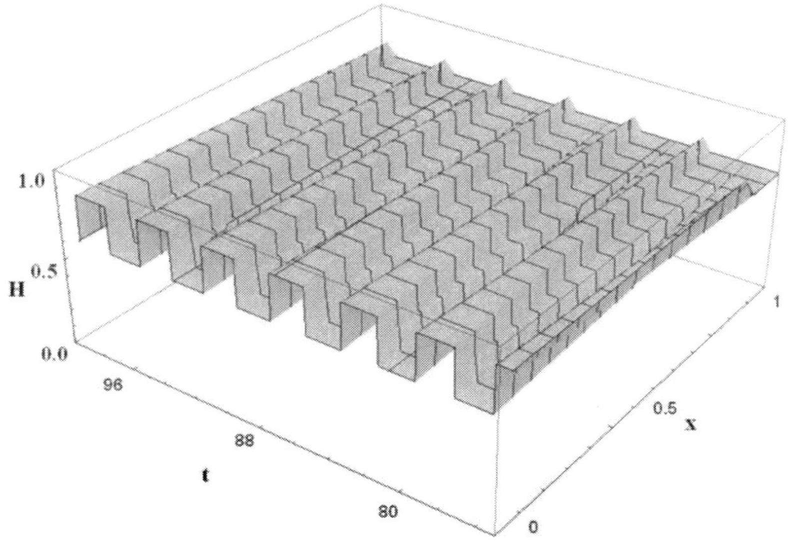

Figure 4. Spatiotemporal evolution of magnetic field, H, in 1D resonator with nonlinear reflecting wall ($x = 1$) for the case of regular periodic solution for Eq. (11) with cubic parabola map (15); cycle of period 2; A= 2.15; time interval {75 – 100}

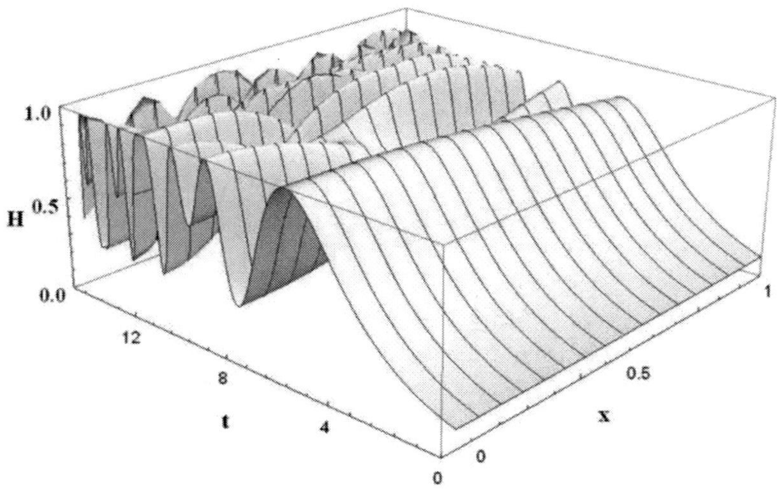

Figure 5. Initial stage of spatiotemporal evolution of magnetic field, H, in 1D resonator with nonlinear reflecting wall ($x = 1$) for the case of "dry turbulence" solution for Eq. (11) with cubic parabola map (15); cycle of period 3; A= 2.46; time interval {0 – 16}

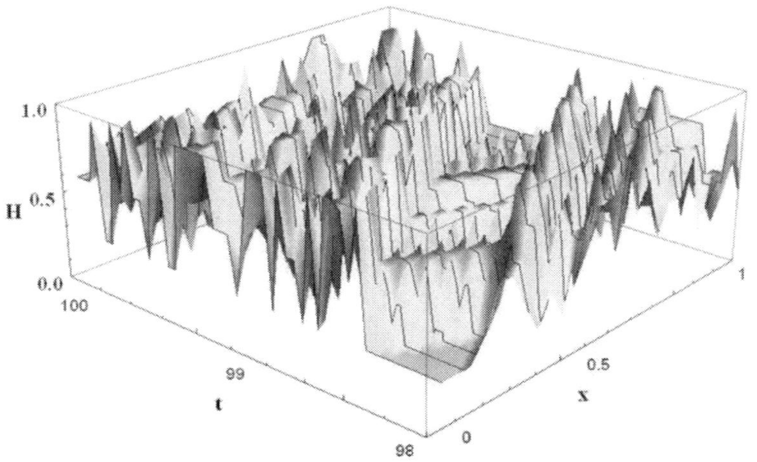

Figure 6. Spatiotemporal evolution of magnetic field, H, in 1D resonator with nonlinear reflecting wall ($x = 1$) at the stage of developed "dry turbulence" solution for Eq. (11) with cubic parabola map (15); cycle of period 3; A= 2.46; time interval $\{98 - 100\}$

bandwidth the following impedance boundary condition can be used:

$$E(0, y, z, t) = R\, H(0, y, z, t) + L\frac{dH(0, y, z, t)}{dt} \tag{16}$$

where $Z = R + i\omega L$ is a complex surface impedance; R is surface resistance and L is surface inductance of the reflecting surface. Let's assume that Ohmic losses are sufficient only in the plane $x = 0$. Then, applying the method of characteristics, the problem of repeated wave scattering by nonlinear reflecting surface with a finite metal conductivity is reduced to the following differential-difference equation

$$\chi\left\{\frac{du(\tau + 2)}{d\tau} + \frac{d\varphi}{du}\frac{du(\tau)}{d\tau}\right\} + \frac{1 - R}{2(1 + R)}u(\tau + 2) = \varphi[u(\tau)] \tag{17}$$

where $\chi \doteq \frac{LD}{2c(1+R)}$.

Eq. (17) is a difference equation of neutral type. However, for sufficiently smooth maps the following condition is valid

$$\left|\frac{d\varphi}{du}\right| \ll |\varphi[u(\tau)]|\,,$$

and we may reduce Eq. (17) to the differential-difference equation with delay argument

$$\chi\frac{du(\tau + 2)}{d\tau} + R_s u(\tau + 2) = \varphi[u(\tau)] \tag{18}$$

where $R_s = \frac{1-R}{2(1+R)}$.

Eq. (18) is suitable for description and analysis of the spatiotemporal dynamics of EM fields in 1D resonators formed by planes with inertial nonlinear reflections and energy dissipation within the frame of the theory of ordinary time-delay difference equations elaborated in [5]). In particular, they have shown that for $\chi \ll 1$ and $\varphi = Au(1 - u)$ the existence of chaotic solutions for this equation has been shown. Important to underline that general properties of the solutions of Eq. (18) are defined by the solutions of the related difference equation (when $\chi = 0$). Due to topological equivalence of two maps $\varphi = Au(1-u)$ and $\tilde{\varphi} = Au(1-u^2)$ they have similar solutions, but for different values of parameter A. For instance, turbulent solutions exist at $A = 3.83$ for function φ while for the function $\tilde{\varphi}$ it appears at $A = 2.46$.

Boundary condition (16) can be also applied in case of EM losses in the system are due to the field leakage through little holes coupling the resonator with outer space, etc. Indeed in this case linear relation between time-harmonic components of the fields E and H on the slot boundary is valid and may be expressed as follows:

$$E(\omega) = Z(\omega)H(\omega), \tag{19}$$

where $Z(\omega)$ is "equivalent impedance" which may be found from solution of the related diffraction problem. For narrow band processes we may assume $\omega = \omega_0 + \delta\omega$ $(\delta\omega \ll \omega_0)$ and applying inverse Fourier transform over variable $\delta\omega$ to the Eq. (19) we obtain the relation

$$E(t) \approx Z(\omega_0)H(t) + \frac{dZ(\omega_0)}{d\omega_0}\frac{dH(t)}{dt} \tag{20}$$

where $H(t) = \tilde{H}(t)\exp(-i\omega_0 t)$ and $E(t) = \tilde{E}(t)\exp(-i\omega_0 t)$, $\tilde{H}(t)$ and $\tilde{E}(t)$ are slowly varying amplitudes of the magnetic and electric fields, respectively. Boundary condition (20) is similar to the impedance boundary condition (16) and may be applied accordingly.

5. Conclusions

Multiple scattering of the electromagnetic waves by the surface with nonlinear reflection is rather common phenomenon in various physical experiments and electronic devices. We have formulated a new class of electromagnetic problems: initial-boundary value problems for linear wave equation with nonlinear boundary conditions. This problem has been reduced to the initial problem for nonlinear difference equation or differential-difference equation with continuous argument, which is much simpler of the primary problem, but still gives a complete description of spatiotemporal dynamics of the EM field in such nonlinear resonator. In spite of its simplicity solutions to this type of problems allow to reveal the main physical phenomenon, which are applicable to realistic EM systems and electronic devices. It has been shown that in such systems spatiotemporal behavior of EM field may be periodic, regular or deterministic, and also chaotic one in spite of major part of the system is formed by linear media. The results obtained, have shown applicability of this approach to analysis of nonstationary process in Diffraction Radiation Generator (DRG) and other open resonator based devices with active elements, such as microwave diodes, transistors, etc. Besides, this approach has been applied for description of complex dynamics of currents formed by electrons and holes in semiconductor PNIPN- reversed biased structure with nonlinear transformation of these currents one each other at the PN-junctions due to avalanche multiplication of the carriers [10] and [11].

References

[1] Shestopalov V P 1997 *Physical foundations of the millimeter and submillimeter waves technique* (The Netherlands: VSP BV)
[2] Lukin K A, Maistrenko Y L, Shestopalov V P and Sharkovsky A N 1989 Method of difference equation in the resonator problems with nonlinear reflections *Soviet Physics Doklady* **34** 977-81
[3] Miloserdova I V and Potapov A N 1980 Elastic oscillations in a bar with nonlinear elastic fastening. *Izv. AN USSR Mechanics of Solids* 178-83 (in Russian)
[4] Bogatyrev Y K, Gorshkov K A and Ostrovskiy L A 1976 Nonlinear waves in periodic structures *Radiotechnika i Electronica* **21** 458-65 (in Russian)
[5] Sharkovsky A N, Maistrenko Y L, Romanenko E Y 1993 *Difference equations and their applications* (New-York: Kluwer Academic)
[6] Sharkovsky A N 1964 Coexistence of cycles of a continuous map of the line into itself *Ukr. Math. Zh.* **16** 61-71 (in Russian)

[7] Feigenbaum H T 1980 Universal behavior in nonlinear systems *Alamos Sci.* **1** 4-27.
[8] Kuznetsov S P 1982 Complex dynamics of the delayed feedback oscillators *Izv. Vuz. Radiofizika* **25** 1410-28 (in Russian)
[9] Mende F F and Spitsyn A I 1985 *Surface impedance of superconductors* (Kiev: Naukova Dumka)
[10] Lukin K A, Cerdeira H A and Colavita A A 1996 Current Oscillations in Avalanche Particle Detectors with p-n-i-p-n-Structure. *IEEE Transactions on Electron Devices* **43** 473-78
[11] Lukin K A, Cerdeira H A and Colavita A A 1997 Chaotic instability of currents in a reverse biased multilayered structure. *Appl. Phys. Lett.* **71** 2484-86

On (ε, δ)-Freudenthal Kantor triple systems and anti-structurable algebras with certain conditions

Noriaki Kamiya[1], **Daniel Mondoc**[2] **and Susumu Okubo**[3]

[1] Center for Mathematical Sciences, University of Aizu, 965-8580 Aizuwakamatsu, Japan
[2] Centre for Mathematical Sciences, Lund University, 22 100 Lund, Sweden
[3] Department of Physics, University of Rochester, Rochester, NY 14627 USA

E-mail: kamiya@u-aizu.ac.jp, daniel.mondoc@math.lu.se, okubo@pas.rochester.edu

Abstract. In this paper we discuss a characterization of anti-structurable algebras in connection with their relation with $(-1,-1)$-Freudenthal Kantor triple systems.

1. Introduction

Freudenthal, Tits [52], I.L. Kantor ([33, 34, 35]) and Koecher ([38, 39]) studied constructions of Lie algebras from nonassociative algebras and triple systems, in particular Jordan algebras, while B.N. Allison ([1, 2]) defined the concept of structurable algebras, containing Jordan algebras. Recently, we have studied constructions of Lie (super)algebras from triple systems and anti-structurable algebras ([24, 26, 29, 30, 31]). As a continuation of [29, 30] we are interested to characterize the structure properties of anti-structurable algebras. Especially, Jordan and Lie (super)algebras ([9, 12]) play an important role in many mathematical and physical subjects ([5, 10, 11, 13, 15, 25, 28, 36, 46, 47, 51, 54, 55]) and the construction and characterization of these algebras can be expressed in terms of triple systems ([19, 22, 23, 27, 37, 48]) by the standard embedding method ([21, 40, 41, 49, 53]).

Summarizing the content of this paper we give an introduction in section §1, definitions and preamble in section §2, while in in section §3 we give properties of anti-structurable algebras satisfying the second order condition, we discuss the notion left neutral pair for (ε, δ)-Freudenthal Kantor triple systems and give examples of anti-structurable algebras with left neutral pair.

2. Definitions and preamble

2.1. (ε, δ)-Freudenthal Kantor triple systems

In this paper triple systems are finite dimensional and defined over a field Φ of characteristic $\neq 2$ or 3. A vector space V over Φ endowed with a trilinear operation $V \times V \times V \to V$, $(x, y, z) \longmapsto (xyz)$ is said to be a *generalized Jordan triple system of second order* (for short GJTS of 2nd order) if the following conditions are fulfilled:

$$(ab(xyz)) = ((abx)yz) - (x(bay)z) + (xy(abz)), \tag{2.1}$$

$$K(K(a, b)x, y) - L(y, x)K(a, b) - K(a, b)L(x, y) = 0, \tag{2.2}$$

for all $a, b, x, y, z \in V$, where $L(a, b)c := (abc)$ and $K(a, b)c := (acb) - (bca), a, b, c \in V$.

A *Jordan triple system* (for short JTS) satisfies (2.1) and $(abc) = (cba)$, for all $a, b, c \in V$.

We can generalize the concept of GJTS of 2nd order as follows (see [13, 14, 17, 18, 19, 20, 21, 53] and the earlier references therein).

For $\varepsilon = \pm 1$ and $\delta = \pm 1$, a triple product that satisfies the identities

$$(ab(xyz)) = ((abx)yz) + \varepsilon(x(bay)z) + (xy(abz)), \tag{2.3}$$

$$K(K(a,b)x,y) - L(y,x)K(a,b) + \varepsilon K(a,b)L(x,y) = 0, \tag{2.4}$$

for all $a, b, x, y, z \in V$, where

$$L(a,b)c := (abc), \quad K(a,b)c := (acb) - \delta(bca), \quad a, b, c \in V, \tag{2.5}$$

is called an (ε, δ)-*Freudenthal Kantor triple system* (for short (ε, δ)-FKTS).

Remark. We note that $K(b,a) = -\delta K(a,b)$.

Remark. The concept of GJTS of 2nd order coincides with that of $(-1,1)$-FKTS. Thus we can construct the corresponding simple Lie algebras by means of the standard embedding method ([6, 13, 14, 15, 16, 17, 21, 24, 26, 35, 53]).

An (ε, δ)-FKTS U is called *unitary* if the identity map Id is contained in $\kappa := K(U,U)$ i.e., if there exist $a_i, b_i \in U$, such that $\Sigma_i K(a_i, b_i) = Id$.

If U is an (ε, δ)-FKTS and $a, b \in U$ then (a,b) is called a *left neutral pair* if $L(a,b) = Id$.

For $\delta = \pm 1$, a triple system $(a,b,c) \mapsto [abc], a, b, c \in V$ is called a δ-*Lie triple system* (for short δ-LTS) if the following three identities are fulfilled

$$[abc] = -\delta[bac],$$
$$[abc] + [bca] + [cab] = 0,$$
$$[ab[xyz]] = [[abx]yz] + [x[aby]z] + [xy[abz]],$$

where $a, b, x, y, z \in V$. An 1-LTS is a *LTS* while a -1-LTS is an *anti-LTS*, by [14].

Proposition 2.1 *([14],[21]) Let $U(\varepsilon, \delta)$ be an (ε, δ)-FKTS. If J is an endomorphism of $U(\varepsilon, \delta)$ such that $J < xyz >=< JxJyJz >$ and $J^2 = -\varepsilon\delta Id$, then $(U(\varepsilon, \delta), [xyz])$ is a LTS (if $\delta = 1$) or an anti-LTS (if $\delta = -1$) with respect to the product*

$$[xyz] :=< xJyz > -\delta < yJxz > +\delta < xJzy > - < yJzx > . \tag{2.6}$$

Corollary 2.1 *Let $U(\varepsilon, \delta)$ be an (ε, δ)-FKTS. Then the vector space $T(\varepsilon, \delta) = U(\varepsilon, \delta) \oplus U(\varepsilon, \delta)$ becomes a LTS (if $\delta = 1$) or an anti-LTS (if $\delta = -1$) with respect to the triple product defined by*

$$\left[\begin{pmatrix} a \\ b \end{pmatrix}\begin{pmatrix} c \\ d \end{pmatrix}\begin{pmatrix} e \\ f \end{pmatrix}\right] = \begin{pmatrix} L(a,d) - \delta L(c,b) & \delta K(a,c) \\ -\varepsilon K(b,d) & \varepsilon(L(d,a) - \delta L(b,c)) \end{pmatrix}\begin{pmatrix} e \\ f \end{pmatrix}. \tag{2.7}$$

Thus we can obtain the standard embedding Lie algebra (if $\delta = 1$) or Lie superalgebra (if $\delta = -1$), $L(\varepsilon, \delta) = D(T(\varepsilon, \delta), T(\varepsilon, \delta)) \oplus T(\varepsilon, \delta)$, associated to $T(\varepsilon, \delta)$, where $D(T(\varepsilon, \delta), T(\varepsilon, \delta))$ is the set of inner derivations of $T(\varepsilon, \delta)$, i.e.

$$D(T(\varepsilon,\delta), T(\varepsilon,\delta)) := \left\{\begin{pmatrix} L(a,b) & \delta K(c,d) \\ -\varepsilon K(e,f) & \varepsilon L(b,a) \end{pmatrix}\right\}_{span}, T(\varepsilon,\delta) := \left\{\begin{pmatrix} x \\ y \end{pmatrix}\middle| x, y \in U(\varepsilon,\delta)\right\}_{span}.$$

Remark. We note that $L(\varepsilon, \delta) := L_{-2} \oplus L_{-1} \oplus L_0 \oplus L_1 \oplus L_2$ is the five graded Lie (super)algebra, such that $L_{-1} \oplus L_1 = T(\varepsilon, \delta)$ and $D(T(\varepsilon, \delta), T(\varepsilon, \delta)) = L_{-2} \oplus L_0 \oplus L_2$ with $[L_i, L_j] \subseteq L_{i+j}$.

2.2. δ-structurable algebras

The motivation for the study of such nonassociative algebras is as follows. The existence of the class of nonassociative algebras called structurable algebras is an important generalization of Jordan algebras giving a construction of Lie algebras. Hence from our concept, by means of triple products, we define a generalization to construct Lie superalgebras as well as Lie algebras.

Our start point briefly described in a historical setting is the construction of Lie (super) algebras starting from a class of nonassociative algebras. Hence within the general framework of (ε,δ)-FKTSs ($\varepsilon,\delta=\pm1$) and the standard embedding Lie (super)algebra construction studied in [6, 7, 13, 14, 15, 26] (see also references therein) we define δ-structurable algebras as a class of nonassociative algebras with involution which coincides with structurable algebras for $\delta=1$ as introduced and studied in [1, 2]. Structurable algebras are a class of nonassociative algebras with involution that include Jordan algebras (with trivial involution), associative algebras with involution, and alternative algebras with involution. They are related to GJTSs 2nd order, or $(-1,1)$-FKTSs, as introduced and studied in [33, 34] and further studied in [3, 4, 32, 42, 43, 44, 45, 50]. Their importance lies with constructions of 5-graded Lie algebras $\mathcal{L}(U):=L(\varepsilon,\delta)=L_{-2}\oplus L_{-1}\oplus L_0\oplus L_1\oplus L_2, [L_i,L_j]\subseteq L_{i+j}$.

For $\delta=-1$ the anti-structurable algebras defined here are a new class of nonassociative algebras that may similarly shed light on the notion of $(-1,-1)$-FKTSs hence (by [6, 7]) on the construction of Lie superalgebras and Jordan algebras as it will be shown.

Let $(\mathcal{A},^-)$ be a finite dimensional nonassociative unital algebra with involution (involutive anti-automorphism, i.e. $\overline{\overline{x}}=x, \overline{xy}=\overline{y}\,\overline{x}, x,y\in\mathcal{A}$) over Φ. The identity element of $\mathcal{A}$ is denoted by 1. Since char$\Phi\neq2$, by [1] we have $\mathcal{A}=\mathcal{H}\oplus\mathcal{S}$, where $\mathcal{H}=\{a\in\mathcal{A}|\overline{a}=a\}$ and $\mathcal{S}=\{a\in\mathcal{A}|\overline{a}=-a\}$.

Suppose $x,y,z\in\mathcal{A}$. Put

$$[x,y]:=xy-yx,\quad [x,y,z]:=(xy)z-x(yz). \tag{2.8}$$

We note that $\overline{[x,y,z]}=-[\overline{z},\overline{y},\overline{x}]$.

Let the operators L_x and R_x be defined by $L_x(y):=xy, R_x(y):=yx, x,y\in\mathcal{A}$ and for $\delta=\pm1$ define

$$^{\delta}V_{x,y}:=L_{L_x(\overline{y})}+\delta(R_xR_{\overline{y}}-R_yR_{\overline{x}}), \tag{2.9}$$

$$^{\delta}B_{\mathcal{A}}(x,y,z):=^{\delta}V_{x,y}(z)=(x\overline{y})z+\delta[(z\overline{y})x-(z\overline{x})y], x,y,z\in\mathcal{A}. \tag{2.10}$$

$^+B_{\mathcal{A}}(x,y,z)$ is called the *triple system obtained from the algebra* $(\mathcal{A},^-)$. We will call $^-B_{\mathcal{A}}(x,y,z)$ the *anti-triple system obtained from the algebra* $(\mathcal{A},^-)$. We write for short

$$V_{x,y}:=^{\delta}V_{x,y},\quad B_{\mathcal{A}}:=(^{\delta}B_{\mathcal{A}},\mathcal{A}). \tag{2.11}$$

Remark. The upper left index notation is chosen in order not to be mixed with the upper right index notation of [1] which has a different meaning.

A unital non-associative algebra with involution $(\mathcal{A},^-)$ is called a *structurable algebra* if the following identity is fulfilled

$$[V_{u,v},V_{x,y}]=V_{V_{u,v}(x),y}-V_{x,V_{v,u}(y)}, \tag{2.12}$$

for $V_{u,v}=^+V_{u,v}, V_{x,y}=^+V_{x,y}, u,v,x,y\in\mathcal{A}$, and we will call $(\mathcal{A},^-)$ an *anti-structurable algebra* if the identity (2.12) is fulfilled for $V_{u,v}=^-V_{u,v}, V_{x,y}=^-V_{x,y}$.

If $(\mathcal{A},^-)$ is structurable then, by [34], the triple system $B_{\mathcal{A}}$ is called a *generalized Jordan triple system* (abbreviated GJTS) and by [8], $B_{\mathcal{A}}$ is a GJTS of 2nd order, i.e. satisfies the identities (2.3) and (2.4). If $(\mathcal{A},^-)$ is anti-structurable then we call $B_{\mathcal{A}}$ an *anti-GJTS*.

Journal of Physics: Conference Series **346** (2012) 012014 doi:10.1088/1742-6596/346/1/012014

3. Main results

3.1. Properties satisfying the second order condition

From now on we assume $\delta = -1$ (unless otherwise specified) and let $(\mathcal{A}, ^{-})$ be an anti-structurable algebra. Define $C(a,b,c) \in End\,\mathcal{A}$ by

$$C(a,b,c)d := [a\bar{b}, \bar{d}, c] - [a, \bar{b}, \bar{d}]c, \quad a,b,c,d \in \mathcal{A}. \tag{3.13}$$

We say that $\mathcal{A}$ satisfies *condition* $\mathfrak{C}$ if

$$C(x,y,w) - C(w,y,x) = C(w,x,y) - C(y,x,w), \quad x,y,w \in \mathcal{A}. \tag{3.14}$$

Theorem 3.1 *Let $(\mathcal{A}, ^{-})$ be an anti-structurable algebra. Then the second order condition (2.4) and condition $\mathfrak{C}$ are equivalent.*

Proof. Suppose first that $\mathcal{A}$ satisfies condition $\mathfrak{C}$. We show then that the second order condition (2.4) is fulfilled for $\delta = -1 = \varepsilon$.

By (3.14) we have $C(x,\bar{y},w)\bar{z} - C(w,\bar{y},x)\bar{z} = C(w,x,\bar{y})\bar{z} - C(\bar{y},x,w)\bar{z}, x,y,z,w \in \mathcal{A}$, that is, by (3.13),

$$[xy,z,w] - [x,y,z]w - [wy,z,x] + [w,y,z]x = [w\bar{x}, z, \bar{y}] - [w,\bar{x},z]\bar{y} - [\bar{y}\,\bar{x}, z, w] + [\bar{y},\bar{x},z]w.$$

Then, by (2.8), the last identity is equivalent to

$$\begin{aligned}((xy)z)w - (xy)(zw) - ((xy)z)w + (x(yz))w - ((wy)z)x + (wy)(zx) + ((wy)z)x - (w(yz))x = \\ ((w\bar{x})z)\bar{y} - (w\bar{x})(z\bar{y}) - ((w\bar{x})z)\bar{y} + (w(\bar{x}z))\bar{y} - ((\overline{yx})z)w + (\overline{yx})(zw) + ((\overline{yx})z)w - (\bar{y}(\bar{x}z))w\end{aligned}$$

and canceling the first with the third term and the fifth with the seventh term both in the left and in the right hand side of the last identity we obtain

$$\begin{aligned}(x(yz))w - (xy)(zw) - (w(yz))x + (wy)(zx) = \\ (w(\bar{x}z))\bar{y} - (w\bar{x})(z\bar{y}) - (\bar{y}(\bar{x}z))w + (\overline{yx})(zw).\end{aligned} \tag{3.15}$$

Set in (3.15) $x = d, y = \bar{c}, z = a\bar{b} + b\bar{a}$, hence $z = \bar{z}$, and then, by using the involution properties, we obtain

$$\begin{aligned}(d(\overline{zc}))w - (d\bar{c})(zw) - (w(\overline{zc}))d + (w\bar{c})(zd) = \\ (w(\overline{zd}))c - (w\bar{d})(zc) - (c(\overline{zd}))w + (c\bar{d})(zw).\end{aligned} \tag{3.16}$$

Now, by definitions (2.5) and (2.10) follows $L(x,y)z = (x\bar{y})z - (z\bar{y})x + (z\bar{x})y, x,y,z \in \mathcal{A}$, so the identity (3.16) can be written

$$(L(d, A(a,b)c) + L(c, A(a,b)d))w = A(c,d)(A(a,b)w), \tag{3.17}$$

for all $w \in \mathcal{A}$, where

$$A(a,b) := L(a,b) + L(b,a), a,b \in \mathcal{A}, \tag{3.18}$$

thus $A(a,b)c = (a\bar{b} + b\bar{a})c = zc$. Then, by (3.17),

$$L(d, A(a,b)c) + L(c, A(a,b)d) = A(c,d)A(a,b). \tag{3.19}$$

By [30] §2, we have $[A(a,b), L(c,d)] = L(A(a,b)c, d) - L(c, A(a,b)d)$, or equivalently,

$$A(a,b)L(c,d) = L(A(a,b)c, d) - L(c, A(a,b)d) + L(c,d)A(a,b). \tag{3.20}$$

Then, by (3.18), the identity (3.20) is equivalent to

$$A(a,b)L(c,d) = L(A(a,b)c,d) - L(c,A(a,b)d) + A(c,d)A(a,b) - L(d,c)A(a,b). \qquad (3.21)$$

Now, by (3.19), $A(c,d)A(a,b) - L(c,A(a,b)d) = L(d,A(a,b)c)$ so (3.21) is equivalent to

$$A(a,b)L(c,d) = L(A(a,b)c,d) + L(d,A(a,b)c) - L(d,c)A(a,b), \qquad (3.22)$$

that is, by (3.18),

$$A(a,b)L(c,d) = A(A(a,b)c,d) - L(d,c)A(a,b). \qquad (3.23)$$

Further, by (2.5) and (2.10), $K(a,b)c = L(a,c)b + L(b,c)a = (a\bar{b} + b\bar{a})c = A(a,b)c$, for all $a,b,c \in \mathcal{A}$, so (3.23) is equivalent to $K(a,b)L(c,d) = K(K(a,b)c,d) - L(d,c)K(a,b)$, that is the second order condition (2.4) is fulfilled.

Conversely, to show that (2.4) implies condition $\mathcal{C}$ it is straightforward and we omit it here. □

Remark. An anti-structurable algebra satisfying the condition $\mathcal{C}$ is a $(-1,-1)$-FKTS.

3.2. Lie admissible structures

In this section we announce results demanding extensive proofs which will to be presented elsewhere.

Theorem 3.2 *Let $(\mathcal{A}, ^-)$ be an anti-structurable algebra such that $^- = Id$. Then $\mathcal{A}$ is a LTS with respect to the new product $[x,y,z] = B_{\mathcal{A}}(x,y,z) - B_{\mathcal{A}}(y,x,z)$, $x,y,z \in \mathcal{A}$.*

Theorem 3.3 *Let $(\mathcal{A}, ^-)$ be an anti-structurable algebra satisfying the second order condition (2.4). Then*

i) $\mathcal{A}$ is a Lie admissible, i.e. the Jacobi identity is fulfilled:

$$[[x,y],z] + [[y,z],x] + [[z,x],y] = 0, \; x,y,z \in \mathcal{A},$$

ii) $[x,y,z] + [z,y,x]$ is totaly symmetric in any exchanges of $x,y,z \in \mathcal{A}$,

iii) $[h,x,y] = [x,h,y] = [x,y,h] = 0$, for all $h \in \mathcal{H}$, $x,y \in \mathcal{A}$.

Theorem 3.4 *Let $(\mathcal{A}, ^-)$ be an anti-structurable algebra satisfying the second order condition (2.4) and let $F(x,y,z) \in End\,\mathcal{A}$ be defined by*

$$F(x,y,z)w := [x\bar{y},w,z] + [z,x\bar{y},w] + ([x,y,w] - [y,x,w])z, \quad x,y,z,w \in \mathcal{A}.$$

Then the following identities are fulfilled:

i) $F(x,y,z) = -F(y,x,z)$, $\quad x,y,z \in \mathcal{A}$,

ii) $F(x,y,z) + F(y,z,x) + F(z,x,y) = 0$, $\quad x,y,z \in \mathcal{A}$.

Remark. We have also $K(u,v)K(x,y) + K(x,y)K(u,v) = K(K(u,v)x,y) + K(x,K(u,v)y$, for $x,y,u,v \in \mathcal{A}$ so the set of $K(x,y), x,y \in \mathcal{A}$, form a Jordan algebra (see [30] for details).

3.3. Left neutral pair and invertible elements

In this section we discuss the notion of left neutral pair for (ε,δ)-FKTSs.

Lemma 3.1 *Let U be an $(\varepsilon,-1)$-FKTS, $\varepsilon = \pm 1$, with product $(abc), a,b,c \in U$, and a left neutral pair $(u,v), u,v \in U$. Then $R(v,u)^2 = Id$, where $R(v,u)w := (wvu), u,v,w \in U$.*

Proof. By (2.5), $K(u,u)v = 2(uvu) = 2u$ and by (2.4), replacing $a = b = u, x = v$ follows $K(K(u,u)v,y) - L(y,v)K(u,u) + \varepsilon K(u,u)L(v,y) = 0$, that is

$$2K(u,y)v - 2L(y,v)u + 2\varepsilon K(u,u)L(v,y)v = 0$$

or equivalent $(uvy) + (yvu) - (yvu) + \varepsilon(u(vyv)u) = 0$ so $y = -\varepsilon(u(vyv)u)$, for all $y \in U$.

On the other hand, by (2.3), replacing $b = y = v, x = z = u$ follows

$$(av(uvu)) = ((avu)vu) + \varepsilon(u(vav)u) + (uv(avu)),$$

hence $(avu) = ((avu)vu) + \varepsilon(u(vav)u) + (avu)$, that is $((avu)vu) = -\varepsilon(u(vav)u)$, for all $a \in U$. Since we have shown above that $a = -\varepsilon(u(vav)u)$ follows from the last identity that $a = ((avu)vu)$, for all $a \in U$, hence $R(v,u)^2 = Id$. □

Lemma 3.2 *Let U be an (ε,δ)-FKTS, $\varepsilon, \delta = \pm 1$, with product $(abc), a, b, c \in U$, and a left neutral pair $(u,v), u, v \in U$. Then*

$$R(v,u)^2 - 2(\delta+1)R(v,u) + (2\delta+1)Id = 0, \quad \text{where} \quad R(v,u)w := (wvu), u, v, w \in U. \quad (3.24)$$

Proof. We remark first that (3.24) is equivalent to $(R(v,u) - Id)(R(v,u) - (2\delta+1)Id) = 0$. Then the proof is clear from Lemma 3.1 for the case $\delta = -1$ while the case corresponding $\delta = 1$ follows from Lemma 1 ([15]). □

Remark. For $\delta = -1$ we note that the following decomposition is valid $U := U(\varepsilon,-1) = U_1(u,v) \oplus U_{-1}(u,v)$, while, for $\delta = 1$ we have $U := U(\varepsilon,1) = U_1(u,v) \oplus U_3(u,v)$, by [15], where

$$U_i(\varepsilon,\delta) = \{x \in U | R(v,u)x = ix\}. \quad (3.25)$$

Lemma 3.3 *Let U be an $(\varepsilon,-1)$-FKTS, $\varepsilon = \pm 1$, with product $(abc), a, b, c \in U$, and a left neutral pair $(u,v), u, v \in U$. Then we have $U_1(u,v) = \{x \in U | K(x,u)v = 2x\}$ and $U_{-1}(u,v) = \{x \in U | K(x,u) = 0\}$, where $U_i(\varepsilon,\delta)$ are defined by (3.25).*

Proof. Consider $x \in U_{-1}(u,v)$. Then, by (3.25), $(xvu) = -x$. Since $(uvx) = x$ then $(xvu) + (uvx) = 0$, hence $K(u,x)v = 0$. Since, by (2.4),

$$K(u,x) = K((uvu),x) = -K(u,(uvx)) - \delta K(u,K(u,x)v) = -K(u,x)$$

then it follows $K(u,x) = 0$.

Conversely, let $K(u,x) = 0$ hence $K(u,x)v = 0$. Thus we get $(xvu) + (uvx) = 0$ hence $(xvu) = -x$, that is $x \in U_{-1}(u,v)$. Hence $U_{-1}(u,v) = \{x \in U | K(x,u) = 0\}$.

On the other hand, let $x \in U_1(u,v)$. Then, by (3.25), $(xvu) = x$. Since $(uvx) = x$ then $(xvu) + (uvx) = 2x$, hence $K(x,u)v = 2x$.

Conversely, if $K(x,u)v = 2x$ then $(xvu) + (uvx) = 2x$, hence $(xvu) = x$, that is $x \in U_1(u,v)$. Hence $U_1(u,v) = \{x \in U | K(x,u)v = 2x\}$. □

Theorem 3.5 *Let U be an (ε,δ)-FKTS, $\varepsilon, \delta = \pm 1$, with product $(abc), a, b, c \in U$, and a left neutral pair $(u,v), u, v \in U$. Then, $U := U(\varepsilon,\delta) = U_1(u,v) \oplus U_{1+2\delta}(u,v)$, where for $\delta = -1$ we have $U_1(u,v) = \{x \in U | K(x,u)v = 2x\}$ and $U_{-1}(u,v) = \{x \in U | K(x,u) = 0\}$, while for $\delta = 1$ we have $U_1(u,v) = \{x \in U | K(x,u) = 0\}$ and $U_3(u,v) = \{x \in U | K(x,u)v = 2x\}$.*

Proof. The proof is a direct consequence of Lemma 3.3 for $\delta = -1$, while for $\delta = 1$ the proof follows from [15]. □

Proposition 3.1 *Let $U := U(\varepsilon,-1)$ be an $(\varepsilon,-1)$-FKTS, $\varepsilon = \pm 1$, with product $(abc), a, b, c \in U$, and a left neutral pair $(u,v), u, v \in U$. Then $U_1(u,v) \simeq \tilde{\kappa} := \{K(a,b) | a, b \in U\}_{span}$ as a JTS with respect to a map $\eta : \tilde{\kappa} \to U_1(u,v)$, $\eta(K(a,b)) = K(a,b)v$.*

Proof. We show that η is a bijection and

$$\eta\{K(a,b)K(c,d)K(e,f)\} = (\eta(K(a,b))\eta(K(c,d))\eta(K(e,f))). \qquad (3.26)$$

Indeed, from (2.4) follows the identity $K((uvx),y) + K(x,(uvy)) - K(u,K(x,y)v) = 0$, hence $K(K(x,y)v,u) = 2K(x,y)$, since (u,v) is a left neutral pair. Then, by Lemma 3.3, the last identity implies $K(x,y)v \in U_1(u,v)$.

Now, if $\theta := \frac{1}{2}K(\cdot,u)$, then $(\theta\circ\eta)(K(x,y)) = \frac{1}{2}K(K(x,y)v,u) = K(x,y)$.

Similarly, for $x \in U_1(u,v)$, $(\eta\circ\theta)(x) = \frac{1}{2}\eta(K(x,u)) = \frac{1}{2}K(x,u)v = x$. Hence η is a bijection with $\eta^{-1} = \theta = \frac{1}{2}K(\cdot,u)$. Furthermore, the left hand side of (3.26) is equal to

$$K(K(a,b)c,K(e,f)d)v + K(K(e,f)c,K(a,b)d)v,$$

by Proposition 2.3 ([30]), and equals the right hand side (3.26), by the definition of η. □

Remark. We note that $U_3(u,v) \simeq \tilde{\kappa} := \{K(a,b)|a,b\in U\}_{span}$ as a JTS under the assumption of existence of a left neutral pair $(u,v), u,v \in U$ and $\delta = 1$, by [15].

Proposition 3.2 *Let U be an $(\varepsilon,-1)$-FKTS, $\varepsilon = \pm 1$, with product $(abc), a,b,c \in U$, and a left neutral pair $(u,v), u,v \in U$. Then the maps U_u, U_v are invertible, where $U_x : U \to U, U_x(y) := (xyx)$.*

Proof. Since $L(u,v) = Id$ then $[L(x,y), L(u,v)] = 0, x,y \in U$.

By (2.3), $L((xyu),v) = -\varepsilon L(u,(yxv))$. Putting $y = v$ in the last identity and applying both sides to u follows $R^2(v,u) = -\varepsilon U_u U_v$. Thus the map U_u is onto and U_v is one to one. Since U is finite dimensional, U_u is one to one, thus invertible so U_v is invertible. □

Remark. An analogous result was obtained for the case of a $(\varepsilon,1)$-FKTS in [15].

Remark. These results can be applied to the $(-1,-1)$-FKTS $U := \mathcal{M}_{k,k}(\Phi)$ of square matrices of order k over Φ with product $(xyz) = xy^{\top}z - zy^{\top}x + zx^{\top}y$ ([29]), where $x^{\top}$ denotes the transposed matrix of x. Also, we emphasize that this triple system is an anti-structurable algebra satisfying the second order condition.

Proposition 3.3 *Let U be an unitary $(-1,-1)$-FKTS, with product $(abc), a,b,c \in U$, and a left neutral pair $(u,v), u,v \in U$. Then $\mathfrak{sl}_2(\Phi) = \{E,F,H\}_{span}$ is a subalgebra of the corresponding Lie superalgebra $\mathfrak{L}(U)$ of U and $L(v,u) = Id$.*

Proof. From the unitary property follows $Id \in \tilde{\kappa} := \{K(a,b)|a,b\in U\}_{span}$.

Let $\mathfrak{L}(U)$ be the corresponding Lie superalgebra of U ([15, 29]). Then $E,F,H \in \mathfrak{L}(U)$ where

$$E := \begin{pmatrix} 0 & Id \\ 0 & 0 \end{pmatrix}, \quad F := \begin{pmatrix} 0 & 0 \\ -Id & 0 \end{pmatrix}, \quad H := [E,F] = \begin{pmatrix} -Id & 0 \\ 0 & Id \end{pmatrix}.$$

On the other hand, since $\begin{pmatrix} L(u,v) & 0 \\ 0 & -L(v,u) \end{pmatrix} \in \mathfrak{L}(U)$ by [15], and $L(u,v) = Id$ it follows $L(v,u) = Id$. □

3.4. Examples of $(-1,-1)$-FKTS with left neutral pairs

We give examples of $(-1,-1)$-FKTS with left neutral pairs and invertible elements. Let

$$GL_k(\Phi) := \{A \in \mathcal{M}_{k,k}(\Phi) | \det A \neq 0\}.$$

If $u \in GL_k(\Phi)$ then set $v = (u^{\top})^{-1}$, where the involution is transposition and so $L(u,v)z = uu^{-1}z - zu^{-1}u + zu^{\top}(u^{\top})^{-1} = z$. Thus there exists a left neutral pair $(u,(u^{\top})^{-1})$. Also we have

$$U_u z = u^{\top}zu - u^{\top}zu + u^{\top}uz, \quad U_{(u^{\top})^{-1}} = (u^{\top})^{-1}((u^{\top})^{-1})(u^{\top})^{\top}z = (u^{\top})^{-1}u^{-1}z,$$

thus by straightforward calculation follows $U_u U_{(u^\top)^{-1}} z = z$. Then the map U_u is invertible. This implies that with any $u \in GL_k(\Phi)$ there can be constructed a left neutral pair $(u, (u^\top)^{-1})$.

Set $O(\Phi) := \{A \in \mathcal{M}_{k,k}(\Phi) | AA^\top = Id\}$. Then in the example above, if any element $u \in O(\Phi)$ it follows that (u, u) is a left neutral pair, i.e. u is a left unit element.

Theorem 3.6 *Let U be a $(-1,-1)$-FKTS. Then (u, v) is a left neutral pair if and only if (v, u) is a left neutral pair.*

Proof. We shall prove that $L(u, v) = Id$ if and only if $L(v, u) = Id$.

If $L(u, v) = Id$ then $[L(u, v), L(v, x)] = 0$ so $L((uvv), x) - L(v, (vux)) = 0$, by (2.3), hence $L(v, x - (vux))v = 0$, since $L(u, v) = Id$. Now, since U_v is invertible follows from the last identity that $(vux) = x$, hence $L(v, u) = Id$.

Conversely, if $L(v, u) = Id$ follows then that $L(u, v) = Id$, by an analogous proof. □

3.5. Almost complex structure and complex structure

Let U be an (ε, δ)-FKTS and $T(\delta)$ be the δ-Lie triple systems associated with U ([21]). Let us set $E := \begin{pmatrix} 0 & Id \\ 0 & 0 \end{pmatrix}, F := \begin{pmatrix} 0 & 0 \\ Id & 0 \end{pmatrix}$ and $J = \delta E - \varepsilon F$. Then it follows $J^2 = -\varepsilon\delta Id$.
We shall call it an (ε, δ)-*almost complex structure*. Thus we can define the following operation

$$N(X, Y) = [JX, JY] - J[JX, Y] - J[X, JY] + J^2[X, Y], \quad X = \begin{pmatrix} x \\ y \end{pmatrix}, Y = \begin{pmatrix} z \\ w \end{pmatrix} \in T(\delta).$$

Then from Proposition 6.1 ([21]) we have that the identity $N(X, Y) = 0$, for $X, Y \in T(\delta)$, is equivalent to $K(x, y) = L(y, x) - \delta L(x, y)$.

Remark. For anti-structurable algebras with product (2.10), $\varepsilon = \delta = -1$. Since then $K(x, y) = L(y, x) + L(x, y)$ we can easily show that there exists a complex structure on the associated anti-LTS $T(-1)$ if $Id \in \tilde{\kappa} := \{K(a, b) | a, b \in U\}_{span}$ (i.e. unitary).

A structurable algebra has no property of complex structure.

Remark. For the $(-1,-1)$-FKTS $U := \mathcal{M}_{k,k}(\Phi)$ of square matrices of order k over Φ with product $(xyz) = xy^\top z - zy^\top x + zx^\top y$ ([29]), since $K(x, y) = L(y, x) + L(x, y)$ we can easily show that there exists a complex structure on the associated anti-LTS $T(-1)$. Moreover, by [29], the standard embedding Lie superalgebra $\mathcal{L}(U)$ corresponding to the $(-1,-1)$-FKTS above is $\mathfrak{osp}(2m|4m)$ or $\mathfrak{osp}(2m+1|4m+2)$ if $k = 2m$ or $k = 2m+1$, respectively.

3.6. Examples of anti-structurable algebras with left neutral pairs

We give examples of anti-structurable algebras with left neutral pairs and invertible elements.

From the fact that $U := \mathcal{M}_{k,k}(\Phi)$ with the product $(xyz) = xy^\top z - zy^\top x + zx^\top y$ is an anti-structurable algebra satisfying the second order condition (2.4) we have the following.

Let $(u, v), u, v \in U$ be a left neutral pair and $GL_k(\Phi) := \{A \in \mathcal{M}_{k,k}(\Phi) | \det A \neq 0\}$.

If $u \in GL_k(\Phi)$ then set $v = (u^\top)^{-1}$, where the involution is transposition and so $L(u, v)z = uu^{-1}z - zu^{-1}u + zu^\top(u^\top)^{-1} = z$. Thus there exists a left neutral pair $(u, (u^\top)^{-1})$. Also we have

$$U_u z = u^\top zu - u^\top zu + u^\top uz, \quad U_{(u^\top)^{-1}} = (u^\top)^{-1}((u^\top)^{-1})(u^\top)^\top z = (u^\top)^{-1}u^{-1}z,$$

thus by straightforward calculation follows $U_u U_{(u^\top)^{-1}} z = z$. Then the map U_u is invertible. This implies that with any $u \in GL_k(\Phi)$ there can be constructed a left neutral pair $(u, (u^\top)^{-1})$.

Set $O(\Phi) := \{A \in \mathcal{M}_{k,k}(\Phi) | AA^\top = Id\}$. Then in the example above, if any element $u \in O(\Phi)$ it follows that (u, u) is a left neutral pair, i.e. u is a left unit element.

Theorem 3.7 *Let $(\mathcal{A}, ^-)$ be an anti-structurable algebra satisfying the second order condition (2.4). Then (u, v) is a left neutral pair if and only if (v, u) is a left neutral pair.*

Proof. It is a direct consequence of Theorem 3.6 and Theorem 3.1 and its remark. □

Acknowledgments
The first two authors, N. Kamiya and D. Mondoc, acknowledge partial support for this research by Grant-in-Aid for Scientific Research No. 19540042 (C),(2), Japan Society for the Promotion of Science. The third author, S. Okubo, acknowledges partial support for this research by U.S. Department of Energy Grant No. DE-FG02-91ER40685.

References
[1] Allison B N 1978 A class of nonassociative algebras with involution containing the class of Jordan algebras *Math. Ann.* **237** 133-56
[2] Allison B N 1979 Models of isotropic simple Lie algebras *Comm. Algebra* **7** 1835-75
[3] Asano H and Yamaguti K 1980 A construction of Lie algebras by generalized Jordan triple systems of second order *Nederl. Akad. Wetensch. Indag. Math.* **42** 249-53
[4] Asano H 1991 Classification of non-compact real simple generalized Jordan triple systems of the second kind *Hiroshima Math. J.* **21** 463-89
[5] Bertram W 2002 Complex and quaternionic structures on symmetric spaces - correspondence with Freudenthal-Kantor triple systems *Theory of Lie Groups and Manifolds* vol 45, ed R Miyaoka and H Tamaru (Tokyo: Sophia Kokyuroku in Math.) 61-80
[6] Elduque A, Kamiya N and Okubo S 2003 Simple $(-1,-1)$ balanced Freudenthal Kantor triple systems *Glasgow Math. J.* **11** 353-72
[7] Elduque A, Kamiya N and Okubo S 2005 $(-1,-1)$ balanced Freudenthal Kantor triple systems and noncommutative Jordan algebras *J. Algebra* **294** 19-40
[8] Faulkner J R 1994 Structurable triples, Lie triples, and symmetric spaces *Forum Math.* **6** 637-50
[9] Frappat L, Sciarrino A and Sorba P 2000 *Dictionary on Lie Algebras and Superalgebras* (San Diego: Academic Press)
[10] Jacobson N 1949 Lie and Jordan triple systems *Amer. J. Math.* **71** 149-70
[11] Jacobson N 1968 *Structure and representations of Jordan algebras* Vol 39 (Providence: Amer. Math. Soc. Colloq. Publ.)
[12] Kac V G 1977 Lie superalgebras *Advances in Math.* **26** 8-96
[13] Kamiya N 1987 A structure theory of Freudenthal-Kantor triple systems *J. Algebra* **110** 108-23
[14] Kamiya N 1988 A construction of anti-Lie triple systems from a class of triple systems *Mem. Fac. Sci. Shimane Univ.* **22** 51-62
[15] Kamiya N 1989 A structure theory of Freudenthal-Kantor triple systems. II. *Comment. Math. Univ. St. Paul.* **38** 41-60
[16] Kamiya N 1991 On (ε,δ)-Freudenthal-Kantor triple systems. *Nonass. alg. and related topics (Hiroshima)* (River Edge: World Sci. Publ.) 65-75
[17] Kamiya N 1991 The construction of all simple Lie algebras over $\mathbb{C}$ from balanced Freudenthal-Kantor triple systems *Contributions to general algebra (Vienna)* vol 7, (Vienna: Hölder-Pichler-Tempsky) 205-13
[18] Kamiya N 1994 On Freudenthal-Kantor triple systems and generalized structurable algebras *Nonass. alg. and its appl. (Oviedo)* Math. Appl. vol 303 (Dordrecht: Kluwer Acad. Publ.) 198-203
[19] Kamiya N 1997 On the Peirce decompositions for Freudenthal-Kantor triple systems *Commun. Algebra* **25** 1833-44
[20] Kamiya N 1998 On a realization of the exceptional simple graded Lie algebras of the second kind and Freudenthal-Kantor triple systems *Bull. Polish Acad. Sci. Math.* **46** 55-65
[21] Kamiya N and Okubo S 2000 On δ-Lie supertriple systems associated with (ε,δ)-Freudenthal-Kantor supertriple systems *Proc. Edinburgh Math. Soc.* **43** 243-60
[22] Kamiya N and Okubo S 2002 A construction of Jordan superalgebras from Jordan-Lie triple systems *Nonass. alg. and its appl.* Lect Notes Pure and Appl. Math. vol 211, ed R Costa, A Grishkov, H Guzzo and L Peresi (New York: Marcel Dekker Inc.) 171-6
[23] Kamiya N and Okubo S 2002 A construction of simple Jordan superalgebra of F type from a Jordan-Lie triple system *Ann. Mat. Pura Appl.* **181** 339-48
[24] Kamiya N and Okubo S 2003 Construction of Lie superalgebras $D(2,1;\alpha), G(3)$ and $F(4)$ from some triple systems *Proc. Edinburgh Math. Soc.* **46** 87-98
[25] Kamiya N and Okubo S 2003 On generalized Freudenthal-Kantor triple systems and Yang-Baxter equations *Proc. XXIV Inter. Colloq. Group Theoretical Methods in Physics (Paris)* vol 173, ed J-P Gazeau, R Kerner, J-P Antoine, S Metens and J-Y Thibon (Paris: IPCS) 815-8
[26] Kamiya N and Okubo S 2004 A construction of simple Lie superalgebras of certain types from triple systems *Bull. Australian Math. Soc.* **69** 113-23.
[27] Kamiya N 2005 Examples of Peirce decomposition of generalized Jordan triple system of second order-

Balanced cases- *Noncommutative geometry and representation theory in mathematical physics* Contemp. Math. vol 391 (Providence: Amer. Math. Soc.) 157-65

[28] Kamiya N and Okubo S 2006 Composition, quadratic, and some triple systems. *Non-associative algebra and its applications* Lect. notes pure appl. math. vol 246 (Boca Raton: Chapman & Hall/CRC) 205-31

[29] Kamiya N and Mondoc D 2008 A new class of nonassociative algebras with involution *Proc. Japan Acad. Ser. A* **84** 68-72

[30] Kamiya N, Mondoc D and Okubo S 2010 A structure theory of $(-1,-1)$-Freudenthal Kantor triple systems *Bull. Australian Math. Soc.* **81** 132-55

[31] Kamiya N and Mondoc D 2009 On anti-structurable algebras and extended Dynkin diagrams *J. Gen. Lie Theory Appl.* **3** 185-92

[32] Kaneyuki S and Asano H 1988 Graded Lie algebras and generalized Jordan triple systems *Nagoya Math. J.* **112** 81-115

[33] Kantor I L 1970 Graded Lie algebras *Trudy Sem. Vect. Tens. Anal.* **15** 227-66

[34] Kantor I L 1972 Some generalizations of Jordan algebras *Trudy Sem. Vect. Tens. Anal.* **16** 407-9

[35] Kantor I L 1973 Models of exceptional Lie algebras *Soviet Math. Dokl.* **14** 254-8

[36] Kantor I L 1998 A generalization of the Jordan approach to symmetric Riemannian spaces *The Monster and Lie algebras (Columbus)* Ohio State Univ. Math. Res. Inst. Publ. vol 7 (Berlin: de Gruyter) 221-34

[37] Kantor I L and Kamiya N 2003 A Peirce decomposition for generalized Jordan triple systems of second order *Commun. Algebra* **31** 5875-913

[38] Koecher M 1967 Embedding of Jordan algebras into Lie algebras I *Amer. J. Math.* **89** 787-816

[39] Koecher M 1968 Embedding of Jordan algebras into Lie algebras II *Amer. J. Math.* **90** 476-510

[40] Lister W G 1952 A structure theory of Lie triple systems *Trans. Amer. Math. Soc.* **72** 217-42

[41] Meyberg K 1972 Lectures on algebras and triple systems *Lecture Notes* (Charlottesville: Virginia Univ. Press)

[42] Mondoc D 2006 Models of compact simple Kantor triple systems defined on a class of structurable algebras of skew-dimension one *Commun. Algebra* **34** 3801-15

[43] Mondoc D 2007 On compact realifications of exceptional simple Kantor triple systems *J. Gen. Lie Theory Appl.* **1** 29-40

[44] Mondoc D 2007 Compact realifications of exceptional simple Kantor triple systems defined on tensor products of composition algebras *J. Algebra* **307** 917-29

[45] Mondoc D 2007 Compact exceptional simple Kantor triple systems defined on tensor products of composition algebras *Commun. Algebra* **35** 3699-712

[46] Neher E 1987 Jordan triple systems by the grid approach *Lecture notes in mathematics* vol 1280 (Berlin: Springer-Verlag)

[47] Okubo S 1995 Introduction to octonion and other non-associative algebras in physics *Montroll Memorial Lect. Ser. in Mathematical Physics* vol 2 (Cambridge: Cambridge Univ. Press)

[48] Okubo S and Kamiya N 1997 Jordan-Lie superalgebra and Jordan-Lie triple system *J. Algebra* **198** 388-411

[49] Okubo S and Kamiya N 2002 Quasi-classical Lie superalgebras and Lie supertriple systems *Commun. Algebra* **30** 3825-50

[50] Okubo S 2005 Symmetric triality relations and structurable algebras *Linear Algebra Appl.* **396** 189-222

[51] Scheunert M 1979 The theory of Lie superalgebras. An introduction *Lecture notes in mathematics* vol 716 (Berlin: Springer-Verlag)

[52] Tits J 1962 Une classe d'algèbres de Lie en relation avec les algèbres de Jordan *Nederl. Acad. Wetensch. Proc. Ser. A* **65** 530-5

[53] Yamaguti K and Ono A 1984 On representations of Freudenthal-Kantor triple systems $U(\varepsilon,\delta)$ *Bull. Fac. School Ed. Hiroshima Univ.* **7** 43-51

[54] Zhevlakov K A, Slinko A M, Shestakov I P and Shirshov A I 1982 *Rings that are Nearly Associative* (New York: Academic Press Inc.)

[55] Zelmanov E 1983 Primary Jordan triple systems *Sibirsk. Mat. Zh.* **4** 23-37

Mal'tsev algebras and triality

Eugen Paal

Department of Mathematics, Tallinn University of Technology,
Ehitajate tee 5, 19086 Tallinn, Estonia

E-mail: `eugen.paal@ttu.ee`

Abstract. A concept of the Mal'tsev pair is presented. This concept is based on the generalized Maurer-Cartan equations of a local analytic Moufang loop. The triality can be seen as a fundamental property of such pairs. Based on triality, the Yamagutian is constructed. Properties of the Yamagutian are studied.

MSC 2010: 17D10

1. Introduction

In [2], the generalized Maurer-Cartan equations for a local analytic Moufang loop were found. In this paper, we propose a concept of the *Mal'tsev pair* which is based on the generalized Maurer-Cartan equations of a local analytic Moufang loop and the structural properties of the Mal'tsev algebras [1]. Such a concept may also be inferred from the theory of *alternative* algebras [3, 4]. Triality can be seen as a fundamental property of such pairs. Based on triality, the Yamagutian is constructed. Properties of the Yamagutian are studied.

2. Mal'tsev pairs

Let $\mathcal{M}$ be an anti-commutative algebra and let $\mathcal{L}$ be a Lie algebra. Throughout this paper we assume that both algebras have the same base field $\mathbb{F}$ ($\doteq \mathbb{R}$ or $\mathbb{C}$; as a matter of fact, only $\operatorname{char}\mathbb{F} \neq 2, 3$ is essential). Denote by (S, T) a pair of the linear maps $S, T : \mathcal{M} \to \mathcal{L}$.

Definition 2.1 (Mal'tsev pair). We call (S, T) a *Mal'tsev pair* if

$$[S_x, S_y] = S_{[y,x]} - 2[S_x, T_y] \tag{2.1a}$$

$$[T_x, T_y] = T_{[x,y]} - 2[T_x, S_y] \tag{2.1b}$$

for all x, y in $\mathcal{M}$.

We call (2.1a,b) the Mal'tsev relations. Note that the same brackets $[\cdot,\cdot]$ are used to denote multiplications in $\mathcal{M}$ and $\mathcal{L}$.

In an anti-commutative algebra $\mathcal{M}$ define the *Jacobi map* $J : \mathcal{M} \times \mathcal{M} \times \mathcal{M} \to \mathcal{M}$ by

$$J(x, y, z) := [x, [y, z]] + [y, [z, x]] + [z, [x, y]]$$

Journal of Physics: Conference Series **346** (2012) 012015 doi:10.1088/1742-6596/346/1/012015

An anti-commutative algebra $\mathcal{M}$ is said to be the *Mal'tsev algebra* [1] if the Mal'tsev identity holds in $\mathcal{M}$:

$$[J(x, y, z), x] = J(x, y, [x, z]), \quad \forall x, y, z \in M$$

Proposition 2.2. *Let (S, T) a Mal'tsev pair. Then*

$$[S_x, T_y] = [T_x, S_y]$$

for all x, y in $\mathcal{M}$.

Proof. Use anti-commutativity in $\mathcal{L}$ and $\mathcal{M}$. □

Corollary 2.3 (minimality conditions). *The Mal'tsev conditions read*

$$2[S_x, T_y] = S_{[y,x]} - [S_x, S_y] = T_{[x,y]} - [T_x, T_y] = 2[T_x, S_y]$$

Remark 2.4. One can see that deviations of S from algebra map and T from anti-algebra may be considered as "minimal" for the Mal'tsev pair.

3. Triality

Proposition 3.1. *Let (S, T) be a Mal'tsev pair. Then $(-T, -S)$ is a Mal'tsev pair as well.*

Proof. By using anti-commutativity in M, rewrite the Mal'tsev conditions of (S, T) as follows:

$$[-T_x, -T_y] = -T_{[y,x]} - 2[-T_x, -S_y]$$
$$[-S_x, -S_y] = -S_{[x,y]} - 2[-S_x, -T_y]$$ □

Consider a triple (S, T, P) of the linear maps $S, T, P : M \to L$, such that

$$S + T + P = 0$$

With a given triple (S, T, P) we can associate the pairs

$$(S, T), \quad (-T, -S), \quad (T, P), \quad (-P, -T), \quad (P, S), \quad (-S, -P) \tag{3.1}$$

Lemma 3.2 (triality). *Let some pair from (3.1) be a Mal'tsev pair. Then all other pairs from (3.1) are the Mal'tsev pairs as well.*

Proof. Assume that (S, T) is a Mal'tsev pair. Then we know from above Proposition 3.1 that $(-T, -S)$ is a Mal'tsev pair. The required minimality conditions for (T, P) and (P, S) read, respectively,

$$2[T_x, P_y] = T_{[y,x]} - [T_x, T_y] = P_{[x,y]} - [P_x, P_y] = 2[P_x, T_y]$$
$$2[P_x, S_y] = P_{[y,x]} - [P_x, P_y] = S_{[x,y]} - [S_x, S_y] = 2[S_x, P_y]$$

As an example, calculate

$$\begin{aligned}-2[S_x, P_y] &= 2[P_x + T_x, P_y] \\ &= 2[P_x, P_y] + 2[P_x, T_y]\end{aligned}$$

$$\begin{aligned}
&= [P_x, P_y] + [P_x, P_y] + 2[P_x, T_y] \\
&= [P_x, P_y] + [S_x + T_x, S_y + T_y] - 2[S_x + T_x, T_y] \\
&= [P_x, P_y] + [S_x, S_y] + [S_x, T_y] + [T_x, S_y] + [T_x, T_y] - 2[S_x, T_y] - 2[T_x, T_y] \\
&= [P_x, P_y] + [S_x, S_y] - [T_x, T_y] \\
&= [P_x, P_y] + S_{[y,x]} - 2[S_x, T_y] - T_{[x,y]} + 2[T_x, S_y] \\
&= [P_x, P_y] - P_{[x,y]}
\end{aligned}$$

All other required equalities in (3.1) can be verified in the same way. The result tells us that (T, P) and (P, S) are the Mal'tsev pairs, which in turn implies that $(-P, -T)$ and $(-S, -P)$ must be the Mal'tsev pairs as well. □

It follows from Lemma 3.2 that a given Mal'tsev pair (S, T) is invariant under the substitutions

$$\begin{aligned}
\sigma &\doteq (S \to T \to P \to S) \\
\tau &\doteq (S \to -T \to S)(P \to -P) \\
\sigma^2 &= (S \to P \to T \to S) \\
\sigma \circ \tau &= (S \to -P \to S)(T \to -T) \\
\sigma^2 \circ \tau &= (T \to -P \to T)(S \to -S)
\end{aligned}$$

which we call the *triality substitutions*. So it is natural to proclaim:

Theorem 3.3 (principle of triality). *All algebraic consequences of the Mal'tsev conditions are triality invariant.*

Such a symmetry we call *triality*. It suggests that we should try to handle the Mal'tsev pairs in triality symmetric manner.

In particular, by using the *triality conjugation*,

$$\begin{aligned}
P^+ &\doteq S - T = P + 2S = -P - 2T \\
S^+ &\doteq T - P = S + 2T = -S - 2P \\
T^+ &\doteq P - S = T + 2P = -T - 2S
\end{aligned}$$

with the evident property

$$S^+ + T^+ + P^+ = 0$$

one can rewrite the Mal'tsev conditions as follows:

$$\begin{aligned}
[S_x, S_y^+] &= [S_x^+, S_y] = S_{[x,y]} \\
[T_x, T_y^+] &= [T_x^+, T_y] = T_{[x,y]} \\
[P_x, P_y^+] &= [P_x^+, P_y] = P_{[x,y]}
\end{aligned}$$

Note that

$$\begin{aligned}
3P &= T^+ - S^+ = P^+ + 2T^+ = -P^+ - 2S^+ \\
3S &= P^+ - T^+ = S^+ + 2P^+ = -S^+ - 2T^+ \\
3T &= S^+ - P^+ = T^+ + 2S^+ = -T^+ - 2P^+
\end{aligned}$$

which means that the triality conjugation is invertible.

4. Yamagutian

We introduced the triple (S,T,P) via the triality symmetric identity (3.1). Following triality, it is natural to search for other but nontrivial triality invariant combinations of the maps from the triple (S,T,P).

Definition 4.1 (Yamagutian [5]). The *Yamagutian* of (S,T) is the skew-symmetric bilinear map $Y: M\otimes M\to L$ defined (cf ([5])) by

$$\begin{aligned}6Y(x;y) &\doteq [S_x,S_y]+[T_x,T_y]+[P_x,P_y]\\ &= -Y(y;x)\end{aligned}$$

We can see the evident but important

Proposition 4.2. *The Yamagutian Y is triality invariant.*

By triality symmetry, the Yamagutian Y can be redefined in several useful ways. In particular,

$$\begin{aligned}6Y(x;y) &= 3[S_x,S_y]-S^+_{[x,y]}\\ &= 3[T_x,T_y]-T^+_{[x,y]}\\ &= 3[P_x,P_y]-P^+_{[x,y]}\end{aligned}$$

and one can also verify that

$$\begin{aligned}6Y(x;y) &= 2P^+_{[x,y]}-6[S_x,T_y]\\ &= 2S^+_{[x,y]}-6[T_x,P_y]\\ &= 2T^+_{[x,y]}-6[P_x,S_y]\end{aligned}$$

Later we shall need the

Proposition 4.3. *Let (S,T) be a Mal'tsev pair. Then*

$$\begin{aligned}6Y(x;y) &= [S^+_x,S^+_y]+S^+_{[x,y]} &\quad (4.1a)\\ &= [T^+_x,T^+_y]+T^+_{[x,y]} &\quad (4.1b)\\ &= [P^+_x,P^+_y]+P^+_{[x,y]} &\quad (4.1c)\end{aligned}$$

for all x,y in M.

Proof. Due to triality, check only the last formula (4.1c):

$$\begin{aligned}[P^+_x,P^+_y] &= [S_x-T_x,S_y-T_y]\\ &= [S_x,S_y]-2[S_x,T_y]+[T_x,T_y]\\ &= \frac{1}{3}Y(x;y)+\frac{1}{3}S^+_{[x,y]}+\frac{1}{3}Y(x;y)-\frac{2}{3}P^+_{[x,y]}+\frac{1}{3}Y(x;y)+\frac{1}{3}T^+_{[x,y]}\\ &= Y(x;y)-P^+_{[x,y]}\end{aligned}$$ □

Remark 4.4. Formulae (4.1a–c) tell us that the Yamagutian Y measures the deviation of S^+, T^+ and P^+ from the anti-algebra maps.

Corollary 4.5. *We have*

$$18Y(x,y) = [S_x^+, S_y^+] + [T_x^+, T_y^+] + [P_x^+, P_y^+]$$

Theorem 4.6. *Let (S, T) be a Mal'tsev pair. Then*

$$\begin{aligned}
[S_x, S_y] &= 2Y(x;y) - \frac{1}{3}S_{[x,y]} - \frac{2}{3}T_{[x,y]} \\
[S_x, T_y] &= -Y(x;y) - \frac{1}{3}S_{[x,y]} + \frac{1}{3}T_{[x,y]} \\
[T_x, T_y] &= 2Y(x;y) + \frac{2}{3}S_{[x,y]} + \frac{1}{3}T_{[x,y]}
\end{aligned}$$

Proof. Evident. □

Corollary 4.7. *By triality, we have*

$$\begin{aligned}
[T_x, T_y] &= 2Y(x;y) - \frac{1}{3}T_{[x,y]} - \frac{2}{3}P_{[x,y]} \\
[T_x, P_y] &= -Y(x;y) - \frac{1}{3}T_{[x,y]} + \frac{1}{3}P_{[x,y]} \\
[P_x, P_y] &= 2Y(x;y) + \frac{2}{3}T_{[x,y]} + \frac{1}{3}P_{[x,y]}
\end{aligned}$$

and

$$\begin{aligned}
[P_x, P_y] &= 2Y(x;y) - \frac{1}{3}P_{[x,y]} - \frac{2}{3}S_{[x,y]} \\
[P_x, S_y] &= -Y(x;y) - \frac{1}{3}P_{[x,y]} + \frac{1}{3}S_{[x,y]} \\
[S_x, S_y] &= 2Y(x;y) + \frac{2}{3}P_{[x,y]} + \frac{1}{3}S_{[x,y]}
\end{aligned}$$

Proposition 4.8. *Let (S, T) be a Mal'tsev pair. Then*

$$6[Y(x;y), S_z] = 3[[S_x, S_y], S_z] - S_{[[x,y],z]} \tag{4.2a}$$
$$6[Y(x;y), T_z] = 3[[T_x, T_y], T_z] - T_{[[x,y],z]} \tag{4.2b}$$
$$6[Y(x;y), P_z] = 3[[P_x, P_y], P_z] - P_{[[x,y],z]} \tag{4.2c}$$

for all x, y, z in M.

Proof. Due to triality, only the first identity must be checked:

$$\begin{aligned}
6[Y(x;y), S_z] &= [3[S_x, S_y] - S_{[x,y]}^+, S_z] \\
&= 3[[S_x, S_y], S_z] - [S_{[x,y]}^+, S_z] \\
&= 3[[S_x, S_y], S_z] - S_{[[x,y],z]}
\end{aligned}$$

□

Corollary 4.9. *Adding formulae (4.2a–c) we obtain (cf (3.1)) the triality symmetric identity*

$$[[S_x, S_y], S_z] + [[T_x, T_y], T_z] + [[P_x, P_y], P_z] = 0$$

Corollary 4.10. *In (4.2a–c) make twicely cyclic permutation of x, y, z and add the resulting equalities with the original ones. Then we obtain*

$$6[Y(x;y), S_z] + 6[Y(y,z), S_x] + 6[Y(z;x), S_y] = S_{J(x,y,z)}$$
$$6[Y(x;y), T_z] + 6[Y(y,z), T_x] + 6[Y(z;x), T_y] = T_{J(x,y,z)}$$
$$6[Y(x;y), P_z] + 6[Y(y,z), P_x] + 6[Y(z;x), P_y] = P_{J(x,y,z)}$$

Proposition 4.11. *Let (S,T) be a Mal'tsev pair. Then*

$$[S^+_{[x,y]}, S^+_z] + [S^+_{[y,z]}, S^+_x] + [S^+_{[z,x]}, S^+_y] = P^+_{J(x,y,z)} \tag{4.3a}$$
$$[T^+_{[x,y]}, T^+_z] + [T^+_{[y,z]}, T^+_x] + [T^+_{[z,x]}, T^+_y] = T^+_{J(x,y,z)} \tag{4.3b}$$
$$[P^+_{[x,y]}, P^+_z] + [P^+_{[y,z]}, P^+_x] + [P^+_{[z,x]}, P^+_y] = P^+_{J(x,y,z)} \tag{4.3c}$$

for all x, y, z in M.

Proof. Subtracting (4.2b) from (4.2a), we obtain

$$6[Y(x,y), P^+_z] = 3[[S_x, S_y], S_z] - 3[[T_x, T_y], T_z] - P^+_{[[x,y],z]}$$

On the other hand, using (4.2c), we have

$$6[Y(x;y), P^+_z] = [[P^+_x, P^+_y], P^+_z] + [P^+_{[x,y]}, P^+_z]$$

and so we obtain

$$[P^+_{[x,y]}, P^+_z] = -P^+_{[[x,y],z]} + 3[[S_x, S_y], S_z] - 3[[T_x, T_y], T_z] - [[P^+_x, P^+_y], P^+_z]$$

Now make twicely cyclic permutation of x, y, z and add the resulting equalities with the original one. Then, using Jacobi conditions (in L) and the definition of J on the right hand-side of the resulting equality we obtain (4.3c). The remaining identities (4.3a,b) are evident from triality. □

Lemma 4.12. *Let (S,T) be a Mal'tsev pair. Then*

$$Y([x,y];z) + Y([y,z];x) + Y([z,x];y) = 0, \quad \forall x, y, z \in \mathcal{M} \tag{4.4}$$

Proof. Use (4.1c) to obtain

$$6Y([x,y];z) = [P^+_{[x,y]}, P^+_z] + P^+_{[[x,y],z]}$$

Make here twicely the cyclic permutation of x, y, z and add the resulting equalities with the original one. Then use (4.3c) and the definition of J to obtain the desired identity (4.4). □

Acknowledgments
The research was in part supported by the Estonian SF, grant ETF9038.

References
[1] Mal'tsev A 1955 Matem. Sb. **36** 569 (in Russian)
[2] Paal 2008 Introduction to Moufang symmetry I. Generalized Lie and Maurer-Cartan equations Preprint 0802.3471
[3] Schafer R D 1952 Trans. Amer. Math. Soc. **72** 1
[4] Schafer R D 1966 An Introduction to Nonassociative Algebras (New York - London: Academic Press)
[5] Yamaguti K 1963 Kumamoto J. Sci. **A6** 9

Algebra, Geometry, and Mathematical Physics 2010 IOP Publishing
Journal of Physics: Conference Series **346** (2012) 012016 doi:10.1088/1742-6596/346/1/012016

Automorphic instanton partition functions on Calabi-Yau threefolds

Daniel Persson[1]
Institüt für Theoretische Physik, ETH Zürich,
CH-8093 Zürich, Switzerland

E-mail: daniel.persson@itp.phys.ethz.ch

Abstract. We survey recent results on quantum corrections to the hypermultiplet moduli space $\mathcal{M}$ in type IIA/B string theory on a compact Calabi-Yau threefold X, or, equivalently, the vector multiplet moduli space in type IIB/A on $X \times S^1$. Our main focus lies on the problem of resumming the infinite series of D-brane and NS5-brane instantons, using the mathematical machinery of automorphic forms. We review the proposal that when the theory in three dimensions exhibits an arithmetic "U-duality" symmetry $G(\mathbb{Z})$ the total instanton partition function arises from a certain unitary automorphic representation of G, whose Fourier coefficients reproduce the BPS-degeneracies. In the case of four-dimensional $\mathcal{N} = 2$ theories on $\mathbb{R}^3 \times S^1$ we argue that the relevant automorphic representation falls in the quaternionic discrete series of G, and that the partition function is a holomorphic section on the twistor space $\mathcal{Z}$ over $\mathcal{M}$.

1. Introduction

U-duality symmetries [1] have provided remarkable insight into the non-perturbative nature of string theory and M-theory (see, e.g., [2]). The moduli space $\mathcal{M}_4$ of string theory on T^6 is protected from quantum corrections and takes form of an arithmetic quotient $E_{7(7)}(\mathbb{Z})\backslash E_{7(7)}/(SU(8)/\mathbb{Z}_2)$ [1, 3]. Moreover, higher derivative couplings in the resulting $\mathcal{N} = 8$ supergravity must be invariant under the U-duality group $E_{7(7)}(\mathbb{Z})$, and are thus naturally described within the mathematical theory of automorphic forms. This provides powerful methods of resumming infinite series of non-perturbative effects, such as D-brane and worldsheet instantons [4, 5, 6, 7, 8, 9, 10, 11]. The Bekenstein-Hawking entropy of BPS black holes is furthermore invariant under U-duality [12], and the associated microscopic degeneracies are given by Fourier coefficients of automorphic forms [13, 14, 15].

The theory becomes even more constrained upon further compactification on S^1 to three dimensions. All dynamical degrees of freedom are then in the form of scalars, which parametrize an enlarged moduli space $E_{8(8)}(\mathbb{Z})\backslash E_{8(8)}/(Spin(16)/\mathbb{Z}_2)$, where the U-duality group is enhanced to $E_{8(8)}(\mathbb{Z})$. In fact, it is a general feature of four-dimensional supergravity theories with symmetry group G_4, that the enhanced group G_3 in three dimensions serves as a spectrum-generating symmetry of the original theory [16, 17, 18, 19, 20].

[1] Also affiliated to *Fundamental Physics, Chalmers University of Technology, SE-412 96 Gothenburg, Sweden.*

Journal of Physics: Conference Series **346** (2012) 012016 doi:10.1088/1742-6596/346/1/012016

For $\mathcal{N} = 2$ supergravities the situation is much less under control. The moduli space is in general not a coset space, but is required by supersymmetry to factorize into a product $\mathcal{M}_V \times \mathcal{M}_H$, respectively corresponding to the vector-multiplet moduli space (special Kähler) and the hypermultiplet moduli space (quaternion-Kähler) [21]. Moreover, the moduli space metric is no longer protected, and may receive both perturbative and non-perturbative quantum corrections. For theories arising from compactifications of string theory on a Calabi-Yau threefold $\mathcal{X}$, there are perturbative corrections both in α' and in the string coupling $g_s = e^{\phi}$, together with non-perturbative effects arising from worldsheet instantons, Euclidean D-branes and NS5-branes wrapping (supersymmetric) cycles in $\mathcal{X}$ [22] (for a review, see [23, 24]).

One can also learn a lot about $\mathcal{N} = 2$ theories by compactification to $D = 3$ [25]. In this case, the hypermultiplet moduli space goes along for the ride, while the vector multiplet moduli space is enhanced to another quaternion-Kähler manifold $\widehat{\mathcal{M}}_V$; this procedure is known as the *c-map* [26, 27]. The enhanced moduli space $\widehat{\mathcal{M}}_V$ holds the key to understanding the wall-crossing behaviour of the BPS-index $\Omega(\gamma)$ (generalized Donaldson-Thomas invariants) [28, 29, 30, 31]. In fact, using a combination of mirror symmetry and T-duality in three dimensions all the moduli spaces of type IIA (resp. IIB) string theory on $\mathcal{X} \times S^1$ (resp. $\hat{\mathcal{X}} \times S^1$) become identified [22, 31, 32]. This unification of moduli has far-reaching consequences that goes beyond classical mirror symmetry [33] as well as homological mirror symmetry [34].

In this survey we shall discuss recent attempts at describing the exact moduli space $\widehat{\mathcal{M}}_V$ in type IIA/B string theory on $\mathcal{X} \times S^1$, or the hypermultiplet moduli space $\mathcal{M}_H$ in type IIB/A on $\mathcal{X}$. We argue that in certain classes of $\mathcal{N} = 2$ supergravities one can use automorphic techniques to constrain quantum corrections. In particular, having identified a "U-duality" group $G_3(\mathbb{Z})$, the BPS-degeneracies $\Omega(\gamma)$ are recovered from the Fourier coefficients of certain automorphic forms. Inspired by earlier related work [35, 15, 36, 37], we revisit the recent results of [38, 39, 40, 41, 42] from a representation theoretic point of view, drawing insight from the "toy models" provided by so called *magic supergravities* [43, 44]. We recall and elaborate upon conjectures stating that the instanton partition function in $\mathcal{N} = 2$ supergravity on $\mathbb{R}^3 \times S^1$ should correspond to an automorphic representation in the quaternionic discrete series of G_3. Along the way, we also make some comparative remarks on similarities with, and differences from, the better understood case of $\mathcal{N} = 2$ field theory [30].

2. Black Hole Partition Functions and Instanton Corrections

In this section we introduce the physical arena in which our analysis takes place. We are mainly focusing on type IIA and IIB string compactifications on a compact Calabi-Yau threefold $\mathcal{X}$. However, occasionally we will also be discussing other classes of $\mathcal{N} = 2$ supergravity theories, which do not necessarily have a known embedding in string theory. We will also be drawing some lessons from supergravities with a higher degree of supersymmetry.

2.1. BPS-States and Partition Functions in $\mathcal{N} = 2$ Theories

Let us for definiteness begin with type IIB on $\mathcal{X}$. The field content of the resulting four-dimensional $\mathcal{N} = 2$ theory splits between vectormultiplets and hypermultiplets. For discussing black holes, it suffices to restrict attention to the vectormultiplet sector, whose bosonic field content consists of the gravity multiplet, containing the 4d spacetime metric $g_{\mu\nu}$ and the graviphoton A_μ, together with the vector multiplets, containing $n_V = h_{2,1}(\mathcal{X})$ abelian vector fields A^I_μ, and $h_{2,1} + 1$ complex scalars X^Λ. The hypermultiplet sector will enter the story in sections 2.4 and 2.5.

We choose a Lagrangian splitting of $H_3(\mathcal{X}, \mathbb{Z}) = \Gamma_e \oplus \Gamma_m \equiv \Gamma$, with an associated symplectic basis $(\mathcal{A}^\Lambda, \mathcal{B}_\Lambda)$, $\Lambda = (0, I)$ of A- and B-cycles. The scalars X^Λ then arise from the periods of the holomorphic 3-form $\Omega_{3,0} \in H^{3,0}(\mathcal{X}, \mathbb{C})$ along $\mathcal{A}^\Lambda$, and thus parametrize the (special Kähler) complex structure moduli space $\mathcal{M}_C(\mathcal{X})$. It is convenient to choose projective coordinates $t^I = X^I/X^0 \in \mathcal{M}_C(\mathcal{X})$, away from the locus $X^0 = 0$.

The BPS-states arise from D3-branes wrapping special Lagrangian submanifolds (sLags) in the homology class $q_\Lambda \mathcal{A}^\Lambda + p^\Lambda \mathcal{B}_\Lambda \in H_3(\mathcal{X}, \mathbb{Z})$. We write the electric-magnetic charge vector as $\gamma = (q_\Lambda, p^\Lambda) \in \Gamma$. The mass M of these states saturate the BPS-bound $M \geq |Z(\gamma;t)|$, where $Z(\gamma;t) \equiv \oint_\gamma \Omega_{3,0}$ is the central charge of the $\mathcal{N} = 2$ supersymmetry algebra. Mathematically, the central charge is a homomorphism $Z : \Gamma \to \mathbb{C}$, referred to as the "stability data" [45, 46].

The black hole partition function $\mathscr{Z}$ is a generating function of BPS-states. It is commonly attached to a "mixed" ensemble [47], where the magnetic charges p^Λ are held fixed, while summing over the electric charges q_Λ weighted by conjugate chemical potentials ζ^Λ. In general the partition function also varies over $\mathcal{M}_C$, hence $\mathscr{Z} = \mathscr{Z}(p; t, \zeta)$. One then arrives at the following expression for the partition function in the mixed ensemble

$$\mathscr{Z}_{\text{mix}}(p; t, \zeta) = \sum_{q_\Lambda \in \Gamma_e} \Omega(q_\Lambda, p^\Lambda; t) e^{2\pi i q_\Lambda \zeta^\Lambda},$$

where $\Omega : \Gamma \to \mathbb{Z}$ is the BPS-index which counts the weighted number of BPS-states of charge vector γ. Physically, the index corresponds to the "second helicity supertrace" which is defined as a trace over the BPS Hilbert space $\mathcal{H}^{\text{BPS}}_{\gamma,t}$ [48, 14]:

$$\Omega(\gamma; t) = \frac{1}{2} \text{Tr}_{\mathcal{H}^{\text{BPS}}_{\gamma,t}} (2J_3)^2 (-1)^{2J_3}, \tag{1}$$

where J_3 is a generator of the little group $SO(3)$ in four dimensions. The insertion of $(2J_3)^2$ is required to obtain a non-vanishing result in the presence of $\mathcal{N} = 2$ supersymmetry. The BPS-index conjecturally counts objects in the derived Fukaya category of $\mathcal{X}$, and by mirror symmetry should be equivalent to the generalized Donaldson-Thomas invariants [28]. The index $\Omega(\gamma; t)$ is furthermore a locally constant function of t, but may jump for special values corresponding to real codimension one submanifolds of $\mathcal{M}_C$, known as walls of marginal stability [49, 28, 29, 30, 31, 50].

For our present purposes it is natural to consider a slightly different generating function, namely one which is covariant with respect to symplectic transformations on the electric-magnetic charges. This corresponds to a partition function attached to a "canonical ensemble" [51], where one introduces chemical potentials $\tilde{\zeta}_\Lambda$ also for the magnetic charges p^Λ and the sum is taken over the full charge lattice. We define such a canonical partition function as

$$\mathscr{Z}_{\text{can}}(t, \zeta, \tilde{\zeta}) = \sum_{\gamma \in \Gamma} \lambda(\gamma) \overline{\Omega}(\gamma; t) e^{2\pi i (q_\Lambda \zeta^\Lambda - p^\Lambda \tilde{\zeta}_\Lambda)}, \tag{2}$$

where we have replaced the integral degeneracies $\Omega(\gamma) \in \mathbb{Z}$ with their rational counterparts $\overline{\Omega}(\gamma) \in \mathbb{Q}$, which are related as follows:

$$\overline{\Omega}(\gamma) = \sum_{d|\gamma} \frac{\Omega(\gamma/d)}{d^2}, \qquad \Omega(\gamma) = \sum_{d|\gamma} \frac{\mu(d) \overline{\Omega}(\gamma/d)}{d^2},$$

with $\mu(d)$ being the Möbius function. The rational degeneracies $\overline{\Omega}(\gamma)$ coincide with the Donaldson-Thomas invariants introduced by Kontsevich-Soibelman [28] and Joyce-Song [29], and are more suited for our purposes. In particular, they are better behaved under wall crossing and S-duality [52, 42, 53]. In (2) we have also inserted a $U(1)$-valued function $\lambda(\gamma)$ which corresponds to a quadratic refinement of the symplectic pairing on Γ. It can be viewed as a generalization of the sign $(-1)^{p^\Lambda q_\Lambda}$ familiar from D0-D2-D4 partition functions [54, 55, 52], and is ultimately required for describing the wall crossing behaviour of $\overline{\Omega}(\gamma)$ [30]. We will discuss properties of $\lambda(\gamma)$ in more detail in section 2.5.

2.2. Instanton Partition Functions in Three Dimensions

We now wish to consider the theory on $\mathbb{R}^3 \times S^1$ and analyze the low-energy physics in $D = 3$. Part of our motivation comes from string compactifications preserving $\mathcal{N} \geq 4$ supersymmetries, for which it is well-known that the 4d effective action is invariant under a discrete U-duality group $G_4(\mathbb{Z})$. In these cases the moduli space $\mathcal{M}_4$ of vacua is protected from quantum corrections, and takes the form of an arithmetic coset space $\mathcal{M}_4 = G_4(\mathbb{Z})\backslash G_4/K_4$, where G_4 is the continuous symmetry of the classical theory, and K_4 is the maximal compact subgroup of G_4, playing the role of a local "R-symmetry". For torus compactifications the symmetry group is $E_{7(7)}(\mathbb{Z})$ [1], while for $K3 \times T^2$ it is $SL(2,\mathbb{Z}) \times SO(6,22;\mathbb{Z})$. By further compactification on a circle of radius $R = e^{-\phi}\ell_{\rm P}$ ($\ell_{\rm P}$ being the 4d Planck length), the U-duality group gets enhanced to a larger group $G_3(\mathbb{Z}) \supset G_4(\mathbb{Z})$. Moreover, in three dimensions all p-form degrees of freedom can be dualized to (pseudo-)scalars which live in the enlarged moduli space[2]

$$\mathcal{M}_3 = G_3(\mathbb{Z})\backslash G_3/K_3.$$

Invariance under the spectrum generating symmetry $G_3(\mathbb{Z})$ provides powerful constraints on quantum corrections in the effective action; in particular, BPS-saturated higher derivative couplings are $G_3(\mathbb{Z})$-invariant Eisenstein series (see, e.g., [5]).

Let us now perform a similar analysis in the $\mathcal{N} = 2$ setting. For definiteness, we still restrict to the type IIB picture. The role of the symmetry group $G_4(\mathbb{Z})$ is played by the monodromy group $\mathrm{M}(\mathcal{X}) \subset Sp(2h_{2,1}+2;\mathbb{Z})$ of the Calabi-Yau threefold $\mathcal{X}$. After compactification to $D = 3$, one finds two new moduli from the gravity multiplet: the radius $e^{-\phi}$ of the circle, together with the "NUT potential" σ, corresponding to the dual of the Kaluza-Klein vector (the off-diagonal component of the 4d metric). In addition, there are $2(n_{\rm V}+1)$ scalars arising from the electric and magnetic Wilson lines of the $n_{\rm V}+1 = h_{2,1}+1$ abelian vector fields A^Λ_μ (including the graviphoton) along the circle:

$$\zeta^\Lambda = \oint_{S^1} A^\Lambda_R dR, \qquad \tilde{\zeta}_\Lambda = \oint_{S^1} (A^\star_R)_\Lambda dR, \tag{3}$$

where A^Λ_R is the component of A^Λ_μ along the circle, while $(A^\star_R)_\Lambda$ is the same component of the dual vector field in $D = 4$. Since Wilson lines couple to electric (or magnetic) charges Q through terms of the form $e^{2\pi i Q \oint_R A dR}$, we may, after comparison with (2), identify the Wilson lines with the chemical potentials, thereby explaining our choice of notation in (3). Thus, one of the virtues of compactification to three dimensions is that the auxiliary chemical potentials $(\zeta^\Lambda, \tilde{\zeta}_\Lambda)$ become promoted to actual physical degrees of freedom. Including also the complex structure moduli $t \in \mathcal{M}_C(\mathcal{X})$, the scalars $\{t, \phi, \zeta^\Lambda, \tilde{\zeta}_\Lambda, \sigma\}$ parametrize the enhanced vector multiplet moduli space in $D = 3$, which we shall denote by $\widehat{\mathcal{M}_C}(\mathcal{X})$. The moduli space $\widehat{\mathcal{M}_C}(\mathcal{X})$ is real $4(h_{2,1}+1)$-dimensional and is equipped with a *quaternion-Kähler* metric [21, 26, 27].[3]

In contrast to $\mathcal{M}_3$, $\widehat{\mathcal{M}_C}(\mathcal{X})$ is in general not a coset space, and is also not protected from quantum corrections. The classical, or "semiflat" [30], metric $\widehat{g}^{\rm sf}$ on $\widehat{\mathcal{M}_C}(\mathcal{X})$ is obtained by naive dimensional reduction on S^1. This metric is singular but it is expected that quantum corrections ensure that the exact moduli space metric $\widehat{g}$ is smooth. These corrections are due to instanton effects which arise from $D = 4$ Euclidean BPS-states wrapping the circle [25]. Microscopically, they correspond to Euclidean D3-branes wrapping $\mathcal{L} \times S^1$, where $\mathcal{L} \subset \mathcal{X}$ is a sLag. The quantum

[2] If the circle S^1 is timelike one obtains instead a pseudo-Riemannian coset space $\mathcal{M}_3^* = G_3/K_3^*$, where K_3^* is non-compact, whose non-positive definite metric can be obtained by analytic continuation from $\mathcal{M}_3$ [17]. We will not distinguish between $\mathcal{M}_3$ and $\mathcal{M}_3^*$ in the following.

[3] Upon reduction on a timelike circle S^1 the resulting moduli space $\widehat{\mathcal{M}^*_C}(\mathcal{X})$ is "para-quaternion-Kähler" [56], which is the quaternionic analogue of the pseudo-Riemannian coset space $\mathcal{M}_3^*$.

deformations of the semiflat metric are weighted by the exponential of minus the radius of the circle times the mass of the BPS-state (or black hole), schematically

$$\widehat{g} \sim \widehat{g}^{\mathrm{sf}} + e^{-2\pi e^{-\phi}|Z(\gamma,t)|}. \tag{4}$$

The explicit form of $\widehat{g}^{\mathrm{sf}}$ is given in eq. (12) below. The instanton corrections are exponentially suppressed in the large radius limit $e^{-\phi} \to \infty$, but contributes with subleading effects for finite values of $e^{-\phi}$. From (3) we learn that the Wilson lines couple to the charges $\gamma = (q_\Lambda, p^\Lambda)$ via terms of the form $e^{-2\pi i(q_\Lambda \zeta^\Lambda - p^\Lambda \tilde{\zeta}_\Lambda)}$, and hence multi-instanton effects are captured by the partition function

$$\mathscr{Z}_{\mathrm{BPS}}(t,\phi,\zeta,\tilde{\zeta}) \sim \sum_{\gamma\in\Gamma} \lambda(\gamma)\overline{\Omega}(\gamma;t) e^{-2\pi \mathcal{S}_\gamma^{\mathrm{BPS}}(t,R,\zeta,\tilde{\zeta})}, \tag{5}$$

where $\mathcal{S}_\gamma^{\mathrm{BPS}}$ is the Euclidean action for a BPS-instanton of charge γ:

$$\mathcal{S}_\gamma^{\mathrm{BPS}}(t,R,\zeta,\tilde{\zeta}) = e^{-\phi}|Z(\gamma;t)| + i(q_\Lambda\zeta^\Lambda - p^\Lambda\tilde{\zeta}_\Lambda).$$

$\mathscr{Z}_{\mathrm{BPS}}$ should be viewed as the instanton analogue of the black hole partition function (2). Charge quantization ("large gauge transformations") requires that $(\zeta^\Lambda, \tilde{\zeta}_\Lambda)$ are angle-valued:

$$\zeta^\Lambda \longmapsto \zeta^\Lambda + m^\Lambda, \qquad \tilde{\zeta}_\Lambda \longmapsto \tilde{\zeta}_\Lambda + n_\Lambda,$$

where $(m^\Lambda, n_\Lambda) \in \mathbb{Z}^{2(n_{\mathrm{V}}+1)}$. Hence, $(\zeta^\Lambda, \tilde{\zeta}_\Lambda)$ parametrize a symplectic torus $\mathcal{T}(\mathcal{X}) = T^{2(n_{\mathrm{V}}+1)}$.

Note that the BPS-index $\overline{\Omega}(\gamma;t)$ reappeared in (5) as the "instanton measure" which encodes multiplicities of instanton bound states in three dimensions. Physically, it corresponds to the path integral measure on the instanton moduli space [57]. The assumption that these two numerical quantities are equal is strongly supported by the behaviour of $\overline{\Omega}(\gamma;t)$ under wall crossing [30] (see [58, 59, 57, 60, 61, 31, 62] for related discussions).

A crucial difference between $\mathcal{N}=2$ field theory and $\mathcal{N}=2$ supergravity is that for large charges the indexed degeneracies of BPS states grow exponentially

$$\overline{\Omega}(\gamma) \sim \exp\left[S_{\mathrm{BH}}(\gamma)\right], \tag{6}$$

where $S_{\mathrm{BH}}(\gamma)$ is the Bekenstein-Hawking entropy of a black hole with charge γ. This implies that the instanton sum in (5) diverges, in contrast to the analogous sum in field theory [30]. A possible resolution to this problem was suggested in [62], by interpreting (5) as an asymptotic series, the ambiguity of which is of the order $\exp[-e^{-2\phi}]$. Such effects indeed exist in $D=3$, and are precisely due to the presence of gravity, as we will now discuss.

In addition to the BPS-instantons, the semiflat metric (12) also receives corrections from gravitational instantons which have no physical analogue in $D=4$ [63, 15, 24, 64, 31]. More precisely, they correspond to four-dimensional Taub-NUT solutions with closed timelike curves. However, in $D=3$ these give rise to physical non-perturbative effects, whose contributions to the metric on $\widehat{\mathcal{M}_C}(\mathcal{X})$ are of the form $e^{-2\pi|k|e^{-2\phi}}$ and are therefore subleading in the large radius limit. The associated gravitational scalar σ couples to the NUT charge $k \in \mathbb{Z}$ through terms of the form $e^{-i\pi k\sigma}$. This requires that σ is periodic modulo 2 (in our conventions), and furthermore that it picks up non-trivial shifts under translations along the torus $\mathcal{T}(\mathcal{X})$ [41, 42]:

$$\sigma \longmapsto \sigma + 2r - n_\Lambda\zeta^\Lambda + m^\Lambda\tilde{\zeta}_\Lambda + 2v(m,n), \tag{7}$$

where $r \in \mathbb{Z}$. Here we have allowed for the presence of a cocycle shift $v(m,n)$ whose appearance is related to the quadratic refinement $\lambda(\gamma)$ as will be explained in section 2.5.

We conclude that the semiflat metric on $\widehat{\mathcal{M}_C}$ is also deformed by gravitational instantons captured by a partition function of the form

$$\mathscr{Z}_{\text{TN}}(t,\phi,\zeta,\tilde{\zeta},\sigma) \sim \sum_{k\neq 0} \mathscr{Z}^{(k)}(t,\zeta,\tilde{\zeta})e^{-2\pi \mathcal{S}_k^{\text{TN}}(\phi,\sigma)}, \tag{8}$$

where the Euclidean action of a charge k Taub-NUT instanton is

$$\mathcal{S}_k^{\text{TN}}(\phi,\sigma) = |k|e^{-2\phi} + \tfrac{1}{2}ik\sigma.$$

We defer a discussion of the coefficients $\mathscr{Z}^{(k)}(t,\zeta,\tilde{\zeta})$ to section 2.5.

Combining the contributions in (5) and (8) we find that the exact metric on the moduli space $\widehat{\mathcal{M}_C}$ should take the schematic form

$$\widehat{g} = \widehat{g}^{\,\text{sf}} + \mathscr{Z}_{\text{tot}}(t,\phi,\zeta,\tilde{\zeta},\sigma),$$

where $\mathscr{Z}_{\text{tot}}$ is the total instanton partition function whose large-radius expansion is given by

$$\mathscr{Z}_{\text{tot}}(t,\phi,\zeta,\tilde{\zeta},\sigma) \sim \sum_{\gamma\in\Gamma} \lambda(\gamma)\overline{\Omega}(\gamma;t)e^{-2\pi \mathcal{S}_\gamma^{\text{BPS}}(t,\phi,\zeta,\tilde{\zeta})} + \sum_{k\neq 0} \mathscr{Z}^{(k)}(t,\zeta,\tilde{\zeta})e^{-2\pi \mathcal{S}_k^{\text{TN}}(\phi,\sigma)}. \tag{9}$$

2.3. Topology of the Moduli Space and the Semiflat Metric

Here we offer some useful remarks on the topology of the moduli space $\widehat{\mathcal{M}_C}(\mathcal{X})$, which will in particular allow for a nice explicit presentation of the semiflat metric $\widehat{g}^{\,\text{sf}}$. By virtue of eq. (7), the shifts of $(\zeta^\Lambda,\tilde{\zeta}_\Lambda)$ do not commute, and the total translation group acting on $(\zeta^\Lambda,\tilde{\zeta}_\Lambda,\sigma)$ is identified with a $(2n_{\text{V}}+3)$-dimensional arithmetic Heisenberg group $N(\mathbb{Z})$. This implies that the coordinates $(\zeta^\Lambda,\tilde{\zeta}_\Lambda,\sigma)$ parametrize the total space of a circle bundle $\mathcal{C}_\sigma \to \mathcal{T}(\mathcal{X})$, with fiber given by the circle S^1_σ parametrized by σ. Moreover, the base $\mathcal{T}(\mathcal{X})$ is itself non-trivially fibered over the complex structure moduli space $\mathcal{M}_C(\mathcal{X})$, and we denote the total space of this torus fibration by $\mathscr{J}_C(\mathcal{X})$:

$$\mathcal{T}(\mathcal{X}) \longrightarrow \mathscr{J}_C(\mathcal{X}) \longrightarrow \mathcal{M}_C(\mathcal{X}). \tag{10}$$

At fixed (large) value of the radius $R >> \ell_{\text{P}}$ $(e^{-\phi}\to\infty)$, the three-dimensional moduli space then exhibits a foliation [41, 42]

$$\widehat{\mathcal{M}_C}(\mathcal{X}) \sim \mathbb{R}_+ \times \mathcal{C}_\sigma(\phi) \tag{11}$$

where the leaves are circle bundles $\mathcal{C}_\sigma(\phi)$ over $\mathscr{J}_C(\mathcal{X})$:

$$S^1_\sigma \longrightarrow \mathcal{C}_\sigma(\phi) \longrightarrow \mathscr{J}_{\mathcal{M}_4}.$$

This represents a supergravity generalization of the Seiberg-Witten (hyperkähler) torus fibration over the Coulomb branch $\mathcal{B}$ of $\mathcal{N}=2$ field theories on $\mathbb{R}^3\times S^1$ [25, 30]. Having identified the topology of $\widehat{\mathcal{M}_C}(\mathcal{X})$ through the foliation (11), we can now write the semiflat metric $\widehat{g}^{\,\text{sf}}$ on $\widehat{\mathcal{M}_C}(\mathcal{X})$ in the following explicit form[4]

$$\widehat{g}^{\,\text{sf}} = 4\mathrm{d}\phi^2 + 4g_{\mathcal{M}_C} + e^{2\phi}g_{\mathcal{T}} + \tfrac{1}{16}e^{4\phi}\mathcal{A}_{\mathcal{C}_\sigma}^2, \tag{12}$$

where $\mathcal{A}_{\mathcal{C}_\sigma} = \mathrm{d}\sigma + \tilde{\zeta}_\Lambda \mathrm{d}\zeta^\Lambda - \zeta^\Lambda \mathrm{d}\tilde{\zeta}_\Lambda$ is the connection on the circle bundle $\mathcal{C}_\sigma(\phi)$, with first Chern class $c_1(\mathcal{C}_\sigma) = \mathrm{d}(\mathcal{A}_{\mathcal{C}_\sigma}/2) = \mathrm{d}\tilde{\zeta}_\Lambda \wedge \mathrm{d}\zeta^\Lambda$ equal to the Kähler form $\omega_{\mathcal{T}}$ on the torus $\mathcal{T}(\mathcal{X})$ [42].[5]

[4] See [26, 27] for the original form of this metric, where it was referred to as the *c-map metric*.
[5] In string theory there is also a one-loop correction to the semiflat metric [65, 66, 67], which modifies the topology of $\mathcal{C}_\sigma$ [42], although this will not concern us in the present work.

2.4. T-Duality and D-Brane Instantons

We now perform a T-duality transformation along the circle S^1_ϕ, which maps the moduli space $\widetilde{\mathcal{M}_C}(\mathcal{X})$ into the hypermultiplet moduli space $\mathcal{M}_{\rm H}^{\rm IIA}(\mathcal{X})$ of type IIA string theory on $\mathcal{X}$. The radius $e^{-\phi}$ of the circle is then mapped to the four-dimensional dilaton, $g_s \equiv e^{\phi}$, in such a way that the large radius limit in the type IIB picture is equivalent to the weak-coupling limit $g_s \to 0$ on the type IIA side. After T-duality the Wilson lines $(\zeta^\Lambda, \tilde{\zeta}_\Lambda)$ correspond to the periods of the type IIA Ramond-Ramond 3-form $C_{(3)}$ over a symplectic basis of $H_3(\mathcal{X}, \mathbb{Z})$:

$$\zeta^\Lambda = \int_{\mathcal{A}^\Lambda} C_{(3)}, \qquad \tilde{\zeta}_\Lambda = \int_{\mathcal{B}_\Lambda} C_{(3)},$$

and the torus $\mathcal{T}(\mathcal{X})$ is thereby identified with the intermediate Jacobian,

$$\mathcal{T}(\mathcal{X}) = H^3(\mathcal{X}, \mathbb{R})/H^3(\mathcal{X}, \mathbb{Z}).$$

The torus fibration (10) thus encodes the fact that the complex structure of $\mathcal{T}(\mathcal{X})$ varies over $\mathcal{M}_C(\mathcal{X})$ and the total space $\mathscr{J}_C(\mathcal{X}) \to \mathcal{M}_C(\mathcal{X})$ represents the entire family of intermediate Jacobians of $\mathcal{X}$. Since $C_{(3)}$ couples to the D2-brane, we conclude that the instanton effects in (5) are T-dual to Euclidean D2-branes wrapping sLags $\mathcal{L} \subset \mathcal{X}$. Indeed, this is consistent with the fact that in the weak-coupling limit, D-instantons are exponentially suppressed by e^{-1/g_s}.

2.5. T-Duality, NS5-Branes and Generalized Theta Functions

On the other hand, after T-duality the Taub-NUT instantons are mapped into non-perturbative effects arising from Euclidean NS5-branes wrapping the Calabi-Yau manifold $\mathcal{X}$ [68, 69, 70]. The form of the coupling (8) is consistent with the fact that NS5-brane effects are weighted by e^{-1/g_s^2} [22]. The scalar σ corresponds to the NS-axion, arising from the dualization of the Kalb-Ramond 2-form $B_{(2)}$ in $D = 4$.

These observations provide information about the structure of the gravitational partition function (8). Invoking the results of [41, 42] we deduce that the coefficient $\mathscr{Z}^{(k)}$ corresponds to the partition function of a charge k Euclidean NS5-brane wrapped on the Calabi-Yau manifold $\mathcal{X}$. Since the worldvolume theory of the type IIA NS5 brane is chiral [71], $\mathscr{Z}^{(k)}$ is not a proper function, but rather a holomorphic section of a line bundle $\mathscr{L}^k$ over $\mathscr{J}_C(\mathcal{X})$ [72, 73, 74].

These results imply that $\mathscr{Z}^{(k)}$ satisfies twisted periodicity relations under translations by a vector $(m^\Lambda, n_\Lambda) \in H^3(\mathcal{X}, \mathbb{Z})$ along the torus $\mathcal{T}$:

$$\mathscr{Z}^{(k)}(t, \zeta + m, \tilde{\zeta} + n) = [\lambda(m,n)]^k e^{\pi i k (m^\Lambda \tilde{\zeta}_\Lambda - n_\Lambda \zeta^\Lambda)} \mathscr{Z}^{(k)}(t, \zeta, \tilde{\zeta}), \tag{13}$$

where $\lambda(m,n)$ is a homomorphism $\lambda : H^3(\mathcal{X}, \mathbb{Z}) \to U(1)$, corresponding to a quadratic refinement of the intersection form on $H_3(\mathcal{X}, \mathbb{Z})$ [72, 74]. It is defined by the cocycle relation:

$$\lambda(H + H') = (-1)^{\langle H, H' \rangle} \lambda(H) \lambda(H'), \tag{14}$$

where $H \in H^3(\mathcal{X}, \mathbb{Z})$. Equivalently, λ describes the holonomies around one-cycles in $\mathcal{T}(\mathcal{X})$. For the choice of Lagrangian decomposition of $H^3(\mathcal{X}, \mathbb{Z})$ corresponding to $H = (m^\Lambda, n_\Lambda)$, one can solve the condition (14) generally in terms of characteristics $\Theta = (\theta^\Lambda, \phi_\Lambda)$ such that [74]

$$\lambda(m,n) = e^{-i\pi n_\Lambda m^\Lambda + 2\pi i (n_\Lambda \theta^\Lambda - m^\Lambda \phi_\Lambda)} \equiv (-1)^{2v(m,n)},$$

where $v(m,n)$ is defined modulo one. It was conjectured in [41, 42]—based on S-duality and other considerations—that the quadratic refinement $\lambda(m,n)$ governing the NS5-partition function is equivalent to the $\lambda(\gamma)$ appearing in (2), after replacing the "flux" $H = (m^\Lambda, n_\Lambda)$ in its argument by the electric-magnetic charge vector $\gamma = (p^\Lambda, q_\Lambda)$.

Sections of $\mathscr{L}$ can be represented by generalized theta series of the form [41]

$$\mathscr{Z}_\Theta^{(1)}(t,\zeta,\tilde\zeta) = \sum_{p^\Lambda\in\Gamma_{\rm m}+\theta^\Lambda} \Psi(t,\zeta^\Lambda - p^\Lambda)e^{2\pi i(\tilde\zeta_\Lambda-\phi_\Lambda)p^\Lambda + i\pi(\theta^\Lambda\phi_\Lambda-\zeta^\Lambda\tilde\zeta_\Lambda)},$$

where $\Psi(\zeta^\Lambda)$ is a "wave function" on $\Gamma_{\rm m}\otimes\mathbb{R}$. When $\Psi(\zeta^\Lambda)$ is a simple Gaussian then $\mathscr{Z}_\Theta^{(1)}$ is a holomorphic section of $\mathscr{L}$ and coincides with the partition function of the self-dual field, as analyzed in [72, 74]. For $k>1$, the bundle $\mathscr{L}^k$ admits $|k|^{2h_{2,1}(\mathcal{X})}$ holomorphic sections corresponding to the elements in $\Gamma_{\rm m}/(|k|\Gamma_{\rm m})$, and the total partition function becomes a sum of vector-valued theta series [42]:

$$\mathscr{Z}_\Theta^{(k)}(t,\zeta,\tilde\zeta) = \sum_{\ell^\Lambda\in\Gamma_{\rm m}/(|k|\Gamma_{\rm m})} \mathscr{Z}_\Theta^{k,\ell^\Lambda}(t,\zeta,\tilde\zeta),$$

where we defined

$$\mathscr{Z}_\Theta^{k,\ell^\Lambda}(t,\zeta,\tilde\zeta) = \sum_{p^\Lambda\in\Gamma_{\rm m}+\theta^\Lambda+\ell^\Lambda} \Psi_{k,\ell^\Lambda}(t,\zeta^\Lambda - p^\Lambda)e^{2\pi i k(\tilde\zeta_\Lambda-\phi_\Lambda)p^\Lambda + i\pi k(\theta^\Lambda\phi_\Lambda-\zeta^\Lambda\tilde\zeta_\Lambda)}. \tag{15}$$

By examining the coupling (8) we deduce that the Heisenberg transformation of σ induced by translations of $(\zeta^\Lambda, \tilde\zeta_\Lambda)$ is compensated by the middle term in the transformation (13) of the partition function $\mathscr{Z}_\Theta^{(k)}$. However, it follows from (13) that this compensating transformation also receives a contribution $[\lambda(m,n)]^k$ from the quadratic refinement. This implies that for the sum (8) to be well-defined from the point of view of the line bundle $\mathscr{L}$, the Heisenberg transformation of the NS-axion σ must contain a term involving the cocycle $v(m,n)$ [41, 42], as was already anticipated in eq. (7). In other words, $e^{i\pi\sigma}$ is valued in the same line bundle $\mathscr{L}$ as the NS5 partition function $\mathscr{Z}_\Theta^{(k)}$.

In the general non-linear case when the wave function $\Psi_{k,\ell^\Lambda}(t,\zeta^\Lambda)$ is non-Gaussian, the partition function $\mathscr{Z}_\Theta^{(k)}$ is no longer a holomorphic section of $\mathscr{L}^k$. In fact, it was shown in [42] that S-duality and mirror symmetry implies that for $k=1$ the wave function $\Psi_{1,0}(\zeta^\Lambda)$ should be identified with the real-polarized partition function $\Psi_{\rm top} = \exp\left(\sum_g \lambda^{2g-2}F_g\right)$ of the topological string on $\mathcal{X}$, consistently with earlier related results [69, 75, 70, 76, 77]. These observations have bearing on the transformation properties of the partition function $\mathscr{Z}_\Theta^{(k)}$ under the monodromy group $\mathrm{M}(\mathcal{X})$, or, more generally, under the 4d duality group $G_4(\mathbb{Z})$. In particular, the wave function $\Psi_{k,\ell^\Lambda}(\zeta^\Lambda)$ transforms in the Schrödinger-Weil representation of $G_4(\mathbb{Z})\ltimes N(\mathbb{Z})$ [37, 42].[6]

2.6. Mirror Symmetry and Unification of Moduli in $D=3$

Let us now also briefly discuss the mirror version of the above story. Mirror symmetry equates the complex structure moduli space $\mathcal{M}_C(\mathcal{X})$ with the complexified Kähler moduli space $\mathcal{M}_K(\hat{\mathcal{X}})$ of the mirror Calabi-Yau threefold $\hat{\mathcal{X}}$. Physically, $\mathcal{M}_K(\hat{\mathcal{X}})$ is the vector multiplet moduli space of type IIA string theory compactified on $\hat{\mathcal{X}}$. Under mirror symmetry, the spectrum of BPS-states states arising from D3-branes wrapping special Lagrangian submanifolds $\mathcal{L}\subset\mathcal{X}$, is mapped to bound states of D0-D2-D4-D6 branes wrapping even cycles in the Calabi-Yau. Such a bound state is accurately described by a stable coherent sheaf $\mathcal{E}$ on $\hat{\mathcal{X}}$, with charges classified by the

[6] The precise dependence of the NS5-partition function on the Calabi-Yau metric, i.e. on the moduli $t\in\mathcal{M}_C$, is subtle [78, 74, 79], and is closely related to the one-loop correction to the semiflat metric $\hat{g}^{\rm sf}$ [41, 42].

K-theory group $K(\hat{\mathcal{X}})$ [80]. Concretely, the charges are described through the Mukai vector $\gamma' : K(\hat{\mathcal{X}}) \to H_{\mathrm{even}}(\hat{\mathcal{X}}, \mathbb{Q})$, which takes the explicit form

$$\gamma' = \mathrm{ch}(\mathcal{E})\sqrt{\mathrm{Td}(\hat{\mathcal{X}})},$$

where $\mathrm{ch}(\mathcal{E})$ is the Chern character of the sheaf $\mathcal{E}$ and $\mathrm{Td}(\hat{\mathcal{X}})$ is the Todd class of $T\hat{\mathcal{X}}$. Since γ' is valued in the rational cohomology group $H_{\mathrm{even}}(\hat{\mathcal{X}}, \mathbb{Q})$ one cannot immediately identify it with its mirror dual $\gamma \in H_3(\mathcal{X}, \mathbb{Z})$. To properly identify the charges under mirror symmetry one must make a symplectic transformation on the electric charges

$$A : \mathbb{Q} \ni q'_\Lambda \mapsto q_\Lambda \equiv q'_\Lambda + A_{\Lambda\Sigma}p^\Sigma \in \mathbb{Z},$$

where $A_{\Lambda\Sigma}$ is a certain integer-valued symplectic matrix (see [42] for more details). Thus, mirror symmetry relates $\gamma \in H_3(\mathcal{X}, \mathbb{Z})$ to a "modified Mukai vector" as follows [42]

$$H_3(\mathcal{X}, \mathbb{Z}) \ni \gamma \quad \overset{\text{mirror map}}{\longleftrightarrow} \quad \gamma = e^A\,\mathrm{ch}(\mathcal{E})\sqrt{\mathrm{Td}(\hat{\mathcal{X}})} \in H_{\mathrm{even}}(\hat{\mathcal{X}}, \mathbb{Z}).$$

This is a realization of the "topological mirror map" discussed in [45]. The electric charges $q_\Lambda = (q_0, q_a)$ now correspond to D0-D2 branes, while $p^\Lambda = (p^a, p^0)$ are D4-D6 charges. After the mirror map, the central charge is given by $Z(\gamma; t) = \int_{\hat{\mathcal{X}}} e^{-\mathcal{J}}\mathrm{ch}(\mathcal{E})\sqrt{\mathrm{Td}(\hat{\mathcal{X}})}$, where $\mathcal{J} = B + iJ$ is the complexified Kähler form on $\mathcal{M}_K(\hat{\mathcal{X}})$.

Compactification to three dimensions promotes the special Kähler moduli space $\mathcal{M}_K(\hat{\mathcal{X}})$ to a quaternion-Kähler manifold $\widehat{\mathcal{M}_K}(\hat{\mathcal{X}})$, whose semiflat metric is deformed by instanton effects arising from Euclidean D0-D2-D4-D6 branes wrapping $\hat{\mathcal{X}} \times S^1$, together with gravitational Taub-NUT instantons. This setup can furthermore be lifted to M-theory on $\mathcal{X} \times T^2$, revealing a hidden $SL(2, \mathbb{Z})$-symmetry corresponding to the mapping class group of the torus. By T-duality, these effects are also equivalent to D(-1)-D1-D3-D5 and NS5-instanton corrections in the hypermultiplet sector of type IIB on $\hat{\mathcal{X}}$, and the $SL(2, \mathbb{Z})$-symmetry is then recognized as the familiar S-duality group.

These considerations imply that by using a combination of mirror symmetry and T-duality, all of the moduli spaces and instanton effects discussed above are in fact equivalent:

$$\mathcal{M}_{\mathrm{H}}^{\mathrm{IIA}}(\mathcal{X}) \quad \overset{\text{T-duality}}{\longleftrightarrow} \quad \widehat{\mathcal{M}_C}(\mathcal{X}) \quad \overset{\text{mirror map}}{\longleftrightarrow} \quad \widehat{\mathcal{M}_K}(\hat{\mathcal{X}}) \quad \overset{\text{T-duality}}{\longleftrightarrow} \quad \mathcal{M}_{\mathrm{H}}^{\mathrm{IIB}}(\hat{\mathcal{X}}).$$

Because of this unification of moduli spaces in three dimensions, it seems that the appropriate language for studying the spectrum of BPS states in type II Calabi-Yau compactifications is that of quaternion-Kähler geometry, rather than special Kähler geometry.

3. Instanton Partition Functions from Automorphic Forms

We now return to the instanton partition function $\mathscr{Z}_{\mathrm{tot}}$ in (9), and assume the theory exhibits a U-duality symmetry $G_3(\mathbb{Z})$. We will argue that the exact partition function should be given by an automorphic form $\mathscr{F}$, whose Fourier expansion reproduces the instanton effects in (9).

What properties should we expect of $\mathscr{F}$? Recall that the partition function must at least be invariant under the semidirect product

$$G^J(\mathbb{Z}) \equiv G_4(\mathbb{Z}) \ltimes N(\mathbb{Z}),$$

which is known as a *Jacobi group* in the mathematical literature.[7] The black hole partition function $\mathscr{Z}_{\mathrm{BPS}}$ in (5) should also be invariant under $G^J(\mathbb{Z})$, while the NS5-wave function Ψ_{k,ℓ^Λ} in (15) transforms in the Schrödinger-Weil representation of $G^J(\mathbb{Z})$.

[7] The group G^J is sometimes also called "Fourier-Jacobi group".

A useful analogue of this situation, first pointed out in [37], is the case of classical holomorphic Siegel modular forms, attached to the metaplectic representation of $Sp(4;\mathbb{Z})$. Fourier expansion around one of the cusps in the Siegel upper half plane yields coefficients attached to the Schrödinger-Weil representation of the Jacobi subgroup $SL(2,\mathbb{Z}) \ltimes N(\mathbb{Z}) \subset Sp(4;\mathbb{Z})$, namely the standard Jacobi theta series [81, 82]. This is known as a *Fourier-Jacobi expansion*.

Inspired by this analogy, we shall now discuss the construction of an automorphic form $\mathscr{F}$ such that the generalization of the Fourier-Jacobi expansion for $G_3(\mathbb{Z}) \supset G^J(\mathbb{Z})$ reproduces the expected non-perturbative contributions to the partition function $\mathscr{Z}_{\text{tot}}$. For the general case when $G^J(\mathbb{Z}) \subset G_3(\mathbb{Z})$ the metaplectic representation is replaced by the so called *minimal representation* of $G_3(\mathbb{Z})$. This suggests that the NS5-partition function (8) has an interpretation as the Fourier-Jacobi expansion of an automorphic form attached to the minimal representation of $G_3(\mathbb{Z})$ [83, 37].

3.1. Constructing Automorphic Forms

We are interested in constructing a function $\mathscr{F}$ on the double coset $G(\mathbb{Z})\backslash G/K$, where G is a real Lie group, K its maximal compact subgroup and $G(\mathbb{Z}) \subset G$ a discrete subgroup. The construction of such functions relies heavily on the representation theory of G. In general, automorphic forms arise from irreducible components in the decomposition of the space $L^2\big(G(\mathbb{Z})\backslash G\big)$, which forms a unitary representation of G. An *automorphic representation* is a pair (π,ρ), where π is an irreducible subspace in the decomposition of $L^2\big(G(\mathbb{Z})\backslash G\big)$, and ρ is its realization. In order to ensure that the associated automorphic form lives on G/K, we look for distinguished K-invariant vectors $\mathcal{W}_0 \in \pi$. Such objects are known as *spherical (Whittaker) vectors*.

To spell out the construction more explicitly, we need the following data:

(i) a unitary (automorphic) representation (π,ρ) of G acting on some Hilbert space $\mathscr{H}$,

(ii) a spherical vector $\mathcal{W}_0 \in \mathscr{H}$,

(iii) a $G(\mathbb{Z})$-invariant *distribution* $\mathcal{W}_{\mathbb{Z}} \in \mathscr{H}^*$.

Using the bilinear pairing $\langle\cdot|\cdot\rangle : \mathscr{H}\times\mathscr{H}^* \to \mathbb{R}$, one may construct $\mathscr{F}$ as the overlap [35, 84]

$$\mathscr{F}(g) = \langle \mathcal{W}_{\mathbb{Z}} | \rho(g)\cdot \mathcal{W}_0\rangle\,, \tag{16}$$

where $g \in G$. This is manifestly invariant under the left action of $\gamma \in G(\mathbb{Z})$ due to the invariance of $\mathcal{W}_{\mathbb{Z}}$, and under the right action of $k \in K$ since $\mathcal{W}_0$ is spherical, and hence defines a function on $G(\mathbb{Z})\backslash G/K$. More generally, one may consider a non-spherical Whittaker vector $\mathcal{W}_K$ ("finite K-type") transforming in some representation of K, for which the associated automorphic *form* $\mathscr{F}_K(g)$ is no longer a function, but rather a section of a vector bundle over G/K.

There is a powerful method to construct the $G(\mathbb{Z})$-invariant distribution $\mathcal{W}_{\mathbb{Z}}$ using p-adic number theory. One may compute $\mathcal{W}_{\mathbb{Z}}$ locally for each prime number p, giving a p-adic spherical vector $\mathcal{W}_p$ invariant under $G(\mathbb{Z}_p) \subset G(\mathbb{Q}_p)$, where $\mathbb{Q}_p$ is the field of p-adic numbers and $\mathbb{Z}_p$ the ring of p-adic integers. It is justified to refer to $\mathcal{W}_p$ as spherical since the discrete group $G(\mathbb{Z}_p)$ is the maximal compact subgroup $K_p \subset G(\mathbb{Q}_p)$, due to the fact that the ring $\mathbb{Z}_p$ is compact in $\mathbb{Q}_p$. By convention, the prime at infinity $p=\infty$ corresponds to the real numbers $\mathbb{Q}_\infty = \mathbb{R}$. Thus, $\mathcal{W}_\infty$ coincides with the standard K-invariant vector $\mathcal{W}_0$.

The distribution $\mathcal{W}_{\mathbb{Z}}$ is recovered by taking the product over all primes, yielding [35, 84]

$$\mathscr{F}(g) = \sum_{x\in\mathbb{Q}^\kappa}\ \prod_{p\,\text{prime}} \rho(g)\cdot\mathcal{W}_p(x), \tag{17}$$

where the sum over rational vectors $x \in \mathbb{Q}^\kappa$, $\kappa \in \mathbb{Z}_+$, encodes the bilinear pairing $\langle\cdot|\cdot\rangle$. The representation-theoretic meaning of the integer κ will be explained momentarily. We will also see below how to evaluate (17) for explicit examples.

3.2. 5-Grading and the Heisenberg Parabolic

Any Lie algebra $\mathfrak{g}$ of rank > 1, with Lie group G, exhibits a 5-grading (vector space direct sums)

$$\mathfrak{g} = \mathfrak{g}_{-2} \oplus \mathfrak{g}_{-1} \oplus \mathfrak{g}_0 \oplus \mathfrak{g}_1 \oplus \mathfrak{g}_2, \tag{18}$$

where the subspaces $\mathfrak{g}_{\pm 2}$ are both one-dimensional and are generated by the highest root of $\mathfrak{g}$, and $\mathfrak{g}_{\pm 1}$ are $2(n_{\text{V}} + 1)$-dimensional vector spaces, endowed with a natural (Kirillov-Kostant) symplectic form. Physically, the parameter n_{V} will play the same role as in section 2, i.e. corresponding to the number of vector multiplets. $\mathfrak{g}_{\pm 1} \oplus \mathfrak{g}_{\pm 2}$ form Heisenberg subalgebras with centers given by the highest and lowest root spaces $\mathfrak{g}_{\pm 2}$.

The 5-grading can be understood as the decomposition of the adjoint representation of $\mathfrak{g}$ with respect to the subalgebra $\mathfrak{sl}(2,\mathbb{R}) = \mathfrak{g}_{-2} \oplus \mathbb{R}h \oplus \mathfrak{g}_2$, where h is the Cartan generator associated to the highest root, and the subscripts in (18) indicate the grade under the action of h. The level zero subspace decomposes further into $\mathfrak{g}_0 = \mathfrak{r} \oplus \mathbb{R}h$, where $\mathfrak{r}$ is the commutant of $\mathfrak{sl}(2,\mathbb{R})$ inside $\mathfrak{g}$. At the level of the Lie group G, the subspace $\mathfrak{g}_0 \oplus \mathfrak{g}_1 \oplus \mathfrak{g}_2$ generates a parabolic subgroup $P \subset G$, with Levi decomposition $P = LN$, where the "unipotent radical" N is the Heisenberg group generated by $\mathfrak{g}_1 \oplus \mathfrak{g}_2$, and L is the Levi factor generated by $\mathfrak{g}_0$. The parabolic subgroup P is known as the *Heisenberg parabolic*. We denote by Z the center of N.

To understand the physical relevance of the grading (18) let us take $\mathfrak{g}$ to be the Lie algebra of the purported symmetry group G_3. The commutant $\mathfrak{r}$ is then identified with the Lie algebra of the duality group G_4 in four dimensions. The Cartan generator h corresponds to rescalings of the radius $e^{-\phi}$, and hence the decomposition (18) mirrors the decompactification limit $e^{-\phi} \to \infty$, which makes the G_4-invariance manifest. The one-dimensional subspace $\mathfrak{g}_2$ corresponds to translations of the NUT-scalar σ, while $\mathfrak{g}_1$ generates shifts of the Wilson lines $(\zeta^\Lambda, \tilde{\zeta}_\Lambda)$. Thus, the unipotent radical N represents the Heisenberg action on $(\zeta^\Lambda, \tilde{\zeta}_\Lambda, \sigma)$ discussed in section 2.2. Finally, the generators of the opposite unipotent radical $\mathfrak{g}_{-2} \oplus \mathfrak{g}_{-1}$ are non-linearly realized and correspond to a Geroch-type solution generating symmetry [16, 17].

3.3. Induced Automorphic Representations

A convenient way to construct an automorphic representation (π, ρ) is through so called parabolic induction. Consider a function $f(n) \in \mathbb{R}$, with $n \in P\backslash G$, where $P \subset G$ is a parabolic subgroup. The group G acts on n from the right, and a compensating transformation of $p \in P$ from the left ensures that $png \equiv n'$ is again an element of $P\backslash G$. A representation ρ of G is obtained by acting on functions $f \in L^2(P\backslash G)$ according to

$$\rho(g) \cdot f(n) = \delta_\rho(p) f(n'),$$

where $\delta_\rho(p) \in \mathbb{R}$ is called the infinitesimal character of ρ. In the mathematical literature such induced representations are commonly denoted by $\text{Ind}_P^G \delta_\rho \equiv I(P)$. By choosing a K-covariant vector $\mathcal{W}_K \in \mathscr{H} = L^2(P\backslash G)$, the representation $I(P)$ extends to an automorphic representation (π, ρ) through the construction in (16). Since G is a non-compact group, the Hilbert space $\mathscr{H}$ is infinite-dimensional. However, one may still associate a dimension to the representation (π, ρ); this is the so called *functional, or Gelfand-Kirillov, dimension* $\kappa = \dim P\backslash G$.

Elements of $I(P)$ are functions (or, more generally, sections) of κ variables, corresponding to the summation variables $x \in \mathbb{Q}^\kappa$ in (17). However, the p-adic Whittaker vector $\mathcal{W}_p$ has only support on $\mathbb{Z}_p \subset \mathbb{Q}_p$, and therefore, after taking the product over all primes, the sum has support on integers $x \in \mathbb{Z}^\kappa$ (see [35, 84]). These integers should be identified with the physical charges of the theory. Hence, physically the automorphic form $\mathscr{F}$ should be attached to a representation (π, ρ) whose functional dimension coincides with number of charges.

An example of particular relevance for what follows is when the parabolic subgroup P coincides with the Heisenberg parabolic. The resulting representation $I(P, s) = \text{Ind}_P^G \delta_\rho(s)$

Journal of Physics: Conference Series **346** (2012) 012016 doi:10.1088/1742-6596/346/1/012016

belongs to the (degenerate) principal continuous series of G and depends on $\operatorname{rank}_{\mathbb{R}} \mathfrak{g}$ complex parameters collectively denoted by s. The functional dimension of $I(P,s)$ is $\kappa = \dim \mathfrak{g}_1 + 1$. The automorphic functions attached to the degenerate principal series are non-holomorphic Eisenstein series

$$\mathscr{E}(g;s) = \sum_{\gamma \in P(\mathbb{Z}) \backslash G(\mathbb{Z})} \delta_\rho(\gamma g; s).$$

When the Heisenberg parabolic P coincides with the Borel subgroup $B \subset G$ then the representation $I(B,s)$ corresponds to the generic (non-degenerate) principal series. This happens in particular for the special cases of $SU(2,1)$ and $SL(3,\mathbb{R})$ to be considered in sections 4 and 5.

Although the BPS partition function (2) only depends on $2(n_V + 1)$ electric and magnetic charges, the 3d theory also exhibits the NUT-charge k, thus providing a total of $2n_V + 3$ charges $(q_\Lambda, p^\Lambda, k)$, which is precisely the functional dimension of $I(P,s)$. A naive first guess is therefore that the complete partition function $\mathscr{Z}_{\rm tot}$ in (9) should be captured by an automorphic function in the degenerate principal series. A characteristic property of the Eisenstein series $\mathscr{E}(g;s)$ is that they are eigenfunctions of the Laplacian on the coset space G/K, with eigenvalue given by the quadratic Casimir operator. Physically, this is precisely the constraint enforced by supersymmetry in $\mathcal{N}=8$ theories [85, 5], although we shall see in section 6 that this constraint is modified for $\mathcal{N}=2$ theories.

One can also obtain smaller ("unipotent") automorphic representations by taking residues of degenerate principal Eisenstein series $\mathscr{E}(g;s)$ along singular loci in the complex s-parameter space [86, 87]. In this way one can in particular construct the *minimal representation* $\pi_{\rm min}$ of smallest functional dimension $\kappa = n_V + 2$ [88]. The minimal representation can be obtained by quantizing the smallest coadjoint orbit $\mathcal{O}_{\rm min}$, i.e. the G-orbit of any root in the Lie algebra $\mathfrak{g}$. It is natural to choose the lowest root, corresponding to the subspace $\mathfrak{g}_{-2}$ in (18). The orbit $\mathcal{O}_{\rm min}$ is then generated by the $(2n_V + 4)$-dimensional subspace $\mathbb{R}h \oplus \mathfrak{g}_1 \oplus \mathfrak{g}_2$. Quantization amounts to choosing a Lagrangian splitting of $\mathfrak{g}_1$ and realizing the generators as differential operators acting on functions of "coordinates" and "momenta" [89, 35, 90]. The functional dimension of the associated representation $\pi_{\rm min}$ is half the dimension of the orbit $\mathcal{O}_{\rm min}$, as advertised. Physically, one may think of these variables as the $n_V + 1$ magnetic charges p^Λ (or, in the opposite polarization, electric charges q_Λ) together with the NUT charge k [15]. Formally, k therefore plays the role of a quantization parameter, by analogy with Planck's constant $\hbar$. The minimal representation generalizes the metaplectic representation for symplectic groups, and its Fourier-Jacobi expansion yields theta series in the Schrödinger-Weil representation of $G^J(\mathbb{Z})$ [83], which is what we expect for the NS5-partition function (8).

3.4. Abelian Fourier Expansion

In order to isolate the dependence on the scalars $(\zeta^\Lambda, \tilde{\zeta}_\Lambda, \sigma)$ as in the instanton sums (9) one must perform a Fourier expansion of $\mathscr{F}$ which diagonalizes the action of $N(\mathbb{Z})$. However, since this group is non-abelian one cannot decompose with respect to all the generators simultaneously. We shall therefore first perform the decomposition along the center Z of N.

The Fourier expansion along Z reads

$$\mathscr{F}(g) = \sum_{\chi_Z \in \hat{Z}} \mathscr{F}_{\chi_Z}(g), \tag{19}$$

where $\chi_Z : Z \to \mathbb{C}^*$ is a unitary character of Z which is trivial on $Z(\mathbb{Z})$. In other words, the sum runs over the unitary irreducible representations of Z, known as the "unitary dual" $\hat{Z}$. The Fourier coefficient $\mathscr{F}_{\chi_Z}(g)$ is formally defined through the integral

$$\mathscr{F}_{\chi_Z} = \int_{Z(\mathbb{Z})\backslash Z(\mathbb{R})} \mathscr{F}(zg)\overline{\chi_Z(z)} dz.$$

Journal of Physics: Conference Series **346** (2012) 012016 doi:10.1088/1742-6596/346/1/012016

More explicitly, in our conventions we may take $\chi_Z(\sigma) = e^{-\pi i k\sigma}$ and write

$$\mathscr{F}_k(t,\zeta,\tilde{\zeta},\phi) = \int_0^2 \mathscr{F}(t,\zeta,\tilde{\zeta},\sigma,\phi)e^{i\pi k\sigma}d\sigma. \tag{20}$$

We restrict first to the case when Z acts trivially, corresponding to the zeroth Fourier coefficient $\mathscr{F}_0$. The abelian expansion of $\mathscr{F}_{\mathrm{A}} \equiv \mathscr{F}_0$ with respect to $N_{\mathrm{a}} = N/Z$ is given by $\sum_{\chi_{N_{\mathrm{a}}}\in\hat{N}_{\mathrm{a}}} \mathscr{F}_{\chi_{N_{\mathrm{a}}}}$, where the $\chi_{N_{\mathrm{a}}}$:th Fourier coefficient is

$$\mathscr{F}_{\chi_{N_{\mathrm{a}}}}(g) = \int_{N_{\mathrm{a}}(\mathbb{Z})\backslash N_{\mathrm{a}}(\mathbb{R})} \mathscr{F}(ng)\overline{\chi_{N_{\mathrm{a}}}(n)}dn.$$

If we choose a Lagrangian splitting of $\mathfrak{g}_1$ such that the characters $\chi_{N_{\mathrm{a}}}$ of $N_{\mathrm{a}}(\mathbb{R})$ takes the explicit form $\chi_{N_{\mathrm{a}}}(\zeta,\tilde{\zeta}) = e^{-2\pi i(q_\Lambda\zeta^\Lambda - p^\Lambda\tilde{\zeta}_\Lambda)}$, where $(q_\Lambda, p^\Lambda) \in \mathbb{Z}^{2n_{\mathrm{V}}+2}$, then we can write the abelian Fourier expansion more explicitly as

$$\mathscr{F}_{\mathrm{A}}(t,\zeta,\tilde{\zeta},\phi) = \sum_{(q_\Lambda,p^\Lambda)\in\mathbb{Z}^{2n+2}} C(q_\Lambda,p^\Lambda)\mathcal{W}_{q_\Lambda,p^\Lambda}(t,\phi)e^{-2\pi i(q_\Lambda\zeta^\Lambda - p^\Lambda\tilde{\zeta}_\Lambda)}, \tag{21}$$

where we have also extracted the purely numerical coefficients $C(q_\Lambda,p^\Lambda)$. The moduli dependent part $\mathcal{W}_{q_\Lambda,p^\Lambda}(t,\phi)$ corresponds to the spherical Whittaker vector $\mathcal{W}_0$ in a "Fourier transformed representation". Similarly, the numerical Fourier coefficients $C(q_\Lambda,p^\Lambda)$ are related to the p-adic spherical vector $\mathcal{W}_p$, after performing a p-adic Fourier transform and evaluating the product over all primes (see [35, 84, 91] for details).

The result (21) bears an strong resemblance with the instanton sum (5). The instanton effects should be reproduced by the leading term in the expansion of the spherical vector $\mathcal{W}_{q_\Lambda,p^\Lambda}(t,\phi)$ around the cusp $e^\phi \to 0$. It is furthermore suggestive to identify the BPS-degeneracies $\overline{\Omega}(\gamma)$ with the numerical Fourier coefficients $C(q_\Lambda,p^\Lambda)$.

3.5. Non-Abelian (Fourier-Jacobi) expansion

We shall now consider the case when the center Z of N acts non-trivially, corresponding to the generic coefficients $\mathscr{F}_k, k\neq 0$, in (20). This provides a generalization of the Fourier-Jacobi expansion of Siegel modular forms as disussed in the beginning of section 3.

For $k \neq 0$ we write the expansion (19) explicitly as

$$\mathscr{F}_{\mathrm{NA}}(t,\zeta,\tilde{\zeta},\sigma,\phi) = \sum_{k\neq 0}\mathscr{F}_k(t,\zeta,\tilde{\zeta},\phi)e^{-i\pi k\sigma}. \tag{22}$$

When the center acts non-trivially one cannot simultaneously diagonalize all of the generators of the abelian subspace $\mathfrak{g}_1$, as was done in (21). Instead, one must choose a *polarization*, i.e. diagonalize either "electric" translations of ζ^Λ or "magnetic" translations of $\tilde{\zeta}_\Lambda$.

For definiteness we consider a magnetic polarization, for which the general form of the non-abelian Fourier coefficients $\mathscr{F}_k$, $k \neq 0$, is given by [92, 38, 39][8]

$$\mathscr{F}_k(t,\zeta,\tilde{\zeta},\phi) = \sum_{\ell^\Lambda\in\Gamma_{\mathrm{m}}/(|k|\Gamma_{\mathrm{m}})}\ \sum_{p^\Lambda\in\Gamma_{\mathrm{m}}+\ell^\Lambda/|k|} \Psi_{k,\ell^\Lambda}(t,\phi,\zeta^\Lambda - p^\Lambda)\, e^{2\pi i k p^\Lambda\tilde{\zeta}_\Lambda - \pi i k\zeta^\Lambda\tilde{\zeta}_\Lambda}. \tag{23}$$

The sum in (22) is then manifestly invariant under magnetic translations $\tilde{\zeta}_\Lambda \mapsto \tilde{\zeta}_\Lambda + n_\Lambda$, while electric shifts of the form $\zeta^\Lambda \mapsto \zeta^\Lambda + n^\Lambda$ require a compensating lattice shift $p^\Lambda \mapsto p^\Lambda + n^\Lambda$. Thus, the non-abelian contribution $\mathscr{F}_{\mathrm{NA}}$ is indeed invariant under the Jacobi group $G^J(\mathbb{Z})$.

[8] For details on the derivation, see for instance section 11.5 of [91].

By comparing (23) with the analysis in section (2.5) it is clear that the non-abelian Fourier expansion corresponds to a sum over vector-valued theta series, with wave function Ψ_{k,ℓ^Λ}. It is therefore suggestive to identify the Fourier coefficient $\mathscr{F}_k$ with the partition function (15) for k NS5-branes. This is consistent with the fact that Ψ_{k,ℓ^Λ} transforms in the Schrödinger-Weil representation of $G^J(\mathbb{Z})$. Moreover, $\mathscr{F}_{\rm NA}$ is a sum over $n_{\rm V}+2$ variables (p^Λ, k) which matches the functional dimension of the minimal representation of $G_3(\mathbb{Z})$. We conclude that (22) is the desired generalization of the Fourier-Jacobi expansion discussed in the beginning of this section.

4. Rigid Calabi-Yau Threefolds and the Universal Sector

We will now put the abstract analysis of the previous section into our first explicit example. We consider a special class of $\mathcal{N}=2$ theories, obtained through compactification of type IIA string theory on a *rigid* Calabi-Yau threefold. A Calabi-Yau threefold is said to be rigid if it has no complex structure deformations; hence $h_{2,1}(\mathcal{X})=0$. We denote such a rigid threefold by $\mathcal{X}_{\rm r}$.

The Hodge decomposition of $H^3(\mathcal{X}_{\rm r},\mathbb{C})$ simplifies considerably:

$$H^3(\mathcal{X}_{\rm r},\mathbb{C}) = H^{3,0}(\mathcal{X}_{\rm r},\mathbb{C}) \oplus H^{0,3}(\mathcal{X}_{\rm r},\mathbb{C}). \tag{24}$$

A consequence of this is that the intermediate Jacobian torus is an elliptic curve (see, e.g., [93])

$$\mathcal{T}(\mathcal{X}) = H^3(\mathcal{X}_{\rm r},\mathbb{R})/H^3(\mathcal{X}_{\rm r},\mathbb{Z}) = H^{3,0}(\mathcal{X}_{\rm r},\mathbb{C})/H^3(\mathcal{X}_{\rm r},\mathbb{Z}) = \mathbb{C}/(\mathbb{Z}+\tau\mathbb{Z}),$$

where τ is the period variable which takes values in the upper half plane $SL(2,\mathbb{R})/U(1)$. Two elliptic curves $\mathbb{C}/(\mathbb{Z}+\tau\mathbb{Z})$ and $\mathbb{C}/(\mathbb{Z}+\tau'\mathbb{Z})$ are equivalent if $\tau'=\gamma\cdot\tau$ where $\gamma\in SL(2,\mathbb{Z})$. For simplicity, we further restrict our attention to those rigid Calabi-Yau threefolds for which the parameter τ takes values in the quadratic number field $\mathbb{Q}(\sqrt{-d})$, where d is a square-free positive integer. Elliptic curves of this type are said to admit complex multiplication, or be of "CM-type". A rigid Calabi-Yau threefold is then of CM-type if and only if the associated Jacobian elliptic curve is of CM-type. The key observation is that in this case, the lattice $\mathbb{Z}+\tau\mathbb{Z}$ corresponds to the ring of integers $\mathcal{O}_d\subset\mathbb{Q}(\sqrt{-d})$.

Physically, rigid Calabi-Yau compactifications are simpler, since the complex structure moduli space $\mathcal{M}_C(\mathcal{X}_{\rm r})$ is trivial. For type IIB on $\mathcal{X}_{\rm r}$, the vector multiplet sector contains a single scalar modulus $X^0=\int_{\mathcal{A}}\Omega_{3,0}$ and the prepotential is quadratic

$$F(X^0) = \tfrac{1}{2}\tau(X^0)^2.$$

The modulus X^0 is non-dynamical and is customarily set to 1; however it is useful to retain it in order to keep track of the homogeneity degree of $F(X^0)$. Since there are no complex structure moduli the BPS-index $\Omega(\gamma)$ is globally constant, and do not suffer wall-crossing. Equivalently, (24) ensures that one is automatically sitting at the end point of the attractor flow [94].

The vector multiplet sector in $D=4$ only contains the gravity multiplet, i.e. a single vector field A_μ (the graviphoton) together with the metric $g_{\mu\nu}$; this is the familiar Maxwell-Einstein theory. Upon dimensional reduction on S^1 it is well-known that this theory exhibits an enhanced symmetry group described by $U(2,1)$, and the classical moduli space in $D=3$ is given by the coset space [16, 17]

$$\mathcal{M}_3^{\rm cl} = U(2,1)/(U(2)\times U(1)).$$

This 4-dimensional coset space is quaternion-Kähler[9] and represents the classical part of the exact moduli space $\widehat{\mathcal{M}_C}(\mathcal{X}_{\rm r})$ in $D=3$. It is parametrized by the radius $e^{-\phi}$, the NUT potential σ, and the two Wilson lines $(\zeta^0,\tilde{\zeta}_0)\equiv(\zeta,\tilde{\zeta})$. Although $\mathcal{M}_3^{\rm cl}$ is described by a coset space,

[9] In fact, it is also Kähler, a curiosity among quaternion-Kähler manifolds.

instanton corrections will deform the metric, and hence the exact moduli space $\mathcal{M}_3^{\text{cl}}$ does not preserve this coset structure.

Upon T-duality, $\mathcal{M}_3^{\text{cl}}$ is mapped to the moduli space of the so called "universal hypermultiplet", parametrized by the dilaton e^{ϕ}, the NS-axion σ and the RR-scalars $(\zeta, \tilde{\zeta})$ arising from the periods of the 3-form along the "universal" 3-cycles $(\mathcal{A}, \mathcal{B}) \in H_3(\mathcal{X}_{\text{r}}, \mathbb{Z})$ [26, 27, 95].

Despite the fact that the quantum moduli space $\widehat{\mathcal{M}_C}(\mathcal{X}_{\text{r}})$ is not a coset, we shall assume that it retains an isometric action of an arithmetic subgroup $G_3(\mathbb{Z}) \subset U(2,1)$. An analogous phenomenon is known to happen for the hypermultiplet sector in type IIB, where the moduli space $\mathcal{M}_{\text{H}}^{\text{IIB}}(\mathcal{X}_{\text{r}})$ exhibits an isometric action of the S-duality group $SL(2,\mathbb{Z})$ [96, 32, 42].

In [39, 40] it was proposed that for rigid Calabi-Yau threefolds of CM-type, the quantum theory should be invariant under the arithmetic U-duality group

$$PU(2,1;\mathcal{O}_d) \equiv U(2,1) \cap PGL(3,\mathcal{O}_d),$$

known as the *Picard modular group*. This group reproduces the correct Heisenberg shifts (7), and is thus consistent with charge quantization For $d = 3 \bmod 4$ the group $PU(2,1;\mathcal{O}_d)$ induces a shift of σ with non-trivial cocycle $v(m,n) = \frac{1}{2}(m+n+mn)$ in (7) [40], while for $d=1$, corresponding to the Gaussian integers $\mathcal{O}_1 = \mathbb{Z} + i\mathbb{Z}$, the cocycle is absent [39]. If taken literally, this has the interesting implication that for each rigid Calabi-Yau compactification there is a unique theta line bundle $\mathscr{L} \to \mathcal{T}(\mathcal{X}_{\text{r}})$, and hence a unique NS5-partition function $\mathscr{Z}_{\Theta}^{(k)}$, as was originally proposed by Witten [72].

To further test this proposal one would like to show that it is compatible with the expected form (9) of quantum corrections to the moduli space metric. Following the discussion in section 3.3, we consider a $PU(2,1;\mathcal{O}_d)$-invariant Eisenstein series in the principal series $I(B,s)$ of $U(2,1)$, which has the same functional dimension $\kappa = 3$ as the number of D2-NS5 charges (q,p,k).[10] Using the techniques of section 3.1 we obtain [39, 40]

$$\mathscr{E}^{PU(2,1;\mathcal{O}_d)}(g;s) = \sum_{\substack{(x_1,x_2)\in\mathbb{Q}(\sqrt{-d})^2 \\ P(x_1,x_2)=0}} \rho(g)\cdot\left[\prod_{p<\infty} \mathcal{W}_p(x_1,x_2;s)\right]\mathcal{W}_0(x_1,x_2;s),$$

where g is a representative in $U(2,1)/(U(2)\times U(1))$ and $P(x_1,x_2)=0$ denotes a certain quadratic constraint on the summation variables (see [39, 40] for details), analogous to the lattice constraints discussed in [5]. The principal series depends on a single parameter $s \in \mathbb{C}$, and the Eisenstein series converges absolutely for $\Re(s) > 2$. However, it is a famous result by Langlands [97] that $\mathscr{E}_{PU(2,1;\mathcal{O}_d)}(g;s)$ can be analytically continued to a meromorphic function in the full complex s-plane, away from the poles at $s = 0, 2$. Indeed, $\mathscr{E}^{PU(2,1;\mathcal{O}_d)}(g;s)$ satisfies a functional equation relating its value at s to the Weyl-transformed value at $2-s$ [39, 40].

The spherical vector $\mathcal{W}_0$ is a function on $B\backslash U(2,1)$, and hence depends on three real variables, or equivalently on 2 variables $(x_1,x_2) \in \mathbb{Q}(\sqrt{-d})^2$ subject to a constraint $P(x_1,x_2)=0$. The explicit form of $\mathcal{W}_0$ is obtained by taking the unitary norm of the first row of a representative $n \in P\backslash U(2,1)$; the norm ensures that $\mathcal{W}_0$ is $U(2)\times U(1)$-invariant, i.e. spherical. The p-adic counterpart $\mathcal{W}_p$ is similarly obtained by taking the corresponding p-adic norm $|z|_p^{\mathbb{Q}(\sqrt{-d})} \equiv \sqrt{|z\bar{z}|_p}$, $z \in \mathbb{Q}(\sqrt{-d})$. For the explicit formulas and more details on this construction, see App. A. of [39], or references [35, 84, 91] for other examples.

By computing the Fourier expansion of $\mathscr{E}^{PU(2,1;\mathcal{O}_d)}(g;s)$ one then obtains explicit candidates for the black hole partition function in (21) and the NS5 wave function $\Psi_{k,\ell}(\phi,\zeta)$ in (23).

[10] This is moreover the most naive extension of previous successful results of summing up D(-1)-instantons in type IIB [4, 96] using an $SL(2,\mathbb{Z})$-invariant Eisenstein series attached to the principal series of $SL(2,\mathbb{R})$.

Extracting first the moduli-dependent part $\mathcal{W}_{q,p}(\phi)$ of $\mathscr{E}^{PU(2,1;\mathcal{O}_d)}_{q,p}(\phi)$ one finds

$$\mathcal{W}_{q,p}(\phi;s) = K_{2s-2}\left(2\pi e^{-\phi}\frac{|q+\tau p|}{\sqrt{\Im(\tau)}}\right), \tag{25}$$

where K_{2s-2} is the modified Bessel function and we recall that τ is the period variable of the elliptic curve $\mathcal{T}(\mathcal{X}) = \mathbb{C}/(\mathbb{Z}+\tau\mathbb{Z})$. For the special value $s=3/2$, the Whittaker vector $\mathcal{W}_{q,p}(\phi;3/2)$ reproduces the known form of the D2-instanton series in type II on a Calabi-Yau threefold [31]. Indeed, upon expanding $\mathcal{W}_{q,p}(\phi;3/2)$ around the weak-coupling cusp $e^{\phi}\to 0$, one finds that the abelian Fourier expansion (21) takes the form

$$\mathscr{E}^{PU(2,1;\mathcal{O}_d)}_{\mathrm{A}}(\zeta,\tilde{\zeta},\phi)\sim \sum_{(p,q)\in\mathbb{Z}^2} C_{\mathrm{A}}(p,q;3/2)\exp\left[-2\pi e^{-\phi}\frac{|q+\tau p|}{\sqrt{\Im\tau}} - 2\pi i(q\zeta - p\tilde{\zeta})\right]. \tag{26}$$

This is indeed of the same form as the BPS-instanton series (5) after noticing that for a rigid Calabi-Yau threefold the D2-brane central charge is [39]

$$Z(\gamma) = \frac{q+\tau p}{\sqrt{\Im(\tau)}}, \qquad \gamma = q\mathcal{A}+p\mathcal{B}\in H_3(\mathcal{X}_{\mathrm{r}},\mathbb{Z}).$$

Note in particular that (26) has the expected suppression e^{-1/g_s} for D-brane instantons.

The numerical coefficients in (26) were evaluated explicitly in [39, 40] with the result

$$C_{\mathrm{A}}(\Lambda;s) = \sum_{\substack{\omega'\in\mathcal{O}_d\\ \Lambda/\omega'\in\mathcal{O}_d^*}}\left|\frac{\Lambda}{\omega'}\right|^{2s-2}\sum_{\substack{z\in\mathcal{O}_d\\ \Lambda/(z\omega')\in\mathcal{O}_d^*}}|z|^{4-4s},$$

where ω is a primitive vector in $\mathcal{O}_d$ and $\Lambda=(q+\tau p)/\Im(\tau)$. As a first consistency check we note that for purely electric or purely magnetic charges, i.e. when either $q=0$ or $p=0$, the Fourier coefficients $C_{\mathrm{A}}(q,0;3/2)$ reproduce the sum over divisors $\sum_{d|q}d^{-2}$ which is known to provide the instanton measure for D(-1)-instantons [4, 96].

Let us now turn to the non-abelian Fourier coefficients (23). The wave function $\Psi_{k,\ell}$ extracted from $\mathscr{E}^{PU(2,1;\mathcal{O}_d)}_{\mathrm{NA}}$ takes the form [39]

$$\Psi^{PU(2,1;\mathcal{O}_d)}_{k,\ell}(\phi,\tilde{\zeta}-p) = \sum_{r=0}^{\infty} C_{\mathrm{NA}}(r,k,\ell;s)\, e^{-\pi|k|(\tilde{\zeta}-p)^2}\, H_r\left(\sqrt{2\pi|k|}(\tilde{\zeta}-p)\right)\, W_{-r-\frac{1}{2},s-1}\left(4\pi|k|e^{-2\phi}\right)$$

where H_r is a Hermite polynomial and $W_{r,s}$ is the Whittaker function. Inserting $s=3/2$ and expanding the Whittaker function for weak coupling yields

$$\Psi^{PU(2,1;\mathcal{O}_d)}_{k,\ell}(\phi,\tilde{\zeta}-p)\sim\sum_{r=0}^{\infty} C_{\mathrm{A}}(r,k,\ell;3/2)\, H_r\left(\sqrt{2\pi|k|}(\tilde{\zeta}-p)\right)\, e^{-2\pi|k|e^{-2\phi}-\pi|k|(\tilde{\zeta}-p)^2}, \tag{27}$$

which is indeed suppressed by e^{-1/g_s^2} as befits NS5-brane instantons.[11] Moreover, for fixed value of r the dependence of $\Psi_{k,\ell}$ on $(\tilde{\zeta}-p)$ matches the general structure of the NS5-brane partition function found in [42], where it was argued that the insertion of a power of $(\tilde{\zeta}-p)$ in the generalized theta series (15) was related to the insertion of $(J_3)^2$ in the second helicity supertrace (1), as is required in order to get a non-vanishing result for $\mathcal{N}=2$ theories. Unfortunately, as of yet there is no explicit expression for the numerical non-abelian Fourier coefficients $C_{\mathrm{NA}}(r,k,\ell;s)$.

[11] To be more accurate, in the special case of the Gaussian integers $\mathcal{O}_1=\mathbb{Z}+i\mathbb{Z}$ the periodicity of the NS-axion is slightly different from (23), forcing the charge k to be a multiple of 4 [39].

Journal of Physics: Conference Series **346** (2012) 012016 doi:10.1088/1742-6596/346/1/012016

5. The Extended Universal Hypermultiplet

For generic Calabi-Yau compactifications there is in general no known candidate for the spectrum generating symmetry $G_3(\mathbb{Z})$. However, it was suggested in [38] that the S-duality symmetry $SL(2,\mathbb{Z})$ of the hypermultiplet moduli space in type IIB extends to an isometric action of $SL(3,\mathbb{Z})$. The motivation behind this can be understood in the T-dual picture of the vector multiplet sector in type IIA on $\mathcal{X}\times S^1$. As already mentioned, the associated moduli space $\widehat{\mathcal{M}_K}(\mathcal{X})$ lifts to M-theory on $\mathcal{X}\times T^2$, thus carrying an action of $SL(2,\mathbb{Z})$. However, it is well-known that upon toroidal compactification to $D=3$ the naive mapping class group $SL(n,\mathbb{Z})$ of an n-torus is enhanced to the "Ehlers symmetry" $SL(n+1,\mathbb{Z})$. This suggests that the moduli space $\widehat{\mathcal{M}_K}(\mathcal{X})$ carries a hidden isometric action of $SL(3,\mathbb{Z})$.

To make this explicit, one can identify a 5-dimensional subspace $SL(3,\mathbb{R})/SO(3)\subset\widehat{\mathcal{M}_K}(\mathcal{X})$ parametrized by the volume modulus t of $\mathcal{X}$, the RR-scalars $(\zeta^0,\tilde{\zeta}_0)\equiv(\zeta,\tilde{\zeta})$ which couple to the (D0, D6)-branes, and the gravity scalars $(e^{-\phi},\sigma)$. In the T-dual picture the role of the modulus t is unchanged, while $(\zeta,\tilde{\zeta})$ couple to D(-1)-D5 instantons, σ to NS5-instantons, while the radius is as usual identified with the 4d string coupling $g_s=e^{\phi}$. By analogy with the standard universal hypermultiplet discussed section 4, the subspace $\{t,\zeta,\tilde{\zeta},\sigma,\phi\}$ is referred to as the "extended universal hypermultiplet".

It was further argued in [38] that in the weak coupling, large volume limit, the full quaternionic moduli space decomposes locally into a product:

$$\widehat{\mathcal{M}_K}(\mathcal{X})\sim SL(3,\mathbb{R})/SO(3)\times\mathcal{R}_K(\mathcal{X})\times(\mathbb{R}^3)^{\otimes h_{1,1}(\mathcal{X})},$$

where $\mathcal{R}_K(\mathcal{X})$ is the vector multiplet moduli space of M-theory on $\mathcal{X}$, of real dimension $h_{1,1}-1$. The action of $SL(3,\mathbb{R})$ on $\widehat{\mathcal{M}_K}(\mathcal{X})$ leaves this factor invariant but acts linearly on the vector space $(\mathbb{R}^3)^{\otimes h_{1,1}(\mathcal{X})}$.

By a similar philosophy as in Section 4 one can now attempt to sum up the contributions from D(-1)-D5 and NS5-brane instantons to the metric on $\widehat{\mathcal{M}_K}(\mathcal{X})$ using an $SL(3,\mathbb{Z})$-invariant Eisenstein series $\mathscr{E}^{SL(3,\mathbb{Z})}(g;s_1,s_2)$, depending on two parameters which are fixed to the values $(s_1,s_2)=(3/2,3/2)$. The Eisenstein series can be constructed using similar methods as in Section 4 (see [84, 98, 91]), and may also be written as a Poincaré series

$$\mathscr{E}^{SL(3,\mathbb{Z})}(t,\phi,\zeta,\tilde{\zeta},\sigma;s_1,s_2)=\sum_{\gamma\in B(\mathbb{Z})\backslash SL(3,\mathbb{Z})}(\gamma\cdot t)^{s_1-s_2}\left(\gamma\cdot e^{-2\phi}\right)^{s_1+s_2},$$

where $t^{s_1-s_2}e^{-2\phi(s_1+s_2)}\equiv\delta_\rho(p;s_1,s_2)$ is the infinitesimal character of the principal series $I(B,s_1,s_2)$, and B is the Borel subgroup. Using old results by Vinogradov-Takhtajan [99] and Bump [100], one may extract the explicit form of the Fourier coefficients of $\mathscr{E}^{SL(3,\mathbb{Z})}(g;s_1,s_2)$. After taking the weak-coupling limit, the abelian contribution is of the form

$$\mathscr{E}_{\text{A}}^{SL(3,\mathbb{Z})}(t,\phi,\zeta,\tilde{\zeta};s_1,s_2)\sim\sum_{(p,q)\in\mathbb{Z}^2}C(p,q;s_1,s_2)\exp\left[-2\pi e^{-\phi}|Z(p,q)|-2\pi i(q\zeta-p\tilde{\zeta})\right].$$

As a non-trivial consistency check we note that the first term in the exponential reproduces the central charge for D0-D6 BPS-bound states [101]:

$$|Z(p,q)|=\left(t|p|^{2/3}+t^{-1}|q|^{2/3}\right)^{3/2}.$$

The numerical coefficients $C(p,q;s_1,s_2)$ are given by a generalized sum over divisors of the charges (p,q) whose explicit form can be found in [38].

The non-abelian Fourier coefficients $\mathscr{E}_{\rm NA}^{SL(3,\mathbb{Z})}$ give rise to a wave function $\Psi_{k,\ell}$ whose weak-coupling expansion reads

$$\Psi_{k,\ell}^{SL(3,\mathbb{Z})} \sim \sum_{\frac{(kn)^3}{d^2}\in\mathbb{Z}} C\left(d, \frac{(nk)^3}{d^2}; s_1, s_2\right) \exp\left[-\frac{2\pi|k|e^{-\phi}\left(e^{-2\phi}+t^3(\tilde{\zeta}-p)^2+tn^2\right)^{3/2}}{e^{-2\phi}+t^3(\tilde{\zeta}-p)^2} + \frac{2\pi i k t^3 n^3(\tilde{\zeta}-p)}{e^{-2\phi}+t^3(\tilde{\zeta}-p)^2}\right] \quad (28)$$

where $d = \gcd(k,\ell)$. We note the interesting fact that the numerical Fourier coefficients in (28) are equivalent to the abelian coefficients $C(p,q;s_1,s_2)$ with the magnetic charge p replaced by d and the electric charge q is replaced by the combination $(nk)^3/d^2$, thus representing an "induced" electric charge. This meshes well with the recent results of [42], which revealed that the instanton measure for bound states of D(-1)-D1-D3 instantons with D5-NS5-instantons is given by the generalized Donaldson-Thomas invariants $\Omega(q_\Lambda, p^\Lambda)$ with p^0 replaced by $\gcd(p,k)$, where $(p,0)$ represents pure D5-branes and $(0,k)$ represents pure NS5-branes. Moreover, the exponential in (28) matches the general NS5-brane instanton action found in [42].

6. Supersymmetry, Representation Theory and Twistors

We shall now outline how to add the final piece of the puzzle, namely the proper implementation of the constraints from supersymmetry. The degree of supersymmetry imposes constraints on the charges $(q_\Lambda, p^\Lambda, k)$ as well as on the geometry of the moduli spaces. We now discuss these constraints in turn, focussing first on charges in sections 6.1-6.3, while treating the moduli space geometry in sections 6.4 and 6.5.

6.1. $\mathcal{N}=8$ Charge Orbits and Representation Theory

Although we are primarily interested in the $\mathcal{N}=2$ case, it is again useful to first draw lessons from theories with a higher degree of supersymmetry. For $\mathcal{N}=8$ supersymmetry in $D=4$ a generic electric-magnetic charge vector $\gamma=(q_\Lambda, p^\Lambda)$ is valued in a 56-dimensional symplectic lattice Γ, invariant under $Sp(56;\mathbb{Z})$ [1]. These charges fit into the **56**-representation of $E_{7(7)}(\mathbb{Z})$. The entropy of a BPS black hole is [12]

$$S_{\rm BH}^{\mathcal{N}=8}(\gamma) = \pi\sqrt{\mathcal{I}_4(\gamma)},$$

where $\mathcal{I}_4(\gamma)$ is the quartic $E_{7(7)}(\mathbb{Z})$-invariant. BPS-solutions require $\mathcal{I}_4 \geq 0$. Generic 1/8-BPS black holes satisfy $\mathcal{I}_4 \neq 0$, while states with $\mathcal{I}_4 = 0$ but $\partial_\gamma \mathcal{I}_4 \neq 0$ correspond to 1/8 BPS black holes with vanishing entropy [102]. States for which $\mathcal{I}_4 = \partial_\gamma\mathcal{I}_4 = 0$ but $\partial_\gamma^2\mathcal{I}_4|_{\rm Ad} \neq 0$ preserve 1/4 supersymmetries, while states with $\partial_\gamma^2\mathcal{I}_4|_{\rm Ad} = 0$ preserve the maximal amount of 1/2 supersymmetries.[12]

The above conditions impose constraints on the BPS-index $\overline{\Omega}(\gamma)$, and consequently, via (6), also on the Fourier coefficients of the automorphic form $\mathscr{F}$. This further translates into a choice of automorphic representation (π,ρ) of the U-duality group $G_3(\mathbb{Z})$, since the number of charges determine the functional dimension κ of π.

As an example, we note that the constraint $\partial_\gamma^2\mathcal{I}_4|_{\rm Ad} = 0$ implies that 1/2 BPS-states in $\mathcal{N}=8$ supergravity has support on 28 electric and magnetic charges. Combined with the NUT-charge k this yields precisely the functional dimension $\kappa = 29$ of the minimal representation $\pi_{\rm min}$ of $E_{8(8)}(\mathbb{Z})$ [19]. Indeed, it was conjectured in [15] that the minimal theta series of $E_{8(8)}$ provides the partition function of 1/2 BPS-states in $\mathcal{N}=8$ supergravity. This is also consistent with recent results revealing that the same theta series captures the non-perturbative

[12] The expression $\partial_\gamma^2\mathcal{I}_4|_{\rm Ad}$ should be understood as the irreducible component of the Hessian of $\mathcal{I}_4$, transforming in the adjoint representation of $E_{7(7)}$ (see [102, 103, 24]).

contributions to the $E_{8(8)}(\mathbb{Z})$-invariant coefficient of the 1/2-BPS saturated R^4-coupling in type II string theory on T^7 [10] (see also [11]). Concretely this implies that when $\mathscr{F}$ is in the minimal representation, the associated Fourier coefficients $C(q_\Lambda, p^\Lambda)$ in (21) vanish unless their arguments satisfy $\partial^2\mathcal{I}_4(q_\Lambda, p^\Lambda)|_{\mathrm{Ad}} = 0$. A simple example of this phenomenon is the minimal representation of $SL(3,\mathbb{R})$ for which the Fourier coefficients vanish unless $pq = 0$ (see, e.g., [38]).

6.2. *Magic $\mathcal{N} = 2$ Supergravities and Quaternionic Representations*

Let us now turn to $\mathcal{N} = 2$ theories. The general form of the Bekenstein-Hawking entropy is given by

$$\mathcal{S}_{\mathrm{BH}}^{\mathcal{N}=2}(\gamma) = \pi\sqrt{\mathcal{Q}_4(\gamma)}$$

where $\mathcal{Q}_4(\gamma)$ is a quartic polynomial in the charges. For example, for 1/2 BPS black holes in M-theory on $\mathcal{X} \times S^1$, corresponding to M2-branes wrapped on 2-cycles with $p^0 = 0$, the polynomial is given by $\mathcal{Q}_4 = d_{ABC}p^Ap^Bp^Cq_0$, $A = 1, \ldots, h_{1,1}$, and $d_{ABC} \sim \int J_A \wedge J_B \wedge J_C$, with $J_A \in H^{1,1}(\mathcal{X},\mathbb{Z})$ a generator of the Kähler cone [13].

Whenever the $\mathcal{N} = 2$ theory exhibits a symmetry G_4 in $D = 4$, the quartic polynomial $\mathcal{Q}_4(\gamma)$ is again identified with the associated quartic invariant $\mathcal{I}_4(\gamma)$. An interesting class of toy models where this happens is provided by the *magic* $\mathcal{N} = 2$ supergravity theories [43, 44], which are classified by degree 3 Jordan algebras J, equipped with a cubic norm N_3 (see [20] for a survey). The classical moduli spaces in $D = 4$ are special Kähler coset spaces,

$$\mathcal{M}_4^{\mathrm{cl}} = \mathrm{Conf}(J)/(U(1) \times M)$$

with cubic prepotentials of the form $F(X) = N_3(X)/X^0$. Here, $\mathrm{Conf}(J)$ is the conformal group leaving the "cubic lightcone" $N_3 = 0$ invariant [18], and $K_4 = U(1) \times M$ is its maximal compact subgroup. The factor M is known as the "reduced structure group" of J.

After compactification to $D = 3$ the moduli space is enhanced to a quaternionic coset space

$$\mathcal{M}_3^{\mathrm{cl}} = \mathrm{QConf}(J)/(SU(2) \times \widetilde{\mathrm{Conf}(J)}),$$

where $\widetilde{\mathrm{Conf}(J)}$ is a compact form of $\mathrm{Conf}(J)$, and $\mathrm{QConf}(J)$ is the quasi-conformal group of J which leaves invariant a "quartic lightcone" $\mathcal{N}_4 = 0$ inside the complex vector space $\mathfrak{g}_1^{\mathbb{C}} \oplus \mathfrak{g}_2^{\mathbb{C}}$. In terms of formal complexified "charges" $(q_\Lambda^{\mathrm{c}}, p_{\mathrm{c}}^\Lambda, k_{\mathrm{c}})$ one can write the $\mathcal{N}_4$ as [104, 37]

$$\mathcal{N}_4(q_\Lambda^{\mathrm{c}}, p_{\mathrm{c}}^\Lambda, k_{\mathrm{c}}) = \mathcal{I}_4\left(q_\Lambda^{\mathrm{c}} - \bar{q}_\Lambda^{\mathrm{c}}, p_{\mathrm{c}}^\Lambda - \bar{p}_{\mathrm{c}}^\Lambda\right) + 2\left(k_{\mathrm{c}} - \bar{k}_{\mathrm{c}} + p_{\mathrm{c}}^\Lambda\bar{q}_\Lambda^{\mathrm{c}} - \bar{p}_{\mathrm{c}}^\Lambda q_\Lambda^{\mathrm{c}}\right)^2, \tag{29}$$

where $\mathcal{I}_4$ is the quartic $\mathrm{Conf}(J)$-invariant. This expression will play a role in section 6.5 below.

In general it is not known whether these magic theories arise from string compactifications. Nevertheless, it has been speculated that they have well-defined quantum completions which are invariant under discrete subgroups $\mathrm{QConf}(J,\mathbb{Z}) \subset \mathrm{QConf}(J)$ (see for instance [20, 36, 37, 104]). An example of particular interest is when $J = \mathbb{R}$ and $\mathrm{Conf}(J) = SL(2,\mathbb{R})$, in which case $\mathcal{M}_3^{\mathrm{cl}} = G_{2(2)}/SO(4)$. This exceptional coset space also arises as the classical vector multiplet moduli space in type IIA compactified on $\mathcal{X} \times S^1$, with $n_{\mathrm{V}} = h_{1,1}(\mathcal{X}) = 1$, or, equivalently, as the hypermultiplet moduli space in type IIB on $\mathcal{X}$ [27, 105].

For the quasiconformal groups $\mathrm{QConf}(J)$ there exist a distinguished class of unitary representations leading to a similar classification of charges as was given above for $\mathcal{N} = 8$. This is the *quaternionic discrete series* π_ν of Gross and Wallach [106, 107]. The representation π_ν depends on a single parameter ν, and for $\nu \geq 2n_{\mathrm{V}} + 3$, it is irreducible and has functional dimension $\kappa = 2n_{\mathrm{V}} + 3$, which is the same as for the principal series. In fact, the quaternionic discrete series can be embedded as a submodule in a degenerate principal series $I(P, s)$, for which the parameter s is restricted to be integer. For smaller values of ν, π_ν is no longer irreducible, but admits unitary irreducible sub-representations π'_ν of smaller functional dimension. For instance, when $\nu = (n_V + 2)/3$ one obtains the quaternionic analogue of the minimal representation $\pi_{\rm min}$, of functional dimension $\kappa = n_V + 2$. Following [106] we denote this minimal quaternionic representation by π_Z, with Z referring to the locus $\partial^2_\gamma \mathcal{I}_4|_{\rm Ad} = 0$.

The physical relevance of the quaternionic discrete series in magic $\mathcal{N} = 2$ theories has been pointed out in a series of works [36, 24, 108, 64, 104, 38]. In particular, it was shown that the BPS Hilbert space $\mathcal{H}^{\rm BPS}$ furnishes a unitary irreducible representation in the quaternionic discrete series of QConf(J). Concretely, this means that physical "BPS wave functions" should have support on $2n_V + 3$ charges $(q_\Lambda, p^\Lambda, k)$. Based on these obervations it is suggestive that the total instanton partition function $\mathscr{Z}_{\rm tot}$ for magic supergravities should be attached to the quaternionic discrete series π_ν of the symmetry group QConf(J), rather than the degenerate principal series $I(P, s)$, as was tentatively assumed in the analysis of sections 4 and 5 following a naive extrapolation of known results in $\mathcal{N} = 8$ theories. These ideas have been advocated in various earlier works [36, 24, 64, 104, 38, 39], although the lack of sufficiently explicit results in the mathematical literature has hindered progress.

6.3. Minimal Theta Series and NS5-Branes

To provide some evidence for the above speculations let us now revisit the analysis of section 5 for the extended universal hypermultiplet. Although the group $SL(3, \mathbb{R})$ does not afford any quaternionic representations, the coset space $SL(3, \mathbb{R})/SO(3)$ embeds naturally into the quaternionic symmetric space $G_{2(2)}/SO(4)$ discussed above. Assuming that the associated Calabi-Yau compactification preserves a discrete subgroup $G_{2(2)}(\mathbb{Z})$, one could hope to sum up the complete series of D(-1)-D1-D3-D5 and NS5-instanton corrections into an automorphic form in the quaternionic discrete series of $G_{2(2)}$.

As a first step in this direction, it was noticed in [38] that $\Psi^{SL(3,\mathbb{Z})}_{k,\ell}$ in fact corresponds to a restriction of the spherical vector $\mathcal{W}_0$ in the minimal quaternionic representation π_Z of $G_{2(2)}$ [109, 104]. It was then conjectured that the automorphic theta series attached to the minimal representation of any group QConf(J) should capture the (vector-valued) non-abelian NS5-brane partition function (15):

$$\mathscr{Z}^{(k,\ell^\Lambda)}(g) = \left\langle \mathcal{W}^{k,\ell^\Lambda}_{\mathbb{Z}} \middle| \rho_Z(g) \cdot \mathcal{W}^{k,\ell^\Lambda}_0 \right\rangle. \tag{30}$$

Indeed, the wave function $\Psi_{k,\ell^\Lambda}(p^\Lambda)$ depends on $n_V + 2$ variables (p^Λ, k), which is precisely the functional dimension of the minimal representation π_Z.

The conjecture (30) is also consistent with the observations in (2.5) that the wave function $\Psi_{1,0}(p^\Lambda)$ should be identified with the topological string amplitude $\Psi_{\rm top}$ on $\mathcal{X}$. For magic supergravities, the holomorphic anomaly equations satisfied by $\Psi_{\rm top}$ are equivalent to the operator equations associated with the Joseph ideal, and thus intimately connected with the minimal representation [37]. For $k > 1$ the minimal wave function $\Psi_{k,\ell^\Lambda}(p^\Lambda)$ provides a candidate for the one-parameter extension of the topological string amplitude proposed in [36, 37].

Although these results are encouraging, the theta series in the minimal quaternionic representation can at most capture part of the quantum corrections to the moduli space metric, notably those states whose charges obey the constraint $\partial^2_\gamma \mathcal{I}_4|_{\rm Ad} = 0$. Physically, this translates into the fact that the associated Fourier coefficients have support on "very small charges", meaning that there are no independent D1-D(-1) charges, but only those induced by the D3-D5 states. To include the contributions from more general BPS states one must consider automorphic representations attached to the generic quaternionic discrete series π_ν, where the Fourier coefficients have support on all charges with $\mathcal{I}_4(\gamma) > 0$ [110, 111, 112]. This is the quaternionic generalization of the familiar fact that the Fourier coefficients $\sum a(n) q^n$ of the holomorphic $SL(2, \mathbb{Z})$-Eisenstein series $E_{2r}(\tau) = \sum (m_1 + m_2\tau)^{-2r}$ vanish unless $n \geq 0$.

6.4. Twistor Space and Deformations of Quaternionic Manifolds

Having identified the quaternionic discrete series π_ν as the physically relevant automorphic representation for $\mathcal{N}=2$ supergravity theories, the next step is to implement the $G_3(\mathbb{Z})$-symmetry while preserving the quaternion-Kähler property of the metric on the moduli space. To this end we shall require some new technology.

A quaternion-Kähler manifold $\mathcal{M}$ is conveniently described through its *twistor space* $\mathcal{Z}$ [113]:

$$\mathbb{P}^1 \longrightarrow \mathcal{Z} \longrightarrow \mathcal{M} \tag{31}$$

where the fiber $\mathbb{P}^1$ corresponds to the S^2 worth of almost complex structures carried by $\mathcal{M}$. If $\mathcal{M}$ is real $4(n_{\rm V}+1)$-dimensional, the twistor space $\mathcal{Z}$ is a complex contact manifold of complex dimension $2n_{\rm V}+3$. The contact structure is encoded in a one-form $\mathcal{X}_\mathcal{Z}$ such that the top form $\mathcal{X}_\mathcal{Z}\wedge(\mathrm{d}\mathcal{X}_\mathcal{Z})^{n_{\rm V}+1}$ is nowhere vanishing. Infinitesimal deformations of $\mathcal{M}$ can be realized as deformations of the complex contact structure of $\mathcal{Z}$, which are classified by $H^1(\mathcal{Z},\mathcal{O}(2))$ [114, 115]. Compatibility with $\mathcal{N}=2$ supersymmetry is therefore ensured if quantum corrections (9) to the metric on the moduli space $\mathcal{M}$ are implemented in terms of deformations of the contact one-form $\mathcal{X}_\mathcal{Z}$ [116, 31]. This situation corresponds to a supergravity generalization of the analysis in [30], where quantum corrections to the hyperkähler metric on the moduli space of $\mathcal{N}=2$ field theory on $\mathbb{R}^3\times S^1$ was studied.

Following [116, 31], one introduces a set of $2n_{\rm V}+3$ local Darboux coordinates $(\xi^\Lambda,\tilde{\xi}_\Lambda,\alpha)$ on some open patch of $\mathcal{Z}$, such that infinitesimal deformations of the contact structure ("contactomorphisms") are realized as holomorphic sections $H(\xi,\tilde{\xi},\alpha)\in H^1(\mathcal{Z},\mathcal{O}(2))$. The Darboux coordinates can formally be viewed as complexifications of the axions $(\zeta^\Lambda,\tilde{\zeta}_\Lambda,\sigma)$ and hence the $N(\mathbb{Z})$ Heisenberg action lifts to an isometric action on $\mathcal{Z}$:

$$\xi^\Lambda\to\xi^\Lambda+m^\Lambda,\qquad \tilde{\xi}_\Lambda\to\tilde{\xi}_\Lambda+n_\Lambda,\qquad \alpha\to\alpha+2r-n_\Lambda\xi^\Lambda+m^\Lambda\tilde{\xi}_\Lambda+2v(m,n).$$

The pair $(\xi^\Lambda,\tilde{\xi}_\Lambda)$ therefore parametrize a complex symplectic torus $\mathcal{T}_\mathbb{C}=\Gamma^*\otimes_\mathbb{Z}\mathbb{C}$ (Γ^* being the dual of the charge lattice Γ). This is the complex torus introduced by Kontsevich and Soibelman [30, 28]. The additional coordinate α then parametrizes the fiber of a "contact line bundle" $\mathscr{L}_\mathbb{C}\to\mathcal{T}_\mathbb{C}$ [42].

When NS5-brane effects are absent the continuous shift symmetry of α is unbroken and the general contactomorphism $H(\xi,\tilde{\xi},\alpha)$ reduces to a symplectomorphism acting on $\mathcal{T}_\mathbb{C}$, which takes the following form [31, 30]:

$$H(\xi,\tilde{\xi})=\sum_{\gamma\in\Gamma}\Omega(t;\gamma)\mathrm{Li}_2\left(\lambda(\gamma)e^{-2\pi i(q_\Lambda\xi^\Lambda-p^\Lambda\tilde{\xi}_\Lambda)}\right)=\sum_{\gamma\in\Gamma}\lambda(\gamma)\overline{\Omega}(t;\gamma)e^{-2\pi i(q_\Lambda\xi^\Lambda-p^\Lambda\tilde{\xi}_\Lambda)}. \tag{32}$$

This is the infinitesimal version of the Kontsevich-Soibelman symplectomorphism $\mathcal{K}_\gamma$ [28]. A key observation for what follows is that $H(\xi,\tilde{\xi})$ can also be viewed as a twistor space realization of the BPS instanton partition function $\mathscr{Z}_{\rm BPS}$ in (5). We now turn to discuss the expected structure of the general contactomorphism $H(\xi,\tilde{\xi},\alpha)$ in the presence of NS5-effects.

6.5. Instanton Partition Function on Twistor Space

Following the logic of section 3, we wish to construct holomorphic sections $H(\xi,\tilde{\xi},\alpha)\in H^1(\mathcal{Z},\mathcal{O}(2))$ which transform nicely under an arithmetic Lie group $G(\mathbb{Z})$, such that the resulting metric on the quaternionic-Kähler base $\mathcal{M}$ is invariant. Some results for $SL(2,\mathbb{Z})$ have been obtained in [32, 42], though a general prescription is lacking.

From the point of representation theory, there is a proposal for how to implement these ideas when the moduli space $\mathcal{M}_3$ is a coset space G/K. In [107] it was shown that the quaternionic discrete series π_ν can be realized as the action of G on holomorphic sections on the twistor space

Journal of Physics: Conference Series **346** (2012) 012016 doi:10.1088/1742-6596/346/1/012016

$\mathcal{Z} \to G/K$. The reason for this can be understood quite naturally by noting that $\mathcal{Z} \cong P_{\mathbb{C}}\backslash G_{\mathbb{C}}$, i.e. it is the complexification of the coset $P\backslash G$ discussed in section 3.3. Morally, this is the complexification of the axions $(\zeta^\Lambda, \tilde{\zeta}_\Lambda, \sigma) \to (\xi^\Lambda, \tilde{\xi}_\Lambda, \alpha)$ mentioned above. The quaternionic discrete series therefore corresponds to the holomorphic quasi-conformal action of G on $\mathfrak{g}_1^{\mathbb{C}} \oplus \mathfrak{g}_2^{\mathbb{C}}$ [104, 117, 118]. Moreover, in terms of the Darboux coordinates, the quartic lightcone (29) provides a Kähler potential for the twistor space according to

$$K_{\mathcal{Z}}(\xi, \tilde{\xi}, \alpha) = \log \mathcal{N}_4(\xi, \tilde{\xi}, \alpha),$$

which makes the G-invariance of $\mathcal{Z}$ manifest.

By identifying the twistor fiber as $\mathbb{P}^1 = SU(2)/U(1)$, the fibration in (31) is such that the twistor space takes the form $\mathcal{Z} = G/(M \times U(1))$. Analogous to the realization of holomorphic modular forms for $SL(2,\mathbb{Z})$ as holomorphic sections of a line bundle over $SL(2,\mathbb{R})/U(1)$, automorphic forms in the quaternionic discrete series of G have a realization in terms of holomorphic sections of G-equivariant vector bundles over $\mathcal{Z}$. Moreover, as a module, one may identify the representation π_ν with the sheaf cohomology group [107]:

$$\pi_\nu = H^1(\mathcal{Z}, \mathcal{O}(-\nu)).$$

In terms of the Darboux coordinates $(\xi^\Lambda, \tilde{\xi}_\Lambda, \alpha) \in \mathcal{Z}$, automorphic forms in the quaternionic discrete series π_ν correspond to holomorphic sections $\mathscr{F}(\xi, \tilde{\xi}, \alpha) \in H^1(\mathcal{Z}, \mathcal{O}(-\nu))$. This is tantalizingly close to what we are after in describing deformations of the complex contact structure of the twistor space $\mathcal{Z}$ itself.

We then propose that the total instanton partition function (9) can be lifted to a holomorphic section $\mathscr{Z}_{\rm tot}(\xi, \tilde{\xi}, \alpha) \in H^1(\mathcal{Z}, \mathcal{O}(-\nu))$ attached to the quaternionic discrete series π_ν of G_3, whose Fourier-Jacobi expansion with respect to $G^J(\mathbb{Z}) = G_4(\mathbb{Z}) \ltimes N(\mathbb{Z})$ exhibits the general structure

$$\begin{aligned} \mathscr{Z}_{\rm tot}(\xi, \tilde{\xi}, \alpha) &= \sum_{(q_\Lambda, p^\Lambda)\in\Gamma} \lambda(q_\Lambda, p^\Lambda)\overline{\Omega}(q_\Lambda, p^\Lambda) e^{-2\pi i(q_\Lambda \xi^\Lambda - p^\Lambda \tilde{\xi}_\Lambda)} \\ &\quad + \sum_{k\neq 0} \sum_{\ell^\Lambda \in \Gamma_m/(|k|\Gamma_m)} \sum_{p^\Lambda\in\Gamma_m+\ell^\Lambda/|k|} \Psi_{k,\ell^\Lambda}(\xi^\Lambda - p^\Lambda)\, e^{2\pi i k p^\Lambda \tilde{\xi}_\Lambda - \pi i k(\alpha - \xi^\Lambda\tilde{\xi}_\Lambda)}. \end{aligned} \tag{33}$$

While the abelian contribution in (33) corresponds to an infinitesimal symplectomorphism of $\mathcal{T}_{\mathbb{C}}$, the non-abelian term provides the generating function of a generic contactomorphism of the twistor space $\mathcal{Z}$ [31, 42].

In the spirit of [108, 64, 42], the partition function $\mathscr{Z}_{\rm tot}(t, \zeta, \tilde{\zeta}, \sigma, \phi)$ on the base $\mathcal{M}$ is defined by quaternionic Penrose transform of (33):

$$\mathscr{Z}_{\rm tot}(t, \zeta, \tilde{\zeta}, \sigma, \phi) \equiv e^{-\phi} \oint_{\mathcal{C}} \frac{d\mathbf{z}}{2\pi i\, \mathbf{z}} \mathscr{Z}_{\rm tot}\Big(\xi(\mathbf{z}), \tilde{\xi}(\mathbf{z}), \alpha(\mathbf{z})\Big),$$

where $\mathbf{z}$ is a coordinate on the twistor fiber $\mathbb{P}^1$, and $\mathcal{C}$ is a suitable contour. Using the explicit form of the Darboux coordinates $(\xi^\Lambda, \tilde{\xi}_\Lambda, \alpha)$ in terms of $(x, \mathbf{z}) \in \mathcal{M} \times \mathbb{P}^1$ found in [108], one can verify that the Penrose transform of the abelian term in (33) reproduces the moduli dependence $K_1(2\pi e^{-\phi}|Z(t;\gamma)|)$ as expected on general grounds for D-brane instantons [31]. Indeed, for rigid Calabi-Yau threefolds, one recovers the Whittaker vector $\mathcal{W}_{q,p}(\phi; 3/2)$ in (25). The saddle-point approximation of the non-abelian term in (33) has furthermore been evaluated in [42], revealing a generalized theta series with insertions as in (27).

7. Concluding Discussion

In this note we have reviewed proposals for implementing quantum corrections to the moduli space metric in $\mathcal{N}=2$ supergravities on $\mathbb{R}^3 \times S^1$, consistently with supersymmetry and $G_3(\mathbb{Z})$-invariance. In particular, we provided arguments that the complete instanton partition function $\mathscr{Z}_{\rm tot}$ is an automorphic form attached to the quaternionic discrete series of G_3.

In light of this it would be interesting to revisit the analysis of section 4 for compactifications on a rigid Calabi-Yau manifold $\mathcal{X}_{\rm r}$. In this case the relevant twistor space is a complex 3-dimensional coset space of the form $\mathcal{Z} = U(2,1)/(U(1) \times U(1))$. Some of the relevant representation theoretic ingredients for constructing $U(2,1;\mathcal{O}_d)$-automorphic forms on $\mathcal{Z}$ in the quaternionic discrete series, such as the K-finite vectors $\mathcal{W}_K^{(\epsilon)}$, have been analyzed in [110, 104]. One would like to construct $\mathscr{Z}_{\rm tot}(\xi,\tilde{\xi},\alpha)$ explicitly and extract the Fourier coefficients $C(p,q)$ which should provide an explicit form for the generalized Donaldson-Thomas invariants of $\mathcal{X}_{\rm r}$.

In order to clarify the relation between the automorphic form $\mathscr{F}(\xi,\tilde{\xi},\alpha) \in H^1(\mathcal{Z},\mathcal{O}(-\nu))$ and the infinitesimal deformation $H(\xi,\tilde{\xi},\alpha) \in H^1(\mathcal{Z},\mathcal{O}(2))$, a possibly useful tool is the *twistor transform* [119, 24, 108, 104]. This is an involutive integral operator $\mathbb{T}$ that takes elements in $H^1(\mathcal{Z},\mathcal{O}(-m))$ to elements in $H^1(\mathcal{Z},\mathcal{O}(m-4))$. Consider for instance the example of $G_{2(2)}/SO(4)$ associated to a type IIB compactification on $\mathcal{X}$ with $n_{\rm V} = h_{1,1} = 1$. The functional dimension of the quaternionic discrete series of $G_{2(2)}$ is $\kappa = 5$, corresponding to the number of D(-1)-D1-D3-D5-NS5 charges $(q_\Lambda, p^\Lambda, k)$. Ideally, one should then construct an automorphic form $\mathscr{F}(\xi,\tilde{\xi},\alpha) \in H^1(\mathcal{Z},\mathcal{O}(-6))$, and take the twistor transform $\mathbb{T}(\mathscr{F}) \in H^1(\mathcal{Z},\mathcal{O}(2))$ as a candidate for the deformation $H(\xi,\tilde{\xi},\alpha)$ of the complex contact structure of $\mathcal{Z} = G_{2(2)}/(U(1) \times SU(2))$. We note that the Fourier coefficients of modular forms on $G_{2(2)}$ generating the quaternionic discrete series have been analyzed in the mathematical literature [87, 111]; it would be very interesting to work out the underlying physics along the lines suggested above.

Finally we offer some remarks on the issue of wall crossing. For the Fourier coefficients in (33) to be identified with the BPS-index (1), one must verify that they are compatible with known wall crossing behaviour [28, 30]. In the absence of NS5-brane effects, $k=0$, this is guaranteed since the abelian terms (32) coincide with infinitesimal Kontsevich-Soibelman symplectomorphisms. However, in the generic case, $k \neq 0$, it is presently unknown how to exponentiate the generating function $H(\xi,\tilde{\xi},\alpha)$ into a finite contactomorphism of the twistor space $\mathcal{Z}$. The non-abelian wall crossing problem would then involve a generalization of the KS formula to an infinite product of such contactomorphisms. Although this formulation is presently out of reach, the recent results of [42] provide a suggestive hint: since the NS5-charge k can be formally identified with a quantization parameter $\hbar$, i.e. with the center of a Heisenberg algebra, one might speculate that these contactomorphisms would act on the quantum torus of [28], and the finite version of (33) would then presumably involve a product of quantum dilogarithms [120].[13]

Acknowledgments

I am grateful to Sergei Alexandrov, Ling Bao, Axel Kleinschmidt, Bengt E.W. Nilsson and Boris Pioline for extremely stimulating collaborations over the last few years on the results presented herein. I also wish to thank Matthias Gaberdiel, Stefan Hohenegger, Pär Kurlberg, Jan Manschot, Yan Soibelman and Stefan Vandoren for related discussions and correspondence. A special thanks goes out to Sergei Alexandrov, Christoffer Petersson and Boris Pioline for very helpful comments on a previous draft. Finally, I thank Eugen Paal and Alexander Stolin for organizing a stimulating workshop in a fantastic location. My research is supported in part by the Swiss National Science Foundation.

[13] Similar speculations on the interpretation of the NS5-brane charge k as a quantization parameter have also been put forward by M. Kontsevich and Y. Soibelman.

References

[1] Hull C M and Townsend P K 1995 *Nucl. Phys.* **B438** 109 (Preprint hep-th/9410167)
[2] Obers N A and Pioline B 1999 *Phys. Rept.* **318** 113 (Preprint hep-th/9809039)
[3] Cremmer E and Julia B 1979 *Nucl. Phys.* **B159** 141
[4] Green M B and Gutperle M 1997 *Nucl. Phys.* **B498** 195 (Preprint hep-th/9701093)
[5] Obers N A and Pioline B 2000 *Commun. Math. Phys.* **209** 275 (Preprint hep-th/9903113)
[6] Lambert N and West P C 2007 *Phys. Rev.* **D75** 066002 (Preprint hep-th/0611318)
[7] Bao L, Cederwall M and Nilsson B E W 2008 *Class. Quant. Grav.* **25** 095001 (Preprint 0706.1183)
[8] Bao L, Bielecki J, Cederwall M, Nilsson B E W and Persson D 2008 *JHEP* **07** 048 (Preprint 0710.4907)
[9] Green M B, Russo J G and Vanhove P 2010 *Phys. Rev.* **D81** 086008 (Preprint 1001.2535)
[10] Pioline B 2010 *JHEP* **03** 116 (Preprint 1001.3647)
[11] Green M B, Miller S D, Russo J G and Vanhove P 2010 *Eisenstein series for higher-rank groups and string theory amplitudes* Preprint 1004.0163
[12] Kallosh R and Kol B 1996 *Phys. Rev.* **D53** 5344 (Preprint hep-th/9602014)
[13] Maldacena J M, Strominger A and Witten E 1997 *JHEP* **12** 002 (Preprint hep-th/9711053)
[14] Dabholkar A, Denef F, Moore G W and Pioline B 2005 *JHEP* **10** 096 (Preprint hep-th/0507014)
[15] Pioline B 2005 *JHEP* **0508** 071 (Preprint hep-th/0506228)
[16] Kinnersley W 1973 *J. Math. Phys.* **14** 651
[17] Breitenlohner P, Gibbons G W and Maison D 1988 *Commun. Math. Phys.* **120** 295
[18] Günaydin M, Koepsell K and Nicolai H 2001 *Commun. Math. Phys.* **221** 57 (Preprint hep-th/0008063)
[19] Günaydin M, Koepsell K and Nicolai H 2002 *Adv. Theor. Math. Phys.* **5** 923 (Preprint hep-th/0109005)
[20] Günaydin M 2005 *AIP Conf. Proc.* **767** 268 (Preprint hep-th/0502235)
[21] Bagger J and Witten E 1983 *Nucl. Phys.* **B222** 1
[22] Becker K, Becker M and Strominger A 1995 *Nucl. Phys.* **B456** 130–152 (Preprint hep-th/9507158)
[23] Aspinwall P S 2000 *Compactification, geometry and duality: $N = 2$* Preprint hep-th/0001001
[24] Pioline B 2006 *Class. Quant. Grav.* **23** S981 (Preprint hep-th/0607227)
[25] Seiberg N and Witten E 1996 *Gauge dynamics and compactification to three dimensions* Preprint hep-th/9607163
[26] Cecotti S, Ferrara S and Girardello L 1989 *Int. J. Mod. Phys.* **A4** 2475
[27] Ferrara S and Sabharwal S 1990 *Nucl. Phys.* **B332** 317
[28] Kontsevich M and Soibelman Y *Stability structures, motivic Donaldson-Thomas invariants and cluster transformations* Preprint 0811.2435
[29] Joyce D and Song Y 2008 *A theory of generalized Donaldson-Thomas invariants* Preprint 0810.5645
[30] Gaiotto D, Moore G W and Neitzke A 2010 *Commun. Math. Phys.* **299** 163 (Preprint 0807.4723)
[31] Alexandrov S, Pioline B, Saueressig F and Vandoren S 2009 *JHEP* **03** 044 (Preprint 0812.4219)
[32] Alexandrov S and Saueressig F 2009 *JHEP* **09** 108 (Preprint 0906.3743)
[33] Candelas P, de la Ossa X C, Green P S and Parkes L 1991 *Nucl. Phys.* **B359** 21
[34] Kontsevich M 1995 *Proceedings of the International Congress of Mathematicians* **1, 2** (Zürich, 1994) 120 (Preprint alg-geom/9411018)
[35] Kazhdan D, Pioline B and Waldron A 2002 *Commun. Math. Phys.* **226** 1 (Preprint hep-th/0107222)
[36] Günaydin M, Neitzke A, Pioline B and Waldron A 2006 *Phys. Rev.* **D73** 084019 (Preprint hep-th/0512296)
[37] Günaydin M, Neitzke A and Pioline B 2006 *JHEP* **12** 070 (Preprint hep-th/0607200)
[38] Pioline B and Persson D 2009 *Commun. Num. Theor. Phys.* **3** 697 (Preprint 0902.3274)
[39] Bao L, Kleinschmidt A, Nilsson B, Persson D and Pioline B 2010 *Commun. Num. Theor. Phys.* **4** 187
[40] Bao L, Kleinschmidt A, Nilsson B E W, Persson D and Pioline B 2010 *Rigid Calabi-Yau threefolds, Picard Eisenstein series and instantons* Preprint 1005.4848
[41] Alexandrov S, Persson D and Pioline B 2011 *Phys.Rev.* **D83** 026001 (Preprint 1009.3026)
[42] Alexandrov S, Persson D and Pioline B 2011 *JHEP* **1103** 111 (Preprint 1010.5792)
[43] Günaydin M, Sierra G and Townsend P K 1983 *Phys. Lett.* **B133** 72
[44] Günaydin M, Sierra G and Townsend P K 1984 *Nucl. Phys.* **B242** 244
[45] Douglas M R *Proceedings of the International Congress of Mathematicians* **3** (Beijing 2002) 395 (Preprint math/0207021)
[46] Bridgeland T 2007 *Ann. of Math. (2)* **166** 317 (Preprint math/0212237)
[47] Ooguri H, Strominger A and Vafa C 2004 *Phys. Rev.* **D70** 106007 (Preprint hep-th/0405146)
[48] Kiritsis E 1997 *Introduction to nonperturbative string theory* Preprint hep-th/9708130
[49] Denef F and Moore G W 2007 *Split states, entropy enigmas, holes and halos* Preprint hep-th/0702146
[50] Alexandrov S 2009 *J. Phys.* **A42** 335402 (Preprint 0902.2761)
[51] de Wit B, Kappeli J, Lopes Cardoso G and Mohaupt T 2006 *JHEP* **03** 074 (Preprint hep-th/0601108)
[52] Manschot J 2010 *Wall-crossing of D4-branes using flow trees* Preprint 1003.1570

[53] Manschot J, Pioline B and Sen A 2011 *JHEP* **1107** 059 (Preprint 1011.1258)
[54] Gaiotto D, Strominger A and Yin X 2007 *JHEP* **08** 070 (Preprint hep-th/0607010)
[55] Manschot J 2010 *Commun. Math. Phys.* **299** 651 (Preprint 0906.1767)
[56] Cortes V, Mayer C, Mohaupt T and Saueressig F 2005 *JHEP* **06** 025 (Preprint hep-th/0503094)
[57] Green M B and Gutperle M 1998 *JHEP* **01** 005 (Preprint hep-th/9711107)
[58] Yi P 1997 *Nucl. Phys.* **B505** 307 (Preprint hep-th/9704098)
[59] Sethi S and Stern M 1998 *Commun. Math. Phys.* **194** 675 (Preprint hep-th/9705046)
[60] Kostov I K and Vanhove P 1998 *Phys.Lett.* **B444** 196 (Preprint hep-th/9809130)
[61] Moore G W, Nekrasov N and Shatashvili S 2000 *Commun.Math.Phys.* **209** 77 (Preprint hep-th/9803265)
[62] Pioline B and Vandoren S 2009 *JHEP* **07** 008 (Preprint 0904.2303)
[63] Behrndt K, Gaida I, Lüst D, Mahapatra S and Mohaupt T 1997 *Nucl. Phys.* **B508** 659
[64] Gunaydin M, Neitzke A, Pioline B and Waldron A 2007 *JHEP* **09** 056 (Preprint 0707.0267)
[65] Antoniadis I, Ferrara S, Minasian R and Narain K S 1997 *Nucl. Phys.* **B507** 571
[66] Günther H, Herrmann C and Louis J 2000 *Fortsch. Phys.* **48** 119 (Preprint hep-th/9901137)
[67] Antoniadis I, Minasian R, Theisen S and Vanhove P 2003 *Class. Quant. Grav.* **20** 5079
[68] Ooguri H and Vafa C 1996 *Nucl. Phys.* **B463** 55 (Preprint hep-th/9511164)
[69] Kapustin A 2004 *JHEP* **09** 034 (Preprint hep-th/0404041)
[70] Dijkgraaf R, Verlinde E P and Vonk M 2002 *On the partition sum of the NS five-brane* Preprint hep-th/0205281
[71] Callan Curtis G J, Harvey J A and Strominger A 1991 *Nucl. Phys.* **B367** 60
[72] Witten E 1997 *J. Geom. Phys.* **22** 103 (Preprint hep-th/9610234)
[73] Henningson M, Nilsson B E W and Salomonson P 1999 *JHEP* **09** 008 (Preprint hep-th/9908107)
[74] Belov D and Moore G W 2006 *Holographic action for the self-dual field* Preprint hep-th/0605038
[75] Nekrasov N, Ooguri H and Vafa C 2004 *JHEP* **0410** 009 (Preprint hep-th/0403167)
[76] Bao L, Cederwall M and Nilsson B E W 2008 *JHEP* **06** 100 (Preprint hep-th/0603120)
[77] Bengtsson V and Cederwall M 2008 *The B-model on the A-model NS5-brane* Preprint 0805.1782
[78] Moore G W 2005 *Comptes Rendus Physique* **6** 251 (Preprint hep-th/0409158)
[79] Monnier S 2010 *Geometric quantization and the metric dependence of the self-dual field theory* Preprint 1011.5890
[80] Minasian R and Moore G W 1997 *JHEP* **11** 002 (Preprint hep-th/9710230)
[81] Eichler M and Zagier D *The theory of Jacobi forms* (*Progress in Mathematics*, vol 55, Birkhäuser Boston 1985) ISBN 0-8176-3180-1
[82] Berndt R and Schmidt R 1998 *Elements of the representation theory of the Jacobi group* **163** (Birkhäuser, Basel, 1st ed. 1998)
[83] Weissmann M H 2003 *Representation Theory, AMS Journal* **7** 275
[84] Pioline B and Waldron A 2003 *Automorphic forms: a physicist's survey* Preprint hep-th/0312068
[85] Green M B and Sethi S 1999 *Phys. Rev.* **D59** 046006 (Preprint hep-th/9808061)
[86] Ginzburg D, Rallis S and Soudry D 1997 *Israel J. Math.* **100** 61, ISSN 0021-2172
[87] Gan W T 2000 *Canad. J. Math.* **52** 737 ISSN 0008-414X
[88] Joseph A 1974 *Comm. Math. Phys.* **36** 325 ISSN 0010-3616
[89] Kazhdan D and Savin G 1990 (*Israel Math. Conf. Proc.* vol 2) (Jerusalem: Weizmann) 209
[90] Kazhdan D and Polishchuk A 2002 *Algebraic groups and arithmetic* (Tata Inst. Fund. Res., Mumbai, 2004) 127 (Preprint math/0209315)
[91] Persson D 2010 *Arithmetic and Hyperbolic Structures in String Theory* Preprint 1001.3154
[92] Ishikawa Y h 1999 *J. Math. Sci. Univ. Tokyo* **6** 477 ISSN 1340-5705
[93] Yui N *Calabi-Yau varieties and mirror symmetry (Toronto, ON, 2001)* (*Fields Inst. Commun.* vol 38) 307
[94] Moore G W 1998 *Arithmetic and attractors* Preprint hep-th/9807087
[95] Becker K and Becker M 1999 *Nucl. Phys.* **B551** 102 (Preprint hep-th/9901126)
[96] Robles-Llana D, Roček M, Saueressig F, Theis U and Vandoren S 2007 *Phys. Rev. Lett.* **98** 211602
[97] Langlands R P 1976 *On the functional equations satisfied by Eisenstein series* Lecture Notes in Mathematics, Vol. 544 (Berlin: Springer-Verlag)
[98] Pioline B and Waldron A 2004 *JHEP* **06** 009 (Preprint hep-th/0404018)
[99] Vinogradov A I and Takhtajan L A 1978 *Zap. Nauchn. Sem. Leningrad. Otdel. Mat. Inst. Steklov. (LOMI)* **76** 5, 216 ISSN 0207-6772
[100] Bump D W 1982 *Automorphic forms on* $GL(3,\mathbb{R})$ PhD Thesis – The University of Chicago
[101] Dhar A and Mandal G 1998 *Nucl.Phys.* **B531** 256 (Preprint hep-th/9803004)
[102] Ferrara S and Maldacena J M 1998 *Class. Quant. Grav.* **15** 749 (Preprint hep-th/9706097)
[103] Ferrara S and Günaydin M 1998 *Int. J. Mod. Phys.* **A13** 2075 (Preprint hep-th/9708025)
[104] Gunaydin M, Neitzke A, Pavlyk O and Pioline B 2008 *Commun. Math. Phys.* **283** 169

[105] Bodner M and Cadavid A C 1990 *Class. Quant. Grav.* **7** 829
[106] Gross B H and Wallach N R *Lie theory and geometry, Progr. Math.* **123** (Birkhäuser, Boston 1994) 289
[107] Gross B H and Wallach N R 1996 *J. Reine Angew. Math.* **481** 73 ISSN 0075-4102
[108] Neitzke A, Pioline B and Vandoren S 2007 *JHEP* **04** 038 (Preprint hep-th/0701214)
[109] Vogan Jr D A 1994 *Invent. Math.* **116** 677 ISSN 0020-9910
[110] Wallach N R 2003 *Comment. Math. Helv.* **78** 266 ISSN 0010-2571
[111] Gan W T, Gross B and Savin G 2002 *Duke Math. J.* **115** 105 ISSN 0012-7094
[112] Weissman M H 2006 *Amer. J. Math.* **128** 849
[113] Salamon S M 1982 *Invent. Math.* **67** 143–171 ISSN 0020-9910
[114] LeBrun C 1988 *Proc. Amer. Math. Soc.* **103** 1205
[115] LeBrun C and Salamon S 1994 *Invent. Math.* **118** 109
[116] Alexandrov S, Pioline B, Saueressig F and Vandoren S 2010 *Commun. Math. Phys.* **296** 353
[117] Gunaydin M and Pavlyk O 2009 *Quasiconformal Realizations of* $E_{6(6)}$, $E_{7(7)}$, $E_{8(8)}$ *and* $SO(n+3, m+3)$, $\mathcal{N}=4$ *and* $\mathcal{N}\geq 4$ *Supergravity and Spherical Vectors* Preprint 0904.0784
[118] Gunaydin M and Pavlyk O 2010 *JHEP* **04** 070 (Preprint 0901.1646)
[119] Eastwood M G and Ginsberg M L 1981 *Duke Math. J.* **48** 177, ISSN 0012-7094
[120] Alexandrov S, Persson D and Pioline B 2011 *Wall-crossing, Rogers dilogarithm, and the QK/HK correspondence* Preprint 1110.0466

Algebra, Geometry, and Mathematical Physics 2010 IOP Publishing
Journal of Physics: Conference Series **346** (2012) 012017 doi:10.1088/1742-6596/346/1/012017

Four ways across the wall

Boris Pioline

Laboratoire de Physique Théorique et Hautes Energies,
CNRS UMR 7589 and Université Pierre et Marie Curie - Paris 6,
4 place Jussieu, 75252 Paris cedex 05, France

E-mail: pioline@lpthe.jussieu.fr

Abstract. An important question in the study of $\mathcal{N} = 2$ supersymmetric string or field theories is to compute the jump of the BPS spectrum across walls of marginal stability in the space of parameters or vacua. I survey four apparently different answers for this problem, two of which are based on the mathematics of generalized Donaldson-Thomas invariants (the Kontsevich-Soibelman and the Joyce-Song formulae), while the other two are based on the physics of multi-centered black hole solutions (the Coulomb branch and the Higgs branch formulae, discovered in joint work with Jan Manschot and Ashoke Sen [1]). Explicit computations indicate that these formulae are equivalent, though a combinatorial proof is currently lacking.

1. Introduction

In quantum field theories and string theory vacua with extended supersymmetry, it is often possible to determine the spectrum of BPS bound states in some weakly coupled region of moduli (or parameter) space $\mathcal{B}$. In extrapolating the BPS spectrum to strong coupling, one usually faces two issues: i) short BPS multiplets may pair up into long multiplets and leave the BPS spectrum and ii) single-particle bound states may decay into the continuum of multi-particle states. The first issue can be avoided by considering a suitable index $\Omega(\gamma, t)$, designed such that contributions from long multiplets cancel. $\Omega(\gamma, t)$ is then a piecewise constant function of the charge vector γ and couplings/moduli $t \in \mathcal{B}$.

The second issue arises at certain loci in moduli space, where the bound state becomes unstable towards decay into a n-particle state with charges $\{\alpha_i\}$ such that $\gamma = \sum_{i=1\dots n} \alpha_i$. In four-dimensional field or string theories with $\mathcal{N} = 2$ supersymmetry, the mass of a BPS bound state $M(\gamma, t)$ is equal to $|Z(\gamma, t)|$, where the central charge Z is a map from $\mathcal{B}$ to $\mathrm{Hom}(\Gamma, \mathbb{C})$, where Γ is the charge lattice. In particular, Z is linear in its first argument γ. The decay is therefore energetically possible only when (and even then, marginally so) the phase of $Z(\gamma, t)$ aligns with the phase of each of the $Z(\alpha_i, t)$'s, so that $M(\gamma) = \sum_{i=1\dots n} M(\alpha_i)$. This alignment takes place in a locus of codimension $p - 1$ in the moduli space $\mathcal{B}$, where p is the dimension of the subspace of Γ spanned by the $\alpha_i's$. The dangerous case is $p = 2$, where the locus defines a codimension one "wall of marginal stability" in $\mathcal{B}$, across which the index $\Omega(\gamma, t)$ may jump. A paradigm of this phenomenon is Seiberg-Witten theory with $SU(2)$ gauge group and no flavors: across the curve $\{a/a_D \in \mathbb{R}^+\}$ in the u-plane, the BPS spectrum jumps from an infinite number of states in the weak coupling region to just two states in the strong coupling region – the monopole and the dyon (see Fig. 1) [2, 3]. An important physical question is therefore to

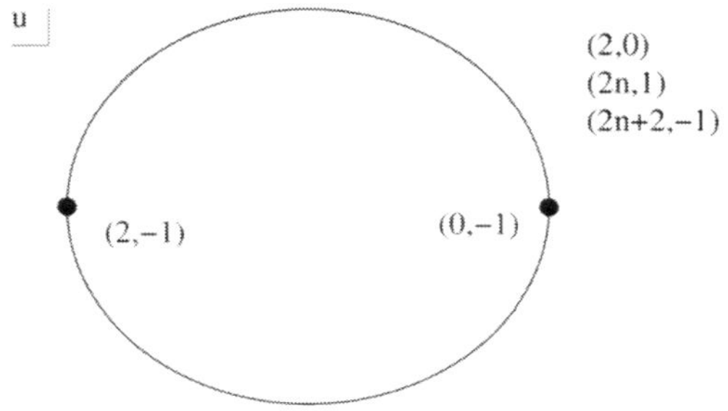

Figure 1. Chamber structure of the u-plane and BPS spectrum in $\mathcal{N}=2, D=4$ SYM theory with $SU(2)$ gauge group and no flavor. The line $\mathrm{Im}\,(a/a_D)=0$ separates the strong and weak coupling chambers. The only stable BPS states in the strong coupling chamber are the monopole and dyons with charges $(q,p)=\pm(0,1), \pm(2,-1)$, in the conventions of [5]. The weak coupling spectrum consists of these same states and their images around the monodromy at infinity, plus the W-boson with charge $(2,0)$.

determine the jump

$$\Delta\Omega(M\gamma_1+N\gamma_2)=\Omega^-(M\gamma_1+N\gamma_2)-\Omega^+(M\gamma_1+N\gamma_2)$$

of the BPS index [1] $\Omega(\gamma,t)\equiv \mathrm{Tr}'(-1)^F$ across the wall of marginal stability

$$W(\gamma_1,\gamma_2)=\{t\in\mathcal{B}/\ \arg[Z(\gamma_1,t)]=\arg[Z(\gamma_2,t)]\}\ , \tag{1}$$

in terms of the BPS indices on side of the wall, say the $\Omega^+(\gamma)'s$. Here we denoted by $\Omega^\pm(\gamma)$ the index in the chamber $c^\pm$ on the side of the wall where $\arg Z(\gamma_1,t)\gtrless \arg Z(\gamma_2,t)$. It will be convenient to choose the basis γ_1,γ_2 such that $\Omega^+(M\gamma_1+N\gamma_2)$ vanishes whenever $MN<0$ (the 'root basis' condition [4]), and denote by $\tilde\Gamma=(\mathbb{Z}^+\gamma_1+\mathbb{Z}^+\gamma_2)\backslash\{0\}$ the positive cone in the two-dimensional sublattice spanned by γ_1,γ_2. We further assume that $\langle\gamma_1,\gamma_2\rangle<0$.

As we shall explain in §4.2, in $\mathcal{N}=2$ supergravity and for suitably large charges, the jump in $\Omega(M\gamma_1+N\gamma_2,t)$ is accounted by the loss or gain of a family n-centered BPS black hole solutions with charges $\alpha_i=M_i\gamma_1+N_i\gamma_2\in\tilde\Gamma$, with $\sum_i(M_i,N_i)=(M,N)$, which exist on the side c_- of the wall, and on this side only. Close to the wall, the n-centered configuration is loosely bound, with the relative distances r_{ij} between the centers diverging at the wall. As a result, its index factorizes into the product of the internal index $\Omega(\alpha_i)$ associated to each center, times the index of the degrees of freedom associated to the relative motion of the centers (with suitable modifications due to Bose-Fermi statistics when some of the (M_i,N_i) coincide) [1]. In §4.2, we shall compute this configurational index using localization techniques, and obtain $\Delta\Omega(M\gamma_1+N\gamma_2)$ for arbitrary values of (M,N).

A closely similar problem arises in the mathematics of Donaldson-Thomas invariants of coherent sheaves on a compact complex manifold $\mathcal{X}$. These invariants, which we shall again denote by $\Omega(\gamma,t)$, are labelled by a class γ in the K-theory lattice $\Gamma=K(\mathcal{X})$, and depend on a choice of stability condition $\phi_t:\Gamma\to S^1$ inside a complex family parametrized by t. $\Omega(\gamma,t)$ is defined as the Euler characteristic [2] of the moduli space $\mathcal{M}(\gamma,t)$ of stable coherent sheaves on $\mathcal{X}$ in the class γ with respect to ϕ_t. As the stability condition is varied, some of the stable sheaves

[1] Here, Tr' denotes the trace in the Hilbert space associated to (γ,t) with the center of motion degrees of freedom removed, and $(-1)^F$ denotes the fermionic parity, equal to $(-1)^{2J_3}$ by the spin-statistics relation, where J_3 is the angular momentum operator along the z axis.

[2] Rather, the Euler characteristic weighted by Behrend's function, see [6, 7] for details.

may become unstable and the DT invariant $\Omega(\gamma, t)$ may jump. This happens on the same walls of marginal stability as in (1), where $\phi_t(\gamma)$ plays the role of $\arg Z(\gamma, t)$. The similarity between these two wall-crossing problems follows from the fact that stable objects in the derived category of coherent sheaves on a Calabi-Yau three-fold $\mathcal{X}$ are realized physically by BPS states in type IIA string theory compactified on $\mathcal{X}$ (see e.g. [8, 9] for reviews).

In two independent pieces of work, Kontsevich-Soibelman [10] and Joyce-Song [7] have determined the variation $\Delta\Omega(\gamma)$ in terms of the DT invariants on one side of the wall. In both works, it was noted that the wall-crossing formula takes a simpler form in terms of the 'rational DT invariants' $\bar{\Omega}(\gamma, t)$, related to the ordinary, integer-valued invariants $\Omega(\gamma, t)$ by the 'multi-cover formula'

$$\bar{\Omega}(\gamma, t) \equiv \sum_{m|\gamma} \Omega(\gamma/m, t)/m^2 ,$$

where the sum runs over all integers $m \geq 1$ such that $\gamma/m \in \Gamma$ (thus, $\bar{\Omega}(\gamma) = \Omega(\gamma)$ if γ is primitive). From a physics point of view, we shall explain in §4.1 that the replacement $\Omega(\gamma, t) \to \bar{\Omega}(\gamma, t)$ effectively converts the Bose-Fermi statistics of the n centers into Boltzmannian statistics, thereby allowing us to treat the centers as distinguishable. In more detail, the KS and JS formulae express the jump $\Delta\bar{\Omega}(\gamma)$ as

$$\Delta\bar{\Omega}(\gamma, t) = \sum_{n\geq 2} \sum_{\substack{\{\alpha_1,\dots\alpha_n\}\in\tilde{\Gamma} \\ \gamma=\alpha_1+\dots+\alpha_n}} \frac{g(\{\alpha_i\})}{\prod m_k!} \prod_{i=1}^{n} \bar{\Omega}^{+}(\alpha_i, t) , \tag{2}$$

where the second sum runs over all (unordered) decompositions of the total charge vector γ into a sum of n vectors $\alpha_i \in \tilde{\Gamma}$. The coefficients $g(\{\alpha_i\})$ (after extracting out a Boltzmann-Gibbs symmetry factor $\prod m_k!$ whenever $\{\alpha_i\}$ contains m_1 copies of β_1, m_2 copies of β_2, etc) are universal functions of the α_i's. As we shall see, the function $g(\{\alpha_i\})$ turns out to be equal to the index $\mathrm{Tr}'(-1)^F$ of the configurational degrees of freedom of an n-centered Boltzmann black hole solution. To compute this index, it is convenient to consider the refined (or equivariant) index $g(\{\alpha_i\}, y) = \mathrm{Tr}'(-y)^{2J_3}$, evaluate the latter by localization methods, and set $y = 1$ at the end. In fact, the refined configurational index $g(\{\alpha_i\}, y)$ for general y enters the wall-crossing formula for so-called motivic Donaldson-Thomas invariants, see (9).

In the remainder of this survey, we first give an executive summary of the KS (§2) and JS (§3) wall-crossing formulae, silencing the subtleties involved in defining the Donaldson-Thomas invariants by themselves. In §4 we then derive the combinatorial factors $g(\{\alpha_i\})$ by quantizing the phase space of multi-centered BPS black holes and evaluating the index by localization. In §5, we give an alternative computation of $g(\{\alpha_i\})$ relying on Reineke's results for quivers without closed loops. We end in §6 with a discussion of some open problems. The material is mostly based on [1, 11], which the reader should consult for more details. Other important references include [12, 13, 14, 15, 16, 10, 7, 17, 18].

2. The Kontsevich-Soibelman wall-crossing formula

We start by reviewing the Kontsevich-Soibelman wall-crossing formula for generalized Donaldson-Thomas invariants. For generality, we first state its motivic (aka refined) version, and then discuss its classical limit. As an application, we use the KS formula to rederive the primitive and semi-primitive wall-crossing formula. Finally, we extract the coefficient $g(\{\alpha_i\})$ appearing in (2) for $n \leq 3$, for comparison with other wall-crossing formulae discussed later.

2.1. The motivic KS formula

The motivic KS formula pertains to 'motivic Donaldson-Thomas invariants' $\Omega_{\mathrm{ref}}(\gamma, t)$ attached to a Calabi-Yau threefold category $\mathcal{X}$, with a stability condition ϕ_t. Informally, $\Omega_{\mathrm{ref}}(\gamma, t)$ is the

$n+\frac{1}{2}d$-th Betti number of the moduli space $\mathcal{M}(\gamma,t)$ of stable degree-γ objects in the triangulated category of coherent sheaves on $\mathcal{X}$, where $d = \dim \mathcal{M}(\gamma,t)$. We define the Poincaré polynomial

$$\Omega_{\rm ref}(\gamma,t,y) = \sum_{n\in\mathbb{Z}} (-y)^n \, \Omega_n(\gamma,t) \,, \tag{3}$$

which is a finite Laurent polynomial in y, symmetric under $y \to 1/y$ (the subscript 'ref' stands for 'refined', which for our purposes is synonymous with 'motivic' [17]).

To state the KS formula, we introduce the Lie algebra $\mathcal{A}$ spanned by abstract generators e_γ, for each $\gamma \in K(\mathcal{X}) = \Gamma$, subject to the commutation rule

$$[e_\gamma, e_{\gamma'}] = \kappa(\langle\gamma,\gamma'\rangle, y)\, e_{\gamma+\gamma'}, \tag{4}$$

where $\langle\gamma,\gamma'\rangle$ is the integer-valued antisymmetric pairing on Γ (physically, the Dirac-Schwinger-Zwanziger product for electromagnetic charge vectors), and

$$\kappa(x,y) \equiv \frac{(-y)^x - (-y)^{-x}}{y-1/y} = (-1)^x \frac{\sinh(\nu\, x)}{\sinh \nu}, \qquad \nu \equiv \ln y \,. \tag{5}$$

It is straightforward to check that (4) satisfies the Jacobi identity for any y.

Now, for a given choice of stability condition t and any $\gamma \in \Gamma$, let $U_\gamma(t)$ be the element in the group $\mathcal{G} = \exp(\mathcal{A})$ defined by

$$U_\gamma(t) = \prod_{n\in\mathbb{Z}} \mathbf{E}\left(y^n (y-1/y)\, e_\gamma\right)^{(-1)^{n+1}\Omega_n(\gamma,t)} \ ,$$

where $\mathbf{E}(x)$ is the quantum dilogarithm function

$$\mathbf{E}(x) \equiv \exp\left[\sum_{k=1}^{\infty} \frac{(xy)^k}{k(1-y^{2k})}\right] \ .$$

We now restrict to $\gamma \in \tilde\Gamma$, where $\tilde\Gamma$ is the positive cone in the two-dimensional sublattice of Γ spanned by two primitive vectors γ_1, γ_2, and assume that the 'root basis' condition stated below (1) holds. The motivic KS wall-crossing formula [10, 17, 18] states that the following two ordered products

$$\prod_{\substack{M\geq 0, N\geq 0,\\ M/N\downarrow}} U^+_{M\gamma_1+N\gamma_2} = \prod_{\substack{M\geq 0, N\geq 0,\\ M/N\uparrow}} U^-_{M\gamma_1+N\gamma_2} \,, \tag{6}$$

where the products are ordered with decreasing (resp., increasing) values of $M/N \in [0,+\infty]$ (such that the argument of $Z(\alpha)$ decreases from left to right on either side). Here, $U^\pm_\gamma$ denote the group element $U_\gamma(t)$ when t lies on the side $c_\pm$ of the wall $\mathcal{W}(\gamma_1,\gamma_2)$. Thus, assuming that the $\Omega^+_n(\gamma)$'s are known for all $\gamma \in \tilde\Gamma$, the $\Omega^-_n(\gamma)$'s can be computed by re-ordering the product on the l.h.s. of (6) in the opposite order, using the commutation rule (4) and the Baker-Campbell-Hausdorff (BCH) formula, and reading off the exponents $\Omega^-_n(\gamma)$.

This procedure is vastly simplified by expressing the products in (6) in terms of the 'rational motivic invariants'

$$\bar\Omega_{\rm ref}(\gamma,t,y) \equiv \sum_{m|\gamma} \frac{(y-y^{-1})}{m(y^m-y^{-m})} \Omega_{\rm ref}(\gamma/m, t, y^m) \,.$$

The relation between $\Omega_{\rm ref}(\gamma, y)$ and $\bar\Omega_{\rm ref}(\gamma, y)$ is easily inverted by means of the Möbius formula,

$$\Omega_{\rm ref}(\gamma, t, y) = \sum_{m|\gamma} \mu(m)\, \frac{(y-y^{-1})}{m(y^m-y^{-m})} \bar\Omega_{\rm ref}(\gamma/m, t, y^m)\,,$$

where $\mu(d)$ is the Möbius function (1 if d is a product of an even number of distinct primes, -1 if d is a product of an odd number of primes, or 0 otherwise). Using the fact that the generators $e_{\ell\gamma}$ commute for all $\ell \in \mathbb{Z}$, we may rewrite (7) as a product of factors labelled by coprime (M, N),

$$\prod_{\substack{M\geq 0, N\geq 0>0,\\ \gcd(M,N)=1, M/N\downarrow}} V^+_{M\gamma_1+N\gamma_2} = \prod_{\substack{M\geq 0, N\geq 0>0,\\ \gcd(M,N)=1, M/N\uparrow}} V^-_{M\gamma_1+N\gamma_2}\,, \tag{7}$$

where

$$V_\gamma(t) = \prod_{\ell\geq 1} U_{\ell\gamma}(t) = \exp\left(\sum_{N=1}^{\infty} \bar\Omega_{\rm ref}(N\gamma, t, y)\, e_{N\gamma}\right)\,. \tag{8}$$

Due to the fact that the algebra (4) is graded by $\tilde\Gamma$, and that every factor of e_α in the logarithm of V_γ is multiplied by a factor of $\Omega_{\rm ref}(\alpha, y)$ for the same vector α, it is clear that the result of the re-ordering procedure outlined below (6) will produce a wall-crossing formula of the form

$$\bar\Omega^-_{\rm ref}(\gamma, y) - \bar\Omega^+_{\rm ref}(\gamma, y) = \sum_{n\geq 2} \sum_{\substack{\{\alpha_1,\dots,\alpha_n\}\in\tilde\Gamma^n\\ \gamma=\alpha_1+\dots+\alpha_n}} \frac{g_{\rm ref}(\{\alpha_i\}, y)}{\prod m_k!} \prod\nolimits_{i=1}^n \bar\Omega^+_{\rm ref}(\alpha_i, y)\,, \tag{9}$$

where the second sum runs over unordered sets of n charge vectors $\alpha_i = M_i\gamma_1 + N_i\gamma_2$ such that $\sum_{i=1\dots n} \alpha_i = \gamma = M\gamma_1 + N\gamma_2$. By the 'root basis' property, only a finite number of terms appear on the right-hand side. In §2.2 below, we shall compute the universal combinatorial factors $g_{\rm ref}(\{\alpha_i\}, y)$ in selected cases.

Before doing so however, we discuss the classical (or numerical) KS wall-crossing formula, which arises from the motivic formula (7) in the limit $y \to 1$. In this limit, the Poincaré polynomial (3) reduces to the Euler-Behrend characteristic of the moduli space $\mathcal{M}(\gamma, t)$

$$\Omega_{\rm ref}(\gamma, t, y) \overset{y\to 1}{\longrightarrow} \chi[\mathcal{M}(\gamma, t)] \in \mathbb{Z}\,,$$

while $\bar\Omega_{\rm ref}(\gamma, y)$ reduces to the 'rational DT invariant'

$$\bar\Omega_{\rm ref}(\gamma, t, y) \overset{y\to 1}{\longrightarrow} \bar\Omega(\gamma, t) \equiv \sum_{m|\gamma} m^{-2}\, \Omega(\gamma/m, t)\,. \tag{10}$$

Moreover, the commutation rule (4) in the Lie algebra $\mathcal{A}$ has a smooth limit

$$[e_\gamma, e_{\gamma'}] = (-1)^{\langle\gamma,\gamma'\rangle} \langle\gamma, \gamma'\rangle\, e_{\gamma+\gamma'}\,,$$

and so does the operator V_γ in (8). As a result, the combinatorial coefficients $g_{\rm ref}(\{\alpha_i\}, y)$ have a smooth limit as $y \to 1$, and the wall-crossing formula for the rational DT invariants is given by the limit of (9) as $y \to 1$, i.e. Eq. (2) with

$$g(\{\alpha_i\}) = \lim_{y\to 1} g_{\rm ref}(\{\alpha_i\}, y)\,.$$

2.2. Primitive and semi-primitive wall-crossing

Despite the fact that either side of the KS wall-crossing formula (7) involves an infinite number of factors, the procedure of re-ordering the product involves only a finite number of operations, for the following reason (already hinted at below (9)): for any $M, N \geq 0$,

$$\mathcal{I}_{M,N} \equiv \{ \sum_{m>M \text{ and/or } n>N} \mathbb{R} \cdot e_{m\gamma_1+n\gamma_2} \}$$

is a two-sided ideal in $\mathcal{A}$, and the quotient $\mathcal{A}_{M,N} = \mathcal{A}/\mathcal{I}_{M,N}$ is a finite dimensional algebra. For the purpose of computing $\Delta\Omega(M\gamma_1 + N\gamma_2)$, it is sufficient to project the relation (7) to $\mathcal{A}_{M,N}$ and use the truncation of the BCH formula at order $\min(M, N)$. E.g. to compute $\Delta\Omega(\gamma_1 + \gamma_2)$, it suffices to re-order the l.h.s of the identity in $\mathcal{A}_{1,1}$

$$\begin{aligned} &\exp(\Omega^+(\gamma_1)e_{\gamma_1}) \exp(\Omega^+(\gamma_1+\gamma_2)e_{\gamma_1+\gamma_2}) \exp(\Omega^+(\gamma_2)e_{\gamma_2}) \\ &= \exp(\Omega^-(\gamma_2)e_{\gamma_2}) \exp(\Omega^-(\gamma_1+\gamma_2)e_{\gamma_1+\gamma_2}) \exp(\Omega^-(\gamma_1)e_{\gamma_1}) \end{aligned} \tag{11}$$

using the truncated BCH formula $e^X e^Y = e^{X+Y+\frac{1}{2}[X,Y]}$, and match the result to the r.h.s. In this way, we find that the motivic invariants $\Omega_{\text{ref}}(\gamma_1, y), \Omega_{\text{ref}}(\gamma_2, y)$ are constant across the wall, while

$$\Delta\Omega_{\text{ref}}(\gamma_1 + \gamma_2, y) = \kappa(\langle\gamma_1, \gamma_2\rangle, y)\, \Omega_{\text{ref}}(\gamma_1, y)\, \Omega_{\text{ref}}(\gamma_2, y) \ . \tag{12}$$

This relation (or its obvious classical limit at $y = 1$) is known as the 'primitive wall crossing formula' [14, 19, 17]. From (12), setting $\alpha_1 = \gamma_2$ and $\alpha_2 = \gamma_1$ so that $\langle\alpha_1, \alpha_2\rangle > 0$, we read off the combinatorial factor

$$g_{\text{ref}}(\alpha_1, \alpha_2, y) = -\kappa(\langle\alpha_1, \alpha_2\rangle, y) = (-1)^{\langle\alpha_1,\alpha_2\rangle+1} \frac{\sinh(\nu\langle\alpha_1, \alpha_2\rangle)}{\sinh\nu} \ . \tag{13}$$

Up to a sign, this is recognized as the character $\mathrm{Tr}\,(-y)^{2J_3}$ of a representation of $SU(2)$ with spin $j = \frac{1}{2}(|\langle\alpha_1, \alpha_2\rangle| - 1)$.

With some more work, one can easily extract the combinatorial coefficients $g_{\text{ref}}(\{\alpha_i\}, y)$ for $n > 2$. E.g, for $n = 3$ and $\alpha_1, \alpha_2, \alpha_3$ three distinct (non necessarily primitive) elements of $\tilde{\Gamma}$ ordered such that $\alpha_{ij} \equiv \langle\alpha_i, \alpha_j\rangle > 0$ for $i < j$, we find

$$g_{\text{ref}}(\alpha_1, \alpha_2, \alpha_3, y) = (-1)^{\alpha_{12}+\alpha_{13}+\alpha_{23}} \frac{\sinh(\nu\alpha_{12})\sinh(\nu(\alpha_{13}+\alpha_{23}))}{\sinh^2\nu} \tag{14}$$

when $\alpha_{12} > \alpha_{23}$, or

$$g_{\text{ref}}(\alpha_1, \alpha_2, \alpha_3, y) = (-1)^{\alpha_{12}+\alpha_{13}+\alpha_{23}} \frac{\sinh(\nu\alpha_{23})\sinh(\nu(\alpha_{12}+\alpha_{13}))}{\sinh^2\nu}$$

when $\alpha_{12} < \alpha_{23}$. The result for $n = 4$ can be found in [1].

While the amount of work necessary to extract the combinatorial factors quickly grows with (M, N), for fixed (small) M it is possible to compute all the jumps for $\gamma \to M\gamma_1 + N\gamma_2$ at once using the Hadamard lemma $\log(e^X Y e^{-X}) = \sum_{n\geq 0} Ad_X^n \cdot Y)/n!$, where $Ad_X \cdot Y \equiv [X, Y]$. Defining

$$Z^{\pm}(M, q, y) = \sum_{N=0}^{\infty} \Omega^{\pm}_{\text{ref}}(M, N, y)\, q^N \ , \quad \Omega^{\pm}_{\text{ref}}(M, N, y) \equiv \Omega^{\pm}_{\text{ref}}(M\gamma_1 + N\gamma_2, y) \ , \tag{15}$$

we find, for $M = 1$ [1], the 'semi-primitive' wall crossing formula [14]

$$Z^-(1, q, y) = Z^+(1, q, y)\, Z_{\text{halo}}(\gamma_1, q, y)$$

where

$$Z_{\rm halo}(\gamma_1,q,y) \equiv \exp\left(\sum_{\ell=1}^{\infty}\kappa(\ell\langle\gamma_1,\gamma_2\rangle,y)\,\bar\Omega_{\rm ref}(\ell\gamma_2,y)\,q^\ell\right) . \tag{16}$$

The reason for the subscript 'halo' will become apparent in §4.1. Re-expressed in terms of the integer motivic invariants, this can be written as an infinite product [18]

$$Z_{\rm halo}(\gamma_1,q,y) = \prod_{\substack{k\geq 1, n\in\mathbb{Z}\\ 1\leq j\leq k|\gamma_{12}|}} \left(1-(-1)^{k|\gamma_{12}|}q^k y^{n+2j-1-k|\gamma_{12}|}\right)^{(-1)^n\,\Omega_n(k\gamma_2)}$$

Generalizations of (16) for $M=2,3$ can be found in [1].

2.3. Exact wall crossing

Finally, we discuss some examples where the re-ordering of the product in (7) can be performed in the full untruncated algebra $\mathcal{A}$. Suppose that in the chamber c_+, the only non-vanishing DT invariants are $\Omega^+(\gamma_1)$ and $\Omega^+(\gamma_2)$. If $\gamma_{12}=-1$, the result of the re-ordering gives[3]

$$U_{\gamma_2}\,U_{\gamma_1} = U_{\gamma_1}\,U_{\gamma_1+\gamma_2}\,U_{\gamma_2}\ , \qquad \gamma_{12}=-1\ , \tag{17}$$

which follows from the pentagonal identity for the quantum dilogarithm function. If instead $\gamma_{12}=-2$, one arrives at [10]

$$U_{(2,-1)}\cdot U_{(0,1)} = U_{(0,1)}\cdot U_{(2,1)}\cdot U_{(4,1)}\dots U_{(2,0)}\dots U_{(3,-1)}\cdot U_{(2,-1)}U_{(1,-1)}\ , \tag{18}$$

where we denoted $\gamma_2=(0,1)$, $\gamma_1=(2,-1)$ to match the usual basis of electromagnetic charges in Seiberg-Witten theory with $G=SU(2)$ and no flavors [5]. As first noted by Denef, Eq. (18) then embodies the BPS spectrum of this gauge theory on the two sides of the curve of marginal stability $\mathrm{Im}\,(a/a_D)=0$, see Fig. 1. Analogues of (18) for $SU(2)$ gauge theories with $0<N_f<4$ flavors can be found in [15, 18]. More general identities of this type can be derived using Y-systems and cluster algebra techniques, see e.g. [20, 21, 22].

3. The Joyce-Song wall-crossing formula

In this section, we briefly review the Joyce-Song wall-crossing formula, which was derived in the context of the Abelian category of coherent sheaves on a Calabi-Yau three-fold $\mathcal{X}$ [7]. Unlike the KS formula, the JS formula only applies to the jump of the classical (or numerical) Donaldson-Thomas invariants. Moreover, it gives a fully explicit formula for the combinatorial factors $g(\{\alpha_i\})$ appearing in (2). The price to pay is that the JS formula is computationally less efficient, as it involves sums over many terms with large denominators and large cancellations.

To state the JS formula, we first introduce S, U and $\mathcal{L}$ factors, which are functions of an ordered list of charge vectors $\alpha_i\in\tilde\Gamma, i=1\dots n$:

- We define $S(\alpha_1,\dots,\alpha_n)\in\{0,\pm1\}$ as follows. If $n=1$, set $S(\alpha_1)=1$. If $n>1$ and, for every $i=1\dots n-1$, either

$$\begin{array}{llll} (a) & \langle\alpha_i,\alpha_{i+1}\rangle\leq 0 & \text{and} & \langle\alpha_1+\dots+\alpha_i,\alpha_{i+1}+\dots+\alpha_n\rangle<0, \quad \text{or}\\ (b) & \langle\alpha_i,\alpha_{i+1}\rangle> 0 & \text{and} & \langle\alpha_1+\dots+\alpha_i,\alpha_{i+1}+\dots+\alpha_n\rangle\geq 0\ , \end{array}$$

  let $S(\alpha_1,\dots,\alpha_n)=(-1)^r$, where r is the number of times option (a) is realized; otherwise, $S(\alpha_1,\dots,\alpha_n)=0$.

[3] In (17) and (18), $\Omega=1$ for each factor, except for the factor $U_{(2,0)}$ in the middle of (18), for which $\Omega=-2$.

Journal of Physics: Conference Series **346** (2012) 012017 doi:10.1088/1742-6596/346/1/012017

- To define the U factor (not to be confused with the operator U of the previous section !), consider all ordered partitions of the n vectors α_i into $1 \le m \le n$ packets $\{\alpha_{a_{j-1}+1}, \cdots, \alpha_{a_j}\}$, $j = 1 \dots m$, with $0 = a_0 < a_1 < \cdots < a_m = n$, such that all vectors within each packet are collinear. Let

$$\beta_j = \alpha_{a_{j-1}+1} + \cdots + \alpha_{a_j}, \qquad j = 1 \dots m$$

be the sum of the charge vectors in each packet. Next, consider all ordered partitions of the m vectors β_j into $1 \le l \le m$ packets $\{\beta_{b_{k-1}+1}, \cdots, \beta_{b_k}\}$, with $0 = b_0 < b_1 < \cdots < b_l = m$, $k = 1 \dots l$, such that the total charge vectors $\delta_k = \beta_{b_{k-1}+1} + \cdots + \beta_{b_k}$, $k = 1 \dots l$ are all collinear. Define the U-factor as the sum

$$U(\alpha_1, \dots, \alpha_n) \equiv \sum_l \frac{(-1)^{l-1}}{l} \cdot \prod_{k=1}^{l}$$
$$\prod_{j=1}^{m} \frac{1}{(a_j - a_{j-1})!} \; S(\beta_{b_{k-1}+1}, \beta_{b_{k-1}+2}, \dots, \beta_{b_k}) \; .$$

over all partitions of α_i and β_j satisfying the conditions above. If none of the α_i are parallel, $S = U$. Contributions with $l > 1$ arise only when $\{\alpha_i\}$ can be split into two (or more) packets with the same total charge, e.g.

$$U[\gamma_1, \gamma_2, \gamma_1, \gamma_2] = S[\gamma_1, \gamma_2, \gamma_1, \gamma_2] - \frac{1}{2} S[\gamma_1, \gamma_2]^2 = 1 - \frac{1}{2}(-1)^2 = \frac{1}{2}$$

- Finally, departing from the notations in [7], define the $\mathcal{L}$ factor by

$$\mathcal{L}(\alpha_1, \dots, \alpha_n) = \sum_{\text{trees}} \prod_{\text{edges(i,j)}} \langle \alpha_i, \alpha_j \rangle$$

where the sum runs over all labelled trees with n vertices labelled $\{1, \dots, n\}$, with edges oriented from i to j if $i < j$. There are n^{n-2} labelled trees with n-vertices, which can be labelled by their Prüfer code, an arbitrary sequence of $n - 2$ numbers in $\{1, \dots n\}$.

With these definitions, the result of [7] can be stated as an explicit formula for the combinatorial factors $g(\{\alpha_i\})$ appearing in (2):

$$g(\{\alpha_i\}) = \frac{1}{2^{n-1}} (-1)^{n-1+\sum_{i<j} \langle \alpha_i, \alpha_j \rangle} \sum_{\sigma \in \Sigma_n} \mathcal{L}\left(\alpha_{\sigma(1)}, \dots \alpha_{\sigma(n)}\right) U\left(\alpha_{\sigma(1)}, \dots \alpha_{\sigma(n)}\right) \; . \tag{19}$$

As an illustration, we now use the JS formula to derive the combinatorial factors $g(\{\alpha_i\})$ for $n = 2, 3$. For $n = 2$ (assuming as before that $\gamma_{12} < 0$), the $U, S, \mathcal{L}$ factors are given in the following table

$\sigma(12)$	S	U	$\mathcal{L}$
12	a	-1	γ_{12}
21	b	1	$-\gamma_{12}$

The JS formula (19) then leads to

$$g(\gamma_1, \gamma_2) = (-1)^{\gamma_{12}} \gamma_{12} \, \Omega(\gamma_1) \, \Omega(\gamma_2) \; ,$$

in agreement with the classical limit of (13).

For $n = 3$, assuming as before that $\alpha_1, \alpha_2, \alpha_3$ are three distinct (non necessarily primitive) elements of $\tilde{\Gamma}$ ordered such that $\alpha_{ij} \equiv \langle \alpha_i, \alpha_j \rangle > 0$ for $i < j$ and moreover that $\alpha_{12} > \alpha_{23}$, we find that the $S, U, \mathcal{L}$ factors are given by

$\sigma(123)$	S	U	$\mathcal{L}$
123	bb	1	$\alpha_{12}\alpha_{13} + \alpha_{13}\alpha_{23} + \alpha_{12}\alpha_{23}$
132	b-	0	$\alpha_{12}\alpha_{13} - \alpha_{13}\alpha_{23} - \alpha_{12}\alpha_{23}$
213	ab	-1	$-\alpha_{12}\alpha_{23} + \alpha_{13}\alpha_{23} - \alpha_{12}\alpha_{13}$
231	-a	0	$\alpha_{12}\alpha_{13} - \alpha_{13}\alpha_{23} - \alpha_{12}\alpha_{23}$
312	ab	-1	$\alpha_{13}\alpha_{23} - \alpha_{12}\alpha_{23} - \alpha_{13}\alpha_{12}$
321	aa	1	$\alpha_{13}\alpha_{23} + \alpha_{12}\alpha_{13} + \alpha_{12}\alpha_{23}$

The JS formula (19) then leads to

$$g(\{\alpha_1, \alpha_2, \alpha_3\}) = (-1)^{\alpha_{12}+\alpha_{23}+\alpha_{13}}\, \alpha_{12}\, (\alpha_{13} + \alpha_{23}) \ ,$$

in agreement with the classical limit of (14). The computation for $n = 4$ can be found in [1], and matches the result from the KS formula.

4. Wall-crossing from multi-centered quantum black holes

In this section, we give a new physical derivation of the wall-crossing formula (2) (in particular, a new formula for the combinatorial factors $g(\{\alpha_i, y\})$) based on the quantum mechanics of multi-centered black hole configuration. Before starting, it should be noted that the KS wall-crossing formula (and to a lesser extent, the JS formula) has already been derived or interpreted in various physical settings [19, 15, 23, 24, 25, 26, 27, 4, 28]. Our derivation is arguably more elementary, as it relies only on the supersymmetric quantum mechanics of point particles interacting by Coulomb and Lorentz-type forces. The down-side is that it does not make the algebra $\mathcal{A}$ manifest and, admittedly, relies on some plausible but not rigorously proven assumptions.

4.1. From Bose-Fermi to Boltzmann statistics

To motivate our approach, let us return to the semi-primitive wall-crossing formula (16) and for simplicity, concentrate on the classical limit $y = 1$. Substituting (10) in (16), one may rewrite (16) as

$$\frac{\sum_{N\geq 0} \Omega^-(1,N)\, q^N}{\sum_{N\geq 0} \Omega^+(1,N)\, q^N} = \prod_{k>0} \left(1 - (-1)^{k\gamma_{12}} q^k\right)^{k\,|\gamma_{12}|\, \Omega^+(k\gamma_2)} \ . \tag{20}$$

E.g. for $\gamma \mapsto \gamma_1 + 2\gamma_2$, we find

$$\begin{aligned}\Delta\Omega(1,2) =& 2\gamma_{12}\, \Omega^+(1,0)\, \Omega^+(0,2) + (-1)^{\gamma_{12}}\, \gamma_{12}\, \Omega^+(1,1)\Omega^+(0,1) \\ &+ \Omega^+(1,0) \left[\frac{1}{2}\gamma_{12}\, \Omega^+(0,1)\left(\gamma_{12}\Omega^+(0,1) + 1\right)\right] \ .\end{aligned} \tag{21}$$

The two contributions on the first line can be interpreted as the index of two-centered black hole solutions, carrying charges $\alpha_1 = \gamma_1$ and $\alpha_2 = 2\gamma_2$ for the first term, or $\alpha_1 = \gamma_1 + \gamma_2$ and $\alpha_2 = \gamma_2$ for the second term. Indeed, for such two-centered solutions, the distance is fixed to [12]

$$r_{12} = \frac{1}{2} \frac{\langle \alpha_1, \alpha_2 \rangle\, |Z(\alpha_1) + Z(\alpha_2)|}{\mathrm{Im}\left[Z(\alpha_1)\bar{Z}(\alpha_2)\right]} \ ,$$

which is positive on the side c_- of the wall only; the unit vector $\vec{r}_{12}/r_{12}$ can be chosen arbitrarily on the unit two-sphere. Such configurations carry angular momentum $\vec{J} = \frac{1}{2}(\alpha_{12} - 1)\vec{r}_{12}/r_{12}$

(similar to the angular momentum carried by an electron in a magnetic monopole background), and therefore have $|\alpha_{12}|$ possible configurational states, with index $g(\alpha_1, \alpha_2) = (-1)^{\alpha_{12}}\alpha_{12}$. Since the distance r_{12} diverges in the vicinity of the wall, the internal degrees of freedom are decoupled from the configurational degrees of freedom, and the total index is the product $g(\alpha_1, \alpha_2)\Omega^-(\alpha_1)\Omega^-(\alpha_2)$, consistently with (21).

The contribution on the second line of (21) is more interesting. Letting $d = \gamma_{12}\Omega^+(0,1)$, the term in bracket is recognized as the index of the symmetric (or, when $d < 0$, antisymmetric) part of the tensor product $\mathcal{H}_1 \otimes \mathcal{H}_1$, where $\mathcal{H}_1$ is the space of quantum states accessible to one particle of charge γ_2 in the field of the particle of charge γ_1. In other words, the second line corresponds to a configuration of two identical centers of charge γ_2 orbiting around a center of charge γ_1, with Bose statistics when $d > 0$, or Fermi statistics when $d < 0$. More generally, (20) can be interpreted as the contribution from halos of particles of charge $k\gamma_2$ orbiting on a fixed shell around a center of charge $\gamma_1 + k_0\gamma_2$, and obeying Bose or Fermi statistics, depending on the sign of $\langle\gamma_1 + k_0\gamma_2, k\gamma_2\rangle\, \Omega^+(k\gamma_2)$.

In contrast, in terms of the rational DT invariants the semi-primitive wall-crossing formula (16) reads

$$\frac{\sum_{N\geq 0}\Omega^-(1,N)\,q^N}{\sum_{N\geq 0}\Omega^+(1,N)\,q^N} = \exp\left(\sum_{\ell=1}^{\infty}(-1)^{\ell\langle\gamma_1,\gamma_2\rangle}\,\ell\langle\gamma_1,\gamma_2\rangle\,\bar{\Omega}(\ell\gamma_2,y)\,q^\ell\right)\,, \tag{22}$$

so that, e.g. for $\gamma \mapsto \gamma_1 + 2\gamma_2$,

$$\Delta\Omega(1,2) = 2\gamma_{12}\,\Omega^+(1,0)\,\bar{\Omega}^+(0,2) + (-1)^{\gamma_{12}}\,\gamma_{12}\,\Omega^+(1,1)\Omega^+(0,1) + \Omega^+(1,0)\left[\frac{1}{2}\left(\gamma_{12}\,\Omega^+(0,1)\right)^2\right]\,, \tag{23}$$

where we combined $\Omega(0,2)$ and $-\frac{1}{4}\Omega(0,1)$ in (21) into $\bar{\Omega}(0,2)$. Unlike (21), the last term in (23) is of the form $\frac{1}{2}d^2$, which would be the result if the two particles of charge $(0,1)$ were distinguishable and obeyed Boltzmann statistics. More generally, (22) can be interpreted as contributions of the same halo of particles with charge $k\gamma_2$ described above, but now satisfying Boltzmann statistics. While (23) is hardly shorter than (21), the reader can easily convince him/herself of the power of this simplification by computing $\Delta\Omega(1,N)$ for higher N.

The lesson to take from this discussion is that, rather than computing the variation of $\Omega(\gamma,t)$ across the wall $\mathcal{W}(\gamma_1,\gamma_2)$, it is advantageous to compute instead the variation of the rational invariants $\bar{\Omega}(\gamma,t)$ defined in (10), and apply the following recipe: treat the centers as distinguishable pointlike particles and compute their configurational index. Whenever m centers carry the same charge α, divide the configuration index by a Bolztmann-Gibbs factor $1/m!$. Finally, multiply the configurational index by the effective (rational) index $\bar{\Omega}(\alpha,t)$ carried by each of the centers. The result of this recipe is the formula (2), where $g(\{\alpha_i\})$ is identified as the configurational index of the quantum mechanics of n distinguishable particles interacting via Coulomb and Lorentz type forces (which we discuss in detail in the next subsection). Although we motivated this recipe by inspecting the classical semi-primitive formula, it in fact holds in full generality and applies to the refined (or motivic) index as well, see [1] for more details.

4.2. The phase space of multi-centered BPS black holes

Let us now review some relevant properties of supersymmetric multi-centered black hole solutions in $\mathcal{N}=2$ supergravity (a similar analysis for multi-centered dyon solution in the low energy limit of $\mathcal{N}=2$ gauge theories can be found in [29, 30]). Such solutions fall into the stationary metric ansatz

$$ds^2 = -e^{2U}\,(dt+\mathcal{A})^2 + e^{-2U}d\vec{r}^2$$

where the scale function U, the Kaluza-Klein one-form $\mathcal{A}$ and the vector multiplet scalars $z^a, a = 1 \dots n_v$ depend on the coordinate $\vec{r}$ on $\mathbb{R}^3$.

For n centers located at $\vec{r}_1, \dots, \vec{r}_n$, carrying electromagnetic charges $\alpha_1, \dots, \alpha_n \in \Gamma$ with total charge $\gamma = \alpha_1 + \dots + \alpha_n$, the values of the vector multiplet scalars t and of the scale factor U are obtained by solving [12],

$$-2\, e^{-U(\vec{r})}\, \mathrm{Im}\left[e^{-\mathrm{i}\phi} Y(t(\vec{r}))\right] = \beta + \sum_{i=1}^{n} \frac{\alpha_i}{|\vec{r} - \vec{r}_i|}, \qquad \phi = \arg Z(\gamma, t_\infty), \tag{24}$$

where $Y(t) = -e^{\mathcal{K}/2}(X^\Lambda(t), F_\Lambda(t))$ is the symplectic section afforded by the special geometry of the vector multiplet moduli space, such that $Z(\gamma, t) = \langle \gamma, Y(t_\infty) \rangle$. The constant vector β on the right-hand side of (24) is determined in terms of the asymptotic values of the moduli at infinity t_∞ by

$$\beta = -2\, \mathrm{Im}\left[e^{-\mathrm{i}\phi}\, Y(t_\infty)\right] .$$

In particular, it follows from (24) that the scale factor U is given by evaluating the Bekenstein-Hawking entropy function $S(\gamma)$ on the harmonic function appearing on the right-hand side of (24) [31],

$$e^{-2U(\vec{r})} = \frac{1}{\pi} S\left(\beta + \sum_{i=1}^{n} \frac{\alpha_i}{|\vec{r} - \vec{r}_i|}\right) .$$

Most importantly for our purposes, the locations $\vec{r}_i$ are subject to the condition of mechanical equilibrium under the Coulomb, Lorentz, Newton and scalar exchange forces (also known as integrability equations) [12]

$$\sum_{\substack{j=1 \\ j \neq i}}^{n} \frac{\alpha_{ij}}{r_{ij}} = c_i\,, \tag{25}$$

where $r_{ij} = |\vec{r}_i - \vec{r}_j|$, $\alpha_{ij} \equiv \langle \alpha_i, \alpha_j \rangle$, and the real constants

$$c_i \equiv 2\, \mathrm{Im}\left[e^{-\mathrm{i}\phi} Z(\alpha_i, t_\infty)\right]$$

depend on the the asymptotic values of the moduli. Since $\phi = \arg Z(\gamma, t_\infty)$, the constants c_i satisfy $\sum_{i=1}^{n} c_i = 0$. The conditions (25) guarantee the existence of a Kaluza-Klein connection $\mathcal{A}$ such that the above configuration is a supersymmetric solution of the equations of motion. In order for the solution to be physical however, one must also require that the scale factor be everywhere positive

$$S\left(\beta + \sum_{i=1}^{n} \frac{\alpha_i}{|\vec{r} - \vec{r}_i|}\right) > 0\,, \qquad \forall\, \vec{r} \in \mathbb{R}^3\,,$$

where $\vec{r}_i$ is the location of the i-th center. For the configurations relevant to the wall-crossing problem, this condition appears to be automatically satisfied.

Now, let $\mathcal{M}_n(\{\alpha_{ij}\}; \{c_i\})$ be the space of of solutions $\{\vec{r}_1, \dots \vec{r}_n\}$ to the equilibrium conditions (25), modulo overall translations of the centers. $\mathcal{M}_n$ is a (possibly disconnected) $2n-2$-dimensional submanifold of $\mathbb{R}^{3n-3} \backslash \Delta$, where Δ is the locus in $\mathbb{R}^{3n-3}$ where two or more of the centers $\vec{r}_i$ coincide. $\mathbb{R}^{3n-3} \backslash \Delta$ is equipped with the closed two-form

$$\omega = \frac{1}{4} \sum_{i<j} \alpha_{ij}\, \frac{\epsilon^{abc} \mathrm{d}r_{ij}^a \wedge \mathrm{d}r_{ij}^b\, r_{ij}^c}{|r_{ij}|^3} .$$

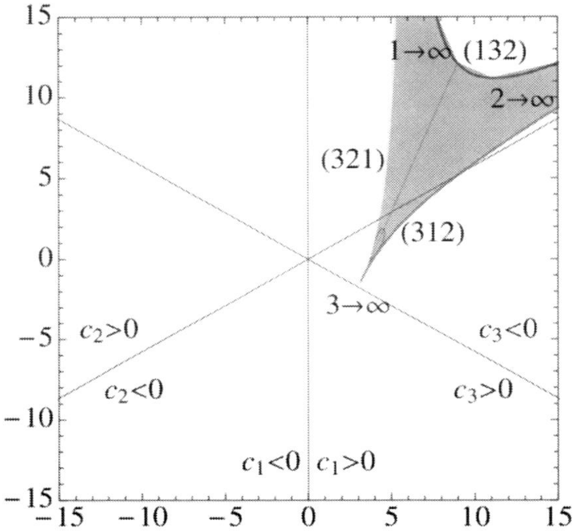

Figure 2. Phase structure of the moduli space $\mathcal{M}_n$ of 3-centered solutions as a function of c_i, for fixed charges such that $\alpha_{12} > 0, \alpha_{23} > 0, \alpha_{13} > 0$, $\alpha_{12} < \alpha_{23}$. The shaded area represent the values of c_i in the two-dimensional section $c_1 + c_2 + c_3 = 0$ which are spanned as the location of the 3rd center is varied, keeping the centers 1 and 2 fixed. Conversely, if the values of c_i is fixed, the range of distances between the centers 1 and 2 can be read off by intersecting the shaded area with a radial line which joins c_i to the origin. Thus, for this choice of charges, 3-centered solutions only exist in the region $c_1 > 0, c_3 < 0$. Inside this region, the range of r_{12} is bounded from below and from above, except on the wall of marginal stability $c_2 = 0$. The boundaries of the shaded region correspond to collinear solutions whose order is indicated. As the wall is crossed, the topology of the collinear solutions changes from $(321), (132)$ to $(321), (312)$.

For generic values of c_i, the restriction of ω to $\mathcal{M}_n$ is non-degenerate and endows $\mathcal{M}_n$ with a symplectic structure [16]. Moreover, the symplectic form ω is invariant under $SO(3)$ rotations. The moment map associated to infinitesimal rotations is the angular momentum

$$\vec{J} = \frac{1}{2} \sum_{i<j} \alpha_{ij} \frac{\vec{r}_{ij}}{|r_{ij}|} .$$

Away from walls of marginal stability, the distances r_{ij} are bounded from above. If it is possible to order the α_i's such that $\langle \alpha_i, \alpha_j \rangle \geq 0$ whenever $i \leq j$, as it is the case when the α_i's lie in a two-dimensional cone $\tilde{\Gamma}$, the distances r_{ij} are also bounded from below by a non-zero $r_{\min} > 0$, and the space $\mathcal{M}_n$ is therefore compact. E.g. for two centers, $\mathcal{M}_2 = S^2$ equipped with $\omega = \frac{1}{2}\alpha_{12} \sin\theta \, d\theta d\phi$ for $\text{sign}(\alpha_{12}) = \text{sign}(c_1)$, and zero otherwise. A representative of the phase structure of $\mathcal{M}_3$ is illustrated in Figure 2.

4.3. Equivariant index and localization

The symplectic space $\mathcal{M}_n \equiv \mathcal{M}_n(\{\alpha_{ij}\}; \{c_i\})$ defines the classical phase space of the configurational degrees of freedom of n-centered BPS black hole solutions. Since such configurations are stationary, the Hamiltonian vanishes and all points in $\mathcal{M}_n$ are degenerate in energy. Quantum mechanically, the Hilbert space consists of sections of $\mathcal{H} = S \otimes \mathcal{L}$, where

$S = S_+ \oplus S_-$ is the total spin bundle[4] over $\mathcal{M}_n$ and $\mathcal{L}$ is a complex line bundle over $\mathcal{M}_n$ with first Chern class ω. BPS states correspond to zero-modes of the Dirac operator D on $\mathcal{M}_n$, and are analogous to states in the lowest Landau level for an electron immersed in a magnetic flux ω. The Dirac operator decomposes as $D = D_+ + D_-$ where D_+ maps $S_+ \otimes \mathcal{L}$ to $S_- \otimes \mathcal{L}$ and vice-versa. The action of $SO(3)$ on $\mathcal{M}_n$ lifts to an action of $SU(2)$ on $S_\pm \otimes \mathcal{L}$, and the refined index is then

$$g_{\rm ref}(\{\alpha_i\}, y) = \text{Tr}_{\,\text{Ker}D_+}(-y)^{2J_3} + \text{Tr}_{\,\text{Ker}D_-}(-y)^{2J_3}$$

where J_3 is the operator representing the rotations along the z axis. Assuming that $\text{Ker}D_- = 0$, it follows that the refined index is equal to the equivariant index of the Dirac operator D,

$$g_{\rm ref}(\{\alpha_i\}, y) = \text{Tr}_{\,\text{Ker}D_+}(-y)^{2J_3} - \text{Tr}_{\,\text{Ker}D_-}(-y)^{2J_3}$$

The assumption that $\text{Ker}D_- = 0$ can be proven when $\mathcal{M}_n$ is Kähler (which is the case when $n = 2, 3$). We do not know how to prove it in general, but it is supported by the fact that it leads to results in agreement with the KS and JS formulae. In §6, we speculate that this assumption may be unnecessary if one were to compute the jump of the protected spin character in the context of $\mathcal{N} = 2$ SYM theories.

Now, by the Atiyah-Bott Lefschetz fixed point formula [32, 33, 34, 35], the equivariant index localizes to the fixed points of the action of J_3 on $\mathcal{M}_n$. Clearly, those correspond to solutions where all centers lie along the z axis, and satisfy the one-dimensional equilibrium conditions

$$\sum_{\substack{j=1\\ j\neq i}}^{n} \frac{\alpha_{ij}}{|z_j - z_i|} = c_i \ , \qquad \sum_{i=1}^{n} z_i = 0 \ , \tag{26}$$

where the last equation fixes the translational zero-mode. For any $\sigma \in \mathcal{S}_n$, where $\mathcal{S}_n$ denotes the set of permutations of $\{1, \ldots, n\}$, we denote by $\mathcal{C}(\sigma)$ the set of solutions to (26) such that $z_{\sigma(i)} < z_{\sigma(j)}$ if $i < j$. The set $\mathcal{C}(\sigma)$ corresponds to the subset of the critical points of the 'superpotential'

$$W(\lambda, \{z_i\}) = -\sum_{i<j} \alpha_{ij}\, \text{sign}(z_j - z_i) \ln|z_j - z_i| - \sum_i (c_i - \lambda/n) z_i \tag{27}$$

which are ordered according to the permutation σ. In the vicinity of a fixed point $p \in \mathcal{C}(\sigma)$, the angular momentum J_3 and the symplectic form ω take the form

$$J_3 = \frac{1}{2}\sum_{i<j} \alpha_{\sigma(i)\sigma(j)} - \frac{1}{4} M_{ij}(p)\,(x_i x_j + y_i y_j) + \cdots \ , \qquad \omega = \frac{1}{2} M_{ij}(p)\, \mathrm{d}x_i \wedge \mathrm{d}y_j + \cdots \ ,$$

where M_{ij} is the Hessian matrix of $W(\lambda, \{z_i\})$ with respect to $z_1, \ldots z_n$, and (x_i, y_i) are coordinates in the plane transverse to the z-axis at the center i, subject to the condition $\sum_i x_i = \sum y_i = 0$. Except for an overall translational zero-mode, the matrix M_{ij} is non-degenerate, and the critical points are isolated, so $\mathcal{C}(\sigma)$ is a finite set (possibly empty).

The Lefschetz fixed point formula of [34] yields an explicit formula for the refined index

$$g_{\rm ref}(\{\alpha_i\}, y) = \frac{(-1)^{\sum_{i<j}\alpha_{ij} + n - 1}}{(y - 1/y)^{n-1}} \sum_{\sigma \in \mathcal{S}_n} s(\sigma)\, y^{\sum_{i<j}\alpha_{\sigma(i)\sigma(j)}} \ , \tag{28}$$

[4] We assume that $\mathcal{M}_n$ is spin, see [11] for a discussion of this issue.

where $s(\sigma)$ counts (with sign) the number of solutions to (26) ordered according to the permutation σ,

$$s(\sigma) = -\sum_{p\in\mathcal{C}(\sigma)} \operatorname{sign}\det \hat{M}\,,$$

where $\hat{M}$ is the Hessian of W as a function of the $n+1$ variables $\lambda, z_1, \dots z_n$, evaluated at the given solution of (26). The factor $(y-1/y)^{n-1}$ in (28) originates form the equivariant $\hat{A}$-genus in the Lefschetz fixed point formula. It is convenient to let the sum in (28) run over all permutations and set $s(\sigma)=0$ when there are no solutions to (26) in the order specified by σ. For reasons that will become clear in §5, we refer to (28) as the 'Coulomb branch wall-crossing formula'.

The formula (28) is fully explicit, yet it depends on our ability to find solutions of the one-dimensional problem (26). While this can be done numerically (approximate solutions are sufficient since the answer depends only on the order σ and the sign of $\det M$), it would be useful to have a general criterium for determining when solutions exist, and if so to compute their Morse index. The answer to these questions is suggested by an alternative approach based on quivers (see §6).

At this point, we can check whether (28) agrees with the answer of the KS or JS formulae. For $n=2$, we find two fixed points with permutations $\sigma(12)=12$ and 21, leading to

$$g(\alpha_1,\alpha_2;y) = (-1)^{\alpha_{12}}\frac{\sinh(\nu\alpha_{12})}{\sinh\nu}\,,$$

in agreement with (13). For $n=3$ (and for the same choice of α_i as above (14)), we find 4 possible orderings, with the following value of $s(\sigma)$,

$$\{1,2,3;+\},\{2,1,3;-\},\{3,1,2;-\},\{3,2,1;+\}\,,$$

leading to

$$\begin{aligned} g_{\rm ref}(\alpha_1,\alpha_2,\alpha_3,y) &= (-1)^{\alpha_{12}+\alpha_{23}+\alpha_{13}}\,(y-y^{-1})^{-2} \\ &\quad\times\left(y^{\alpha_{12}+\alpha_{13}+\alpha_{23}} - y^{\alpha_{13}+\alpha_{23}-\alpha_{12}} - y^{\alpha_{12}-\alpha_{23}-\alpha_{13}} + y^{-\alpha_{12}-\alpha_{13}-\alpha_{23}}\right), \end{aligned}$$

in agreement with the result (14). Further checks for $n>3$ can be found in [1]. It is an open problem to show by combinatorial means that the result (28) agrees with the KS formula in general.

5. Wall-crossing from Abelian quivers

Finally, we discuss an alternative computation of the combinatorial factors $g_{\rm ref}(\{\alpha_i\},y)$ based on Reineke's formula for invariants of quivers without closed loop.

The physical idea is that the supersymmetric quantum mechanics of multi-centered black holes admits two branches: the Coulomb branch, where the centers are far separated and well described by the supergravity solution described in §4.2, and the Higgs branch, where the centers are very close to each other and are better represented as D-branes, with open strings stretched between them. At large string coupling, the wave function is mainly supported on the Coulomb branch, while at small string coupling it is mainly supported on the Higgs branch [13]. However, the BPS index $\mathrm{Tr}'(-1)^F$ is independent of the string coupling (and other hypermultiplet fields), and should be computable in both. It is less clear that the same should be true of the refined index $\mathrm{Tr}'(-y)^{2J_3}$, since this quantity is in general non-protected in string theory [27], but for what concerns the jump of the refined index across walls of marginal stability, we shall find strong evidence that this is the case.

Journal of Physics: Conference Series **346** (2012) 012017 doi:10.1088/1742-6596/346/1/012017

On the Higgs branch, the D-brane system is described at low energy by a supersymmetric quiver quantum mechanics, with gauge group $\prod_{i=1}^n U(N_i)$, where N_i is the number of coinciding D-branes at point i, and $\langle\alpha_i, \alpha_j$ fields in the bifundamental representation $(N_i, \bar{N}_j)$ when $\langle\alpha_i, \alpha_j\rangle$ is positive, or $-\langle\alpha_i, \alpha_j\rangle$ fields in the bifundamental representation $(\bar{N}_i, N_j)$ in the opposite case. Thanks to the Bose-Fermi/Boltzmann correspondence described in §4.1, we can treat all centers as distinguishable and assume that $N_i = 1$ for any i, provided that we attach an effective rational index $\bar{\Omega}_{\rm ref}(\alpha_i, y)$ to each center, and perturb the charge vectors so that none of them coincide. Moreover, since the α_i can be ordered such that $\langle\alpha_i, \alpha_j\rangle \geq 0$ whenever $i < j$, the quiver admits no closed loop.

For arbitrary quivers without loops, Reineke has computed the Poincaré polynomial of the moduli space of quiver representations [36]. In the special case of Abelian quivers ($N_i = 1$), Reineke's formula gives

$$g_{\rm ref}(\{\alpha_i\}, y) = \frac{(-y)^{-\sum_{i<j}\alpha_{ij}}}{(y-1/y)^{n-1}} \sum_{\rm partitions} (-1)^{s-1} y^{2\sum_{a\leq b}\sum_{j<i}\alpha_{ji}\, m_i^{(a)}\, m_j^{(b)}}\,, \tag{29}$$

where the sum runs over all ordered partitions of $\gamma = \alpha_1 + \cdots + \alpha_n$ into s vectors $\beta^{(a)}$ ($1 \leq a \leq s$, $1 \leq s \leq n$) such that

(i) $\beta^{(a)} = \sum_i m_i^{(a)} \alpha_i$ with $m_i^{(a)} \in \{0, 1\}$, $\sum_a \beta^{(a)} = \gamma$

(ii) $\langle \sum_{a=1}^b \beta^{(a)}, \gamma\rangle > 0 \quad \forall \quad b \quad \text{with} \quad 1 \leq b \leq s-1$

We refer to (29) as the 'Higgs branch wall-crossing formula'.

Let us illustrate how this formula works by computing $g_{\rm ref}(\{\alpha_i\}, y)$ for $n = 2, 3$. For $n = 2$ case with $\alpha_{12} > 0$, there are two possible ordered partitions satisfying the conditions stated above:

$$\{\alpha_1 + \alpha_2\}, \quad \{\alpha_1, \alpha_2\}\,.$$

The first term contributes $y^{2\alpha_{12}}$ and the second term contributes -1 to the sum. In total,

$$g_{\rm ref}(\alpha_1, \alpha_2, y) = (-y)^{1-\alpha_{12}}\,(y^2-1)^{-1}\,(y^{2\alpha_{12}}-1) = (-1)^{\alpha_{12}+1}\,\frac{y^{\alpha_{12}} - y^{-\alpha_{12}}}{y - y^{-1}}\,, \tag{30}$$

in agreement with (13). For $n = 3$, assuming the same conditions on α_i as in (14), we find 6 possible ordered partitions

$$\{\alpha_1+\alpha_2+\alpha_3\}, \quad \{\alpha_1, \alpha_2+\alpha_3\}, \quad \{\alpha_1+\alpha_2, \alpha_3\}, \quad \{\alpha_1+\alpha_3, \alpha_2\}, \quad \{\alpha_1, \alpha_2, \alpha_3\}, \quad \{\alpha_1, \alpha_3, \alpha_2\}\,.$$

The second and the last contribution cancel, leaving

$$\begin{aligned} g_{\rm ref}(\alpha_1, \alpha_2, \alpha_3, y) =& (-1)^{\alpha_{12}+\alpha_{13}+\alpha_{23}}\,(y-y^{-1})^{-2} \\ & \Big(y^{\alpha_{12}+\alpha_{13}+\alpha_{23}} - y^{\alpha_{12}-\alpha_{23}-\alpha_{13}} - y^{\alpha_{13}+\alpha_{23}-\alpha_{12}} + y^{-\alpha_{12}-\alpha_{13}-\alpha_{23})}\Big)\,, \end{aligned}$$

in agreement with (14).

6. Conclusion and open problems

In this survey, we have described four apparently different wall-crossing formulae, the KS formula (7), the JS formula (19), the 'Coulomb branch' formula (28) and the 'Higgs branch' formula (29). In [1] we have checked that these formulae agree among each other for $n \leq 5$. Unfortunately, we do not have a mathematical proof of their equivalence at this point. The equivalence of the KS and JS formula appears to be on solid ground [10, 7]. The Reineke formula (or rather, its specialization at $y = 1$) is known to follow from the JS formula [37] (although I do not know of an explicit combinatorial proof). The underlying mathematical structure of these formulae involves Ringel-Hall algebras (and generalizations thereof [38]), which might provide a realization of the long sought-after algebra of BPS states [39, 40].

The Coulomb branch formula (28) appears to be new, and it would be desirable to derive it e.g. from the Higgs branch formula (29). Indeed, the equivalence between these two formulae could be viewed as a toy model of open/closed string duality. The comparison of the Higgs and Coulomb branch formulae suggests that the permutations σ for which $s(\sigma)$ does not vanish are those whose maximal increasing subsequence[5] satisfies the condition ii) below (29), while none of its non-maximal increasing subsequence does, in which case $s(\sigma) = (-1)^{\#\{i;\sigma(i+1)<\sigma(i)\}}$. It would be interesting to derive these conditions by applying Morse theory to the 'superpotential' W in (27).

While the Coulomb and Higgs branch formulae have been derived in the context of $\mathcal{N} = 2$ supergravity, the fact that they appear to agree with the KS and JS formulae suggests that they should work just as well in the context of $\mathcal{N} = 2$ SYM theories. For such field theories, unlike in supergravity, the refined index (or rather a variant thereof, known as the protected spin character, and constructed out of the spatial rotation generator J_3 and a $SU(2)_R$ symmetry generator I_3 [27]) receives contributions from short multiplets only. It would be desirable to derive the jump of the protected spin character by quantizing multi-centered dyonic solutions of the low energy Abelian gauge theory, along the lines of [29, 30], and see if it agrees with the Coulomb formula (presumably this jump will coincide with the equivariant index of the Dirac operator D, without the need to assume that $\mathrm{Ker}D_- = 0$). It would also be interesting to clarify the relation with other realizations of BPS dyons in $\mathcal{N} = 2$ gauge theories based on string webs or non-Abelian monopoles [41, 42, 43, 44, 45].

More generally, localization techniques appear to be a powerful way of quantizing multi-centered black hole solutions at fixed values of the moduli, not only in the vicinity of walls of marginal stability. In contrast to the situation studied here, the phase space $\mathcal{M}_n$ is in general no longer be compact, due to the presence of scaling solutions [13, 46, 47, 14, 48]. We refer the reader to [11] for an application of localization techniques to this problem.

Acknowledgments
I wish to thank J. Manschot and A. Sen for collaboration on the results reported in this note, and I. Bena, D. Joyce, D. Persson, S. El-Showk, M. Vergne for valuable discussions. I am also grateful to the organizers of the AGMP-06 meeting in Tjärnö for the invitation to present these results in a stimulating and scenic environment.

References
[1] Manschot J, Pioline B and Sen A 2011 *JHEP* **1107** 059 (Preprint `1011.1258`)
[2] Seiberg N and Witten E 1994 *Nucl. Phys.* **B426** 19–52 (Preprint `hep-th/9407087`)
[3] Ferrari F and Bilal A 1996 *Nucl. Phys.* **B469** 387–402 (Preprint `hep-th/9602082`)
[4] Andriyash E, Denef F, Jafferis D L and Moore G W 2010, *Wall-crossing from supersymmetric galaxies* (Preprint `1008.0030`)
[5] Seiberg N and Witten E 1994 *Nucl. Phys.* **B431** 484–550 (Preprint `hep-th/9408099`)
[6] Behrend K 2009 *Ann. of Math. (2)* **170** 1307–38
[7] Joyce D and Song Y 2008 *A theory of generalized Donaldson-Thomas invariants*, Memoirs of the AMS, 2011 (Preprint `0810.5645`)

[5] Rather than defining the notion of (maximal) increasing subsequence, we illustrate it on an example: for the permutation $\sigma(1234) = 3142$, the increasing subsequences are $\{\{3\}, \{14\}, \{2\}\}$ and $\{\{3\}, \{1\}, \{4\}, \{2\}\}$, associated to the ordered decompositions $\{\alpha_2, \alpha_1+\alpha_4, \alpha_3\}$ and $\{\alpha_2, \alpha_4, \alpha_1, \alpha_3\}$, respectively. The first increasing subsequence is maximal, the second is not. More details can be found in [1], §3.3.

[8] Douglas M R 2002 *Dirichlet branes, homological mirror symmetry, and stability*, proceedings of ICM 2002 (Preprint math/0207021)
[9] Aspinwall P S 2003, *D-branes on Calabi-Yau manifolds*, talk at TASI 2003 (Preprint hep-th/0403166)
[10] Kontsevich M and Soibelman Y 2008 *Stability structures, motivic Donaldson-Thomas invariants and cluster transformations* (Preprint 0811.2435)
[11] Manschot J, Pioline B and Sen A 2011 *JHEP* **1105** 057 (Preprint 1103.1887)
[12] Denef F 2000 *JHEP* **08** 050 (Preprint hep-th/0005049)
[13] Denef F 2002 *JHEP* **10** 023 (Preprint hep-th/0206072)
[14] Denef F and Moore G W 2007, *Split states, entropy enigmas, holes and halos* (Preprint hep-th/0702146)
[15] Gaiotto D, Moore G W and Neitzke A 2010 *Commun. Math. Phys.* **299** 163–224 (Preprint 0807.4723)
[16] de Boer J, El-Showk S, Messamah I and Van den Bleeken D 2009 *JHEP* **05** 002 (Preprint 0807.4556)
[17] Dimofte T and Gukov S 2010 *Lett. Math. Phys.* **91** 1 (Preprint 0904.1420)
[18] Dimofte T, Gukov S and Soibelman Y 2011 *Lett. Math. Phys.* **95** 1 (Preprint 0912.1346)
[19] Diaconescu E and Moore G W 2007, *Crossing the wall: Branes versus bundles* (Preprint 0706.3193)
[20] Gliozzi F and Tateo R 1995 *Phys. Lett.* **B348** 84–8 (Preprint hep-th/9411203)
[21] Frenkel E and Szenes A 1995 *Math. Res. Lett.* **2** 677–93
[22] Nakanishi T 2010, *Periodicities in cluster algebras and dilogarithm identities*, Proceedings of ICRA XIV, Tokyo (Preprint 1006.0632)
[23] Alexandrov S, Pioline B, Saueressig F and Vandoren S 2009 *JHEP* **03** 044 (Preprint 0812.4219)
[24] Jafferis D L and Moore G W 2008 *Wall crossing in local Calabi Yau manifolds* (Preprint 0810.4909)
[25] Cecotti S and Vafa C 2009 *BPS Wall Crossing and Topological Strings* (Preprint 0910.2615)
[26] Cecotti S, Neitzke A and Vafa C 2010 *R-Twisting and 4d/2d Correspondences* (Preprint 1006.3435)
[27] Gaiotto D, Moore G W and Neitzke A 2010, *Framed BPS states* (Preprint 1006.0146)
[28] Manschot J 2010 *Commun. Math. Phys.* **299** 651–676 (Preprint 0906.1767)
[29] Denef F 1999 *Nucl. Phys.* **B547** 201–220 (Preprint hep-th/9812049)
[30] Lee S and Yi P 2011 *JHEP* **1104** 098 (Preprint 1102.1729)
[31] Bates B and Denef F 2003 *Exact solutions for supersymmetric stationary black hole composites* (Preprint hep-th/0304094)
[32] Atiyah M F and Bott R 1967 *Ann. Math. (2)* **86** 374–407
[33] Atiyah M F and Bott R 1968 *Ann. Math. (2)* **88** 451–91
[34] Berline B and Vergne M 1985 *Amer. J. Math.* **107** 1159–1190
[35] Berline B, Getzler E and Vergne M 1992 *Heat kernels and Dirac operators*, Springer-Verlag.
[36] Reineke M 2003 *Invent. Math.* **152** 349–68
[37] Joyce D 2008 *Adv. Math.* **217** 125–204
[38] Kontsevich M and Soibelman Y 2010, *Cohomological Hall algebra, exponential Hodge structures and motivic Donaldson-Thomas invariants* (Preprint 1006.2706)
[39] Harvey J A and Moore G W 1998 *Commun. Math. Phys.* **197** 489–519 (Preprint hep-th/9609017)
[40] Fiol B and Marino M 2000 *JHEP* **07** 031 (Preprint hep-th/0006189)
[41] Bergman O 1998 *Nucl. Phys.* **B525** 104–116 (Preprint hep-th/9712211)
[42] Lee K M and Yi P 1998 *Phys. Rev.* **D58** 066005 (Preprint hep-th/9804174)
[43] Gauntlett J P, Kim N, Park J and Yi P 2000 *Phys. Rev.* **D61** 125012 (Preprint hep-th/9912082)
[44] Ritz A, Shifman M A, Vainshtein A I and Voloshin M B 2001 *Phys. Rev.* **D63** 065018 (Preprint hep-th/0006028)
[45] Argyres P C and Narayan K 2001 *JHEP* **03** 047 (Preprint hep-th/0101114)
[46] Bena I, Wang C W and Warner N P 2006 *JHEP* **11** 042 (Preprint hep-th/0608217)
[47] Bena I, Wang C W and Warner N P 2008 *JHEP* **07** 019 (Preprint 0706.3786)
[48] de Boer J, El-Showk S, Messamah I and Van den Bleeken D 2010 *JHEP* **02** 062 (Preprint 0906.0011)

New approach for solving Maxwell equations with strong singularity

V A Rukavishnikov and A O Mosolapov

Computing Center of Far-Eastern Branch Russian Academy of Sciences, Kim-Yu-Chen Str. 65, Khabarovsk, 680000, Russia

E-mail: vark@mail.redcom.ru

Abstract. In this note we introduce a notion of R_ν-generalized solution to Maxwell equations with strong singularity in a 2D nonconvex polygonal domain. We develop a new weighted edge FEM. Results of numerical experiments prove the efficiency of this method.

1. Introduction

Maxwell equations are used in mathematical models of electromagnetic fields, for example, in plasma physics, electrodynamics and engineering of high-frequency devices. As a rule, in practical problems the computational domain is nonconvex with reentrant corners or edges on its boundary. Such geometry singularities lead to strong electromagnetic fields in their neighborhood, and a solution of Maxwell equations is strongly singular, i.e. it does not belong to the Sobolev space H^1 (W_2^1).

For numerical solution of such problems P. Ciarlet Jr., F. Assous and others developed the singular complement method (SCM). This method consists in splitting of the space of solutions into a two-term direct sum, where the first term contains the regular part of solution and the second contains the singular part (see [1, 2, 3, 4]).

Another approach to solving Maxwell equations with strongly singular solution was developed by M. Costabel and M. Daug. This method is based on regularization of the initial equation with a weighted divergence term. The resulting elliptic problem considered in a special weighted space is equivalent to the original one (see [5, 6]).

For boundary value problems with strongly singular solutions we developed the theory of numerical methods based on the conception of R_ν-generalized solution (see, for example, [7, 8, 9, 10, 11]. This conception allows us, depending on the singularity of input data (coefficients and right hand sides of equations and boundary conditions) and geometry of the boundary, to define a weighted space or a set containing the solution. It is also possible to build a regularizator which bounds the influence of singularity on the accuracy of numerical solution.

In the present paper we develop the weighted edge finite element method (FEM) based on the conception of R_ν-generalized solution of the Maxwell equations with strong singularity due to a reentrant corner on the boundary. Numerical experiments of model problems showed that the rate of convergence of the numerical solution to the exact one is more than one and a half times better in comparison with the results established in [3, 6, 12]. Another advantage of this method is the simplicity of the solution determination which is an additional benefit for numerical experiments.

2. Notations and preliminaries

We denote two- and three-dimensional Euclidian spaces by R^2 and R^3, respectively, with the Cartesian coordinates $x = (x_1, x_2)$, $x = (x_1, x_2, x_3)$. Let Ω be a bounded simply connected domain in these spaces with the boundary $\partial\Omega$. Let $\mathbf{n}$ be the unit outward normal to $\partial\Omega$. In the case $\Omega \subset R^2$ $\boldsymbol{\tau}$ means a unite tangent vector to $\partial\Omega$. For vector-functions $f : R^3 \to R^3$ or $f : R^3 \times [0, T] \to R^3$ we use the notation $\bar{E}(x) = (E_1(x), E_2(x), E_3(x))$. Vector-functions $f : R^2 \to R^2$ are denoted by boldface latin letters: $\mathbf{E}(x) = (E_1(x), E_2(x))$.

In the 2D case the notations curl and **curl** distinguish between scalar and vector curl operators:

$$\operatorname{curl} \mathbf{E} = \frac{\partial E_2}{\partial x_1} - \frac{\partial E_1}{\partial x_2}, \quad \mathbf{curl}\, f = \left(\frac{\partial f}{\partial x_2}, -\frac{\partial f}{\partial x_1} \right).$$

Let us recall the definitions of some functional spaces associated with Maxwell equations:

$$\begin{aligned} &H_0(\operatorname{curl}, \Omega) = \{\mathbf{E} \in H(\operatorname{curl}, \Omega) \mid \mathbf{E} \cdot \boldsymbol{\tau} = 0,\ x \in \partial\Omega\}, \\ &H(\operatorname{div}, \Omega) = \{\mathbf{E} \in \mathbf{L}_2(\Omega) \mid \operatorname{div} \mathbf{E} \in \mathrm{L}_2(\Omega)\}, \\ &H(\operatorname{div}0, \Omega) = \{\mathbf{E} \in H(\operatorname{div}, \Omega) \mid \operatorname{div} \mathbf{E} = 0\}, \quad V = H_0(\operatorname{curl}, \Omega) \cap H(\operatorname{div}0, \Omega). \end{aligned}$$

Let the domain $\Omega \subset R^2$ have one special point $O(0,0)$ on its boundary. Let $B_\delta(O)$ be an open ball centered at point $O(0,0)$ with radius δ. We introduce a weight function $\rho(x)$ which coincides with the distance to the point $O(0,0)$ in $B_\delta(O)$ and it is equal to δ outside $B_\delta(O)$.

We introduce a weighted Sobolev space $H^k_{2,\nu}(\Omega)$ with the norm

$$\|f\|_{H^k_{2,\nu}(\Omega)} = \left(\sum_{|\lambda| \le k} \int_\Omega \rho^{2(\nu+|\lambda|-k)}(x) \left| D^\lambda f \right|^2 dx \right)^{1/2},$$

where $D^\lambda = \frac{\partial^{|\lambda|}}{\partial x_1^{\lambda_1} \partial x_2^{\lambda_2}}$, ν is nonnegative real, $\lambda = (\lambda_1, \lambda_2)$, λ_1,λ_2 are nonnegative integers and $|\lambda| = \lambda_1 + \lambda_2$. When $k = 0$ $H^0_{2,\nu}(\Omega) = L_{2,\nu}(\Omega)$. For the corresponding spaces of vector-functions we use notations $\mathbf{H}^k_{2,\nu}(\Omega)$, $\mathbf{L}_{2,\nu}(\Omega)$.

A subspace of $\mathbf{H}^k_{2,\nu}(\Omega)$, composed of the vector-functions with zero tangential component on $\partial\Omega$, is denoted by $\overset{\circ}{\mathbf{H}}{}^k_{2,\nu}(\Omega)$:

$$\overset{\circ}{\mathbf{H}}{}^k_{2,\nu}(\Omega) = \{\mathbf{E} \in \mathbf{H}^k_{2,\nu}(\Omega) \mid \mathbf{E} \cdot \boldsymbol{\tau} = 0 \text{ for almost all } x \in \partial\Omega\}.$$

3. Problem statement

Let $\Omega \subset R^3$. We suppose (see [1, 6]) that following conditions are satisfied:

- $\partial\Omega$ is a perfect conductor;
- the physical environment is vacuum;
- free charges are absent;
- electromagnetic field and current density are monochromatic with circular frequency ω:

$$\bar{E}(x,t) = \bar{E}(x) \exp(-i\omega t), \quad \bar{B}(x,t) = \bar{B}(x) \exp(-i\omega t), \quad \bar{J}(x,t) = \bar{J}(x) \exp(-i\omega t).$$

Then the Maxwell equations with appropriate boundary conditions read:

$$\begin{aligned} & i\omega\bar{E} + c^2\mathbf{curl}\,\bar{B} = \frac{1}{\varepsilon_0}\bar{J}, \\ & -i\omega\bar{B} + \mathbf{curl}\,\bar{E} = 0, \\ & \operatorname{div}\bar{E} = 0,\ x \in \Omega, \\ & \bar{E} \times \mathbf{n} = 0, \\ & \bar{B}\cdot\mathbf{n} = 0,\ x \in \partial\Omega, \end{aligned} \tag{1}$$

where $\bar{E}$ and $\bar{B}$ are, respectively, the electric and magnetic fields, $\bar{J}$ is the current density, and c and ε_0 are the speed of light and dielectric permittivity.

In what follows we assume that the data does not depend on variable x_3. As noted in [1], we can decouple the initial system (1) in two systems of unknowns $\bar{E} = (E_1, E_2, 0) = \mathbf{E}$, $\bar{B} = (0, 0, B_3) = B$ (so called TE mode) and $\bar{E} = (0, 0, E_3) = E$, $\bar{B} = (B_1, B_2, 0) = \mathbf{B}$ (TM mode).

Let Ω be a L-shaped domain with one reentrant corner on its boundary. Later in this paper we assume that $\Omega = (-1,1)\times(-1,1)\backslash[0,1]\times[-1,0]$.

Using a standard technique (see [1]), we reduce the original problem for the TE mode to the first boundary value problem of the second order:

$$\mathbf{curl}\,\mathrm{curl}\,\mathbf{E} - k^2\mathbf{E} = \mathbf{f}, \tag{2}$$

$$\operatorname{div}\mathbf{E} = 0,\ x \in \Omega, \tag{3}$$

$$\mathbf{E}\cdot\boldsymbol{\tau} = 0,\ x \in \partial\Omega. \tag{4}$$

It is well known that for a non-convex domain the solution of (2)–(4) belongs to V and does not belong to $\mathbf{H}^1(\Omega)$ ([1, 5, 6]). In other words, problem (2)–(4) has a strongly singular solution.

Let us introduce the notion of R_ν-generalized solution for problem (2)–(4). Denote by

$$a(\mathbf{E},\mathbf{F}) = (\mathrm{curl}\,\mathbf{E}, \mathrm{curl}(\rho^{2\nu}\mathbf{F})) - k^2(\mathbf{E}, \rho^{2\nu}\mathbf{F}), \quad b(\mathbf{F}) = (\mathbf{f}, \rho^{2\nu}\mathbf{F})$$

the bilinear and linear forms, respectively.

Definition 3.1 *A vector-function $\mathbf{E} \in \mathring{\mathbf{H}}^1_{2,\nu}(\Omega)$ is called R_ν-generalized solution of problem (2)–(4) if it satisfies the condition (3) and for any $\mathbf{F} \in \mathring{\mathbf{H}}^1_{2,\nu}(\Omega)$ the identity $a(\mathbf{E},\mathbf{F}) = b(\mathbf{F})$ holds.*

4. Construction of a scheme of the finite element method

We construct a scheme of the weighted edge FEM for the determination of an approximate R_ν-generalized solution of the problem (2)–(4). This method is based on the $H(\mathrm{curl})$-conforming edge FEM proposed by J.-C. Nédélec in 1980-s. Moreover, the finite element subspace is formed by solenoidal basis functions so that condition (3) holds automatically.

We divide $\bar{\Omega}$ into the set $\{K\} = \{K_i\}_{i=1}^{N_h}$ of the closed squares K_i by means of vertical and horizontal straights $x_1 = -1 + jh$, $x_2 = -1 + lh$, $j, l = \overline{0, N}$, where N is a positive even integer, $h = \frac{2}{N}$ and $N_h = \frac{3N^2}{4}$. Square K_i we name as element.

Let $S_1, \ldots, S_{S_h}$ be the set of sides of the squares K which do not belong to $\partial\Omega$,

$$S = \{S_i,\ i = 1, \ldots, S_h,\ S_i \notin \partial\Omega,\ S_h = \frac{3N^2}{2} - 2N\}.$$

We denote by M_i the center of the side S_i, $i = \overline{1, S_h}$.

For every element K_i we denote by $O^{K_i} = (o_1^{K_i}, o_2^{K_i})$ its center, $S_j^{K_i}$ are its sides and $M_j^{K_i} = (m_{1j}^{K_i}, m_{2j}^{K_i})$ are the centers of $S_j^{K_i}$, $j = \overline{1,4}$. The rule of the local numeration of the $S_j^{K_i}$, $j = \overline{1,4}$ is depicted on figure 1.

Journal of Physics: Conference Series **346** (2012) 012018 doi:10.1088/1742-6596/346/1/012018

We introduce form functions of the element K_i associated with its sides:

$$\boldsymbol{\psi}_1^{K_i} = \frac{1}{h}\rho^{-\nu}(m_{11}^{K_i}, x_2)\left(\frac{h}{2} + o_2^{K} - x_2\right)\mathbf{i},$$

$$\boldsymbol{\psi}_2^{K_i} = \frac{1}{h}\rho^{-\nu}(m_{12}^{K_i}, x_2)\left(\frac{h}{2} + x_2 - o_2^{K_i}\right)\mathbf{i},$$

$$\boldsymbol{\psi}_3^{K_i} = \frac{1}{h}\rho^{-\nu}(x_1, m_{23}^{K_i})\left(\frac{h}{2} + o_1^{K_i} - x_1\right)\mathbf{j},$$

$$\boldsymbol{\psi}_4^{K_i} = \frac{1}{h}\rho^{-\nu}(x_1, m_{24}^{K_i})\left(\frac{h}{2} + x_1 - o_1^{K_i}\right)\mathbf{j},$$

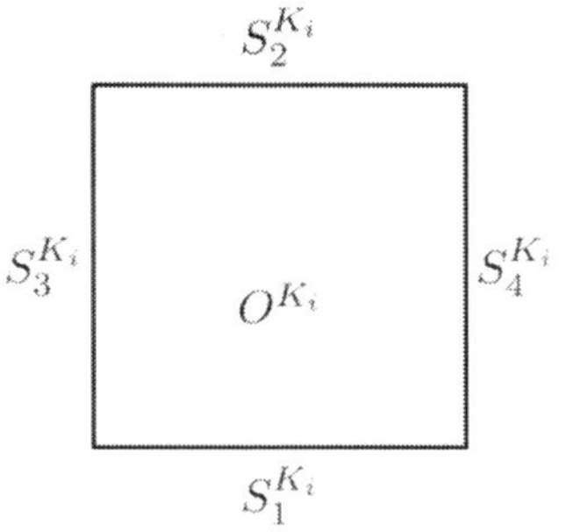

Figure 1. The rule of the local numeration.

where $\mathbf{i}$ and $\mathbf{j}$ are the standard basis vectors of R^2.

For every side $S_i \in S$ we introduce the weighted basis function $\boldsymbol{\psi}_i$ in the following way. Let K_m and K_n be two finite elements with common side S_i, and let $\boldsymbol{\psi}_j^{K_m}$, $\boldsymbol{\psi}_l^{K_n}$ be the form functions associated with side S_i. Then the corresponding basis function is given by

$$\boldsymbol{\psi}_i = \begin{cases} \boldsymbol{\psi}_j^{K_m}, \; x \in K_m, \\ \boldsymbol{\psi}_l^{K_n}, \; x \in K_n, \\ 0, \; x \in \bar{\Omega} \setminus (K_m \cup K_n). \end{cases}$$

Denote by $\mathbf{V}_h$ the linear span of $\{\boldsymbol{\psi}_i\}_{i=1}^{S_h}$. It is obvious that $\mathbf{V}_h \subset \mathring{\mathbf{H}}^1_{2,\nu}(\Omega)$. We shall approximate the R_ν-generalized solution of problem (2)–(4) in this subspace.

Definition 4.1 *A function $\mathbf{E}_\nu^h \in \mathbf{V}_h$ satisfying the identity*

$$a(\mathbf{E}_\nu^h, \mathbf{v}_h) = b(\mathbf{v}_h) \quad \forall \mathbf{v}_h \in \mathbf{V}_h$$

is called the approximate R_ν-generalized solution of problem (2)–(4).

An approximate solution $\mathbf{E}_\nu^h$ will be found in the form

$$\mathbf{E}_\nu^h = \sum_{i=1}^{S_h} d_i \boldsymbol{\psi}_i,$$

where $d_i = \rho^\nu(M_i) g_i$, $g_i = \text{const}$.

The unknowns d_i are defined from the system of equations

$$A\mathbf{d} = \mathbf{b}, \tag{5}$$

where

$$\mathbf{d} = (d_1, \ldots, d_{S_h})^T, \quad A = (A_{ij}), \quad i, j = \overline{1, S_h},$$
$$A_{ij} = a(\boldsymbol{\psi}_i, \boldsymbol{\psi}_j), \quad \mathbf{b} = (b_1, \ldots, b_{S_h})^T, \quad b_i = b(\boldsymbol{\psi}_i).$$

Journal of Physics: Conference Series **346** (2012) 012018 doi:10.1088/1742-6596/346/1/012018

5. Results of numerical experiments

We developed a program "Proba-III" for carrying out a set of numerical tests using our FEM described in section 4 and GMRES-method for solving system (5). The errors of numerical approximations to the R_ν-generalized solution were computed as module between approximate and exact solutions in the points M_i and in the norm of space $\mathbf{L}_{2,\nu}(\Omega)$. We present here the results of numerical experiments for one model problem. To do this, introduce an auxiliary function $\varphi = (x_1x_2(x_1^2-1)(x_2^2-1))^2\left(\sqrt{x_1^2+x_2^2}\right)^{\beta+2}$. Define $\mathbf{E} = \mathbf{curl}\,\varphi$. Substituting $\mathbf{E}$ in equation (2), we get the right hand side $\mathbf{f}$ for the different coefficients (wave numbers) $k_1 = 220$ and $k_2 = 300$ (frequencies are 10.497 GHz and 14.314 GHz respectively). Numerical experiments were carried out on meshes with different step sizes h.

Table 1. Dependence of error $\|\mathbf{E}-\mathbf{E}_\nu^h\|_{\mathbf{L}_{2,\nu}(\Omega)}$ on the number of segments of fragmentation N, $\beta = -6.5$, $\delta = 0.06$, $\nu = 3.5$.

N	32	64	128
$\|\mathbf{E}-\mathbf{E}_\nu^h\|_{\mathbf{L}_{2,\nu}(\Omega)}$, $k=k_1$	0.0001249	0.0000650	0.0000333
$\|\mathbf{E}-\mathbf{E}_\nu^h\|_{\mathbf{L}_{2,\nu}(\Omega)}$, $k=k_2$	0.0001249	0.0000646	0.0000274

Results presented in table 1 are also visually depicted on figure 2.

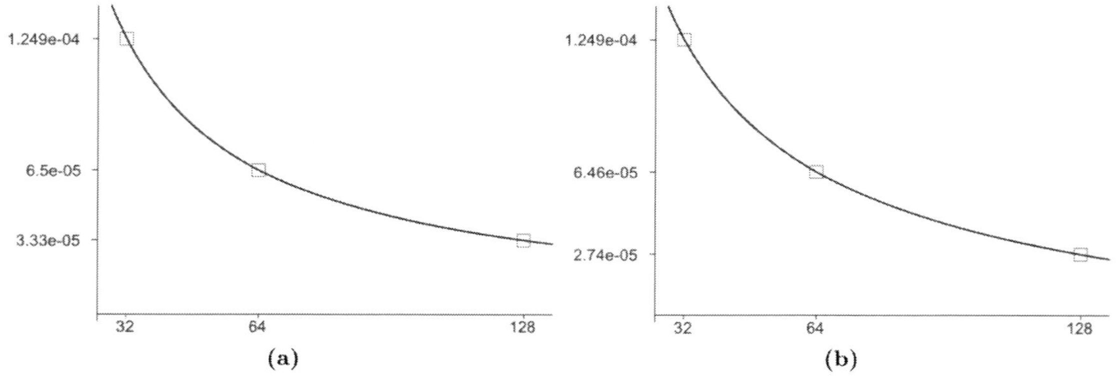

Figure 2. Dependence of error $\|\mathbf{E}-\mathbf{E}_\nu^h\|_{\mathbf{L}_{2,\nu}(\Omega)}$ on the number of segments of fragmentation N, $k = 220$ **(a)**, $k = 300$ **(b)**.

In table 2 for the determined approximate R_ν-generalized solutions we present the numbers of points n_1 and n_2 (in percentage of their total number) where the errors $\delta_{1i} = |E_1(M_i) - E_{1\nu}^h(M_i)|$, $\delta_{2i} = |E_2(M_i) - E_{2\nu}^h(M_i)|$, $i = \overline{1, S_h}$, are less than the given limit value $\Delta = 0.001$.

On figure 3 we depict the distribution of the points M_i with error δ_{1i} for the component $E_{1\nu}^h$ of the approximate R_ν-generalized solution on different meshes. The figures for the component $E_{2\nu}^h$ are analogous.

Journal of Physics: Conference Series **346** (2012) 012018 doi:10.1088/1742-6596/346/1/012018

Table 2. The numbers of points (in percentage of their total number) where the errors are less than the given limit value $\bar{\Delta} = 0.001$, $\beta = -6.5$, $\delta = 0.06$, $\nu = 3.5$.

N	32	64	128
n_1, $k = k_1$	42.93%	63.30%	83.57%
n_2, $k = k_2$	42.93%	69.28%	83.40%

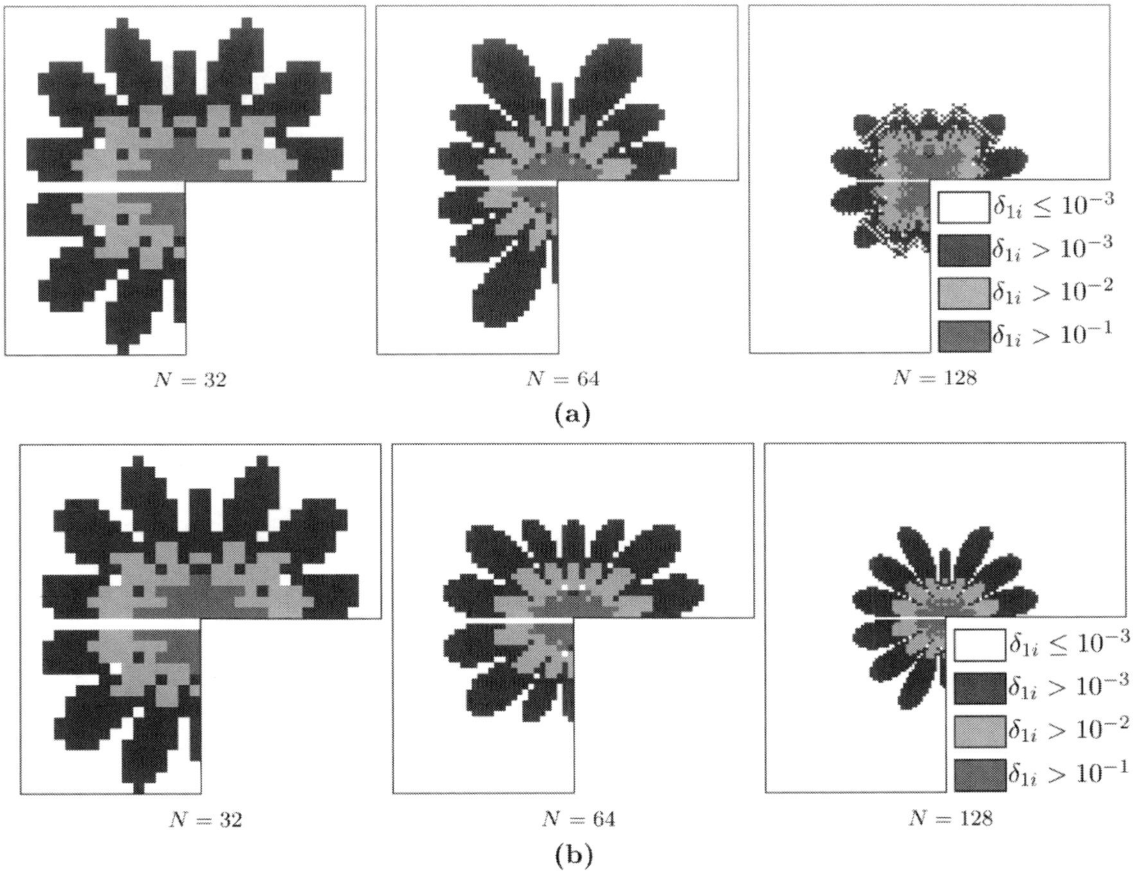

Figure 3. Distribution of the points M_i with error δ_{1i} for the component $E^h_{1\nu}$ of the approximate R_ν-generalized solution, $k = 220$ **(a)**, $k = 300$ **(b)**.

The series of numerical experiments showed that:

(i) approximate solution by a proposed FEM converges to an exact R_ν-generalized solution in the weight space $\mathbf{L}_{2,\nu}(\Omega)$ with the first rate (see table 1 and figure 2), that is more than one and a half times better in comparison with the results obtained with some another methods (see [3], [6] and [12]);

(ii) the number of points where the errors are greater than the given limit value $\bar{\Delta}$ decreases with mesh refinement; the radius of the neighborhood containing these points decreases when h decreases.

Acknowledgments
This work was supported by Russian Foundation of Basic Research (grant 10-01-00060) and by Presidium of Far Eastern Branch of Russ. Acad. Sci. (project 09-II-SB-01-001)

References
[1] Assous F, Ciarlet P and Segré J 2000 Numerical solution to the time-dependent Maxwell equations in two-dimensional singular domain: the singular complement method *J. Comput. Phys.* **161** 218
[2] Assous F, Ciarlet P and Garcia E 2005 Singular electromagnetic fields: inductive approach *C. R. Acad. Sci. Paris, Ser. I.* **341** 605
[3] Assous F, Ciarlet P, Garcia E and Segré J 2006 Time-dependent Maxwell's equations with charges in singular geometries *Comput. Methods Appl. Mech. Eng.* **196** 665
[4] Buffa A, Ciarlet P and Jamelot E 2009 Solving electromagnetic eigenvalue problems in polyhedral domains *Numer. Math.* **113** 497
[5] Costabel M, Daug M and Schwab C 2005 Exponential convergence of hp-FEM for Maxwell's equations with Weighted Regularization in polygonal domains *Mathematical Models and Methods in Applied Sciences* **15** 575
[6] Costabel M and Daug M 2002 Weighted regularization of Maxwell equations in polyhedral domains *Numer. Math.* **93** 239
[7] Rukavishnikov V A 1990 On differentiability properties of an R_ν-generalized solution of the Dirichlet problem *Sov. Math. Dokl.* **40** 653
[8] Rukavishnikov V A and Rukavishnikova H I 1995 The finite element method for the first boundary value problem with compatible degeneracy of the initial data *Russ. Acad. Sci., Dokl. Math.* **50** 335
[9] Rukavishnikov V A and Bespalov A Y 2000 Exponential Convergency Rate in a Finite Element Method for a Dirichlet Problem with Singularity of Solution *Dokl. Math.* **62** 266
[10] Rukavishnikov V A 2009 Methods of numerical analysis for boundary value problem with strong singularity *Russ. J. Numer. Anal. Math. Model.* **24** 565
[11] Rukavishnikov V A and Rukavishnikova H I 2010 The finite element method for a boundary value problem with strong singularity *J. Comput. Appl. Math.* **234** 2870
[12] Jamelot E 2004 A nodal finite element method for Maxwell's equations *C. R. Acad. Sci. Paris, Ser. I.* **339** 809
[13] Jin J M 2002 *The Finite Element Method in Electromagnetics (2nd Edition)* (New York: John Wiley & Sons, Inc.)

Algebra, Geometry, and Mathematical Physics 2010 IOP Publishing
Journal of Physics: Conference Series **346** (2012) 012019 doi:10.1088/1742-6596/346/1/012019

Inverse scattering in guides

Yury Shestopalov[1] **and Yury Smirnov**[2]
[1]Karlstad University, Karlstad, 65188 Sweden
[2]Penza State University, ul. Krasnaya 40, Penza, 440017 Russia
E-mail: youri.shestopalov@kau.se, smirnovyug@mail.ru

Abstract. We present statements and a method of solution to the inverse scattering problem of reconstructing permittivity of a dielectric inclusion in a 2D or 3D waveguide from the transmission characteristics. The approach employs a volume singular integral equation (VSIE) method. The unique solvability of VSIE is established. The inverse problem is solved by the method of iterations applied to VSIE; the convergence of the method is proved

1. Introduction

The methods of reconstructing the shape of the scatterer or its permittivity are developed by Colton and Kress [3] mainly for the cases when the obstacles are supposed to be perfectly conducting or dielectric bodies in two- or three-dimensional space. Among alternative techniques, note an approach proposed by Shestopalov and Lozhechko [11] for cylindrical scatterers whose (two-dimensional) cross sections are formed by domains with infinite noncompact boundaries. The uniqueness for such problems stated in the whole space or in the half-space is proved when the data in the inverse problem of finding the shape of the scatterer or permittivity of the inclusion consist of the far-field patterns of the scattered field given for the plane wave irradiating the obstacle from all directions, and are available for all frequencies varying in a certain interval.

The direct and inverse problems associated with the acoustic and electromagnetic wave propagation in parallel-plane guides were studied by many authors (see e.g. [2], [1], [4], [6], [7] and the bibliography therein). Most of the results were obtained for parallel-plane waveguides. However, when a dielectric body is situated in a waveguide of a bounded cross-section (as in [2] where the so called waveguide method is presented), similar results concerning the unique solvability and efficient solution techniques for the inverse scattering problems of reconstructing permittivity of the scatterer are not available. This fact becomes a driving force of our effort in developing a new approach to the solution of both forward and inverse scattering problems in waveguides. The present work is devoted to the development of the methods of reconstructing permittivity of the scatterer in a 2D-waveguide and a 3D-waveguide of rectangular cross section. We develop solution techniques elaborated in [10], [9], [5], [20], [12], [16], and [17] for the analysis of forward and inverse boundary value problems (BVPs) for Maxwell's and Helmholtz equations associated with the wave propagation in waveguides with dielectric inclusions. The specific methods applied in this study were proposed and justified in [15] on the basis of volume singular integral equation (VSIE) method [8] recently developed in [14], [18], and [13].

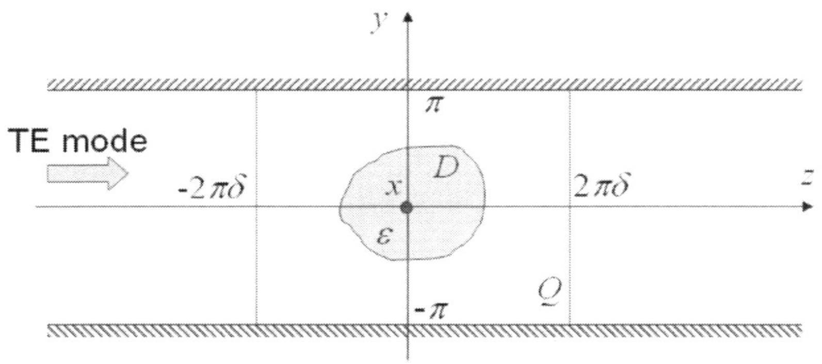

Figure 1. TE-mode diffraction by a dielectric inclusion in a parallel-plane waveguide

2. 2D-guide

2.1. Forward (diffraction) problem

We consider a parallel-plane waveguide $S = \{(y, z) : -\pi < y < \pi, |z| < \infty\}$ containing a nonmagnetic, isotropic, and inhomogeneous dielectric inclusion having the cross section $D \subset \mathcal{Q} = \{(y, z) : -\pi < y < \pi, -2\pi\delta < z < 2\pi\delta\}$ bounded by a piecewise smooth closed contour ∂D (Figure 1), where $\mathcal{Q} \subset S$ denotes the so-called transition domain. The permittivity function $\varepsilon = \varepsilon(y, z)$ is assumed to be such that $\operatorname{supp} m(y, z) \subset \mathcal{Q}$, where $m(y, z) = 1 - \varepsilon(y, z)$. We also assume that $\varepsilon\,(y, z)$ is a complex-valued function of two real arguments y, z piecewise continuously differentiable and bounded in S and denote

$$\varepsilon\,(y, z) = \varepsilon_1\,(y, z) \exp\left[i\varepsilon_2\,(y, z)\right] = g_1\,(y, z) + i\,g_2\,(y, z)\,. \tag{1}$$

In accordance with physical assumptions of the model, the real and imaginary parts of (1) are positive, piecewise continuously differentiable, and bounded satisfying $g_1\,(y, z) \geq 1$, so that the modulus ε_1 and argument ε_2 of the $\varepsilon\,(y, z)$ are also positive functions that are piecewise continuous and bounded on the line and satisfy $0 \leq \varepsilon_2\,(y, z) < \pi/2$ and $\varepsilon_1\,(y, z) \geq 1$.

We further introduce the complex magnitude of the stationary electric and magnetic field, $\mathbf{E}(\mathbf{r}, t)$ and $\mathbf{H}(\mathbf{r}, t)$, respectively, where $\mathbf{r} = (x, y, z)$, and consider the problem of diffraction of the TE mode (assumed to be linearly polarized)

$$\mathbf{E}(\mathbf{r}, t) = \mathbf{E}(\mathbf{r}) \exp\left(-i\omega t\right), \quad \mathbf{H}(\mathbf{r}, t) = \mathbf{H}(\mathbf{r}) \exp\left(-i\omega t\right),$$
$$\mathbf{E}(\mathbf{r}) = (E_x, 0, 0), \quad \mathbf{H}(\mathbf{r}) = \left(0, \frac{1}{i\omega\mu_0}\frac{\partial E_x}{\partial z}, -\frac{1}{i\omega\mu_0}\frac{\partial E_x}{\partial y}\right),$$

by a dielectric inclusion D.

The total field $u(y, z) = E_x\,(y, z) = E_x^{inc}\,(y, z) + E_x^{scat}\,(y, z) = u^i(y, z) + u^s(y, z)$ of the diffraction by the D of the unit-magnitude TE wave with the only nonzero component is the solution to the BVP [10]

$$[\Delta + \kappa^2\varepsilon\,(y, z)]u\,(y, z) = 0 \text{ in } S, \quad u(\pm\pi, z) = 0, \tag{2}$$

$$u(y, z) = u^i(y, z) + u^s(y, z), \quad u^s(y, z) = \sum_{n=1}^{\infty} a_n^{\pm} \exp(i\Gamma_n z) \sin(ny), \tag{3}$$

where $\Delta = \frac{\partial^2}{\partial y^2} + \frac{\partial^2}{\partial z^2}$ is the Laplace operator, superscripts $^+$ and $^-$ correspond, respectively, to the domains $z > 2\pi\delta$ and $z < -2\pi\delta$, $\omega = \kappa c$ is the dimensionless circular frequency, $\kappa = \omega/c = 2\pi/\lambda$ is the dimensionless frequency parameter (λ is the free-space wavelength), $c = (\varepsilon_0 \mu_0)^{-1/2}$ is the speed of light in vacuum, and $\Gamma_n = (\kappa^2 - n^2)^{1/2}$ is the transverse wavenumber satisfying the conditions

$$Im\,\Gamma_n \geq 0, \quad \Gamma_n = i|\Gamma_n|, \quad |\Gamma_n| = Im\,\Gamma_n = (n^2 - \kappa^2)^{1/2}, \quad n > \kappa. \tag{4}$$

It is also assumed that the series in (3) converges absolutely and uniformly and allows for double differentiation with respect to y and z. If $\varepsilon(y,z)$ is a piecewise continuous function, the continuity and transmission conditions [3, Ch. 2], should be added on the discontinuity lines.

Note that $u^i(y,z)$ satisfies (2) in S, the boundary condition, and radiation condition (3) only in the positive direction, so that the electromagnetic field with the x-component $u^i(y,z)$ may be interpreted as *a mode* coming from the domain $z < -2\pi\delta$.

2.2. Integral equation

We solve the (forward) problem (2)—(4) in the transition domain Q. To this end, make use of Green's function (see [9])

$$G_0(y,z;\, y_0,z_0) = \frac{1}{\pi}\sum_{n=1}^{\infty} \exp[i\Gamma_n|z - z_0|]\sin(ny)\sin(ny_0)\Gamma_n^{-1}.$$

of problem (2)—(4) defined at $\varepsilon = 1$ in the domain S.

The IE with respect to the sought for scattered field $u(y,z)$ has the form

$$u(y,z) = -\int_{-2\pi\delta}^{2\pi\delta}\int_{-\pi}^{\pi} G_0(y,z;\, y_0,z_0)\,[1 - \varepsilon(y_0,z_0)]\,u(y_0,z_0)\,dy_0 dz_0 + u^i(y,z), \quad (y,z) \in Q. \tag{5}$$

We have obtained a Lippmann—Schwinger IE which can be written in the operator form $(I - K)u = f$, where I denotes the identity operator. The following statement [15] is valid.

Theorem 1. *Suppose that* $m(y,z) \equiv 1 - \varepsilon(y,z) = 0$ *for* $\sqrt{y^2 + z^2} \geq a$ *with some* $a > 0$ *(so that the inclusion* D *is contained in a circle* B_a^O *of radius* a *centered at the origin) and* $B_a^O \subset S$ *and let* $\kappa^2 < 2/Ma^2$*, where* $M = \max_{B_a^O}|m(y,z)|$*. Then there exists a unique solution to IE (5).*

2.3. Asymptotic representation of the field

If only one TE mode propagates in the waveguide without the inclusion, (that is, $1 < \kappa < 2$), then the solution to (2)—(4) can be represented as a superposition of the reflected (in the domain $z < -2\pi\delta$) and transmitted (in the domain $z > 2\pi\delta$) fields and exponentially decreasing terms.

This form of the solution is crucial for developing a method of solution to both direct and inverse scattering problems.

Namely, under the condition $1 < \kappa < 2$, the asymptotic representation of the total field has the form

$$u(y,z) = \begin{cases} [e^{i\Gamma_1 z} + R(\kappa)\, e^{-i\Gamma_1 z}]\sin(y) + O(e^{|\Gamma_2|z}) = u^i(y,z) + R(\kappa)u^i(y,-z) + O(e^{|\Gamma_2|z}), \\ \qquad z < -2\pi\delta, \\ T(\kappa)\sin(y)e^{i\Gamma_1 z} + O(e^{-|\Gamma_2|z}) = T(\kappa)\,u^i(y,z) + O(e^{-|\Gamma_2|z}), \\ \qquad z > 2\pi\delta, \end{cases}$$

where $R(\kappa)$ and $T(\kappa)$ are, respectively, the reflection and transmission coefficients.

For $z < -2\pi\delta$ the scattered field

$$u^s(y,z) = -\sin(y)\exp(-i\Gamma_1 z)u^{(1)}_{\infty,-}(\kappa) - \sum_{n=2}^{\infty}\frac{i}{\pi}\frac{\exp(|\Gamma_n|z)}{|\Gamma_n|}u^{(n)}_{\infty,-}(y;\kappa)$$

The partial far field patterns

$$u^{(n)}_{\infty,-}(y,\kappa) = \sin(ny)u^{(n)}_{\infty,-}(\kappa), \quad n = 2,3,\ldots$$

where

$$u^{(n)}_{\infty,-}(\kappa) = \int\int_Q \exp(-|\Gamma_n|z_0)\sin(ny_0)m(y_0,z_0)\,u(y_0,z_0)\,dy_0dz_0.$$

An asymptotic expression for the scattered field in terms of partial far field patterns has the form

$$\begin{aligned} u^s(y,z) &= -\sin(y)\exp(i\Gamma_1 z)u^{(1)}_{\infty,+}(\kappa) + O(e^{-|\Gamma_2|z}) \\ &= -u^i(y,z)u^{(1)}_{\infty,+}(\kappa) + O(e^{-|\Gamma_2|z}), z > 2\pi\delta, \\ u^s(y,z) &= -\sin(y)\exp(-i\Gamma_1 z)u^{(1)}_{\infty,-}(\kappa) + O(e^{|\Gamma_2|z}) \\ &= -u^i(y,-z)u^{(1)}_{\infty,-}(\kappa) + O(e^{|\Gamma_2|z}), z < -2\pi\delta. \end{aligned}$$

We will use this representation to justify the method of permittivity reconstruction based on the knowledge of one complex number, the transmission coefficient.

2.4. Statement and a method of solution to the inverse problem of permittivity reconstruction

Statement. Assume that domain D occupied by the dielectric inclusion in the waveguide is fixed and there are given the sets of the partial far field patterns

$$U_{\infty,\pm}(y,\kappa) = U_{\infty,\pm}(y,\kappa;\varepsilon;u) = \left\{u^{(n)}_{\infty,\pm}(y,\kappa)\right\}_{n=1}^{\infty} = \left\{\sin(ny)u^{(n)}_{\infty,\pm}(\kappa)\right\}_{n=1}^{\infty},$$

for all $y \in [-\pi,\pi]$ and (i) for one frequency or (ii) for all values in an interval $\kappa \in \mathrm{K} = (\kappa_1,\kappa_2) \subset (1,2)$. It is necessary to determine permittivity $\varepsilon_1(y,z)$ of the inclusion. Variable y here plays the role of the observation angle of the far field pattern. However, the angle of incidence of the plane wave which enters the far field pattern representation in the case of scattering by an obstacle in the space has no counterpart in this statement. Therefore, it is not possible to apply the solution technique and prove the uniqueness of finding the permittivity of the inclusion directly using the methods developed e.g. in [3].

Solution. From the asymptotic representations

$$u^s_j(y,z) = -\sin(y)\exp(i\Gamma_1 z)u^{(1),j}_{\infty,+}(y,\kappa) - \sum_{n=2}^{\infty}\frac{i}{\pi}\frac{\exp(-|\Gamma_n|z)}{|\Gamma_n|}u^{(n),j}_{\infty,+}(y,\kappa), \quad j = 1,2, \tag{6}$$

for $z > 2\pi\delta$ and

$$u^s_j(y,z) = -\sin(y)\exp(-i\Gamma_1 z)u^{(1),j}_{\infty,-}(y,\kappa) - \sum_{n=2}^{\infty}\frac{i}{\pi}\frac{\exp(|\Gamma_n|z)}{|\Gamma_n|}u^{(n),j}_{\infty,-}(y,\kappa), \quad j = 1,2, \tag{7}$$

for $z < -2\pi\delta$ for the transmitted and reflected fields generated by partial far field patterns $U^j_{\infty,\pm}(y,\kappa)$, it follows that in the single-mode (when the condition $1 < \kappa < 2$ holds) parallel-plane waveguide excited by a unit-amplitude principal mode, the knowledge of two complex numbers $R(\kappa) = u^{(1)}_{\infty,-}(\kappa)$ and $T(\kappa) = u^{(1)}_{\infty,+}(\kappa)$ is virtually sufficient to solve the inverse problem of reconstructing permittivity of an inhomogeneous dielectric inclusion in the guide.

Let us briefly describe the solution technique for the inverse problem under study in the case of a homogeneous inclusion when the permittivity $\varepsilon =$const. Using conditions at infinity (3) and asymptotic representations (6) and (7), we can write

$$u(y,z) = u^i(y,z) + a_1^+ \exp(i\Gamma_1 z)\sin(y) + \sum_{n=2}^{\infty} a_n^{\pm} \exp(i\Gamma_n z)\sin(ny), \tag{8}$$

$$a_1^+ = \iint_Q \exp(-i\Gamma_1 z_0)\sin(y_0)(1-\epsilon(y_0,z_0))u(y_0,z_0)\,dy_0dz_0,$$

$$u(y,z) = T(\kappa)u^i(y,z) + O(e^{-|\Gamma_2|z}). \tag{9}$$

Discarding the exponentially decaying terms in the limit $z \to \infty$, taking into account that $\operatorname{supp}(1-\varepsilon(y,z)) = D$ and

$$\iint_D \exp(-i\Gamma_1 z_0)\sin(y_0)u(y_0,z_0)\,dy_0dz_0 = (u, \overline{u^i}),$$

where brackets denote the inner product in the space $L_2(D)$ and the bar complex conjugation, and equating the left-hand sides of (8) and (9), we obtain

$$(1-\varepsilon)(u,\overline{u^i}) = \hat{T}; \tag{10}$$

here it is assumed that $\hat{T} = T - 1$ is a known (measured) quantity.

If solution u were known, then (10) can be used for the direct computation of the sought-for ε. However, u is a quantity that must be determined. To this end, taking into notice that the integral representation for the scattered field has the same form as IE (5),

$$u(y,z) = -\int_{-2\pi\delta}^{2\pi\delta}\int_{-\pi}^{\pi} G_0(y,z;\,y_0,z_0)\,[1-\varepsilon(y_0,z_0)]\,u(y_0,z_0)\,dy_0dz_0 + u^i(y,z), \quad (y,z) \in S\setminus Q, \tag{11}$$

we substitute (10) into (11) to obtain a nonlinear volume IE with respect to u

$$(u-u^i)\frac{(u,\overline{u^i})}{\hat{T}} = \mathcal{G}u, \quad \mathcal{G}u = \int_D G_0(y,z;\,y_0,z_0)\,u(y_0,z_0)\,dy_0dz_0, \quad (y,z)\in D. \tag{12}$$

Setting $\hat{u} = u - u^i$ and $\hat{T}_0 = \dfrac{\hat{T}}{||u^i||^2}$, we can rewrite nonlinear volume IE (12) in the form suitable for the application of contraction mapping

$$\hat{u} = \mathcal{B}(\hat{u}), \tag{13}$$

$$\mathcal{B}(\hat{u}) = -\frac{\hat{u}(\hat{u},\overline{u^i})}{||u^i||^2} + \hat{T}_0\mathcal{G}\hat{u} + \hat{T}_0\mathcal{G}u^i.$$

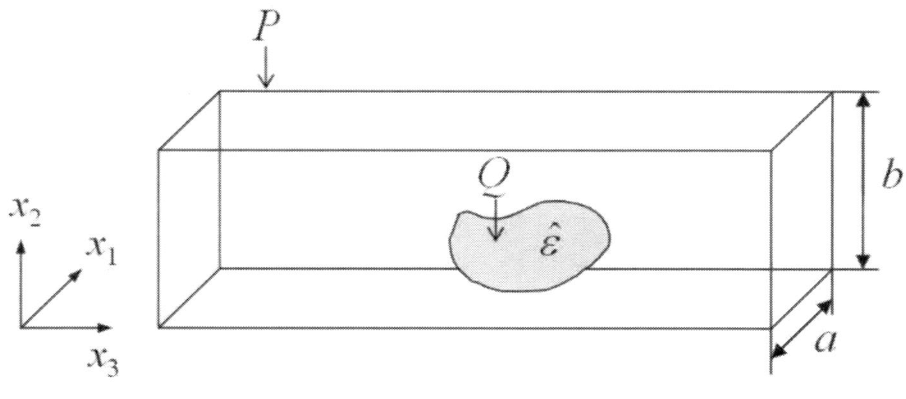

Figure 2. Scattering by a dielectric body in a 3D waveguide

Equation (13) can be solved by the method of fixed-point iterations. The justification of the method, including the proof that operator $\mathcal{B}$ is a contraction and the convergence of the fixed-point iterations based on equation (13), is performed for the 3D case in Section 3.5. In fact, the solution technique for the inverse problem considered in a 3D-guide is of practical importance. We show that when the inclusion is homogeneous (the permittivity is a complex constant), one can determine the permittivity solely from one complex number, the transmission coefficient.

We note that the 2D case is mainly of theoretical character and is studied as the first step towards the practically important 3D case considered in the next sections.

3. 3D-guide

3.1. *Forward (diffraction) problem*

Assume that a waveguide

$$P := \{x : 0 < x_1 < a,\ 0 < x_2 < b,\ -\infty < x_3 < \infty\}$$

with the perfectly conducting boundary surface ∂P is given in the cartesian coordinate system. A three-dimensional body Q ($Q \subset P$ is a domain) with a constant magnetic permeability μ_0 and a positive (3×3) matrix (tensor) permittivity $\hat{\epsilon}(x)$ is placed in the waveguide. The components of $\hat{\epsilon}$ are bounded functions in $\bar{Q}$, $\hat{\epsilon} \in L_\infty(Q)$, and $\hat{\epsilon}^{-1} \in L_\infty(Q)$. The boundary ∂Q of Q is piecewise smooth and Q does not touch the walls of the waveguide.

We look for the electromagnetic field $\mathbf{E}, \mathbf{H} \in L_2^{loc}(P)$ that is induced in the waveguide by the external field with the time dependence $e^{-i\omega t}$. The external field is induced by the electric current $\mathbf{j}_E^0 \in L_2^{loc}(P)$. The differential operators grad, div, and curl are interpreted in the sense of distributions.

We will seek weak (generalized) solutions to Maxwell's system of equations

$$\mathrm{rot}\mathbf{H} = -i\omega\hat{\epsilon}\mathbf{E} + \mathbf{j}_E^0, \quad \mathrm{rot}\mathbf{E} = i\omega\mu_0\mathbf{H}, \tag{14}$$

$$\mathbf{E}_\tau|_{\partial P} = 0, \quad \mathbf{H}_\nu|_{\partial P} = 0, \tag{15}$$

admitting for $|x_3| > C$ and sufficiently large $C > 0$ the representations (+ corresponds to $+\infty$

and $-$ to $-\infty$)

$$\begin{pmatrix}\mathbf{E}\\ \mathbf{H}\end{pmatrix} = \sum_p R_p^{(\pm)} e^{-i\gamma_p^{(1)}|x_3|}\begin{pmatrix}\lambda_p^{(1)}\Pi_p e_3 - i\gamma_p^{(1)}\nabla_2\Pi_p\\ -i\omega\varepsilon_0(\nabla_2\Pi_p)\times e_3\end{pmatrix}+$$
$$+\sum_p Q_p^{(\pm)} e^{-i\gamma_p^{(2)}|x_3|}\begin{pmatrix} i\omega\mu_0(\nabla_2\Psi_p)\times e_3\\ \lambda_p^{(2)}\Psi_p e_3 - i\gamma_p^{(2)}\nabla_2\Psi_p\end{pmatrix}. \tag{16}$$

Here, $\gamma_p^{(j)} = \sqrt{k_0^2 - \lambda_p^{(j)}}$, $\operatorname{Im}\gamma_p^{(j)} < 0$ or $\operatorname{Im}\gamma_p^{(j)} = 0$, $k_0\gamma_p^{(j)} \geq 0$, and $\lambda_p^{(1)}$, $\Pi_p(x_1,x_2)$ and $\lambda_p^{(2)}$, $\Psi_p(x_1,x_2)$ ($k_0^2 = \omega^2\varepsilon_0\mu_0$) are the complete system of eigenvalues and orthogonal and normalized in $L_2(\Pi)$ eigenfunctions of the two-dimensional Laplace operator $-\Delta$ in the rectangle $\Pi := \{x' = (x_1,x_2) : 0 < x_1 < a,\ 0 < x_2 < b\}$ with the Dirichlet and the Neumann conditions, respectively; and $\nabla_2 \equiv e_1\,\partial/\partial x_1 + e_2\,\partial/\partial x_2$.

We assume that $\mathbf{E}^0$ and $\mathbf{H}^0$ are solutions of BVP under consideration in the absence of the nonhomogeneous body Q, $\hat\epsilon(x) = \epsilon_0\hat I$, $x \in P$ ($\hat I$ is the identity tensor):

$$\operatorname{rot}\mathbf{H}^0 = -i\omega\epsilon_0\mathbf{E}^0 + \mathbf{j}_E^0, \quad \operatorname{rot}\mathbf{E}^0 = i\omega\mu_0\mathbf{H}^0, \tag{17}$$

$$\mathbf{E}_\tau^0|_{\partial P} = 0, \quad \mathbf{H}_\nu^0|_{\partial P} = 0. \tag{18}$$

These solutions can be expressed in an analytical form in terms of $\mathbf{j}_\mathbf{E}^\mathbf{0}$ using Green's tensor of domain P. These solutions should not satisfy the conditions at infinity (16). For example, $\mathbf{E}^0$ and $\mathbf{H}^0$ can be TM- or TE-mode of this waveguide.

3.2. Green's tensor function for the waveguide of rectangular cross-section

One can construct explicitly (see [13]) the diagonal Green's tensor $\hat G_E = \operatorname{diag}(G_E^1, G_E^2, G_E^3)$ whose components are the fundamental solutions to the Helmholtz equation in P with the coefficient k_0^2 and satisfy the boundary conditions of the first or second kind on ∂P, which ensure that the tangential components of the intensity of the electric field vanish on the walls of the waveguide. The components of this tensor are

$$G_E^1 = \frac{2}{ab}\sum_{n=0}^{\infty}\sum_{m=1}^{\infty}\frac{1}{(1+\delta_{0n})\gamma_{nm}}\cos(\frac{\pi n}{a}x_1)\sin(\frac{\pi m}{b}x_2)\cos(\frac{\pi n}{a}y_1)\sin(\frac{\pi m}{b}y_2)e^{-\gamma_{nm}|x_3-y_3|},$$

$$G_E^2 = \frac{2}{ab}\sum_{n=1}^{\infty}\sum_{m=0}^{\infty}\frac{1}{(1+\delta_{0m})\gamma_{nm}}\sin(\frac{\pi n}{a}x_1)\cos(\frac{\pi m}{b}x_2)\sin(\frac{\pi n}{a}y_1)\cos(\frac{\pi m}{b}y_2)e^{-\gamma_{nm}|x_3-y_3|},$$

$$G_E^3 = \frac{2}{ab}\sum_{n=1}^{\infty}\sum_{m=1}^{\infty}\frac{1}{\gamma_{nm}}\sin(\frac{\pi n}{a}x_1)\sin(\frac{\pi m}{b}x_2)\sin(\frac{\pi n}{a}y_1)\sin(\frac{\pi m}{b}y_2)e^{-\gamma_{nm}|x_3-y_3|},$$

where δ_{nm} is the Kronecker delta. Here, $\gamma_{nm} = \sqrt{\left(\frac{\pi n}{a}\right)^2 + \left(\frac{\pi m}{b}\right)^2 - k_0^2}$ and the branch of the square root is chosen such that $\operatorname{Re}\gamma_{nm} \geq 0$ and $\operatorname{Im}\gamma_{nm} \geq 0$.

Write G_E^j in the form in which the singularity at $x = y$ is singled out:

$$G_E^j = \frac{1}{4\pi}\frac{e^{ik_0|x-y|}}{|x-y|} + g^j(x,y), \quad x,y \in P.$$

Journal of Physics: Conference Series **346** (2012) 012019 doi:10.1088/1742-6596/346/1/012019

Here, the function $g^j \in C^\infty(\bar{Q} \times \bar{P})$ (see [5, p. 85]). Note that Green's function is not connected with a particular domain Q; our requirement however is that Q cannot touch the walls which yields the smoothness of g^j. Due to the symmetry of Green's functions $G^j_E(x,y) = G^j_E(y,x)$ $(j = 1,2,3)$, we have the following proposition (see [8], [19]).

Proposition. *Green's tensor $\hat{G}_E$ can be represented by*

$$\hat{G}_E = \frac{1}{4\pi}\frac{e^{ik_0|x-y|}}{|x-y|}\hat{I} + \hat{g}(x,y), \quad x, y \in P,$$

where the matrix function (tensor) $\hat{g} \in C^\infty(\bar{Q} \times \bar{P})$ and $\hat{g} \in C^\infty(\bar{P} \times \bar{Q})$.

Note that Green's functions have a single singularity of the form $\frac{1}{4\pi}\frac{e^{ik_0|x-y|}}{|x-y|}$ and have no other singularities due to the assumption that the body does not touch the waveguide walls, i.e., $\bar{Q} \subset P$.

3.3. Volume singular integral equation

Let BVPs (14)–(15) and (17), (18) have unique solutions. Rewrite Eq. (14) in the equivalent form

$$\mathrm{rot}\mathbf{H} = -i\omega\epsilon_0\mathbf{E} + \mathbf{j}_E, \quad \mathrm{rot}\mathbf{E} = i\omega\mu_0\mathbf{H}, \tag{19}$$

where

$$\mathbf{j}_E = \mathbf{j}^0_E + \mathbf{j}^p_E. \tag{20}$$

In the last equation, $\mathbf{j}^p_E = -i\omega(\hat{\epsilon}(x) - \epsilon_0\hat{I})\mathbf{E}$ is the polarization current.

The solution of BVP (19), (15) is given by

$$\mathbf{E} = i\omega\mu_0\mathbf{A}_E - \frac{1}{i\omega\epsilon_0}\mathrm{grad}\ \mathrm{div}\mathbf{A}_E, \quad \mathbf{H} = \mathrm{rot}\mathbf{A}_E \tag{21}$$

where

$$\mathbf{A}_E = \int_P \hat{G}_E(r)\mathbf{j}_E(y)dy, \ r := |x-y|, \tag{22}$$

is the vector potential of the electric current that satisfies the equation

$$\triangle\mathbf{A}_E + k_0^2\mathbf{A}_E = -\mathbf{j}_E.$$

Formulas (21) do not give an explicit solution of problem (19),(15) because the current $\mathbf{j}_E$ depends on $\mathbf{E}$. Relations (20)–(22) for $\mathbf{E}$ imply the integro-differential equation which can be reduced to a VSIE

$$\begin{aligned}\mathbf{E}(x) &+ \frac{1}{3}\left[\frac{\hat{\epsilon}(x)}{\epsilon_0} - \hat{I}\right]\mathbf{E}(x) - v.p.\int_Q \hat{\Gamma}_1(x,y)\left\{\left[\frac{\hat{\epsilon}(y)}{\epsilon_0} - \hat{I}\right]\mathbf{E}(y)\right\}dy \\ &- \int_Q \hat{\Gamma}(x,y)\left\{\left[\frac{\hat{\epsilon}(y)}{\epsilon_0} - \hat{I}\right]\mathbf{E}(y)\right\}dy - \int_Q \hat{\Gamma}_2(x,y)\left\{\left[\frac{\hat{\epsilon}(y)}{\epsilon_0} - \hat{I}\right]\mathbf{E}(y)\right\}dy = \mathbf{E}^0(x),\end{aligned} \tag{23}$$

where the tensors $\hat{\Gamma}$, $\hat{\Gamma}_1$, and $\hat{\Gamma}_2$ are given by

$$\hat{\Gamma}(x,y) = k_0^2\hat{G}_E(r) + (\cdot, \mathrm{grad})\mathrm{grad}G_0(r), \quad \hat{\Gamma}_1(x,y) = (\cdot, \mathrm{grad})\mathrm{grad}G_1(r),$$

$$(\hat{\Gamma}_2(x,y))_{ij} = \frac{\partial^2 g^j(r)}{\partial x_i \partial x_j}, \quad \hat{G}_E(r) = G_0(r)\hat{I} + G_1(r)\hat{I} + \hat{G}_2(r),$$

$$G_0(r) = \frac{e^{ik_0 r} - 1}{4\pi r}, \quad G_1(r) = \frac{1}{4\pi r}, \quad \hat{G}_2(r) = \mathrm{diag}\,\{g^1, g^2, g^3\}.$$

Journal of Physics: Conference Series **346** (2012) 012019 doi:10.1088/1742-6596/346/1/012019

The solvability of Eq. (23) and the equivalence of the boundary value diffraction problem and the singular integral equation are established by the following theorems (see [8], [19]).

Theorem 2. *Suppose the body Q has the piecewise smooth boundary ∂Q and the positive permittivity tensor $\hat{\epsilon} \in L_\infty(Q)$ and $\hat{\epsilon}^{-1} \in L_\infty(Q)$. Suppose that $\mathbf{E}, \mathbf{H}$ and $\mathbf{E}^0$, $\mathbf{H}^0$ are unique solutions of BVPs (14), (15), (17), (18) respectively. Then, the integral equation (23) has a unique solution $\mathbf{E} \in L_2(Q)$. Conversely, if $\mathbf{E} \in L_2(Q)$ is a solution to the integral equation (23), then formulas (20)–(22) give a solution of BVP for Maxwell's equation system (14) satisfying condition (15).*

Rewrite integral equation (23) for the electric field in the form

$$(I + S - K)\,\mathbf{E} = \mathbf{E}^0, \tag{24}$$

where operators S and K are defined according to (23):

$$(S\,\mathbf{E})\,(x) = \frac{1}{3}\left[\frac{\hat{\epsilon}(x)}{\epsilon_0} - \hat{I}\right]\mathbf{E}(x) - v.p.\int\limits_Q \hat{\Gamma}_1(x,y)\left\{\left[\frac{\hat{\epsilon}(y)}{\epsilon_0} - \hat{I}\right]\mathbf{E}(y)\right\}dy,$$

$$(K\,\mathbf{E})\,(x) = \int\limits_Q \hat{\Gamma}(x,y)\left\{\left[\frac{\hat{\epsilon}(y)}{\epsilon_0} - \hat{I}\right]\mathbf{E}(y)\right\}dy + \int\limits_Q \hat{\Gamma}_2(x,y)\left\{\left[\frac{\hat{\epsilon}(y)}{\epsilon_0} - \hat{I}\right]\mathbf{E}(y)\right\}dy\,.$$

Theorem 3. *Assume that homogeneous integral equation (24) has only a trivial solution and the permittivity tensor is such that*

$$ess\ \sup\left[\sum_{l,n=1}^{3}\left|\frac{\epsilon_{ln}(x)}{\epsilon_0} - \delta_{ln}\right|^2\right]^{1/2} < \left(1 + \sqrt{\frac{1}{2}}\right)^{-1}.$$

Then equation (24) is uniquely solvable for any right-hand side $\mathbf{E}^0 \in L_2(Q)$.

3.4. Statement of the inverse problem

Assumptions. Let Q be an isotropic dielectric body and $\hat{\epsilon}(x) = \epsilon$, where the effective permittivity ϵ of the medium filling Q is an unknown constant to be determined Assume that

$$\frac{\pi}{a} < k_0 < \frac{\pi}{b}. \tag{25}$$

Then the only one wave (mode) can propagate in the waveguide because $\operatorname{Im}\gamma_1^{(2)} = 0$,

$$\gamma_1^{(2)} = \sqrt{k_0^2 - \frac{\pi^2}{a^2}} > 0,$$

and $\operatorname{Im}\gamma_p^{(j)} < 0$ for all p, j except for $p = 1$ and $j = 2$. Assume that

$$\mathbf{E}^0(x) = \mathbf{e_2}A^{(+)}i\omega\mu_0\frac{\pi}{a}\sin\frac{\pi x_1}{a}e^{-i\gamma_1^{(2)}x_3}\,.$$

Here $A^{(+)} \neq 0$ is the (known) amplitude of the propagating wave and $\Psi_1 = \cos\dfrac{\pi x_1}{a}$ (see formulas (16)).

Statement of inverse problem. Assume that it is given:

(i) a waveguide P of rectangular cross-section with the perfectly conducting boundary surface ∂P, numbers a, b, and $k_0 = \omega\sqrt{\mu_0\epsilon_0}$ satisfy the condition (25);
(ii) a homogeneous three-dimensional dielectric inclusion, a body Q, which is placed in the waveguide, does not touch its walls ($Q \subset P$ is a domain), and has a constant magnetic permeability μ_0 (given) and a constant (unknown) permittivity ϵ;
(iii) a positive number ω, frequency of the initial electromagnetic field $\mathbf{E}^0(x)$ and two (complex) numbers $Q_1^{(+)}$ and $A^{(+)}$, the amplitudes of transmitted and incident waves, respectively.

It is necessary to determine (complex) permittivity ϵ of dielectric inclusion Q.

3.5. Solution to the inverse problem

From (25), it follows that $G_E^1 \to 0$ and $G_E^3 \to 0$ uniformly with respect to $y \in Q$ for $x_3 \to +\infty$. We have

$$G_E^2 - \frac{1}{ab\gamma_{10}} \sin\frac{\pi x_1}{a} \sin\frac{\pi y_1}{a} e^{-i\gamma_1^{(2)}|x_3-y_3|} \to 0$$

uniformly with respect to $y \in Q$ for $x_3 \to +\infty$. Also $\operatorname{div}\hat{G_E} \to 0$ uniformly with respect to $y \in Q$ for $x_3 \to +\infty$ because this condition holds for $\dfrac{\partial G_E^2}{\partial x_2}$.

Calculating the limit for $x_3 \to +\infty$ in the representation formula for the electric field

$$\mathbf{E}(x) = \mathbf{E}^0(x) + k_0^2 \int_Q \hat{G}_E \left[\frac{\hat{\epsilon}(y)}{\epsilon_0} - \hat{I}\right] \mathbf{E}(y)dy + \operatorname{grad}\operatorname{div} \int_Q \hat{G}_E \left[\frac{\hat{\epsilon}(y)}{\epsilon_0} - \hat{I}\right] \mathbf{E}(y)dy, \quad x \in P\backslash Q, \tag{26}$$

and taking into account the condition at infinity (16) we have

$$\begin{aligned} \mathbf{e_2} Q_1^{(+)} e^{-i\gamma_1^{(2)}x_3} i\omega\mu_0 \frac{\pi}{a} \sin\frac{\pi x_1}{a} &= \mathbf{e_2} A^{(+)} e^{-i\gamma_1^{(2)}x_3} i\omega\mu_0 \frac{\pi}{a} \sin\frac{\pi x_1}{a} + \\ &+ \left(\frac{\epsilon}{\epsilon_0} - 1\right) \frac{k_0^2 \mathbf{e_2}}{ab\gamma_{10}} \int_Q \sin\frac{\pi x_1}{a} \sin\frac{\pi y_1}{a} e^{-i\gamma_1^{(2)}(x_3-y_3)} E_2(y)dy\,, \end{aligned}$$

which yields

$$Q_1^{(+)} = A^{(+)} + k_0^2 \left(\frac{\epsilon}{\epsilon_0} - 1\right) \frac{1}{b\gamma_{10} i\pi\omega\mu_0} \int_Q \sin\frac{\pi y_1}{a} e^{i\gamma_1^{(2)}y_3} E_2(y)dy\,. \tag{27}$$

We can rewrite (27) as

$$\frac{\epsilon}{\epsilon_0} - 1 = \frac{C}{(\mathbf{E},\, \mathbf{f})}, \tag{28}$$

where

$$C = \frac{i\pi\omega\mu_0 b\gamma_{10}(Q_1^{(+)} - A^{(+)})}{k_0^2}, \qquad \mathbf{f} = \mathbf{e_2} \sin\frac{\pi y_1}{a} e^{-i\gamma_1^{(2)}y_3}, \tag{29}$$

$$(\mathbf{E},\, \mathbf{f}) = \int_Q \mathbf{E}(y)\overline{\mathbf{f}(y)}dy\,.$$

$Q_1^{(+)}$ is a known (measured) quantity. Therefore, C is also a known quantity.

Journal of Physics: Conference Series **346** (2012) 012019 doi:10.1088/1742-6596/346/1/012019

Substituting (28) and (29) into the representation formula (26) for the electric field (considered at $x \in Q$) we obtain a nonlinear volume IE

$$\frac{(\mathbf{E},\, \mathbf{f})}{C}(\mathbf{E}(x) - \mathbf{E}^0(x)) = k_0^2 \int\limits_Q \hat{G_E}(x,y)\mathbf{E}(y)dy + \operatorname{grad}\operatorname{div} \int\limits_Q \hat{G_E}(x,y)\mathbf{E}(y)dy, \quad x \in Q. \quad (30)$$

Introduce a linear integral operator

$$\mathcal{T}_0\mathbf{E} = k_0^2 \int\limits_Q \hat{G_E}(x,y)\mathbf{E}(y)dy + \operatorname{grad}\operatorname{div} \int\limits_Q \hat{G_E}(x,y)\mathbf{E}(y)dy \quad (31)$$

and use (31) to rewrite equation (30)

$$\frac{(\mathbf{E},\, \mathbf{f})}{C}(\mathbf{E} - \mathbf{E}^0) = \mathcal{T}_0\mathbf{E}.$$

Performing some normalizations, we can further rewrite (30) in the form

$$\mathbf{U} = \mathcal{T}(\mathbf{U}) \equiv (-\mathbf{U}, \tilde{\mathbf{f}})\mathbf{U} + \tilde{\mathbf{f}}_0 + \hat{A}_0\mathbf{U}, \qquad \mathbf{U} = \tilde{\mathbf{E}} - \mathbf{f}. \quad (32)$$

Here

$$\tilde{\mathbf{f}} = \frac{\mathbf{f}}{||\mathbf{f}||^2}, \quad \tilde{\mathbf{f}}_0 = \frac{\mathbf{f}_0}{||\mathbf{f}||^2}, \quad \mathbf{f}_0 = \tilde{A}_0\mathbf{f}, \quad \hat{A}_0 = \frac{\tilde{A}_0}{||\mathbf{f}||^2},$$

$$\tilde{A}_0 = A_0\frac{C}{\tilde{A^{(+)}}}, \qquad \tilde{A^{(+)}} = A^{(+)}i\omega\mu_0\frac{\pi}{a}$$

Let $\overline{S_r(0)}$ be a closed ball of a radius r centered at 0 in the space $L_2(Q)$. Assume that $\mathbf{U} \in \overline{S_r(0)}$; then the following estimate is valid

$$||\mathcal{T}(\mathbf{U})|| \le \left(r||\tilde{\mathbf{f}}|| + ||\hat{A}_0||\right) r + ||\tilde{\mathbf{f}}_0||, \quad \forall \mathbf{U} \in \overline{S_r(0)}.$$

If

$$\left(r||\tilde{\mathbf{f}}|| + ||\hat{A}_0||\right) r + ||\tilde{\mathbf{f}}_0|| \le r, \quad (33)$$

then operator $\mathcal{T}$ acts from the ball $\overline{S_r(0)}$ to ball $\overline{S_r(0)}$.

Now, assuming that $\mathbf{U}, \mathbf{V} \in \overline{S_r(0)}$ and using the Cauchy–Bunyakovski inequality, we obtain

$$||\mathcal{T}(\mathbf{U}) - \mathcal{T}(\mathbf{V})|| \le \left(2r||\tilde{\mathbf{f}}|| + ||\hat{A}_0||\right) ||\mathbf{U} - \mathbf{V}||, \quad \forall \mathbf{U}, \mathbf{V} \in \overline{S_r(0)}.$$

Thus, if

$$2r||\tilde{\mathbf{f}}|| + ||\hat{A}_0|| < 1, \quad (34)$$

operator $\mathcal{T}(\mathbf{U})$ is a contraction.

Choose radius r of the ball so that both conditions (33) and (34) hold (below we will show when it is possible). Taking into account the equivalence of transformations leading from equation (30) to equation (32) and applying the contraction principle we arrive at the main results proved in [13] and [14].

Theorem 4. *Assume that conditions (33) and (34) hold. Then equation (32)* $\mathbf{U} = \mathcal{T}(\mathbf{U})$ *(and, respectively, equation (30)) is uniquely solvable. Also, there exists one and only one solution to the inverse problem given by formula (28).*

$$\epsilon = \epsilon_0 \left(1 + \frac{C}{(\mathbf{E}, \mathbf{f})} \right). \tag{35}$$

Approximate solutions to equation (32) can be determined using the iterations $\mathbf{U}_{n+1} = \mathcal{T}(\mathbf{U}_n)$ *which converge for any initial approximation* $\mathbf{U}_0 \in \overline{S_r(0)}$ *with the rate of a geometrical progression.*

Theorem 5. *Assume that the condition*

$$|\tilde{C}| < F = ||\mathbf{f}||^2 \frac{1 - \frac{2r^*}{||\mathbf{f}||}}{||\mathcal{T}_0||} \tag{36}$$

holds. Then VSIE (30) is uniquely solvable. Also, there exists one and only one solution to the inverse problem given by formula (35). Approximate solutions $\mathbf{E}_n \in \overline{S_{r^0}(\mathbf{E}^0)}$ *to equation (30) can be determined using the iterations*

$$\mathbf{E}_{n+1} = \mathbf{E}_n - \frac{1}{\tilde{A^{(+)}}||\mathbf{f}||^2} \left[(\mathbf{E}_n, \mathbf{f}) \left(\mathbf{E}_n - \mathbf{E}^0 \right) - C(\mathcal{T}_0 \mathbf{E}_n) \right], \tag{37}$$

which converge to exact solution $\mathbf{E} \in \overline{S_{r^0}(\mathbf{E}^0)}$ *for any initial approximation* $\mathbf{E}_0 \in \overline{S_{r^0}(\mathbf{E}^0)}$ *with the rate of a geometrical progression, where* $r^0 = |\tilde{A^{(+)}}| r^*$ *and* r^* *is determined according to the formula*

$$r^* = \left(\frac{||\mathcal{T}_0 \mathbf{f}||^2}{||\mathcal{T}_0||^2} + \frac{||\mathcal{T}_0 \mathbf{f}|| ||\mathbf{f}||}{||\mathcal{T}_0||} \right)^{1/2} - \frac{||\mathcal{T}_0 \mathbf{f}||}{||\mathcal{T}_0||}.$$

Condition (36) is fulfilled if the quantity $|Q_1^{(+)} - A_1^{(+)}|$ is sufficiently small. From the physical viewpoint, it means that the amplitude of the transmitted wave does differ substantially from the amplitude of the incident wave.

Consider the iteration procedure (37) for the solution of the inverse problem under study; that is, for the solution of integral equation (30). One has to calculate, at $n = 0, 1, 2, \ldots$, the image of integral operator $\mathcal{T}_0$ defined by (31); the calculation algorithm is proposed and justified by Shestopalov and Smirnov [13] and Shestopalov, Smirnov and Mironov [14] using the VSIE technique [8]. As an initial approximation, one can choose $\mathbf{E}_0 = \mathbf{E}^0$. After solving iteratively equation (30) by (37) with a prescribed accuracy, one can find the sought-for permittivity ϵ by formulas (28), (29).

Acknowledgments

This work was partially supported by the Swedish Institute within the frames of the Visby Program.

References

[1] Buchanan J, Gilbert R, Wirgin A and Xu Y 2004 *Marine Acoustics. Direct and Inverse Problems* (Philadelphia: Society for Industrial and Applied Mathematics)
[2] Chen H, Zhang J, Bai Y, Luo Y, Ran L, Jiang Q and Kong J-A 2006 Experimental retrieval of the effective parameters of metamaterials base don a waveguide method *Opt. Express* **14** 12944–9
[3] Colton D and Kress R 1998 *Inverse Acoustic and Electromagnetic Scattering Theory* (Berlin: Springer-Verlag)

[4] Ikehata M, Makrakis G and Nakamura G 2004 Inverse boundary value problem for ocean acoustics using point sources *Math. Methods Appl. Sci.* **27** 1367–84
[5] Ilinski A and Smirnov Y 1998 *Electromagnetic Wave Diffraction by Conducting Screens* (VSP Int. Science Publishers: Utrecht, Boston, Köln, Tokyo)
[6] Nakamura G and Sini M 2004 On the near field measurement for the inverse scattering problem for ocean acoustics *Inverse Problems* **20** 1387–92
[7] Ramm A 1992 *Multidimensional Inverse Scattering Problems* (New York: John Wiley)
[8] Samokhin A 2001 *Integral Equations and Iteration Methods in Electromagnetic Scattering* ed Y Shestopalov (VSP Int. Science Publishers: Utrecht, Boston, Köln, Tokyo)
[9] Shestopalov V and Shestopalov Y 1996 *Spectral Theory and Excitation of Open Structures* (London: Peter Peregrinus)
[10] Shestopalov V and Sirenko Y 1989 *Dynamical Theory of Gratings* (Kiev: Naukova Dumka)
[11] Shestopalov Y and Lozhechko V 2003 Direct and inverse problems of the wave diffraction by screens with arbitrary finite inhomogeneities *J. Inverse Ill-Posed Problems* **11** 643–53
[12] Shestopalov Y, Kobayashi K and Smirnov Y 2009 Investigation of electromagnetic diffraction by a dielectric body in a waveguide using the method of volume singular integral equation *SIAM J. Appl. Maths.* **70** 969–83
[13] Shestopalov Y and Smirnov Y 2010 Existence and uniqueness of solution to the inverse problem of complex permittivity reconstruction of a dielectric body in a waveguide *Inverse Problems* **26** 105002
[14] Shestopalov Y, Smirnov Y and Mironov D 2010 Analysis of Inverse Scattering in a Waveguide Using the Method of Volume Singular Integral Equation *2010 International Symposium on Electromagnetic Theory EMTS 2010, Berlin, Germany, August 15–9, 2010* pp 532-534
[15] Shestopalov Y, Smirnov Y and Yakovlev V 2008 Volume singular integral equation method for determination of permittivity of meta- and nano materials *Proc. Progress in Electromagnetics Research Symposium July 2–6, 2008* (Cambridge MA) 291–2
[16] Smirnov Y 2009 Inverse boundary value problem for determination of permittivity of a dielectric body in a waveguide using the method of volume singular integral equation *IEEJ Fundamentals and Materials* **129** 675–80
[17] Smirnov Y 2008 Application of GRID technologies to the solution of a nonlinear volume singular integral equation for the determination of effective permittivity of nanomaterials *Izv. Vyssh. Uchebn. Zav. Povolzhsk. Region* **3** 39–55
[18] Smirnov Y, Shestopalov Y and Yakovlev V 2010 A Volume Singular Integral Equation Technique for Reconstruction of Complex Per- mittivity of a Body in a Rectangular Waveguide *26th International Review of Progress in Applied Computational Electromagnetics ACES 2010, Tampere, Finland, April 25-29, 2010* pp 426–430
[19] Smirnov Y and Tsupak A 2004 Investigation of electromagnetic diffraction by a dielectric body using the volume singular integral equation *Comp. Maths. Math. Phys.* **44** 2143–58
[20] Smirnov Y, Schürmann H W and Shestopalov Y 2004 Integral equation approach for the propagation of TE-waves in a nonlinear dielectric cylindrical waveguide *J. Nonlin. Math. Phys.* **11** 256–68

Some formulas for Legendre functions related to the Poisson transform and Lorentz group representation

I A Shilin[1,2] **and A I Nizhnikov**[3]

[1] Department of Higher Mathematics, Sholokhov Moscow State University for the Humanities, V. Radishevskaya 16 – 18, Moscow 109240, Russia
[2] Department of Mathematical Simulation, Moscow Aviation Institute , Volokolamskoe shosse 4, Moscow 125993, Russia

E-mail: ilyashilin@li.ru
[3] Department of Mathematical Physics, Moscow Pedagogical State University, M. Pirogovskaya 1, Moscow 119991, Russia

E-mail: tensit@mail.ru

Abstract. We consider the representation of Lorentz group in the spaces of homogeneous functions on cone and hyperboloid. Using the Poisson transform, which maps a function from the first space into the corresponding function from the second space, we derive some formulas containing Legendre functions. Using the restriction of the representation in the first space on a one-parameter subgroup of hyperbolic rotations, we obtain another formula for Legendre functions. All obtained formulas are new.

1. Introduction

Let the linear space $\mathbb{R}^{n+1}$ is endowed with the quadratic form $q(x) := x_0^2 - x_1^2 - \ldots - x_n^2$ and $\hat{q}$ be a polar bilinear form for q. We will deal with two kinds of orbits of the Lorentz group $SO(n,1)$. One of them is cone $O_0 := \{x : q(x) = 0\}$. The second kind of orbits consists of two-sheet hyperboloids $O_r := \{q(x) = r\}$ for any $r > 0$.

Let G be a connected component of $SO(n,1)$, which contains the identical element. For any $\sigma \in \mathbb{C}$, let $\mathfrak{C}_\sigma$ be a linear subspace in $C^\infty(O_0)$ consisting of σ-homogeneous functions on the O_1. It is useful to suppose throughout this paper that $1-n < \operatorname{re}\sigma < 0$. We define the G-representations T_σ in $\mathfrak{C}_\sigma$ and T_σ in $\mathfrak{D}_\sigma$ by left shifts: $T_\sigma(g)[f(x)] := f(g^{-1}x)$ and $S_\sigma(g)[f(x)] := f(g^{-1}x)$.

Suppose that γ is a contour on O_0 intersecting all generatrices. Every point $x \in O_0$ can be represented as $x_i = \{tF_i(\xi_1, \ldots, \xi_{n-1}),\ i = 0, 1, \ldots, n$, where $\xi_1, \ldots \xi_{n-1}$ are parameters of the contour γ. For the subgroup H_γ of G which acts transitively on γ, we have$dx = t^{n-2}\,\mathrm{d}t\,\mathrm{d}\gamma$, where $\mathrm{d}x$ is a G-invariant measure on O_0 and $\mathrm{d}\gamma$ is a H_γ-invariant measure on γ.

For any pair $(\mathfrak{C}_\sigma, \mathfrak{C}_{\sigma^*})$, we define the bilinear functionals

$$\mathsf{F}_\gamma : (\mathfrak{C}_\sigma, \mathfrak{C}_{\sigma^*}) \longrightarrow \mathbb{C},\ (f_1, f_2) \longmapsto \int_\gamma f_1(x)\, f_2(x)\, \mathrm{d}\gamma.$$

The functional F_γ does not depend on γ if $\sigma^* = 1 - n - \sigma$ because of the homogeneity of f_1 and f_2. It follows from the relation between $\mathrm{d}x$ and $\mathrm{d}\gamma$.

Let $f \in \mathfrak{C}_\sigma$ and $y \in O_1$. The integral map $\Pi(f)(y) := \mathsf{F}_\gamma(f, (\hat{q}(x,y))^{1-n-\sigma})$ is said to be the Poisson transform (see, for instance, page 194 in [5]). This map intertwines the representations T_σ and S_σ, moreover, $\operatorname{Im}\Pi \subset \operatorname{Ker}\Box_{1n}$, where $\Box_{n1} = \frac{\partial^2}{\partial x_0^2} - \sum_{i=1}^{n} \frac{\partial^2}{\partial x_i^2}$ is the Laplace – Beltrami operator (see, e.g. [3]). The asymptotical properties of the Poisson transform was researched in [1]. The aim of our paper is to use the above transform for the obtaining of formulae for Legendre functions.

Let us consider the spherical, parabolic and hyperbolic contours on O_0. Each of them intersects any or almost any generatrix. Let γ_1 be the intersection of cone O_0 and plane $x_0 = 1$. Then $H_{\gamma_1} \simeq SO(n)$ and $\mathrm{d}\gamma_1 = \frac{\mathrm{d}x_{\zeta(1)}\ldots\mathrm{d}x_{\zeta(n-1)}}{|x_{\zeta(n)}|}$, where $\zeta \in \mathbf{S}_n$(symmetric group). We denote the intersection of O_0 and plane $x_0 + x_n = 1$ by symbol γ_2. The subgroup H_{γ_2} consists of matrices

$$n(b) = \begin{pmatrix} \operatorname{diag}(\underbrace{1,\ldots,1}_{n-1}) & & b^{\mathrm{T}} & b^{\mathrm{T}} \\ & -b & 1-b^* & -b^* \\ & b & b^* & b^* \end{pmatrix},$$

where $b = (b_1, \ldots, b_{n-1})$ and $b^* = \frac{1}{2}(b_1^2 + \ldots + b_{n-1}^2)$. Here $\mathrm{d}\gamma_2 = \mathrm{d}x_1 \ldots \mathrm{d}x_{n-1}$. Let $\gamma_{3,\pm}$ be the intersection of cone O_0 and plane $x_0 + x_n = 1$. If $\gamma_3 = \gamma_{3+} \cup \gamma_{3-}$, then $H_{\gamma_3} = SO(n-1,1)$ and

$$\mathrm{d}\gamma_3 = \frac{\mathrm{d}x_0\,\mathrm{d}x_{\zeta(1)}\ldots\mathrm{d}x_{\zeta(n-2)}}{|x_{\zeta(n-1)}|},$$

where $\zeta \in \mathbf{S}_{n-1}$.

For any contour, we define a basis in the space of restrictions of functions $f \in \mathfrak{C}_\sigma$ to this contour. This basis can be continued on O_0 via a homogeneity. Finally, we have the following bases (named as spherical, parabolic, and hyperbolic basis, respectively): $\{f_{k,K} : k \geq k_1\}$, $\{f_{\lambda,L} : \lambda \geq 0\}$, $\{f_{\mu,M,\pm} : \mu \in \mathbb{R}\}$, where the short notation T denotes $T = (t_1, \ldots, t_{n-2})$, integers t_i satisfy the conditions $t_i \geq t_{i+1} \geq |t_{n-2}|$, and

$$f_{k,K}(x) = N_{(k,K)}\, x_0^{\sigma-k}\, \Xi_{(k,K)}(x),$$

$$f_{\lambda,L}(x) = N^*_{\lambda,L}\,(x_0+x_n)^{\sigma-\frac{n-3}{2}} \left(\frac{\lambda}{2}\right)^{l_1} \left(\frac{\lambda r_{n-1}}{2}\right)^{\frac{3-n}{2}-l_1} J_{\frac{n-3}{2}+l_1}\left(\frac{\lambda r_{n-1}}{x_0+x_n}\right) \Xi_L(x),$$

$$f_{\mu,M,\pm}(x) = N^{**}_{\mu,M}\,(x_n)_\pm^{\sigma+\frac{n-3}{2}}\, r_{n-1}^{\frac{3-n}{2}-m_1}\, P^{\frac{3-n}{2}-m_1}_{-\frac{1}{2}+\mathrm{i}\mu}\left(\frac{x_0}{x_n}\right) \Xi_M(x),$$

$$\Xi_{(t_1,\ldots,t_s)}(x) = \prod_{i=1}^{s-3} r_{s-i}^{t_i-t_{i+1}}\, C^{\frac{s-i-2}{2}+t_{i+1}}_{t_i-t_{i+1}}\left(\frac{x_{s-i}}{r_{s-i}}\right)(x_2+\mathbf{i}x_1)^{t_{s-2}},$$

where [4]

$$t_1 \geq \ldots \geq t_{s-3} \geq |t_{s-2}| \geq 0, \quad r_j^2 = x_1^2 + \ldots + x_j^2,$$

and

$$(x_n)_\pm^\omega = \begin{cases} |x_n|^\omega, & \text{for } x \in \gamma_{3\pm}, \\ 0, & \text{for } x \in \gamma_{3\mp}, \end{cases}$$

Above, $J_{\frac{n-3}{2}+l_1}\left(\frac{\lambda r_{n-1}}{x_0+x_n}\right)$ is the Bessel function, $P^{\frac{3-n}{2}-m_1}_{-\frac{1}{2}+\mathrm{i}\mu}\left(\frac{x_0}{x_n}\right)$ is the Legendre function, $C^{\frac{s-i-2}{2}+t_{i+1}}_{t_i-t_{i+1}}\left(\frac{x_{s-i}}{r_{s-i}}\right)$ is the Gegenbauer polynomial, and $N_{(k,K)}, N^*_{\lambda,L}, N^{**}_{\mu,M}$ are normalizing constants (the exact description of this constants follows from the same formulas 9.4.1.7, 11.3.3.5, 10.3.4.9 in [4]). Omitting all details, let us note, that each of the above bases consists of the associated spherical functions of the subrepresentation to the subgroup H_{γ_i} (respectively).

2. Formulae induced by the Poisson transform

Let us consider the case $SO(3,1)$ of Lorentz group. Moving to this group, let us note that length of the multi-indexes K, L, M is equal to 1, therefore, $K \equiv k_1$ and so on. We have the distribution

$$f_{\lambda,l_1}(x) = \sum_{k=0}^{\infty} \sum_{k_1=-|k|}^{|k|} c_{\lambda,l_1,k,k_1} f_{k,k_1}(x).$$

If $x \in \gamma_1$, then $x = (1, \sin\alpha \sin\beta, \sin\alpha\cos\beta, \cos\alpha)$, where $\alpha \in [0;\pi)$ and $\beta \in [0;2\pi)$. For any $x \in \gamma_2$, we can write $x = \left(\frac{1+u^2}{2}, u\sin\iota, u\cos\iota, \frac{1-u^2}{2}\right)$. Here $u \geq 0$, $\iota \in [0;2\pi)$. We will deal with the restrictions

$$f_{k,k_1}(\alpha,\beta) = C_{k-k_1}^{k_1+\frac{1}{2}}(\cos\alpha)\,\sin^{k_1}\alpha\,\mathrm{e}^{\mathrm{i}k_1\beta},$$
$$f_{\lambda,l_1}(u,\iota) = 2^{-l_1}\,\mathrm{e}^{\mathrm{i}l_1\iota}\,J_{l_1}(\lambda u).$$

From the formula

$$\int_{-1}^{1} (1-s^2)^{\nu-\frac{1}{2}}\,C_a^\nu(s)\,C_b^\nu(s)\,\mathrm{d}s = 0,$$

where $a \neq b, \operatorname{re}\nu > -\frac{1}{2}$ (see, e.g. 7.313(1) in [2]), we derive

$$f_{\lambda,l_1}(x) = \sum_{k=0}^{\infty} c_{\lambda,l_1,k,-l_1}\,f_{k,-l_1}(x).$$

Let $l_1 = 0$. Then

$$\Pi(f_{\lambda,0}) = \sum_{k=0}^{\infty} c_{\lambda,0,k,0}\,\Pi(f_{k,0}).$$

We choose γ_2 in the left side of this equality and γ_1 in the right side. Let $y(v) = \left(\frac{v^{-1}+v}{2}, 0, \ldots, 0, \frac{v^{-1}-v}{2}\right)$ in the left side and $y(p) = (\cosh p, 0, \ldots, 0, \sinh p)$ in the opposite side, so $v = \mathrm{e}^{-p}$. In addition, $c_{\lambda,l_1,k,k_1} = \mathsf{F}_\gamma(f_{\lambda,l_1}, f_{k,k_1})$, where we choose $\gamma = \gamma_2$. Finally, we have

$$G_{13}^{21}\left(\left(\frac{\lambda}{2\mathrm{e}^p}\right)^2 \middle| \begin{matrix} 0 \\ -\sigma,\,0,\,0 \end{matrix}\right) = 2^{-\sigma-\frac{1}{2}}\,\pi^{\frac{3}{2}}\,\mathrm{e}^{\sigma p}\,\sinh^{\frac{1}{2}} p\,\sin^{-1}[\pi(\sigma+1)]$$
$$\cdot \sum_{k=0}^{\infty} (-1)^k\,(k!)^{-1}\,c_{\lambda,0,k,0}\,\Gamma(k+1)\,\Gamma^{-1}(\sigma-k+1)\,P_{\sigma+\frac{1}{2}}^{-k-\frac{1}{2}}(\cosh p).$$

Here G_{pq}^{mn} means the Meijer G-function.

By analogy, from the distribution

$$f_{k,k_1}(x) = \int_0^{+\infty} c_{k,k_1,\lambda,-k_1}\,f_{\lambda,-k_1}(x)\,\mathrm{d}\lambda,$$

we obtain

$$P_{-\sigma-\frac{3}{2}}^{-\frac{1}{2}}(\cosh p) = 2\pi^{-\frac{3}{4}}\,\sinh^{\frac{1}{2}} p\,\Gamma^{-1}(-\sigma)$$
$$\cdot \int_0^{+\infty} G_{13}^{21}\left(\frac{\lambda^2}{4} \middle| \begin{matrix} 0 \\ -\sigma-1,\,0,\,0 \end{matrix}\right) G_{13}^{21}\left(\left(\frac{\lambda}{2\mathrm{e}^p}\right)^2 \middle| \begin{matrix} 0 \\ \sigma-1,\,0,\,0 \end{matrix}\right) \mathrm{d}\lambda.$$

In the same way,

$$f_{\mu,m_1,\pm}(x) = \int\limits_0^{+\infty} c_{\mu,m_1,\pm,\lambda,-m_1}\, f_{\lambda,-m_1}(x)\,\mathrm{d}\lambda$$

and, consequently, for instance,

$$\Pi(f_{\mu,m_1,+}) = \int\limits_0^{+\infty} c_{\mu,m_1,+,\lambda,-m_1}\, \Pi(f_{\lambda,-m_1})\,\mathrm{d}\lambda.$$

We choose γ_4 (in fact, γ_{3+}) in the left side of this equality and γ_2 in the right side. Let us choose the following parametrization in γ_3: $x = (\cosh\xi, \sinh\xi\sin\eta, \sinh\xi\cos\eta, \pm 1)$. Then

$$f_{\mu,m_1,\pm}(\xi,\eta) = \begin{cases} P^{-m_1}_{-\frac{1}{2}+\mathrm{i}\mu}(\cosh\xi)\,\sinh^{m_1}\xi\,\mathrm{e}^{\mathrm{i}m_1\eta}, & \text{for } x(\xi,\eta)\in\gamma_{3\pm},\\ 0, & \text{for } x(\xi,\eta)\in\gamma_{3,\mp}. \end{cases}$$

It follows to

$$\begin{aligned} P^{\sigma+1}_{-\frac{1}{2}+\mathrm{i}\mu}(\tanh p) &= \sqrt{2}\,\pi^{\frac{3}{2}}\,\cosh p\,\sin^{-1}[\pi(\sigma+1)]\\ &\quad\cdot\Gamma^{-1}\left(\mathrm{i}\mu-\sigma-\frac{1}{2}\right)\Gamma^{-1}\left(-\frac{3}{2}-\sigma-\mathrm{i}\mu\right)\sum_{k=0}^{\infty}(-1)^k\,(k!)^{-1}\\ &\quad\cdot N_{(k,k_1)}\,\sinh^{\frac{1}{2}}p\,\Gamma(k+1)\,\Gamma^{-1}(\sigma-k+1)\,P^{-\frac{1}{2}-k}_{\sigma+\frac{1}{2}}(\cosh p). \end{aligned}$$

In the simplest case $SO(2,1)$ of Lorentz group, we have $x(\alpha) = (1,\cos\alpha,\sin\alpha)\in\gamma_1$ and $x(z) = \left(\frac{1+z^2}{2}, \frac{1-z^2}{2}, z\right)\in\gamma_2$, and, for the decomposition

$$\Pi(f_\lambda) = \sum_{k\in\mathbb{Z}} c_{\lambda k}\,\Pi(f_k),$$

we obtain, if $\lambda\neq 0$,

$$\begin{aligned} c_{\lambda k} = \mathsf{F}_{\gamma_2}(f_\lambda, f_k) &= 2^{2\sigma+1}\,\mathrm{e}^{-|\lambda|}\,\mathrm{sign}\,\lambda\,\Gamma^{-1}(-\sigma-k\mathrm{sign}\,\lambda)\\ &\quad\cdot\Gamma^{-1}(\sigma+k\mathrm{sign}\,\lambda+1)\,\Gamma^{-1}(k\mathrm{sign}\,\lambda-\sigma-1)\,G^{21}_{12}\left(2|\lambda|\,\middle|\,\begin{matrix}-\sigma-k\mathrm{sign}\,\lambda\\ 0,\ -2\sigma-1\end{matrix}\right). \end{aligned}$$

Further,

$$\Pi(f_\lambda) = \mathsf{F}_{\gamma_2}(f_\lambda,\hat{q}^{-\sigma-1}) = 2^{-\sigma-\frac{1}{2}}\,\pi^{-1}\,\cosh^\sigma p\int\limits_{-\infty}^{+\infty}(z^2-2z\tanh p+1)^{-\sigma-1}\,\mathrm{e}^{\mathrm{i}\lambda z}\,\mathrm{d}z,$$

$$\Pi(f_k) = \mathsf{F}_{\gamma_1}(f_k,\hat{q}^{-\sigma-1}) = \int\limits_0^{2\pi}(\cosh p-\sin\alpha\sinh p)^{-\sigma-1}\,\mathrm{e}^{\mathrm{i}k\alpha\,\mathrm{sign}\,\lambda}\,\mathrm{d}\alpha.$$

It follows to

$$K_{\sigma+\frac{1}{2}}\left(\frac{|\lambda|}{\tanh p}\right) = \sqrt{\frac{\pi\cosh p}{2}}\,\mathrm{e}^{-\mathrm{i}\lambda\cosh p}\sum_{k\in\mathbb{Z}}(-\mathrm{i}\mathrm{sign}\,\lambda)^k\,P^k_\sigma(\cosh p)\,W_{-k\mathrm{sign}\,\lambda,\sigma+\frac{1}{2}}(2|\lambda|).$$

In this formula, $K_{\sigma+\frac{1}{2}}\left(\frac{|\lambda|}{\tanh p}\right)$ denotes the modified Bessel function of the second kind.

3. Formulae induced by the subrepresentation

For example, let us consider the subgroup of hyperbolic rotations in plane x_0x_{n-1}. This one-parameter subgroup consists of matrices

$$h(\varrho) = \begin{pmatrix} \cosh\varrho & 0 & \sinh\varrho & 0 \\ 0 & \mathrm{diag}(\underbrace{1,\dots,1}_{n-2}) & 0 & 0 \\ \sinh\varrho & 0 & \cosh\varrho & 0 \\ 0 & 0 & 0 & 1 \end{pmatrix}.$$

Let $t_{\lambda,L,\tilde{\lambda},\tilde{L}}(\varrho)$ be the matrix element of the restriction of T_σ to above subgroup with respect to basis $\{f_{\lambda,L} \,:\, \lambda \geq 0\}$. This matrix element can be obtained via the functionals F_γ, i.e.

$$t_{\lambda,L,\tilde{\lambda},\tilde{L}}(\varrho) = \mathsf{F}_\gamma(T_\sigma(h(\varrho))[f_{\tilde{\lambda},\tilde{L}}], f_{-\lambda,-L}).$$

Let us choose $x \in O_0$ such that $x_i = 0$ for $1 \leq i \leq n-2$. Then $f_{\lambda,L}(x) = 0$ under the condition $l_i \neq 0$ for $i \leq 2$. We choose $\gamma = \gamma_2$ and use the addition theorem for Bessel functions (see, e.g. 9.3.4(6) in [4]).

It is possible to prove that F_γ is invariant with respect to any pair $(T_\sigma(h(\varrho)), T_{1-\sigma-n}(h(\varrho))$, i.e.

$$t_{\lambda,L,\tilde{\lambda},\tilde{L}}(\varrho) = \mathsf{F}_\gamma(f_{\tilde{\lambda},\tilde{L}}, T_{1-\sigma-n}(h(-\varrho))[f_{-\lambda,-L}]).$$

It follows from the decomposition $G = KAK$, where K is a maximal compact subgroup (indeed, $K = H_{\gamma_1}$) and A is an maximal Abelian subgroup (see, for instance, [6]). Let us express the functions $f_{\tilde{\lambda},\tilde{L}}$ and $f_{-\lambda,-L}$ via functions of the basis $\{f_{\mu,M,\pm} \,:\, \mu \in \mathbb{R}\}$. We need the denotation $E := c_{\lambda,L,\mu,-M,+}c_{-\tilde{\lambda},-\tilde{L},\mu,\tilde{M},+} + c_{\lambda,L,\mu,-M,-}c_{-\tilde{\lambda},-\tilde{L},\mu,\tilde{M},-}$. Suppose $n > 3$, $\varrho > 0$ and $0 < \lambda\cosh^2\frac{\varrho}{2} < \tilde{\lambda}$. From the comparison of two approaches for $t_{\lambda,L,\tilde{\lambda},\tilde{L}}(\varrho)$, we have

$$\begin{aligned} &\int_{-\infty}^{+\infty} E\,\Gamma^2\left(\frac{4-n}{2} - \mu\mathbf{i}\right) P^{\frac{3-n}{2}}_{-\frac{1}{2}+\mu\mathbf{i}}(\cosh\varrho)\,\mathrm{d}\mu \\ &\quad = 2^{\frac{5-n}{2}}\,\mathbf{i}^{n-3}\,[\pi(n-3)]^{-1}\,\tilde{\lambda}^{\frac{1-n}{2}}\,\sinh^{\frac{n-3}{2}}\varrho\,\cosh^{1-n-2\sigma}\frac{\varrho}{2} \\ &\quad \cdot \tanh^{\frac{3-n}{2}}\frac{\varrho}{2}\,\Gamma^{-1}(n-2)\,\Gamma^{-1}\left(\frac{n-3}{2}\right)\sum_{s=0}^{\infty}(-1)^s\,(s!)^{-1} \\ &\quad \cdot \left(s + \frac{n-3}{2}\right)\Gamma(n+s-3)\,\Gamma\left(\frac{s+1}{2}\right)\Gamma^{-1}\left(\frac{n-s}{2} - 1\right) \\ &\quad \cdot J_{\frac{n-3}{2}+s}\left(\frac{\tilde{\lambda}\sinh\varrho}{\cosh^4\frac{\varrho}{2}}\right){}_2F_1\left(\frac{s+1}{2}, 2 - \frac{n+3}{2}; \frac{n-1}{2}; \left(\frac{\lambda\cosh^2\frac{\varrho}{2}}{\tilde{\lambda}}\right)^2\right). \end{aligned}$$

Here ${}_2F_1$ denotes the Gauss hypergeometric function.

Acknowledgments

The authors wish to acknowledge the Ministry of Education and Science of the Russian Federation: the research described in this paper is supported by grant NK 586-P30.

Appendix

The logical continuation of this paper follows to the consideration of bilinear functionals of more general type. For instance, recently, for the rudimentary Lorentz group $SO(2,1)$, the authors study the functionals

$$\mathsf{D}_{\gamma_i} : \mathfrak{C}_\sigma^2 \longrightarrow \mathbb{C},\ (u,v) \longmapsto \int_{\gamma_i}\int_{\gamma_i} k(x,\hat{x})\, u(x)\, v(\hat{x})\, \mathrm{d}x\, \mathrm{d}\hat{x},$$

where the kernel k provides that D_{γ_i} is invariant with respect to T_σ, i.e., for any $g \in SO(2,1)$,

$$\mathsf{D}_{\gamma_i}(T_\sigma(g)[u], T_\sigma(g)[v]) = \mathsf{D}_{\gamma_i}(u,v).$$

The authors obtained recently the restrictions of the kernel k to the direct products $\gamma_i \times \gamma_i$ of contours. Using these functionals, the authors derived some new formulas for Whittaker functions $M_{\mu,\nu}$ and $W_{\mu,\nu}$. At present time, the corresponding paper is preparing for publication.

References

[1] Artemov A A 2004 The Poisson transform for a hyperboloid of one sheet *Sb. Math.* **195** 643
[2] Gradshteyn I S and Ryzhik I M 2007 *Table of Integrals, Series, and Products* (New York: Academic Press)
[3] Shtepina T V 2004 Generalization of the Funk-Hecke theorem to the case of a hyperbolic space *Izv. Math.* **68** 1051
[4] Vilenkin N Ja 1968 *Special Functions and Theory of Group Representations* (Providence, R. I.: AMS)
[5] Vilenkin N Ja and Klimyk A U 1993 *Representation of Lie Groups and Special Functions* vol 2 (Dordrecht: Kluwer Academic Publishers)
[6] Wehrhahn R F, Smirnov Yu F and Shirokov A M 1992 Symmetry scattering on the hyperboloid $SO(2,1)/SO(2)$ in different coordinate systems *J. Math. Phys.* **33** 2384

Burchnall-Chaundy annihilating polynomials for commuting elements in Ore extension rings

Johan Richter[1], **Sergei D. Silvestrov**[1,2]

[1]Centre for Mathematical Sciences, Lund University, Box 118, 22100 Lund, Sweden
[2]Division of Applied Mathematics, The School of Education, Culture and Communication, Mälardalen University, Box 883, 721 23 Västerås, Sweden

E-mail: `johanr@maths.lth.se`, `ssilvest@maths.lth.se`, `sergei.silvestrov@mdh.se`

Abstract. In this article further progress is made in extending the Burchnall-Chaundy type determinant construction of annihilating polynomial for commuting elements to broader classes of rings and algebras by deducing an explicit general formula for the coefficients of the annihilating polynomial obtained by the Burchnall-Chaundy type determinant construction in Ore extension rings. It is also demonstrated how this formula can be used to compute the annihilating polynomials in several examples of commuting elements in Ore extensions. Also it is demonstrated that additional properties which may be possessed by the endomorphism, such as for example injectivity, may influence strongly the annihilating polynomial.

MSC 2010: 16S99, 81S05, 39A13
PACS: 02.10.Hh, 03.65.Fd
Keywords: commuting elements, annihilating polynomial, algebraic dependence, Burchnall-Chaundy determinant construction, Ore extensions

1. Introduction

It is a classical result, going back to [1, 2, 3], that all pairs of commuting elements in the Heisenberg (Weyl) algebra are algebraiclly dependent over $\mathbb{C}$. This result was later rediscovered and applied to the study of non-linear partial differential equations [9, 10, 14].

In 1994, one of the authors of the present paper, S. Silvestrov, based on consideration of the previous literature and a series of trial computations, made the following three part conjecture.

- The first part of the conjecture stated that the Burchnall–Chaundy type result on algebraic dependence of commuting elements can be proved in greater generality, that is for much more general classes of non-commutative algebras and rings than the Heisenberg algebra and related algebras of differential operators treated by Burchnall and Chaundy and in subsequent literature [1, 2, 3, 5, 9, 10, 14].
- The second part stated that the Burchnall–Chaundy eliminant construction of annihilating polynomials formulated in determinant (resultant) form works after some appropriate modifications for most or possibly all classes of algebras where the Burchnall–Chaundy type result on algebraic dependence of commuting elements can be proved.
- Finally, the third part of the conjecture stated that the proof of the vanishing of the corresponding determinant polynomial on the commuting elements can be performed in a purely algebraic way for all classes of algebras or rings where this fact is true, that is

using only the internal structure and calculations with the elements in the corresponding algebras or rings and the algebraic combinatorial expansion formulas for the corresponding determinants, that is, without any need of passing to operator representations and use of analytic methods as in the Burchnall–Chaundy type proofs.

In the first and the second part of the conjecture more progress has been made. In [6], the key Burchnall–Chaundy type theorem on algebraic dependence of commuting elements in q-deformed Heisenberg algebras (and thus as a corollary for q-difference operators as operators representing q-deformed Heisenberg algebras) was obtained. The result and the methods have been extended to more general algebras and rings generalizing q-deformed Heisenberg algebras (generalized Weyl structures and graded rings) in [7]. The proof in [6] is totally different from the Burchnall–Chaundy type proof. It is an argument based only on the intrinsic properties of the elements and internal structure of q-deformed Heisenberg algebras, thus supporting the first part of the conjecture. It can be used successfully for an algorithmic implementation for computing the corresponding annihilating polynomial for given commuting elements. However, it does not give any specific information on the structure or properties of such polynomials or any general formulae. It is thus important to have a way of describing such annihilating polynomials by some explicit formulae, as for example those obtained using the Burchnall–Chaundy eliminant construction for the $q = 1$ case, i.e., for the classical Heisenberg algebra. In [11], a step in that direction was taken by offering a number of examples, all supporting the claim that the eliminant determinant method should work in the general case. However, no general proof for this was provided. The complete proof following the Burchnall–Chaundy approach in the case of q not a root of unity has been recently obtained [8], by showing that the determinant eliminant construction, properly adjusted for the q-deformed Heisenberg algebras, gives annihilating polynomials for commuting elements in the q-deformed Heisenberg algebra when q is not a root of unity, thus confirming the second part of the conjecture for these algebras. That proof was obtained by adapting the Burchnall–Chaundy eliminant determinant method of the case $q = 1$ of differential operators to the q-deformed case, after passing to a specific faithful representation of the q-deformed Heisenberg algebra on Laurent series and then performing a detailed analysis of the kernels of arbitrary operators in the image of this representation. While exploring the determinant eliminant construction of the annihilating polynomials, we also obtained some further information on such polynomials and some other results on dimensions and bases in the eigenspaces of the q-difference operators in the image of the chosen representation of the q-deformed Heisenberg algebra. In the case of q being a root of unity the algebraic dependence of commuting elements holds only over the center of the q-deformed Heisenberg algebra [6], and it is still unknown how to modify the eliminant determinant construction to yield annihilating polynomials for this case.

This third part of the conjecture remains widely open for most classes of non-commutative algebras and rings, even in the case of the usual Heisenberg algebra and differential operators, and with only a series of examples calculated for the Heisenberg algebra, q-Heisenberg algebra and some more general algebras, all supporting the conjecture. An interesting partial progress in this direction has been made in the recent work by Daniel Larsson on extension of Burchnall–Chaundy theory to Ore extensions [12]. An algebraic proof has been given in this article in case of Ore extension algebras. However it uses some general properties of resultants in Ore extension rings in order to deduce that Burchnall–Chaundy type determinant polynomial for commuting elements is annihilating. This proof does not involve explicit combinatorics of computations in Ore extensions based on their defining commutation relations, and thus still does not reveal explicit formulas for annihilating polynomials and does not add to understanding of why it again works for such broader class of algebras defined by this more general class of commutation relations.

In this article we make further progress in this third part of the conjecture by deducing an

Journal of Physics: Conference Series **346** (2012) 012021 doi:10.1088/1742-6596/346/1/012021

explicit general formula for the coefficients of annihilating polynomial obtained by the Burchnall–Chaundy type determinant construction in Ore extensions, and we also demonstrate how this formula can be used to compute the annihilating polynomials in several examples of commuting elements.

2. Extension of Burchnall–Chaundy theory to Ore extensions

In this section we describe an extension of Burchnall–Chaundy theory to Ore extensions developed by Daniel Larsson in [12].

Given a commutative ring $\mathbf{R}$ with an endomorphism σ and another additive map δ that satisfies the σ-Leibniz' rule:

$$\delta(ab) = \sigma(a)\delta(b) + \delta(a)b,$$

we define a multiplication on $\mathbf{R}[x]$ by

$$xa = \sigma(a)x + \delta(a).$$

As usual we write multiplication as juxtaposition. This is known as an Ore polynomial ring or an Ore extension of $\mathbf{R}$. The multiplication in an Ore polynomial ring is associative and distrubitive, but typically not commutative. There is a normal form in $\mathbf{R}[x]$ where we can write all elements uniquely as $\sum_{j\in\mathbb{N}} a_j x^j$, where $\mathbb{N} = \{n \in \mathbb{Z}\colon n \geq 0\}$. We can also write explicitly how the multiplication operation acts in this normal form. For this purpose we introduce, following [12], the functions π_i^n defined as the sum of all possible compositions of i copies of σ and $n-i$ copies of δ. Thus for example $\pi_0^1(a) = \sigma(a)$ and $\pi_1^2 = \sigma(\delta(a)) + \delta(\sigma(a))$. Also it is convenient to define that $\pi_i^j(a) = 0$ if $i < 0$ or $i > j$ or $x = 0$. Using these functions we get the following expression for the multiplication

$$(\sum_{i=0}^{n} a_i x^i)\cdot(\sum_{j=0}^{m} b_j x^j) = \sum_{z=0}^{n+m}\sum_{i=0}^{n}\sum_{j=0}^{m} a_i \pi_{z-j}^{i}(b_j) x^z.$$

2.1. Determinant polynomial

Let M be a $r \times c$ matrix with entries in $\mathbf{R}$ where $r \leq c$. Then we define the determinant polynomial of M, denoted by $|M|$, as follows

$$|M| = \sum_{i=0}^{c-r} \det(M_i) x^i$$

where M_i is the $r \times r$ matrix whose first $r-1$ colummns coincide with M and whose last column equals the $(c-i)th$ column of M.

Larsson and Li show how to rewrite the determinant polynomial as a determinant of a $r \times r$ matrix, where the entries in the last column are elements of $\mathbf{R}[x]$ and all other entries lie in $\mathbf{R}$. The expansion of such a determinant is defined by the convention that the element of $\mathbf{R}[x]$ is placed on the right-hand side of the entries in $\mathbf{R}$, in the usual expansion that defines the determinant.

The following two useful Propositions easily follow from basic properties of the determinant.

Proposition 1. *Let M be a $r \times c$ matrix, $r \leq c$, with determinant polynomial $|M|$. Let H_i denote the polynomial*

$$m_{i,1}x^{c-1} + \ldots + m_{i,r}x^{c-r} + \ldots + \ldots + m_{i,c}.$$

Then

$$|M| = \det \begin{pmatrix} m_{1,1} & \dots & m_{1,r-1} & H_1 \\ m_{2,1} & \dots & m_{2,r-1} & H_2 \\ . & \dots & . & . \\ m_{r,1} & \dots & m_{r,r-1} & H_r \end{pmatrix}$$

Now assume we have a sequence $A_1, A_2, \dots, A_r$ of polynomials in $\mathbf{R}[x]$ and let d be the maximum degree of the polynomials. We assume that $d \geq r$. We form a $r \times (d+1)$ matrix, denoted $\mathrm{mat}(A)$, whose entry in the ith row and jth column is the coefficient of x^{d+1-j} in A_i. The determinant polynomial of A is defined as $|\mathrm{mat}(A)|$ and denoted $|A|$.

Proposition 2. *Let $A_1, A_2, \dots, A_r$ be a sequence of elements of $\mathbf{R}[x]$ of maximum degree d. Then*

$$|A| = \det \begin{pmatrix} a_{1,d} & \dots & a_{1,d-r+1} & A_1 \\ a_{2,d} & \dots & a_{2,d-r+1} & A_2 \\ . & \dots & . & . \\ a_{r,d} & \dots & a_{r,d-r+1} & A_r \end{pmatrix}$$

where $a_{i,j}$ is the coefficient of x^j in A_i.

2.2. The resultant

Let P and Q be two elements of $\mathbf{R}[x]$, of degree m and n respectively. We define their resultant, $\mathrm{Res}(P,Q)$, as the determinant polynomial of the sequence

$$P, xP, \dots, x^{n-1}P, Q, xQ, \dots, x^{m-1}Q.$$

It is easy to see that this will be a determinant of size $m+n$.

Proposition 3. *For all P and Q in $\mathbf{R}[x]$ there exists element S, T in $\mathbf{R}[x]$ such that*

$$\mathrm{Res}(P,Q) = SP + TQ.$$

Proof. By the definition and the previous proposition we know that for some $m_{i,j} \in \mathbf{R}$

$$\mathrm{Res}(P,Q) = \det \begin{pmatrix} m_{n,1} & \dots & m_{n,n+m-1} & P \\ . & \dots & . & . \\ m_{2,1} & \dots & m_{2,n+m-1} & x^{n-2}P \\ m_{1,1} & \dots & m_{1,n+m-1} & x^{n-1}P \\ m_{n+m} & \dots & m_{n+m,n+m-1} & Q \\ . & \dots & . & . \\ m_{n+1} & \dots & m_{n+1,n+m-1} & x^{m-1}Q \end{pmatrix}.$$

If we expand this using the definition of the determinant we see that the theorem is true. □

Now let P and Q be commuting elements, and let s and t be variables that commute with each other and everything in $\mathbf{R}[x]$. (So they can be seen as elements in the larger algebra $\mathbf{R}[x][s,t]$.) Then $\mathrm{Res}(P-s, Q-t) = S(P-s) + T(Q-t)$ for some S, T in $\mathbf{R}[x][s,t]$. We note that $\mathrm{Res}(P-s, Q-t)$ is a element of $\mathbf{R}[s,t]$, a fact that follows from the definition. It is also true that $\mathrm{Res}(P-s, Q-t)$ will depend polynomially on s and t, once again by the definition of the resultant. Finally if we formally replace s by P and t by Q the resultant becomes zero. Putting it all together we have proven

Theorem 1. *If P and Q are commuting elements of $\mathbf{R}[x]$ then*

$$f(s,t) = \mathrm{Res}(P-s, Q-t)$$

is a polynomial in two commuting variables such that $f(P,Q) = 0$.

Journal of Physics: Conference Series **346** (2012) 012021 doi:10.1088/1742-6596/346/1/012021

3. Recursive construction of the matrix of the resultant

We can also construct the matrix used for computing the resultant in a recursive way. For the first row we simply take the coefficients of $P - s$. Let $d_{i,j}$ denote the element in row i and column j of the resultant. If i or j is non-positive we agree to set $d_{i,j} = 0$. Further we set $\sigma(s) = s$ and $\delta(s) = 0$. This means that $xs = \sigma(s)x + \delta(s)$ in accordance with the σ-Leibniz rule. For $1 < i \leq n$ and $n + 1 < i \leq n + m$ we then have the recursive formula

$$d_{i,j} = \delta(d_{i-1,j}) + \sigma(d_{i-1,j-1}).$$

This formula simply expresses the definition of the resultant and the σ-Leinbiz rule.

3.1. The Heisenberg algebra case

We illustrate the result in the classical Heisenberg (Weyl) algebra case.

Assume that P and Q are commuting elements in the Heisenberg algebra, of degree m and n respectively. We change notation and write $P = \sum_j a_j D^j$ and $Q = \sum_i b_i D^i$, where the a_i and b_i are polynomials over $\mathbb{C}$ in one variable, which we denote by y.

We want to give a more explicit expression for the resultant of P and Q. To this end we use Leibniz' rule. In the context of Heisenberg algebras this takes the form

$$D^n p = \sum_{k=0}^{n} \binom{n}{k} p^{(k)} D^{n-k}$$

for any polynomial p in y.

Thus

$$D^e P = D^e \sum_{j=0}^{m} a_j D^j = \sum_{j=0}^{m} (\sum_{k=0}^{n} \binom{e}{k} a_j^{(k)} D^{e-k}) D^j =$$
$$\sum_{j=0}^{m} (\sum_{l=0}^{n+m} \binom{e}{n+j-l} a_j^{(n+j-l)} D^l) = \sum_{l=0}^{n+m} \sum_{j=0}^{m} \binom{e}{n+j-l} a_j^{(n+j-l)} D^l.$$

We can now write down the expression for $\mathrm{Res}(P, Q)$. At place (e, f), when $e \leq n$, we get

$$\sum_{j=0}^{m} \binom{e-1}{e-1+j-n-m+f} a_j^{(e-1+j-n-m+f)}$$

and similarly for the last m rows.

The expression for the resultant becomes

$$\mathrm{Res}(P, Q) = \begin{vmatrix} a_{n+m} & \dots & a_1 & P \\ a'_{n+m} + a_{n+m-1} & \dots & a'_1 + a_0 & DP \\ . & \dots & . & . \\ a_m & \dots & \sum_{j=0}^{m} \binom{n-1}{n+j-2} a_j^{(n+j-2)} & D^{n-1}P \\ b_{n+m} & \dots & b_1 & Q \\ b'_{n+m} + b_{n+m-1} & \dots & b'_1 + b_0 & DQ \\ . & \dots & & . \\ b_n & \dots & \sum_{i=0}^{n} \binom{m-1}{m+j-2} b_i^{(m+i-2)} & D^{m-1}Q \end{vmatrix}.$$

Journal of Physics: Conference Series **346** (2012) 012021 doi:10.1088/1742-6596/346/1/012021

As an example take $P = yD$ and $Q = y^2D^2$. Then

$$\operatorname{Res}(P-s, Q-t) = \begin{vmatrix} 0 & y & yD - s \\ y & (1-s) & yD^2 + (1-s)D \\ y^2 & 0 & y^2D^2 - t \end{vmatrix}.$$

Expanding this we get

$$\operatorname{Res}(P-s,Q-t) = y^3(yD^2 + (1-s)D) - y^2(y^2D^2 - t) - y^2(1-s)(yD-s) =$$
$$y^4D^2 + y^3D - sy^3D - y^4D^2 + y^2t + sy^2 - y^3D - y^2s^2 + sy^3D = (t + s - s^2)y^2.$$

We note that $Q + P - P^2 = 0$.

4. Necessity of injectivity

We give an example to show what can happen if we do not require σ to be injective. Let k be a field and set $\mathbf{R} = k[y]$, the polynomials over k in the variable y.

Set $\delta(f) = 0$ and $\sigma(f) = f(0)$, for all polynomials f. Then σ is an endomorphism and δ is a σ-derivation. So $\mathbf{R}[x]$ is well-defined.

Now set $P = yx^2$ and $Q = P$. Then P and Q commute. But the resultant we get is

$$\operatorname{Res}(P-s, Q-t) = \begin{vmatrix} 0 & y & 0 & -s \\ 0 & 0 & -s & 0 \\ 0 & y & 0 & -t \\ 0 & 0 & -t & 0 \end{vmatrix}.$$

This resultant is zero. We note that as long as the leading coefficient of both P and Q belong to the kernel of σ the resultant will be zero.

Theorem 2. *Let P, Q be commuting elements in some Ore extension $\mathbf{R}[x; \sigma, \delta]$ of degrees m and n respectively in x. Suppose the highest coefficients a_m and b_n both belong to the kernel of σ. Then* $\operatorname{Res}(P-s, Q-t) = 0$.

Proof. Form the matrix of the resultant according to the definition. Consider the first column of the matrix. The only potentially non-zero elements in it are the elements in row n and $n+m$ where the elements $\sigma^{n-1}(a_m)$ respectively $\sigma^{m-1}(b_n)$ appear. But since a_m and b_n belong to the kernel of σ these elements must also be zero and thus the determinant is zero. □

5. The case $\delta = 0$

When we set $\delta(r) = 0$, for all $r \in \mathbf{R}$ the formulae for the elements in the determinant simplify. We transpose the columns of the resultant to simplify the formulae further.

Let $D_{i,j}$ denote the element in place i, j of $\operatorname{Res}(P-s, Q-t)$, where $P = \sum_{i=0}^m a_i x^i$ has degree m and $Q = \sum_{j=0}^n b_j x^j$ has degree n. Multiplying by x repeatedly and using the commutation rule $xa = \sigma(a)x$ we find that if $i \le n$ then

$$D_{i,j} = \sigma^{i-1}(a_{j-i}) - \Delta_{i,j}s$$

where Δ is the Kronecker delta-function. If $i > n$ then

$$D_{i,j} = \sigma^{i-n-1}(b_{j-i-n}) - \Delta_{i-n,j}t.$$

The formula is in some sense simpler than it looks. The first row simply has the same coefficients as $P - s$. Then you apply σ to the coefficients and shift them one step to the left to get the second row. (Where we use the rule $\sigma(s) = s$.) This gives us the first n rows. For the last m rows we do the same with $Q - t$. We note that it is a generalization of the classical resultant for polynomials which we recover if we set $\sigma \equiv 1$.

We can expand the determinant along the last column to get

$$\mathrm{Res}(P-s, Q-t) = (-1)^{m+n+1} \begin{vmatrix} \sigma(a_{n+m-1}) & & \dots & \sigma(a_1) & \sigma(a_0) - s \\ \sigma^2(a_{n+m-2}) & & \dots & \sigma^2(a_0) - s & 0 \\ \cdot & & \dots & & \cdot \\ b_{n+m} & & \dots & b_2 & b_1 \\ \cdot & & \dots & \cdot & \cdot \\ \sigma^{m-1}(b_n) & \sigma^{m-1}(b_{n-1}) & & \dots & 0 \end{vmatrix} (a_0 - s) +$$

$$+ (-1)^{m+2n+1} \begin{vmatrix} a_{n+m} & & \dots & a_2 & a_1 \\ \sigma(a_{n+m-1}) & & \dots & \sigma(a_1) & \sigma(a_0) - s \\ \cdot & & \dots & \cdot & \cdot \\ \sigma(b_{n+m-1}) & & \dots & \sigma(b_1) & \sigma(b_0) - t \\ \cdot & & \dots & \cdot & \cdot \\ \sigma^{m-1}(b_n) & \sigma^{m-1}(b_{n-1}) & & \dots & 0 \end{vmatrix} (b_0 - t).$$

Now in every row there is only one element that contains s or t. This helps us to determine the coefficient of s^n and t^m which are the highest possible powers of s and t. Start with s. In the precedcing expression for $\mathrm{Res}(P - s, Q - t)$ it is easy to see that we need only expand the first determinant. Further, when we have expanded this determinant we see that there is only one relevant subdeterminant and so on. It help with the presentation of the result to make a new definition. Set p_k to be the coefficient of x^k in $P - s$. (So p_k equals b_k plus possibly a $(-s)$-term.) We set $\sigma(b_k - s) = \sigma(b_k) - s$.

Doing this we get the cofficient of s^n as

$$(-1)^{m+n+1} \begin{vmatrix} 0 & \dots & p_n \\ \cdot & \dots & \cdot \\ \sigma^{m-1}(p_n) & \dots & \sigma^{m-1}(p_{n-m}) \end{vmatrix}.$$

Similarly we get (with q_k the coefficient of x^k in $Q - t$) that the coefficient of t^m is

$$(-1)^{m+2n+1} \begin{vmatrix} 0 & \dots & q_m \\ \cdot & \dots & \cdot \\ \sigma^{n-1}(q_m) & \dots & \sigma^{n-1}(q_{m-n}) \end{vmatrix}.$$

These matrices are of rather special form ("triangular with respect to the wrong diagonal") so we can compute them explicitly. The coefficient of s^n becomes $(-1)^{n+1} b_n \sigma(b_n) \cdot \ldots \cdot \sigma^{m-1}(b_n)$ and the coefficient of t^m becomes $(-1)^{m+n+1} a_m \sigma(a_m) \cdot \ldots \cdot \sigma^{n-1}(a_m)$.

5.1. The leading coefficients in general

We can extend our computation of the leading coefficients to the general case. So let P and Q be two Ore-polynomials of degree m and n respectively. If we form the determinant $\mathrm{Res}(P-s, Q-t)$ we notice that all elements containing s lie on the anti-diagonal. We also note that there are n such elements so the highest possible power of s that can occur is s^n.

We have the following expression for the resultant

$$\mathrm{Res}(P - s, Q - t) = \sum_{\alpha} p(\alpha) \prod_{i=1}^{m+n} d_{i,\alpha(i)}$$

where α runs over all the permutations of $m+n$ elements, $p(\alpha)$ denotes the sign of the permutation and $d_{i,j}$ denotes the element in place (i,j) of the resultant.

Now if $i > n$ and $i < m+n-j$ then $d_{i,j} = 0$. This helps us compute the coefficient of s^n. The only element of the sum given earlier that contains s^n is the one that comes from multiplication of all the elements on the anti-diagonal. Thus up to some sign we get that s^n has the coefficient $\prod_{k=0}^{m-1} \sigma^k(b_n)$, where the exponent denotes functional iteration. Similarly we get that the coefficient of t^m is $\prod_{k=0}^{n-1} \sigma^k(a_m)$. As a corollary we obtain

Theorem 3. *$\mathrm{Res}(P-s, Q-t)$ is non-zero if R is an integral domain and σ is injective.*

Proof. We have just shown that s^n has the coefficient $\prod_{k=0}^{m-1} \sigma^k(b_n)$. b_n is non-zero and if σ is injective so must all the $\sigma^k(b_n)$ be. Finally, if R is an integral domain the product of non-zero elements is non-zero. □

The reason that the formulae for the leading coefficients remain the same in the general case as when $\delta = 0$ is that the highest term when you multiply a Ore-polynomial by x from the left is determined by σ alone.

6. The lower-order coefficients

One of our main goals in this article is to obtain formula for calculating the annihilating polynomial. In other words we want to derive a formula not only for the highest coefficient but also for the lower-order coefficients of $\mathrm{Res}(P-s, Q-t)$. To do this we start with formulating a formula for the k-derivative of a determinant in our general context of Ore extension rings. This formula has been obtained in the special case of determinants of ordinary enough times differentiable real valued functions in [4].

Theorem 4. *Let A be a square matrix of size m with entries in $R[t]$ for some commutative ring R. We denote the determinant of A by $|A|$. Then*

$$\frac{\partial^k |A|}{\partial t^k} = \sum_{k_1+k_2+\ldots+k_m=k} \frac{k!}{k_1!k_2!\ldots k_m!} \begin{vmatrix} a_{11}^{(k_1)} & a_{12}^{(k_1)} & \ldots & a_{1m}^{(k_1)} \\ a_{21}^{(k_2)} & a_{22}^{(k_2)} & \ldots & a_{2m}^{(k_2)} \\ \ldots & \ldots & \ldots & \ldots \\ a_{m1}^{(k_m)} & a_{m2}^{(k_m)} & \ldots & a_{mm}^{(k_m)} \end{vmatrix}$$

with the sum taken over all combinations of non-negative integers $k_1, k_2 \ldots k_m$ that sum to k. Multiplication by a positive integer here denotes the obvious and usual way that $\mathbb{N}$ acts on $R[t]$.

Proof. We give a proof by induction on both k and m. The formula is clearly true when $k = 0$ for any m. It is also easy to check that it is true when $k = 1$ and $m = 2$. So we assume it is true for $k = 1$ and m and try to prove it is true for $m+1$ and $k = 1$. Let A be any $m+1$-sized square matrix. Then

$$\frac{\partial |A|}{\partial t} = \frac{\partial}{\partial t}\Big(\sum_{i=1}^{m+1}(-1)^{i+1} a_{1i}|A_{1i}|\Big) = \sum_{i=1}^{m+1} a'_{1i}|A_{1i}| + \sum_{i=1}^{m+1} a_{1i}|A'_{1i}|$$

$$= \sum_{i=1}^{m+1} \begin{vmatrix} a_{1,1} & a_{1,2} & \ldots & a_{1,m+1} \\ a_{2,1} & a_{2,2} & \ldots & a_{2,m+1} \\ \ldots & \ldots & \ldots & \ldots \\ a'_{i,1} & a'_{i,2} & \ldots & a'_{i,m+1} \\ \ldots & \ldots & \ldots & \ldots \\ a_{m+1,1} & a_{m+1,2} & \ldots & a_{m+1,m+1} \end{vmatrix}.$$

This proves that the claim is valid for $k = 1$ and m arbitrary. So we now assume the theorem is true for k and try to prove it for $k + 1$. If A is a square matrix of size m, then

$$\frac{\partial^{k+1}|A|}{\partial t^{k+1}} = \frac{\partial}{\partial t}\left(\frac{\partial^k |A|}{\partial t^k}\right) = \frac{\partial}{\partial t} \sum_{k_1+k_2+\ldots+k_m=k} \frac{k!}{k_1!k_2!\ldots k_m!} \begin{vmatrix} a_{1,1}^{(k_1)} & a_{1,2}^{(k_1)} & \ldots & a_{1,m}^{(k_1)} \\ a_{2,1}^{(k_2)} & a_{2,2}^{(k_2)} & \ldots & a_{2,m}^{(k_2)} \\ \ldots & \ldots & \ldots & \ldots \\ a_{m,1}^{(k_m)} & a_{m,2}^{(k_m)} & \ldots & a_{m,m}^{(k_m)} \end{vmatrix}$$

$$= \sum_{k_1+k_2+\ldots+k_m=k} \frac{k!}{k_1!k_2!\ldots k_m!} \sum_{i=1}^{m} \begin{vmatrix} a_{1,1}^{(k_1)} & a_{1,2}^{(k_1)} & \ldots & a_{1,m}^{(k_1)} \\ a_{2,1}^{(k_2)} & a_{2,2}^{(k_2)} & \ldots & a_{2,m}^{(k_2)} \\ \ldots & \ldots & \ldots & \ldots \\ a_{i,1}^{(k_i+1)} & a_{i,2}^{(k_i+1)} & \ldots & a_{i,m}^{(k_i+1)} \\ \ldots & \ldots & \ldots & \ldots \\ a_{m,1}^{(k_m)} & a_{m,2}^{(k_m)} & \ldots & a_{m,m}^{(k_m)} \end{vmatrix}$$

$$= \sum_{k_1+k_2+\ldots+k_m=k+1} \frac{(k+1)!}{k_1!k_2!\ldots k_m!} \begin{vmatrix} a_{1,1}^{(k_1)} & a_{1,2}^{(k_1)} & \ldots & a_{1,m}^{(k_1)} \\ a_{2,1}^{(k_2)} & a_{2,2}^{(k_2)} & \ldots & a_{2,m}^{(k_2)} \\ \ldots & \ldots & \ldots & \ldots \\ a_{m,1}^{(k_m)} & a_{m,2}^{(k_m)} & \ldots & a_{m,m}^{(k_m)} \end{vmatrix}$$

where we have used a well-known property of multinomial coefficients. □

We now apply Proposition 4 to get a formula for the lower order coefficients of $\mathrm{Res}(P-s, Q-t)$. We note that any time we differentiate a row of the determinant twice we end up with zero and that only the last m rows contains any t-terms. Set $A = \mathrm{Res}(P-s, Q-t)$ and let $\mathbf{a}_i$ denote the vector consisting of the elements in row i of A.

We find that

$$\frac{\partial^k |A|}{\partial t^k} = \sum_{\substack{k_1+k_2+\ldots+k_m=k \\ 0\le k_i \le 1}} k! \begin{vmatrix} \mathbf{a}_1 \\ \vdots \\ \mathbf{a}_n \\ \frac{\partial^{k_1}\mathbf{a}_{n+1}}{\partial t} \\ \frac{\partial^{k_2}\mathbf{a}_{n+2}}{\partial t} \\ \vdots \\ \frac{\partial^{k_m}\mathbf{a}_{n+m}}{\partial t} \end{vmatrix}.$$

We differentiate this expression with respect to s, and get for non-negative integers k, q,

$$\frac{\partial^{k+q} |A|}{\partial s^q \partial t^k} = \sum_{\substack{k_1+k_2+\ldots+k_m=k \\ q_1+q_2+\ldots+q_n=q \\ 0\le k_i \le 1 \\ 0 \le q_i \le 1}} k!q! \begin{vmatrix} \frac{\partial^{q_1}\mathbf{a}_1}{\partial s} \\ \vdots \\ \frac{\partial^{q_n}\mathbf{a}_n}{\partial s} \\ \frac{\partial^{k_1}\mathbf{a}_{n+1}}{\partial t} \\ \frac{\partial^{k_2}\mathbf{a}_{n+2}}{\partial t} \\ \vdots \\ \frac{\partial^{k_m}\mathbf{a}_{n+m}}{\partial t} \end{vmatrix}.$$

Journal of Physics: Conference Series **346** (2012) 012021 doi:10.1088/1742-6596/346/1/012021

This gives almost immediately an expression for the coefficient of $s^q t^k$. (At least if $\mathbf{R}$ torsion-free, seen as a module over $\mathbb{Z}$.)

The coefficient in front of $s^q t^k$ is the same as the derivative of $\text{Res}(P-s, Q-t)$ with respect to s and t taken respectively q and k times and then evaluated at $t=0$ and $s=0$. Thus we get the following theorem describing the formula for all coefficients of $\text{Res}(P-s, Q-t)$.

Theorem 5. *Let* $\mathbf{R}$ *be torsion-free as a module over* $\mathbb{Z}$ *with the natural action of* $\mathbb{Z}$. *Let* P *and* Q *be two commuting elements of* $\mathbf{R}[x]$. *Denote the coefficient of* $s^q t^k$ *in* $\text{Res}(P-s, Q-t)$ *by* $c_{q,k}$. *Denote row number* i *in the resultant matrix by* $\mathbf{a}_i$.

Then

$$c_{q,k} = \sum_{\substack{k_1+k_2+\ldots+k_m=k \\ q_1+q_2+\ldots+q_n=q \\ 0\leq k_i\leq 1 \\ 0\leq q_i\leq 1}} \begin{vmatrix} \frac{\partial^{q_1}\mathbf{a}_1}{\partial s}\big|_{s=0,t=0} \\ \vdots \\ \frac{\partial^{q_n}\mathbf{a}_n}{\partial s}\big|_{s=0,t=0} \\ \frac{\partial^{k_1}\mathbf{a}_{n+1}}{\partial t}\big|_{s=0,t=0} \\ \frac{\partial^{k_2}\mathbf{a}_{n+2}}{\partial t}\big|_{s=0,t=0} \\ \vdots \\ \frac{\partial^{k_m}\mathbf{a}_{n+m}}{\partial t}\big|_{s=0,t=0} \end{vmatrix}.$$

We will illustrate now the preceding theorem with several examples.

Example 1. We choose the Heisenberg algebra setting we considered earlier for our example. So we set $\mathbf{R} = \mathbb{C}[y]$ and denote the Ore extension by $\mathbf{R}[D]$.

Set $P = yD$ and $Q = y^2D^2$. We have already computed that $\text{Res}(P-s, Q-t) = (t+s-s^2)y^2$. We want to check that we get the same result using the theorem. We compute

$$c_{0,0} = \begin{vmatrix} 0 & y & 0 \\ y & 1 & 0 \\ y^2 & 0 & 0 \end{vmatrix} = 0, \quad c_{0,1} = \begin{vmatrix} 0 & y & 0 \\ y & 1 & 0 \\ 0 & 0 & -1 \end{vmatrix} = y^2,$$

$$c_{1,0} = \begin{vmatrix} 0 & 0 & -1 \\ y & 1 & 0 \\ y^2 & 0 & 0 \end{vmatrix} + \begin{vmatrix} 0 & y & 0 \\ 0 & -1 & 0 \\ y^2 & 0 & 0 \end{vmatrix} = y^2,$$

$$c_{1,1} = \begin{vmatrix} 0 & 0 & -1 \\ y & 1 & 0 \\ 0 & 0 & -1 \end{vmatrix} + \begin{vmatrix} 0 & y & 0 \\ 0 & -1 & 0 \\ 0 & 0 & -1 \end{vmatrix} = 0,$$

$$c_{2,0} = \begin{vmatrix} 0 & 0 & -1 \\ 0 & -1 & 0 \\ y^2 & 0 & 0 \end{vmatrix} = -y^2.$$

Thus we get the same result as before.

Example 2. We now give an example using the quantum plane. This can be defined by setting $\mathbf{R} = \mathbb{C}[y]$, setting $\sigma(y) = qy$ for some constant q and $\sigma(a) = a$ for all $a \in \mathbb{C}$ and finally defining $\delta \equiv 0$. This defines an Ore extension that is known as the quantum plane.

In this algebra $P = yx$ and $Q = (yx)^2 = qy^2x^2$ commute. We compute $\text{Res}(P-s, Q-t)$ in two ways. Using the definition we get

$$\text{Res}(P-s, Q-t) = \begin{vmatrix} 0 & y & -s \\ qy & -s & 0 \\ qy^2 & 0 & -t \end{vmatrix} = (t-s^2)qy^2.$$

We now compute the same thing using Theorem 5:

$$c_{0,0} = \begin{vmatrix} 0 & y & 0 \\ qy & 0 & 0 \\ qy^2 & 0 & 0 \end{vmatrix} = 0, \quad c_{0,1} = \begin{vmatrix} 0 & y & 0 \\ qy & 0 & 0 \\ 0 & 0 & -1 \end{vmatrix} = qy^2,$$

$$c_{1,1} = \begin{vmatrix} 0 & 0 & -1 \\ qy & 0 & 0 \\ 0 & 0 & -1 \end{vmatrix} + \begin{vmatrix} 0 & y & 0 \\ 0 & -1 & 0 \\ 0 & 0 & -1 \end{vmatrix} = 0,$$

$$c_{1,0} = \begin{vmatrix} 0 & 0 & -1 \\ qy & 0 & 0 \\ qy^2 & 0 & 0 \end{vmatrix} + \begin{vmatrix} 0 & y & 0 \\ 0 & -1 & 0 \\ qy^2 & 0 & 0 \end{vmatrix} = 0,$$

$$c_{2,0} = \begin{vmatrix} 0 & 0 & -1 \\ 0 & -1 & 0 \\ qy^2 & 0 & -0 \end{vmatrix} = -qy^2, \quad c_{2,1} = \begin{vmatrix} 0 & 0 & -1 \\ 0 & -1 & 0 \\ 0 & 0 & -1 \end{vmatrix} = 0.$$

Example 3. We now describe a larger example, using the q-Heisenberg algebra. We generate this as an Ore extension by setting $\mathbf{R} = \mathbb{C}[y]$, $\sigma(y) = qy$ and $\delta(y) = 1$ for some $q \in \mathbb{C}$.

Set $P = Q = (yx)^2 = qy^2x^2 + yx$. Then

$$\mathrm{Res}(P-s, Q-t) = \begin{vmatrix} 0 & qy^2 & y & -s \\ q^3y^2 & (2q+q^2)y & 1-s & 0 \\ 0 & qy^2 & y & -t \\ q^3y^2 & (2q+q^2)y & 1-t & 0 \end{vmatrix}.$$

On computing the determinant we find that

$$\mathrm{Res}(P-s, Q-t) = q^4y^4t^2 - 2q^4y^4st + q^4y^4s^2.$$

We now wish to compute the resultant using our formula. With the same notation as in the previous examples we get

$$c_{0,0} = \begin{vmatrix} 0 & qy^2 & y & 0 \\ q^3y^2 & (2q+q^2)y & 1 & 0 \\ 0 & qy^2 & y & 0 \\ q^3y^2 & (2q+q^2)y & 1 & 0 \end{vmatrix} = 0,$$

$$c_{0,1} = \begin{vmatrix} 0 & qy^2 & y & 0 \\ q^3y^2 & (2q+q^2)y & 1 & 0 \\ 0 & 0 & 0 & -1 \\ q^3y^2 & (2q+q^2)y & 1 & 0 \end{vmatrix} + \begin{vmatrix} 0 & qy^2 & y & 0 \\ q^3y^2 & (2q+q^2)y & 1 & 0 \\ 0 & qy^2 & y & 0 \\ 0 & 0 & -1 & 0 \end{vmatrix} = 0 + 0 = 0,$$

$$c_{1,0} = \begin{vmatrix} 0 & 0 & 0 & -1 \\ q^3y^2 & (2q+q^2)y & 1 & 0 \\ 0 & qy^2 & y & 0 \\ q^3y^2 & (2q+q^2)y & 1 & 0 \end{vmatrix} + \begin{vmatrix} 0 & qy^2 & y & 0 \\ 0 & 0 & -1 & 0 \\ 0 & qy^2 & y & 0 \\ q^3y^2 & (2q+q^2)y & 1 & 0 \end{vmatrix} = 0 + 0 = 0,$$

$$c_{1,1} = \begin{vmatrix} 0 & 0 & 0 & -1 \\ q^3y^2 & (2q+q^2)y & 1 & 0 \\ 0 & 0 & 0 & -1 \\ q^3y^2 & (2q+q^2)y & 1 & 0 \end{vmatrix} + \begin{vmatrix} 0 & 0 & 0 & -1 \\ q^3y^2 & (2q+q^2)y & 1 & 0 \\ 0 & qy^2 & y & 0 \\ 0 & 0 & -1 & 0 \end{vmatrix} +$$
$$\begin{vmatrix} 0 & qy^2 & y & 0 \\ 0 & 0 & -1 & 0 \\ 0 & 0 & 0 & -1 \\ q^3y^2 & (2q+q^2)y & 1 & 0 \end{vmatrix} + \begin{vmatrix} 0 & qy^2 & y & 0 \\ 0 & 0 & -1 & 0 \\ 0 & qy^2 & y & 0 \\ 0 & 0 & -1 & 0 \end{vmatrix} = 0 - q^4y^4 - q^4y^4 + 0 = -2q^4y^4,$$

$$c_{2,0} = \begin{vmatrix} 0 & 0 & 0 & -1 \\ 0 & 0 & -1 & 0 \\ 0 & qy^2 & y & 0 \\ q^3y^2 & (2q+q^2)y & 1 & 0 \end{vmatrix} = q^4y^4,$$

$$c_{0,2} = \begin{vmatrix} 0 & qy^2 & 1 & 0 \\ q^3y^2 & (2q+q^2)y & 1 & 0 \\ 0 & 0 & 0 & -1 \\ 0 & 0 & -1 & 0 \end{vmatrix} = q^4y^4,$$

$$c_{2,1} = \begin{vmatrix} 0 & 0 & 0 & -1 \\ 0 & 0 & -1 & 0 \\ 0 & 0 & 0 & -1 \\ q^3y^2 & (2q+q^2)y & 1 & 0 \end{vmatrix} + \begin{vmatrix} 0 & 0 & 0 & -1 \\ 0 & 0 & -1 & 0 \\ 0 & qy^2 & y & 0 \\ 0 & 0 & -1 & 0 \end{vmatrix} = 0 + 0 = 0,$$

$$c_{1,2} = \begin{vmatrix} 0 & 0 & 0 & -1 \\ q^3y^2 & (2q+q^2)y & 1 & 0 \\ 0 & 0 & 0 & -1 \\ 0 & 0 & -1 & 0 \end{vmatrix} + \begin{vmatrix} 0 & qy^2 & y & 0 \\ 0 & 0 & -1 & 0 \\ 0 & 0 & 0 & -1 \\ 0 & 0 & -1 & 0 \end{vmatrix} = 0 + 0 = 0,$$

$$c_{2,2} = \begin{vmatrix} 0 & 0 & 0 & -1 \\ 0 & 0 & -1 & 0 \\ 0 & 0 & 0 & -1 \\ 0 & 0 & -1 & 0 \end{vmatrix} = 0.$$

As expected we get the same result using both calculations.

Example 4. We give a second example in the q-Heisenberg algebra. Take $P = (y^2x)^2 + y^2x$ and $Q = y^2x$. Then $P = q^2y^4x^2 + (y^3 + qy^3 + y^2)x$ and

$$\operatorname{Res}(P - s, Q - t) = \begin{vmatrix} y^4 & qy^3 + y^3 + y^2 & -s \\ 0 & y^2 & -t \\ q^2y^2 & qy + y - t & 0 \end{vmatrix} = -q^2y^4t^2 - q^2y^4t + q^2y^4s.$$

As before we also compute the coefficients directly using our formula:

$$c_{0,1} = \begin{vmatrix} y^4 & qy^3 + y^3 + y^2 & 0 \\ 0 & y^2 & 0 \\ 0 & -1 & 0 \end{vmatrix} + \begin{vmatrix} y^4 & qy^3 + y^3 + y^2 & 0 \\ 0 & 0 & -1 \\ q^2y^2 & qy + y & 0 \end{vmatrix} = -q^2y^4,$$

$$c_{0,2} = \begin{vmatrix} y^4 & qy^3 + y^3 + y^2 & 0 \\ 0 & 0 & -1 \\ 0 & -1 & 0 \end{vmatrix} = -q^2y^4,$$

$$c_{1,0} = \begin{vmatrix} 0 & 0 & -1 \\ 0 & y^2 & 0 \\ q^2y^2 & (1+q)y & 0 \end{vmatrix} = q^2y^4,$$

$$c_{1,1} = \begin{vmatrix} 0 & 0 & -1 \\ 0 & 0 & -1 \\ q^2y^2 & (1+q)y & 0 \end{vmatrix} + \begin{vmatrix} 0 & 0 & -1 \\ 0 & y^2 & 0 \\ 0 & -1 & 0 \end{vmatrix} = 0.$$

7. Annihilating polynomials for elements in a specific commutative subalgebra

Suppose that $\mathbf{R} = S[y]$ for some commutative ring S, by which we mean simply the ordinary polynomial ring over S. In this case it is natural to suppose that σ and δ are S-linear functions. The important case of the Heisenberg algebra is one such example as are the more general q-deformed Heisenberg algebras. Under the preceding assumptions we can specify σ and δ completely by just giving $\sigma(y)$, respectively $\delta(y)$.

The elements of the form $(yx)^n$, for natural numbers n, generate a subalgebra of $S[y][x;\sigma,\delta]$. It will be a commutative sub-algebra, as is easy to see. If we assume that P and Q come from this subalgebra we can give a formula for the elements in the determinant, using the function π_i^n introduced in section 2. We first have that

$$x^n(yx)^m = \sum_{i_1=1}^{n+m}\sum_{i_2=1}^{n+m-1}\cdots\sum_{i_m=1}^{n+1}\pi_{i_m-1}^{n}(y)\pi_{i_{m-1}-1}^{i_m}(y)\dots\pi_{i_1}^{i_2}(y)x^{i_1}$$

which is proven by induction. If $P = \sum_{i=0}^{m} a_i(yx)^m$ and $Q = \sum_{j=0}^{n} b_j(yx)^n$ we can from this get an expression for the elements of the matrix whose determinat is $\text{Res}(P-s,Q-t)$. Setting $e_{i,j}$ to be the element of row i and column $n+m-j$ and letting $d_{i,j}$ denote the Kronecker delta-function we have that

$$d_{i,j}s + e_{i,j} = \sum_{k=0}^{m} a_k \sum_{i_2=1}^{k+i-2}\cdots\sum_{i_k=1}^{i}\pi_{i_k-1}^{i-1}(y)\dots\pi_{j-2}^{i_2}(y)$$

if $j \le n$ and similarily if $j > n$

$$d_{i,j+n}t + e_{i,j} = \sum_{k=0}^{n} b_k \sum_{i_2=1}^{k+i-n-2}\cdots\sum_{i_k=1}^{i-n}\pi_{i_k-1}^{i-n-1}(y)\dots\pi_{j-2}^{i_2}(y).$$

Assume now that $\mathbf{R} = k[y]$ for some field k, and that $\delta \equiv 0$.

Set $P = \sum_{i=0}^{m} a_i(yx)^m$ and $Q = \sum_{j=0}^{n} b_j(yx)^n$. In this special case we can prove the following

Theorem 6. *Let* $\mathbf{R} = k[y]$ *for some field* k. *Let* σ *be any endomorphism and assume* δ *is identically zero. If* $P\sum_{i=0}^{m} a_i(yx)^m$ *and* $Q = \sum_{j=0}^{n} b_j(yx)^n$ *then* P *and* Q *commute and* $\text{Res}(P-s,Q-t) = G(s,t)\prod_{j=0}^{n-1}\prod_{i=0}^{j+m-1}\sigma^i(y)$ *where* $G(s,t)$ *does not contain any non-zero power of* y.

Proof. Denote the element in row i and column $n+m-j+1$ by $r_{i,j}$. Then it can be seen that

$$r_{i,j} = C_{i,j}\prod_{k=1}^{j-1}(\sigma^{i+k-2}(y))$$

if $i \le n$, where $C_{i,j}$ does not depend on y and we interpret the empty product as 1. If $i > n$ we get that

$$r_{i,j} = C_{i,j}\prod_{k=1}^{j+n-1}\sigma^{i+k-2-n}(y).$$

If $i \le n$ and $i > j$ we must have that $C_{i,j} = 0$. Similarily if $i > n$ and $i - n > j$ we have that $C_{i,j} = 0$.

We know that

$$\text{Res}(P-s,Q-t) = \sum_{\alpha\in S_{n+m}}\text{sign}(\alpha)\prod_{i=1}^{m+n} r_{i,\alpha(n+m-i+1)}$$

Journal of Physics: Conference Series **346** (2012) 012021 doi:10.1088/1742-6596/346/1/012021

where the sum is over all possible permutations of $n+m$ elements. We let γ denote the permutation that maps $n+m-j+1$ to j for all $1 \le j \le n+m$. Then we can write

$$\mathrm{Res}(P-s, Q-t) = \sum_{\alpha \in S_{n+m}} \mathrm{sign}(\alpha \circ \gamma) \prod_{i=1}^{m+n} r_{i,\alpha((\gamma(n+m-i+1))} = \sum_{\alpha \in S_{n+m}} \mathrm{sign}(\alpha \circ \gamma) \prod_{i=1}^{m+n} r_{i,\alpha(i)}.$$

Now, we compute one term of this expansion, ignoring the sign:

$$\begin{aligned}\prod_{i=1}^{n+m} r_{i,\alpha(i)} &= \prod_{i=1}^{n} C_{i,\alpha(i)}(\prod_{k=1}^{\alpha(i)-i} \sigma^{k+i-2}(y))^{-1}) \prod_{i=n+1}^{n+m} C_{i,\alpha(i)} \prod_{k=1}^{\alpha(i)+n-i} \sigma^{i+k-2-n} \\ &= \prod_{i=1}^{n+m} C_{i,\alpha(i)} \prod_{i=1}^{n}(\prod_{k=1}^{\alpha(i)-i} \sigma^{k+i-2}(y)) \prod_{i=n+1}^{n+m}(\prod_{k=1}^{\alpha(i)+n-i} \sigma^{i+k-2-n}(y)).\end{aligned}$$

We count the number of times a factor σ^p appears in this expression. It can be seen that for any natural number p, $\sigma^p(y)$ occurs

$$S = |A_1 \cap C_1| + |A_2 \cap C_1|$$

number of times in the factorization, where

$$\begin{aligned} A_1 &= \{i \mid 1 \le i \le n \text{ and } i-1 \le p\} \\ A_2 &= \{i \mid n+1 \le i \le n+m \text{ and } i \le p+n+1\} \\ C_1 &= \{i \mid \alpha(i) \ge p+2\} \end{aligned}$$

and $|F|$ denotes the number of elements of set F. We are not interested in those α that make any $C_{i,\alpha(i)}$ zero since the claim is trivially true for them. We will use this fact. So we assume that $\alpha(i) \ge i$ if $i \le n$ and $\alpha(i) \ge i-n$ otherwise.

Define

$$\begin{aligned} B_1 &= \{i \mid 1 \le i \le n \text{ and } p \le i-2\} \\ B_2 &= \{i \mid n+1 \le i \le n+m \text{ and } p+n+2 \le i\} \\ C_2 &= \{i \mid \alpha(i) \le p+1\} \end{aligned}$$

We can see that the A_i and B_i form one partition of the set $\{1, \ldots n+m\}$ and the C_i another. We further note that if $i \in C_2 \cap B_1$ we must have that $i > \alpha(i)$ and $i \le n$ which we have assumed can not happen. Similarly $C_2 \cap B_2$ is empty. We use this to rewrite S as

$$S = |A_1 \cap C_1| + |A_2 \cap C_1| - |B_1 \cap C_2| - |B_2 \cap C_2|.$$

Setting $A = A_1 \cup A_2$ and $B = B_1 \cup B_2$ we then compute S as follows

$$\begin{aligned} S &= |A \cap C_1| - |B \cap C_2| = |A \cap C_1| - |A^c \cap C_1^c| = |A \cap C_1| - |(A \cup C_1)^c| \\ &= |A \cap C_1| - (m+n-|A \cup C_1|) = |A \cap C_1| + |A \cup C_1| - m - n \\ &= |A \cap C_1| + |A| + |C_1| - |A \cap C_1| - m - n = |A| + |C_1| - m - n \end{aligned}$$

where we have used the principle of inclusion-exlusion in the last step. This result is independent of the permuation α which almost proves the proposition. Simply see that if we multiply all the elements along the anti-diagonal we get the claimed formula to finish the proof.

□

As a special case of the theorem we get the following

Corollary 1. *Assume that* $\mathbf{R} = k[y]$. *If* P,Q *are linear combinations of elements of the form* $(yx)^i$, *where* P *is of degree* m *and* Q *of degree* n *in* x, δ *is identically zero and* $\sigma(y) = qy^p$, *where* q *is an element of the field* k *and* p *is a positive integer, then* P *and* Q *commute and*

$$\operatorname{Res}(P - s, Q - t) = G(s,t) \cdot y^{(\sum_{i=0}^{n-1} p^i)(\sum_{i=0}^{m-1} p^i)}.$$

For general commuting P and Q we can write the resultant as

$$\operatorname{Res}(P - s, Q - t) = \sum G_i(s,t) y^i$$

where the G_i do not contain any power of y. It would be interesting to find conditions guaranteeing that $G_i(P,Q) = 0$ for all i. This will not be true in general but if $\mathbf{R}[x] = S[y][x;\sigma,0]$ where S is an integral domain and σ is injective and P and Q are of the form considered in this section then $G_i(P,Q) = 0$ for all i, since all the G_i will in fact be equal.

Example 5. We include an example to illustrate Corollary 1. Set $P = (yx)^2$ and $Q = yx$. Then $P = qy^{k+1}x^2$ and $xQ = qy^k x^2$ so

$$\operatorname{Res}(P - s, Q - t) = \begin{vmatrix} qy^{k+1} & 0 & -s \\ 0 & y & -t \\ qy^k & -t & 0 \end{vmatrix} = q(s^2 - t) \cdot y^{k+1}.$$

With the notation of the theorem $G(s,t) = q(s - t^2)$ in this case.

Acknowledgments

This research was supported in part by The Swedish Foundation for International Cooperation in Research and Higher Education (STINT), The Swedish Research Council (grant 2007-6338), The Swedish Royal Academy of Sciences and The Crafoord Foundation. The second author also is grateful to Institut Mittag-Leffler for support of his participation in Quantum Information Theory program in Autumn 2010, which has been beneficial for completion of this work.

References

[1] Burchnall J L and Chaundy T W 1922 Commutative ordinary differential operators *Proc. London Math. Soc. Ser. 2* **21** 420
[2] Burchnall J L and Chaundy T W 1928 Commutative ordinary differential operators *Proc. Roy. Soc. London A* **118** 557
[3] Burchnall J L and Chaundy T W 1932 Commutative ordinary differential operators. II. — The Identity $P^n = Q^m$ *Proc. Roy. Soc. London A* **134** 471
[4] Christiano J G and Hall J E 1964 On the n-th Derivative of a Determinant of the j-th Order *Math. Mag.* **37** 215
[5] Goodearl K R 1983 Centralizers in differential, pseudodifferential, and fractional differential operator rings *Rocky Mountain J. Math.* **13** 573
[6] Hellström L and Silvestrov S 2000 *Commuting Elements in q-Deformed Heisenberg Algebras* (River Edge, NJ: World Scientific Publishing Co., Inc.)
[7] Hellström L and Silvestrov S 2007 Ergodipotent maps and commutativity of elements in non-commutative rings and algebras with twisted intertwining *J. Algebra* **314** 17
[8] de Jeu M, Svensson C and Silvestrov S 2009 Algebraic curves for commuting elements in the q-deformed Heisenberg algebra *J. Algebra* **321** 1239
[9] Krichever I M 1977 Integration of non-linear equations by the methods of algebraic geometry *Funktz. Anal. Priloz.* **11** 15
[10] Krichever I M 1977 Methods of algebraic geometry in the theory of nonlinear equations *Uspekhi Mat. Nauk* **32** 183

[11] Larsson D and Silvestrov S D 2003 Burchnall-Chaundy theory for q-difference operators and q-deformed Heisenberg algebras *J. Nonlin. Math. Phys.* **10** Suppl 2 95

[12] Larsson D 2008 Burchnall-Chaundy theory, Ore extensions and σ-differential operators *Preprint U.U.D.M. Report 2008:45, Department of Mathematics, Uppsala University*

[13] Li Z 1998 *A subresultant theory for Ore polynomials with applications* Proc. of the 1998 Int. Symp. on Symbolic and Algebraic Computation (Rostock) (ACM, New York) pp 132–139

[14] Mumford D 1978 *An algebro-geometric construction of commuting operators and of solutions to the Toda lattice equation, Korteweg deVries equation and related nonlinear equation* Proc. of the International Symposium on Algebraic Geometry (Kyoto Univ., Kyoto, 1977) (Kinokuniya Book Store, Tokyo) pp 115–153

[15] Richter J and Silvestrov S 2009 On algebraic curves for commuting elements in q-Heisenberg algebras *J. Gen. Lie Theory Appl.* **3** 321

Algebra, Geometry, and Mathematical Physics 2010 IOP Publishing
Journal of Physics: Conference Series **346** (2012) 012022 doi:10.1088/1742-6596/346/1/012022

Some generalizations of the Hewitt extension

Jaan Vajakas
Institute of Mathematics, University of Tartu, J. Liivi 2, Tartu 50409, Estonia
E-mail: jv@ut.ee

Abstract. We introduce two extensions of a Tychonoff space that generalize the notions of the Hewitt extension and the "generalized Hewitt extension" introduced by Bachman, Beckenstein, Narici and Warner. The extensions are determined by sets of continuous functions. In particular, we study the extensions determined by all continuous functions and the extension determined by all continuous bounded functions and show that these can be regarded as epireflective functors. Finally we discuss the possibility of using these extensions to describe the characters of function algebras.

1. Introduction

A topological space Y is called an *extension* of another topological space X if X is a dense subspace of Y. An extension that is a compact Hausdorff space is called a *compactification*. It is a well known fact that every Tychonoff space X has a compactification βX such that for every compact Hausdorff space Y, every continuous map $f : X \to Y$ has a continuous extension $f^\beta : \beta X \to Y$ (uniquely determined since Y is Hausdorff). Such an extension βX is unique up to topological equivalence and is called the *Stone-Čech compactification* of X [3].

Let us denote for two topological spaces X and Y by $C(X,Y)$ the set of all continuous functions from X to Y, by $C_C(X,Y)$ the set of all continuous functions with relatively compact image, and if Y is a topological vector space then by $C_b(X,Y)$ the set of all continuous functions with bounded image. Unless noted otherwise, by *topological vector space* we mean a topological vector space over $\mathbb{R}$ or $\mathbb{C}$.

Another well-known extension of a Tychonoff space X is its *Hewitt extension*, also called the *realcompactification* or *repletion* and denoted by υX. It is defined as follows:

$$\upsilon X := \{p \in \beta X \mid \hat{f}(p) \neq \infty \quad \forall f \in C(X,\mathbb{R})\},$$

where $\hat{f} : \beta X \to \mathbb{R} \cup \{\infty\}$ is the continuous extension of f whose existence is provided by the definition of the Stone-Čech compactification. A Tychonoff space X is called *replete* (or *realcompact*) if $X = \upsilon X$. There are surprisingly many replete spaces: e. g. all Lindelöf spaces and complete uniform spaces of non-measurable cardinality [3].

In [2], G. Bachman, E. Beckenstein, L. Narici and S. Warner generalized the Stone–Čech compactification and Hewitt extension: if X and Y are Hausdorff spaces then X is called Y*-completely regular* if the topology of X is the weak topology determined by $C_C(X,Y)$. For a Hausdorff space Y and a Y-completely regular space X, they define the compactification $\beta_Y X$ as the completion of X with respect to the weak uniformity determined by $C_C(X,Y)$ (we regard each $f \in C_C(X,Y)$ as a function into $\operatorname{cl} f[X]$ and since $\operatorname{cl} f[X]$ is a compact Hausdorff space, it admits exactly one uniformity that is compatible with its topology [4, TG II.27, Théorème 1]; $\beta_Y X$ is an extension since the weak uniformity generated by a set of functions is compatible with the weak topology generated by the functions [4, TG II.8, Corollaire]). Every Y-completely regular Hausdorff space is completely regular since a topological space is completely regular iff it is uniformizable [3, p. 5, 0.2-3].

For every compact Y-completely regular topological space Z, each $f \in C(X, Z)$ has an extension $f^{\beta_Y} \in C(\beta_Y X, Z)$. If Y is a complete Hausdorff uniform space and the topology of X is the weak topology generated by $C(X,Y)$ (e. g. if X is Y-completely regular) then the extension $\upsilon_Y X$ is defined in [2] as the completion of X with respect to the weak uniformity determined by $C(X,Y)$. If Y is Y-completely regular then

$$\upsilon_Y X = \{p \in \beta_Y X \mid f^{\beta_Y}(p) \in Y\}.$$

It is easy to see that if $Y = \mathbb{R}$ or, more generally, if Y is a Hausdorff space containing a subspace homeomorphic to $\mathbb{R}$, then a Hausdorff space is Y-completely regular if and only if it is completely regular. For such Y and a Tychonoff space X we have $\beta_Y X = \beta X$, and if additionally Y is a uniform space then $\upsilon_Y X = \upsilon X$.

Generalizing the definition of υX, given Tychonoff spaces X and Y and a set of functions $A \subset C(X,Y)$, let us introduce the extension

$$\alpha_A X := \{p \in \beta X \mid f^\beta(p) \in Y \quad \forall f \in A\}$$

where $f^\beta : \beta X \to \beta Y$ is the continuous extension of f.

Likewise, generalizing the definition of $\upsilon_Y X$, we set

$$\gamma_A X := \{p \in \beta_Y X \mid f^{\beta_Y}(p) \in Y \quad \forall f \in A\}$$

if Y is a Y-completely regular complete Hausdorff uniform space, X a Y-completely regular Hausdorff space, A a set of functions such that

$$C_C(X,Y) \subset A \subset C(X,Y)$$

and $f^{\beta_Y} \in C(\beta_Y X, \beta_Y Y)$ denotes the continuous extension of f.

In addition, we denote

$$\alpha_Y^C X := \alpha_{C(X,Y)} X, \quad \alpha_Y^{C_b} X := \alpha_{C_b(X,Y)} X, \quad \gamma_Y^C X := \gamma_{C(X,Y)} X, \quad \gamma_Y^{C_b} X := \gamma_{C_b(X,Y)} X.$$

2. Some properties of $\alpha_A X$ and $\gamma_A X$

It is not hard to see that $\alpha_A X$ (or $\gamma_A X$) may be alternatively defined as the largest subset in βX (or, respectively, $\beta_Y X$) that contains X and where every function in A can be continuously extended or as the set of points p in βX (or, respectively, $\beta_Y X$) such that all functions in A transform every net in X converging to p into one that converges in Y.

As special cases we have

$$\alpha_{\mathbb{R}}^{C_b} X = \beta X, \quad \alpha_{\mathbb{R}}^C X = \upsilon X, \quad \gamma_{C_C(X,Y)} X = \beta_Y X, \quad \gamma_Y^C X = \upsilon_Y X.$$

If the extension $\gamma_A X$ is defined (i. e. if Y is a Y-completely regular complete Hausdorff uniform space, X is a Y-completely regular Hausdorff space and $C_C(X,Y) \subset A \subset C(X,Y)$) and the extensions $\beta_Y X$ and βX are topologically equivalent, then we can identify $\alpha_A X = \gamma_A X$. In particular, since every Hausdorff topological vector space with more than one element contains a subspace homeomorphic to $\mathbb{R}$ (for it obviously has a one-dimensional subspace and every finite-dimensional vector space admits only one Hausdorff topology [9, p. 151]) and Y-complete regularity is just complete regularity for a space Y containing $\mathbb{R}$, we have that $\alpha_A X = \gamma_A X$ if Y is a complete Hausdorff topological vector space with more than one element, X is a Tychonoff space and $C_C(X,Y) \subset A \subset C(X,Y)$.

Proposition 2.1. *Let Y be a Y-completely regular complete Hausdorff uniform space, X a Y-completely regular Hausdorff space and $C_C(X.Y) \subset A \subset C(X, Y)$. Then the uniform space $(\gamma_A X, \mathcal{U}_\gamma)$ is the completion of the uniform space $(X, \mathcal{U}_X)$ where $\mathcal{U}_\gamma$ is the weak uniformity determined by the functions in $A^\gamma := \{f^{\gamma_A} \mid f \in A\}$ (where $f^{\gamma_A} \in C(\gamma_A X, Y)$ is the continuous extension of f) and $\mathcal{U}_X$ is the weak uniformity determined by the functions in A.*

Proof. Analogous to [2, p. 22, Theorem 1.5-1]. □

The following properties can be proven in a straightforward manner.

Proposition 2.2. *Let X, Y and Z be Tychonoff spaces. Then the following properties hold.*

(a) *For $A \subset C(X, Y)$, the equality $\alpha_A X = \beta X$ holds if and only if $A \subset C_C(X, Y)$.*

(b) *Let $V \subset C(X, Z)$ and $W \subset C(Z, Y)$. If $\alpha_W Z = Z$ and*

$$\{w \circ v \mid w \in W, \ v \in V\} \subset U \subset C(X, Y)$$

then $\alpha_U X \subset \alpha_V X$.

(c) *If $h : Z \to Y$ is an embedding with closed image and $A \subset C(X, Y)$ then $\alpha_A X \subset \alpha_B X$, where $B := \{h^{-1} \circ f \mid f \in A, \ f[X] \subset h[Z]\}$.*

3. The extension determined by all continuous functions

Proposition 3.1. *Let X, Y and Z be Tychonoff spaces.*

(a) *If $\alpha_Y^C Z = Z$ then $\alpha_Y^C X \subset \alpha_Z^C X$.*

(b) *If $\upsilon Y = Y$ then $\upsilon X \subset \alpha_Y^C X$.*

(c) *If Y contains a closed subspace homeomorphic to $\mathbb{R}$ then $\alpha_Y^C X \subset \upsilon X$.*

(d) *If Y contains a closed subspace homeomorphic to $\mathbb{R}$ and at least one of X and Y is replete, then $\upsilon X = \alpha_Y^C X$.*

(e) *If Y is a Hausdorff topological vector space with more than one element and at least one of X and Y is replete then $\upsilon X = \alpha_Y^C X$.*

(f) *Let X_1 and X_2 be Tychonoff spaces and K_1 and K_2 their compactifications, correspondingly. If there exists $\phi \in C(K_1, K_2)$ such that $\phi^{-1}[X_2] = X_1$ then $\alpha_{X_2}^C X_1 = X_1$.*

Proof. 2.2(b) $\Rightarrow$ (a) $\Rightarrow$ (b). 2.2(c) $\Rightarrow$ (c). (b) $\wedge$(c) $\Rightarrow$ (d) $\Rightarrow$ (e) (since every finite-dimensional subspace in a topological vector space is closed [9, p. 152]).

(f): Since any compactification of X_1 is a continuous image of βX_1 under a mapping that leaves points of X_1 fixed [13, p. 10], there exists $h \in C(\beta X_1, K_1)$ such that $h^{-1}[X_1] = X_1$. Then $(\phi|_{X_1})^\beta = \phi \circ h$ and $(\phi \circ h)^{-1}[X_2] = h^{-1}[X_1] = X_1$, so for any $p \in K_1 \setminus X_1$ we have $(\phi|_{X_1})^\beta(p) \in K_2 \setminus X_2$. □

4. The extension determined by continuous bounded functions

Let Y be a Hausdorff topological vector space with one or more element, and X a completely regular Hausdorff space. Since $\beta_Y X = \beta X$ for a topological Hausdorff vector space Y with more than one element, $\gamma_Y^{C_b} X = \alpha_Y^{C_b} X$ whenever $\gamma_Y^{C_b} X$ is defined (i. e. if Y is complete).

The extension $\alpha_Y^{C_b} X$ equals βX if and only if the image of every continuous bounded function is relatively compact. Thus, in case every bounded set in Y is relatively compact (e. g. if Y is finite-dimensional or, more generally, a Montel space), we have $\alpha_Y^{C_b} X = \beta X$. In what follows we shall find a sufficient condition for the equality $\alpha_Y^{C_b} X = \upsilon X$.

Journal of Physics: Conference Series **346** (2012) 012022 doi:10.1088/1742-6596/346/1/012022

Proposition 4.1. *Let Y be a Hausdorff topological vector space and X a Tychonoff space. Then*

$$\alpha_Y^{C_b} X = \bigcap_{U \in \mathcal{B}_Y} \alpha_U^C X,$$

where $\mathcal{B}_Y$ is the set of all closed bounded sets of the space Y.

Proof. The inclusion $\subset$ follows from 2.2(c). The inclusion $\supset$ is also easy to see. □

Corollary 4.2. *Let Y be a replete topological vector space. If Y contains a closed bounded subset K such that $\alpha_K^C \mathbb{R} = \mathbb{R}$ then $\alpha_Y^{C_b} X = \upsilon X$ for every Tychonoff space X.*

Proof. By 3.1(b) $\upsilon X \subset \alpha_Y^C X$; since $\alpha_Y^C X \subset \alpha_Y^{C_b} X$, we have $\upsilon X \subset \alpha_Y^{C_b} X$. By 4.1 we have $\alpha_Y^{C_b} X \subset \alpha_K^C X$ and by 3.1(a), $\alpha_K^C X = \alpha_{\mathbb{R}}^C X \subset \upsilon X$. □

Proposition 4.3. *Every infinite-dimensional locally bounded Hausdorff vector space contains a closed bounded subset homeomorphic to the half-open interval $[0, \infty)$.*

Proof. The proof is similar to the proof of non-compactness of the unit sphere in infinite-dimensional normed space.

Let Y be an infinite-dimensional locally bounded Hausdorff vector space. Then the topology of Y is induced by a quasinorm $\|\cdot\|$, i. e. a functional satisfying all the norm axioms except that the triangle equality is replaced with $\|x+y\| \leq k(\|x\| + \|y\|)$, where $k \geq 1$ [9, p. 159]. Let S bet the corresponding unit sphere, i. e. $S = \{y \in Y \mid \|y\| = 1\}$. For every finite subset $\{x_1, \ldots, x_n\} \subset Y$, let $L(x_1, \ldots, x_n)$ denote the linear span of the subset, and for each point $y \in Y$ and subset $T \subset Y$, let $d(y, T) := \inf_{z \in T} \|y - z\|$.

As in the case of normed spaces, one can prove that the Riesz lemma holds: for every closed proper subspace $T \subset Y$ and $\alpha < 1$ there exists $y \in S$ such that $d(y, T) > \alpha$.

To construct the homeomorphism, we begin by choosing an arbitrary $y_0 \in S$. $L(y_0)$ is closed since it is finite-dimensional, and since Y is infinite-dimensional, $L(y_0) \neq Y$. Hence, by the Riesz lemma there exists $y_1 \in S$ such that $d(y_1, L(y_0)) > \frac{1}{2}$. By applying Riesz lemma again we get $y_2 \in S$ such that $d(y_2, L(y_0, y_1)) > \frac{1}{2}$. Continuing this way, we get a sequence $(y_n)_{n \in \mathbb{N}}$, where $y_n \in S$ and $d(y_n, L(y_0, \ldots, y_{n-1})) > \frac{1}{2}$ for all $n \in \mathbb{N}$.

Let us define for each $n \in \mathbb{N}$ a map $f_n : [n-1, n] \to Y$ as follows:

$$f_n(t) := (n-t)y_{n-1} + (t-n+1)y_n \quad \forall t \in [n-1, n].$$

The maps f_n are obviously continuous and agree on the intersections of their domains, i. e. $f_n(n) = f_{n+1}(n)$ for all $n \in \mathbb{N}$. Let $f : [0, \infty) \to Y$ be the map obtained by gluing together the functions f_n (i. e. $f|_{[n-1,n]} = f_n$ for all $n \in \mathbb{N}$). Since a function that is continuous on each element of a locally finite closed cover of its domain is continuous, f is continuous.

The map f is one-to-one because if $x \neq y$ then $f(x) - f(y)$ is a non-trivial linear combination of the vectors y_n and by the construction of the sequence (y_n), the elements y_n are linearly independent. The inverse map of f can be obtained by gluing together the inverse maps $f_n^{-1} : f_n[[n-1, n]] \to [0, \infty)$.

Since the domains of the functions f_n, the intervals $[n-1, n]$, are compact and Y is Hausdorff, the functions f_n map closed sets to closed sets and therefore their inverses f_n^{-1} are continuous and their images $f_n[[n-1, n]]$ are closed. To show that f is a homeomorphism onto its image it suffices to prove that the images $f_n[[n-1, n]]$ form a locally finite cover in $f[[0, \infty)]$. For that we show that every point in Y has a neighbourhood that intersects with at most two of the images $f_n[[n-1, n]]$.

Journal of Physics: Conference Series **346** (2012) 012022 doi:10.1088/1742-6596/346/1/012022

Choose $\beta > 0$ such that $1 - \beta > 2k\beta$ and define $d := \min\left\{\frac{\beta}{2}, \frac{1-\beta}{2k} - \beta\right\}$. Let $y \in Y$ and U the open ball with center y and radius $r := \frac{d}{2k}$. We will show that there exists $N \in \mathbb{N}$ such that $U \cap f_n[[n-1,n]] = \emptyset$ for all $n \in \mathbb{N} \setminus \{N,\ N+1\}$. Suppose the opposite. Then there would exist $m, n \in \mathbb{N}$ such that $n > m+1$, $U \cap f[[m, m+1]] \neq \emptyset$ and $U \cap f[[n, n+1]] \neq \emptyset$. Hence there would exist $s \in [m-1, m]$ and $t \in [n-1, n]$ such that $f(s), f(t) \in U$. So we have

$$\|f(t) - f(s)\| \leq k(\|f(t) - y\| + \|y - f(s)\|) < d.$$

Let $\lambda := s - m + 1, \mu := t - n + 1$. Then

$$f(s) = (1-\lambda)y_{m-1} + \lambda y_m, \quad f(t) = (1-\mu)y_{n-1} + \mu y_n.$$

If $\mu \geq \beta$ then, using the inequality $d(y_n, L(y_0, \ldots, y_{n-1})) > \frac{1}{2}$, we would get

$$\begin{aligned}\|f(t) - f(s)\| &= \|\mu y_n + (1-\mu)y_{n-1} - (1-\lambda)y_{m-1} - \lambda y_m\| \\ &= \mu \left\| y_n + \frac{1-\mu}{\mu} y_{n-1} - \frac{1-\lambda}{\mu} y_{m-1} - \frac{\lambda}{\mu} y_m \right\| \\ &\geq \mu\, d(y_n, L(y_0, \ldots, y_{n-1})) \geq \frac{\beta}{2} \geq d\end{aligned}$$

— a contradiction. If $\mu < \beta$ then

$$\begin{aligned}\|f(t) - f(s)\| &= \|\mu y_n + (1-\mu)y_{n-1} - (1-\lambda)y_{m-1} - \lambda y_m\| \\ &\geq \frac{1-\mu}{k} \left\| y_{n-1} - \frac{1-\lambda}{1-\mu} y_{m-1} - \frac{\lambda}{1-\mu} y_m \right\| - \mu\|y_n\| \\ &\geq \frac{1-\beta}{k} d(y_{n-1}, L(y_0, \ldots, y_{n-2})) - \beta \geq d\end{aligned}$$

— again a contradiction. So our assumption must be false, i. e. there must exist $N \in \mathbb{N}$ such that $U \cap f_n[[n-1, n]] = \emptyset$ for all $n \notin \{N,\ N+1\}$.

We have shown that the images of the functions f_n form a locally finite closed cover of the image of f and therefore f is a homeomorphism onto its image. Since the union of a locally finite set of closed sets is closed, the image of f is closed. As the values of f are convex combinations of pairs of elements of the unit sphere, we have $f[[0, \infty)] \subset kB$, where $B := \{x \in Y \mid \|x\| \leq 1\}$ is the closed unit ball, so the image of f is bounded. Hence, the image of f is the desired subset. □

Theorem 4.4. *Let Y be a replete topological vector space that contains an infinite-dimensional closed subspace which is homeomorphically isomorphic to a locally bounded vector space. Then $\alpha_Y^{C_b} X = \upsilon X$ for every Tychonoff space X.*

Proof. Let $Z \subset Y$ be the closed subspace mentioned in the assumption. By Proposition 4.3 there exists a closed bounded subset $K \subset Z$ homeomorphic to $[0, \infty)$. The set K is also closed and bounded in Y. From 3.1(f) we get that $\alpha_{[0,\infty)}^{C} \mathbb{R} = \mathbb{R}$ (by considering the map $\phi : x \mapsto |x|,\ [-\infty, \infty] \to [0, \infty]$). Apply Corollary 4.2. □

Example 4.5. *If Y is the space $C(\mathbb{R}, \mathbb{R})$ with compact-open topology, then $\alpha_Y^{C_b} X = \upsilon X$ for every Tychonoff space X.*

Proof. The set $\{f \in C(\mathbb{R}, \mathbb{R}) \mid f[\mathbb{R} \setminus (0,1)] = \{0\}\}$ is a closed infinite-dimensional subspace of $C(\mathbb{R}, \mathbb{R})$ and the subspace topology is induced by the norm $\|f\| = \sup_{x \in [0,1]} |f(x)|$. □

5. α_Y^C, $\alpha_Y^{C_b}$ and γ_Y^C as epireflective functors

For every topological space X there exist a Tychonoff space ρX and a surjection $h : X \to \rho X$ such that for every Tychonoff space Y each $f \in C(X, Y)$ factors as $f = f^\rho \circ h$ where $f^\rho \in C(\rho X, Y)$ [13, pp. 5–7]. It is well known [13, p. 261] that the functors $X \mapsto \rho X$, $X \mapsto \beta\rho X$ and $X \mapsto \upsilon\rho X$ are epireflective functors on the category of Hausdorff spaces. We shall see that α_Y^C, $\alpha_Y^{C_b}$ and γ_Y^C generate similar epireflective functors. We shall use the following definition of epireflectivity (from [13, pp. 248–250]).

Definition 5.1. A functor r from a category $\mathcal{A}$ to its subcategory $\mathcal{B}$ is called a *reflective functor* if for every object A in $\mathcal{A}$ there exists a morphism $\eta_A : A \to r(A)$ such that for every object B in $\mathcal{B}$, each morphism $f : A \to B$ factors uniquely as $f = g \circ \eta_A$ where $g : r(A) \to B$ is a morphism in $\mathcal{B}$. A subcategory $\mathcal{B}$, for which such a functor r exists, is called a *reflective subcategory* of $\mathcal{A}$ and η_A is called the *reflection* of A. The reflective functor r and the reflective subcategory $\mathcal{B}$ are called *epireflective* if η_A is always an epimorphism.

Lemma 5.2. *Let X and Y be Tychonoff spaces. Then $\alpha_Y^C\alpha_Y^C X = \alpha_Y^C X$. If Y is a topological vector space, then $\alpha_Y^{C_b}\alpha_Y^{C_b} X = \alpha_Y^{C_b} X$. If $\gamma_Y^C X$ is defined then $\gamma_Y^C\gamma_Y^C X = \gamma_Y^C X$.*

Proof. It follows from the definition of the Stone-Čech compactification that we may identify $\beta X = \beta Z$ whenever $X \subset Z \subset \beta X$. Thus $\beta\alpha_Y^C X = \beta X$. It is easy to see that the map $f \mapsto f^{\alpha_Y^C X}$, $C(X, Y) \to C(\alpha_Y^C X, Y)$, taking a function to its continuous extension, is bijective. Since $f^\beta = (f^{\alpha_Y^C X})^\beta$ for all $f \in C(X, Y)$, we have

$$\{f^\beta \mid f \in C(\alpha_Y^C X, Y)\} = \{f^\beta \mid f \in C(X, Y)\}$$

and thus, by the definition of α^C, we get $\alpha_Y^C\alpha_Y^C X = \alpha_Y^C X$.

The statement $\alpha_Y^{C_b}\alpha_Y^{C_b} X = \alpha_Y^{C_b} X$ can be proven in a similar way, by using the bijection $f \mapsto f^{\alpha_Y^{C_b}}$, $C_b(X, Y) \to C_b(\alpha_Y^{C_b} X, Y)$ that takes a function to its continuous extension (that this map is correctly defined and bijective follows from the fact that closures and subsets of bounded sets are bounded). The statement about γ_Y^C has an analogous proof (one must additionally verify that $\gamma_Y^C\gamma_Y^C X$ is defined whenever $\gamma_Y^C X$ is defined, but that is easy to do). □

Note that we do not talk about $\gamma_Y^{C_b} X$ since whenever it is defined, it coincides with $\alpha_Y^{C_b} X$.

Lemma 5.3. *Let X_1, X_2 and Y be Tychonoff spaces. Then*

(a) *each $f \in C(X_1, X_2)$ can be extended to a continuous map from $\alpha_Y^C X_1$ to $\alpha_Y^C X_2$;*

(b) *if Y is a topological vector space, then each $f \in C(X_1, X_2)$ can be extended to a continuous map from $\alpha_Y^{C_b} X_1$ to $\alpha_Y^{C_b} X_2$;*

(c) *if the extensions $\gamma_Y^C X_1$ and $\gamma_Y^C X_2$ are defined, then each $f \in C(X_1, X_2)$ can be extended to a continuous map from $\gamma_Y^C X_1$ to $\gamma_Y^C X_2$.*

Proof. Let us prove (b); (a) and (c) are analogous.

Let $f \in C(X_1, X_2)$. For each $g \in C_b(X_2, Y)$ then $g \circ f \in C_b(X_1, Y)$ and $g^\beta \circ f^\beta = (g \circ f)^\beta$. Thus $g^\beta(f^\beta(p)) = (g \circ f)^\beta(p) \in Y$ for every $p \in \alpha_Y^{C_b} X_1$. Consequently $f^\beta(p) \in \alpha_Y^{C_b} X_2$, as g was arbitrary. Thus $f^\beta|_{\alpha_Y^{C_b} X_1}$ is the desired extension. □

Proposition 5.4. *Let Y be a Tychonoff space and let* Haus *denote the category of all Hausdorff spaces.*

(a) *Let* $\alpha_Y^C\mathrm{Repl}$ *denote the full subcategory of* Haus *consisting of all Tychonoff spaces* X *for which* $X = \alpha_Y^C\rho X$. *The map* Haus $\to \alpha_Y^C\mathrm{Repl}$, $X \mapsto \alpha_Y^C\rho X$ *together with the maps* $C(X_1, X_2) \to C(\alpha_Y^C\rho X_1, \alpha_Y^C\rho X_2)$, *that map each* $f \in C(X_1, X_2)$ *to the continuous extension of* $f^\rho : \rho X_1 \to \rho X_2$ *(provided by lemma 5.3), forms an epireflective functor from* Haus *to* $\alpha_Y^C\mathrm{Repl}$.

(b) *Let* Y *be a topological vector space. Let* $\alpha_Y^{C_b}\mathrm{Repl}$ *be the full subcategory of* Haus *consisting of all Tychonoff spaces* X *for which* $X = \alpha_Y^{C_b}X$. *The map* Haus $\to \alpha_Y^{C_b}\mathrm{Repl}$, $X \mapsto \alpha_Y^{C_b}\rho X$ *and the maps* $C(X_1, X_2) \to C(\alpha_Y^{C_b}\rho X_1, \alpha_Y^{C_b}\rho X_2)$, *that take a function* $f \in C(X_1, X_2)$ *to the continuous extension of* f^ρ, *form an epireflective functor from* Haus *to* $\alpha_Y^{C_b}\mathrm{Repl}$.

(c) *Let* Y *be a* Y*-completely regular complete Hausdorff uniform space. Let* $\gamma_Y^C\mathrm{Repl}$ *denote the full subcategory of* Haus *consisting of all* Y*-completely regular Hausdorff spaces* X *such that* $X = \gamma_Y^C\rho_Y X$. *The map* Haus $\to \gamma_Y^C\mathrm{Repl}$, $X \mapsto \gamma_Y^C\rho_Y X$ *and the maps* $C(X_1, X_2) \to C(\gamma_Y^C\rho_Y X_1, \gamma_Y^C\rho_Y X_2)$, *taking each* $f \in C(X_1, X_2)$ *to the continuous extension of* f^{ρ_Y}, *form an epireflective functor from* Haus *to* $\gamma_Y^C\mathrm{Repl}$.

Proof. Let us consider only case (a) since (b) and (c) are analogous. Let r be the functor from the category of Tychonoff spaces to $\alpha_Y^C\mathrm{Repl}$ that maps every space X to $\alpha_Y^C X$ and every function $f \in C(X_1, X_2)$ to its extension in $C(\alpha_Y^C X_1, \alpha_Y^C X_2)$.

By 5.2 ja 5.3, r is correctly defined, and r is indeed a functor, i. e. preserves identity functions and satisfies $r(g) \circ r(f) = r(g \circ f)$, since continuous functions to a Hausdorff space coincide if they coincide on a dense subset of their domain. It is also easy to see that r is reflective (in the definition of reflectivity we take $\eta_A : X \to r(X)$ to be the inclusion embedding).

As r and ρ are reflective, their composition $r\rho$, the functor the statement is about, is also reflective. It is also an epimorphism since it maps X to a dense subset in $\alpha_Y^C\rho X$. □

As a corollary, products and closed subspaces of spaces belonging to $\alpha_Y^C\mathrm{Repl}$ (or $\alpha_Y^{C_b}\mathrm{Repl}$ or $\gamma_Y^C\mathrm{Repl}$) also belong to $\alpha_Y^C\mathrm{Repl}$ (or, correspondingly, $\alpha_Y^{C_b}\mathrm{Repl}$ or $\gamma_Y^C\mathrm{Repl}$), since a full replete subcategory of Haus is epireflective if and only if it is closed under products and taking closed subspaces [13, p. 259].

6. Characters of function algebras

By *topological algebra* we mean a topological vector space over a field F (with F assumed to be $\mathbb{R}$ or $\mathbb{C}$, unless specified otherwise) endowed with associative distributive multiplication such that the maps $x \mapsto xa$ and $x \mapsto ax$ are continuous for all a. If Y is a topological algebra and A is a subset of $C(X,Y)$ that is closed under multiplication then we regard it as an algebra with operations defined pointwise (note that e. g. $C(X,Y)$ might not be closed under multiplication if multiplication on Y is not jointly continuous [11]). For a topological algebra A, let $\mathrm{Hom}\,A$ denote the set of all characters on A (i. e. multiplicative linear functions with values in F) and $\mathrm{hom}\,A$ the set of all continuous characters.

Let X be a Tychonoff space. Every $f \in C(X,\mathbb{R})$ has a continuous extension $\hat{f} : \upsilon X \to \mathbb{R}$ and thus for each $p \in \upsilon X$ we may consider the functional $\delta_p : f \mapsto \hat{f}(p)$ on $C(X,\mathbb{R})$. The map $p \mapsto \delta_p$ is a homeomorphism between υX and $\mathrm{Hom}\,C(X,\mathbb{R})$ (equipped with the Gelfand topology) [3, p. 30]. Similarly, for $p \in \beta X$ and $f \in C_b(X,\mathbb{R})$, we can define $\delta_p(f) = \hat{f}(p)$ where $\hat{f}$ is the continuous extension of f; then $p \mapsto \delta_p$ is a homeomorphism from βX to $\mathrm{Hom}\,C_b(X,\mathbb{R})$ (actually this is a corollary of the previous sentence since $C_b(X,\mathbb{R}) = C(\beta X,\mathbb{R})$). If F is a complete ultraregular topological field and X is an ultraregular topological space then $\mathrm{Hom}\,C(X,F)$ and $\mathrm{Hom}\,C_C(X,F)$ can be described analogously using $\upsilon_Y X$ and $\beta_Y X$ [2].

A natural approach for describing the characters of a subalgebra of some function algebra $A \subset C(X,Y)$ where Y is a topological algebra would be to look for characters χ in the form

$\chi(f) = \tau(f^\beta(p))$, where $p \in \beta X$, $\tau \in \operatorname{Hom} Y$ and $f^\beta \in C(\beta X, \beta Y)$ is the extension of f. For that to make any sense, $f^\beta(p)$ must belong to Y for all $f \in A$, i. e. $p \in \alpha_A X$. This motivates our definition of the extensions $\alpha_A X$ and $\gamma_A X$.

Unfortunately, such description of characters is not always possible. For example, W. Govaerts has shown in [7] that the algebra $C_b(X, Y)$ may have continuous characters χ that cannot be represented as $\chi(f) = (\tau \circ f)^\beta(p)$, where $p \in \beta X$ and $\tau \in \operatorname{hom} Y$. Since in his example $\operatorname{Hom} Y = \operatorname{hom} Y$ (he considers a certain Banach algebra Y and all characters of a Banach algebra are continuous), we also see that there are characters χ that cannot be represented as $\chi(f) = \tau(f^\beta(p))$, as $(\tau \circ f)^\beta(p) = \tau(f^\beta(p))$ for any $\tau \in C(Y, F)$, $f \in C(X, Y)$ and $p \in \beta X$ such that $f^\beta(p) \in Y$.

However, in some cases results can be obtained. Many authors have gotten descriptions for the characters of subalgebras of $C(X, Y)$, with Y a topological algebra, in the form $\chi(f) = \tau(f^\beta(p))$ where $p \in X$ or $p \in \upsilon X$ or $p \in \beta X$ (and in the situations considered, $\alpha_A X$ usually coincides with X or υX or βX), see e. g. [8] for the case when Y is a commutative unital Q-algebra with continuous inversion (in addition to the real and complex cases, the case of an algebra over an ultraregular field F is also considered there, and $\upsilon_F X$ and $\beta_F X$ are used in those cases), [6] for the case when Y is metrizable, [1] for a generic approach for describing continuous characters of vector-valued function algebras. Some cases when the space of possible values of p might be some other subspace of βX have been considered by L. Oubbi. In what follows we shall state some of his results using the extension $\alpha_A X$ and fix a small error in a lemma in his article [12].

In [12], subalgebras $A \subset C(X, Y)$, where Y is an algebra with continuous multiplication over $F = \mathbb{C}$, are considered and the following subsets of βX are defined:

$$\begin{aligned}\upsilon_A X &:= \left\{p \in \alpha_A X \mid \exists f \in A : f^\beta(p) \notin r(Y)\right\},\\ \beta_A X &:= \left\{p \in \alpha_B X \mid \exists f \in A\ \exists \tau \in \operatorname{hom}_A Y : (\tau \circ f)^\beta(p) \neq 0\right\},\end{aligned}$$

(in the right hand sides we have made use of our own notation, i. e. the extension $\alpha_A X$) where

$$\begin{aligned}r(Y) &:= \bigcap_{\tau \in \operatorname{Hom} Y} \operatorname{Ker} \tau,\\ B &:= \{\tau \circ f \mid f \in A, \tau \in \operatorname{hom}_A Y\},\\ \operatorname{hom}_A Y &:= \{\tau \in \operatorname{Hom} Y \mid \tau \circ f \in C(X, F) \quad \forall f \in A\}.\end{aligned}$$

It is claimed in the first part of Lemma 2.1 in [12] that if A is a Y-bimodule and a $C_b(X, \mathbb{R})$-module (i. e. closed under pointwise multiplication by elements of Y and $C_b(X, \mathbb{R})$), then the maps

$$\begin{aligned}G_1 &: \upsilon_A X \times \operatorname{Hom} Y \to \operatorname{Hom} A,\ (p, \tau) \mapsto \tau \circ \delta_p,\\ G_2 &: \beta_A X \times \operatorname{hom}_A Y \to \operatorname{Hom} A,\ (p, \tau) \mapsto \gamma_{p,\tau},\end{aligned}$$

are one-to-one and relatively open (in other words: their inverses are continuous), where

$$\delta_p(f) := f^\beta(p), \qquad \gamma_{p,\tau}(f) := (\tau \circ f)^\beta(p).$$

This is false. Indeed, let $\tau, \tau' \in \operatorname{hom} Y$ be two different continuous characters and consider the subalgebra $A := \{f \in C(X, Y) \mid \tau(f(x)) = 0 \quad \forall x \in X\}$. The kernel of a character is a maximal two-sided ideal, so $\operatorname{Ker} \tau \setminus \operatorname{Ker} \tau' \neq \emptyset$. If $y \in \operatorname{Ker} \tau \setminus \operatorname{Ker} \tau'$ then for the constant function $f(x) = y$ we have $f \in A$ with $f^\beta(x) \notin r(Y)$ and $(\tau' \circ f)^\beta(x) \neq 0$ everywhere, hence $\upsilon_A X = \alpha_A X$ and $\beta_A X = \alpha_B X$. On the other hand, $G_1(x, \tau) = 0$ for each $x \in \alpha_A X$ and $G_2(x, \tau) = 0$ for each

$x \in \alpha_B X$. So, if the algebra Y has at least two continuous characters then the values of G_1 and G_2 are not always characters, and if, in addition, the space X has at least two points then they are not one-to-one either.

The statement can be fixed and the proof in [12] works, if the maps G_1 and G_2 are regarded with different domains (the proof also works if $F = \mathbb{R}$ or if multiplication on Y is not jointly continuous):

Proposition 6.1. *Let X be a Tychonoff space, Y a Haudorff topological algebra over $F \in \{\mathbb{R}, \mathbb{C}\}$ and $A \subset C(X,Y)$ a function algebra that is a Y-bimodule and a $C_b(X,\mathbb{R})$-module. Then the maps*

$$G_1 : U \to \operatorname{Hom} A,\ (p,\tau) \mapsto \tau \circ \delta_p, \quad G_2 : V \to \operatorname{Hom} A,\ (p,\tau) \mapsto \gamma_{p,\tau}$$

are one-to-one and relatively open, where

$$\begin{aligned} U &:= \{(p,\tau) \mid p \in \alpha_A X,\ \tau \in \operatorname{Hom} Y,\ \exists f \in A \text{ such that } \tau(f^\beta(p)) \neq 0\}, \\ V &:= \{(p,\tau) \mid p \in \alpha_{D(\tau)} X,\ \tau \in \operatorname{hom}_A Y,\ \exists f \in A \text{ such that } (\tau \circ f)^\beta(p) \neq 0\}, \\ D(\tau) &:= \{\tau \circ f \mid f \in A\}. \end{aligned}$$

An analogous statement with the α-extension replaced by the γ-extension holds, if Y is a Y-completely regular complete Hausdorff algebra over an arbitrary topological field F, X is a Y-completely regular Hausdorff space, $\beta_Y X$ is an ultraregular space (i. e. a Hausdorff space whose topology has a base consisting of closed sets), and $A \subset C(X,Y)$ a function algebra that is a Y-bimodule and a $C(X,\{0,1\})$-module (where 0 and 1 are, respectively, the zero and identity of F). The proof is similar, except that instead of the condition of complete regularity it uses the fact that the points of $\beta_Y X$ are distinguished by $\{0,1\}$-valued continuous functions. The assumptions on X, Y and $\beta_Y X$ hold if F is an ultraregular non-discrete field with valuation, Y is a complete topological algebra over F with more than one element and X is an ultraregular space — indeed, then Y contains a subspace homeomorphic to F [5, I.13] and it is easy to see from the definition of Y-complete regularity that thus every F-completely regular space is Y-completely regular; as said in [2], for any ultraregular space F with at least two points, ultraregular spaces are F-completely regular.

The second part of Lemma 2.1 in [12] says that G_1 is continuous if $\operatorname{Hom} Y$ is locally equicontinuous; this works for algebras over any topological field, too.

In [12], Oubbi also gives a sufficient condition for describing the characters of a subalgebra of $C(X,Y)$ as $\chi(f) = (\tau \circ f)^\beta(p)$ in the case Y is an algebra with involution over $\mathbb{C}$ having jointly continuous multiplication (actually his proof also works in the case of algebras over $\mathbb{R}$). In his earlier article [10] he stated a special case of this result for the situation when $Y = F \in \{\mathbb{R}, \mathbb{C}\}$, obtaining a sufficient condition for representing the characters in the form $\chi(f) = \hat{f}(p)$, where $\hat{f} : \beta X \to Y \cup \{\infty\}$ is the continuous extension of f. In particular, it is shown in [10] that if $A \subset C(X,F)$ is closed under pointwise involution and multiplication by elements of $C_b(X,\mathbb{R})$ then $\operatorname{Hom} A$ is homeomorphic to $\alpha_A X \setminus \bigcap_{f \in A} \hat{f}^{-1}[\{0\}]$.

Finding more sufficient or necessary conditions for using the extensions $\alpha_A X$ and $\gamma_A X$ to describe the characters of function algebras remains an open research topic.

Acknowledgments
The author thanks Prof. Mati Abel for the idea of generalizing the Hewitt extension and the Estonian Doctoral School in Mathematics and Statistics and the Estonian Science Foundation (grant ETF7427) for financial support.

References

[1] Abel M 2004 *Proc. of the 3rd Int. Conf. on Topological Algebra and Applications, Oulu 2–6 July, 2001* (Acta Universitatis Ouluensis A vol 408) ed J Arhippainen and M Filali (Oulu: Oulun Yliopisto) pp 25–42
[2] Bachman G, Beckenstein E, Narici L and Warner S 1975 Rings of continuous functions with values in a topological field *Trans. Amer. Math. Soc.* **204** 91
[3] Beckenstein E, Narici L and Suffel C 1977 *Topological Algebras* (Amsterdam–New York–Oxford: North-Holland Publ. Co.)
[4] Bourbaki N 1971 *Éléments de mathématique. Topologie générale. Chapitres 1 à 4* (Paris: Hermann)
[5] Bourbaki N 1981 *Éléments de mathématique. Espaces vectoriels topologiques. Chapitres 1 à 5* (Berlin–Heidelberg–New York: Springer)
[6] Dierolf S, Schröder K H and Wengenroth J 1999 Characters on certain function algebras *Funct. Approx. Comment Math.* **26** 53
[7] Govaerts W 1978 Homomorphisms of weighted algebras of continuous functions *Ann. Mat. Pura Appl.* **116** 151
[8] Hery W J 1976 Maximal ideals in algebras of topological algebra valued functions *Pacific J. Math.* **65** 365-
[9] Köthe G 1969 *Topological Vector Spaces I* (New York: Springer-Verlag)
[10] Oubbi L 1999 On different algebras contained in $CV(X)$ *Bull. Belg. Math. Soc. Simon Stevin* **6** 111
[11] Oubbi L 2000 Weighted algebras of vector-valued continuous functions *Math. Nachr.* **212** 117
[12] Oubbi L 2007 Characters on algebras of vector-valued continuous functions *Rocky Mountain J. Math.* **37** 947
[13] Walker R C 1974 *The Stone-Čech Compactification* (Berlin–Heidelberg–New York: Springer-Verlag)

Algebra, Geometry, and Mathematical Physics 2010 IOP Publishing
Journal of Physics: Conference Series **346** (2012) 012023 doi:10.1088/1742-6596/346/1/012023

Tangent structures: sector-forms, jets and connections

Maido Rahula[1], **Petr Vašík**[2], **Nicoleta Voicu**[3]
[1]University of Tartu, J.Liivi 2, Tartu, 50409, Estonia
[2]Brno University of Technology, Technická 2, Brno, 616 69, Czech Republic
[3]University of Braşov, Iuliu Maniu Str., 50, 500091, Braşov, Romania

E-mail: rahula@ut.ee, vasik@fme.vutbr.cz, nico.brinzei@unitbv.ro

Abstract. To define a higher order connection on a fibered manifold one can use the sections of nonholonomic jet prolongations. However, a more natural approach seems to be the one assuming the structure of a higher-order tangent bundle and using White's sector-forms on these bundles.

MSC2010: 58A05, 58A20, 53C05

1. Iterations of a tangent functor

The tangent functor T assigns to a smooth manifold M a tangent bundle $TM \to M$ (first order tangent bundle), and to a mapping $\varphi : M_1 \to M_2$ between two smooth manifolds M_1, M_2 its differential (the tangent mapping $T\varphi : TM_1 \to TM_2$ as a fibered morphism). If the functor T is applied k times, then the k-th iteration assigns to a manifold M its *k-th order tangent bundle* $T^kM \to M$ and to a mapping φ its *k-th differential* $T^k\varphi : T^kM_1 \to T^kM_2$. In short we have

$$T^k : \quad \begin{cases} M \rightsquigarrow T^kM\,, \\ \varphi \rightsquigarrow T^k\varphi\,. \end{cases}$$

The total space T^kM of a fibre bundle $T^kM \to M$, as a manifold, has a dimension of 2^kn where $n = \dim M$. Each time the tangent functor T is applied, the dimension is doubled.

Proposition 1. *The bundle $T^kM \to T^{k-1}M$ is equipped with the structure of a k-fold vector bundle. Particularly, T^kM admits k different projections to $T^{k-1}M$,*

$$\boxed{\rho_s := T^{k-s}\pi_s : T^kM \to T^{k-1}M,}$$

where π_s is the natural projection $T^sM \to T^{s-1}M\,, s = 1, 2, \ldots, k$. Each projection defines a vector bundle with basis $T^{k-1}M$ and the total space composed of $2^{k-1}n$-dimensional vector spaces as fibers.

For every order l, $l = 2, .., k$, the fibers of $T^lM \to T^{l-1}M$ have common n-dimensional intersections as equalizers of all possible projections. Using this fact, on an order k tangent bundle $T^kM \to M$ one can define a $(k-1)n$-dimensional subbundle, *the osculating bundle* of the manifold M, denoted by $\operatorname{Osc}^{k-1}M\,, k > 1$.

Definition 1.1. The subbundle $\mathrm{Osc}^{k-1}M \subset T^kM$ is defined by the equality of projections

$$\rho_1 = \rho_2 = \ldots = \rho_k .$$

This means that the osculating bundle $\mathrm{Osc}^{k-1}M$ consists of exactly those elements of T^kM for which all k projections on $T^{k-1}M$ coincide.

2. Local coordinates

The local coordinates on the neighborhoods

$$T^sU \subset T^sM, \quad \text{where} \quad T^{s-1}U = \pi_s(T^sU), \ s = 1, 2, \ldots, k,$$

are derived from coordinates, or coordinate mappings, (u^i), which are given on the neighborhood $U \subset M$:

U: (u^i), $i = 1, 2, \ldots, n$,
TU: (u^i, u^i_1), where $u^i := u^i \circ \pi_1$, $u^i_1 := du^i$,
T^2U: $(u^i, u^i_1, u^i_2, u^i_{12})$,
where $u^i := u^i \circ \pi_1\pi_2$, $u^i_1 := du^i \circ \pi_2$, $u^i_2 := d(u^i \circ \pi_1)$, $u^i_{12} := d^2u^i$,
etc.

Proposition 2. *Coordinate mappings given on the neighborhood $T^{s-1}U$ can be transformed into coordinate mappings defined on the neighborhood T^sU with respect to the projection π_s by adding the differentials of these mappings.*

Local coordinates are obtained by the following principle:
to the coordinates of a point of a manifold we attach the coordinates of the vector tangent to the manifold at that point. We use the following notation: the coordinates of a neighborhood T^kU consist of two copies of local coordinates on $T^{k-1}U$ where the second copy is equipped with an additional subscript k. This principle is suitable in the sense that the coordinates with index s are recognized as the coordinates of the kernels of projections ρ_s, $s = 1, 2, \ldots, k$, i.e. the coordinates with index s disappear after the application of projection ρ_s.

The coordinate form of three projections $\rho_s : T^3U \to T^2U$, $s = 1, 2, 3$, is given by the following diagram:

$$(u^i, u^i_1, u^i_2, u^i_{12}, u^i_3, u^i_{13}, u^i_{23}, u^i_{123})$$

$$\rho_1 \swarrow \qquad\qquad \rho_2 \downarrow \qquad\qquad \searrow \rho_3$$

$$(u^i, u^i_2, u^i_3, , u^i_{23}) \qquad (u^i, u^i_1, u^i_3, , u^i_{13}) \qquad (u^i, u^i_1, u^i_2, , u^i_{12}).$$

Proposition 3. *The osculating bundle $Osc^{k-1}M$ is defined on the neighborhood $T^kU \subset T^kM$ by equality of the coordinates with the same number of subscripts.*

For the osculating bundle $\mathrm{Osc}^{k-1}M$ we use the coordinate notation

$$(u^i, du^i, d^2u^i, \ldots, d^ku^i).$$

Thus for $k = 3$ the osculating bundle Osc^2M in third order tangent bundle $T^3M \to M$ is defined by the equality of the projections

$$\rho_1 = \rho_2 = \rho_3 ,$$

and on the neighborhood $T^3U \subset T^3M$ this leads to the equality of osculating coordinates and to the notation: $du^i := u_1^i = u_2^i = u_3^i$, $d^2u^i := u_{12}^i = u_{13}^i = u_{23}^i$.

We describe the embedding $\mathrm{Osc}^{k-1}M \hookrightarrow T^kM$. In case $k = 3$ this embedding is constructed by means of the immersion $\zeta : \mathrm{Osc}M \to T^2M$ and its tangent mapping $T\zeta : T\mathrm{Osc}M \to T^3M$ described as follows:

$$\begin{pmatrix} u^i \\ u_1^i \\ u_2^i \\ u_{12}^i \end{pmatrix} \circ \zeta = \begin{pmatrix} u^i \\ du^i \\ du^i \\ d^2u^i \end{pmatrix}, \quad \begin{pmatrix} u_3^i \\ u_{13}^i \\ u_{23}^i \\ u_{123}^i \end{pmatrix} \circ T\zeta = \begin{pmatrix} du^i \\ d^2u^i \\ d^2u^i \\ d^3u^i \end{pmatrix},$$

$$T\zeta\Big(\frac{\partial}{\partial u^i}, \frac{\partial}{\partial(du^i)}, \frac{\partial}{\partial(d^2u^i)}\Big) = \Big(\frac{\partial}{\partial u^i}, \frac{\partial}{\partial u_1^i} + \frac{\partial}{\partial u_2^i}, \frac{\partial}{\partial u_{12}^i}\Big).$$

The fibers of $\mathrm{Osc}M$ are the integral surfaces of a distribution

$$\langle \partial_i^1 + \partial_i^2, \partial_i^{12} \rangle, \text{ where } \partial_i^1 + \partial_i^2 := \frac{\partial}{\partial u_1^i} + \frac{\partial}{\partial u_2^i}, \ \partial_i^{12} := \frac{\partial}{\partial u_{12}^i}.$$

The functions $(u_1^i - u_2^i)$ vanish on $\mathrm{Osc}M$,

$$(\partial_i^1 + \partial_i^2)(u_1^i - u_2^i) = \partial_i^{12}(u_1^i - u_2^i) = 0.$$

In case $k = 3$ the appropriate distribution is $\langle \partial_i^1 + \partial_i^2 + \partial_i^3,\ \partial_i^{12} + \partial_i^{23} + \partial_i^{13},\ \partial_i^{123} \rangle$, etc.

Historically, osculating bundles were investigated much earlier than higher order tangent bundles. The work begun by V.V. Vagner, [9], 60 years ago and has recently been finalized in the theory of Miron-Atanasiu, [2]. Recently, more attention has been given to iterations of the functor T, i.e. to higher order tangent bundles and sector forms. It is fully justified, since the above concepts form the base of differential and global analysis. The structural approaches given by Ehresmann and Pradin, [5, 7], although important, are not efficient. The geometric structure of higher order tangent bundles T^kM suggests applications in mechanics, such as higher order movements. From this point of view, sector calculus and the generalization of Cartan differential forms seem to be suitable tools. By White [10], a sector-form is a scalar function on T^kM, linear on the fibers of all k projections $T^kM \to T^{k-1}M$. Sector-forms include not only Cartan forms but, as coefficients, they also use generally non-holonomic jets. Bertram in [3] presented an original approach to these structures. In our recently published book [1], we tried to unify these approaches.

3. Sector forms

Definition 3.1. A general k-sector form is defined as a scalar function on T^kM that is linear and homogenous on the fibers of all projections $\rho, \dots, \rho_k : T^kM \to T^{k-1}M$, with arbitrary coefficients defined by the coordinate maps of manifold M. The coefficients are formed by a non-holonomic jet.

Proposition 4. *k-sector forms, same as covectors and 1-forms, form a vector space, i.e. an arbitrary linear combination of k-sector forms is again a k-sector form.*

Examples of sector forms.

1. Any 1-form on a manifold M is a 1-sector form on the tangent bundle TM.
2. A sector form Φ on a neighborhood T^2U and a sector form Ψ on a neighborhood T^3U can generally be written as:

$$\Phi = \varphi_{ij}u_1^iu_2^j + \varphi_iu_{12}^i,$$
$$\Psi = \psi_{ijk}u_1^iu_2^ju_3^k + \psi_{ij}^1u_1^iu_{23}^j + \psi_{ij}^2u_2^iu_{13}^j + \psi_{ij}^3u_3^iu_{12}^j + \psi_iu_{123}^i,$$

with arbitrary coefficients φ_{ij}, φ_i and ψ_{ijk}, ψ^1_{ij}, ψ^2_{ij}, ψ^3_{ij}, ψ_i. Note that, in each expression, the lower subscript 1 and 2, or 1, 2 and 3, respectively, occur exactly once, which means that these functions are really linear on the fibers of projections ρ_1 and ρ_2 or ρ_1, ρ_2 and ρ_3, respectively.

3. Locally, on the neighborhoods $U, TU, T^2U, \ldots$, the differentials of a function f, defined on manifold M, can be written as a sector form:

$$\begin{aligned} f_1 &:= df = f_i u_1^i , \\ f_{12} &:= d^2 f = f_{ij} u_1^i u_2^j + f_i u_{12}^i , \\ f_{123} &:= d^3 f = f_{ijk} u_1^i u_2^j u_3^k + f_{ij}(u_1^i u_{23}^j + u_2^i u_{13}^j + u_3^i u_{12}^j) + f_i u_{123}^i , \\ &\ldots \qquad \ldots \qquad \ldots \end{aligned}$$

The coefficients are formed by the holonomic jet of the function f,

$$f_i := \frac{\partial f}{\partial u^i}, \quad f_{ij} := \frac{\partial^2 f}{\partial u^i \partial u^j}, \quad f_{ijk} := \frac{\partial^3 f}{\partial u^i \partial u^j \partial u^k}, \quad \ldots$$

4. The differentials of a 1-form $\Phi_1 = \varphi_i u_1^i$ defined on M, as scalar functions on TM, can be written as sector forms:

$$\begin{aligned} \Phi_2 &:= d\Phi = \varphi_{i,j} u_1^i u_2^j + \varphi_i u_{12}^i , \\ \Phi_{23} &:= d^2\Phi = \varphi_{i,jk} u_1^i u_2^j u_3^k + \varphi_{i,j}(u_{13}^i u_2^j + u_1^i u_{23}^j + u_3^j u_{12}^i) + \varphi_i u_{123}^i , \\ &\ldots \qquad \ldots \qquad \ldots \end{aligned}$$

having as coefficients partial derivatives of functions φ_i,

$$\varphi_{i,j} := \frac{\partial \varphi_i}{\partial u^j}, \quad \varphi_{i,jk} := \frac{\partial^2 \varphi_i}{\partial u^j \partial u^k}, \ldots$$

The coefficients form a semi-holonomic (pseudo-, quasi-) jet.

Let us now list several important properties of sector forms.

Proposition 5. *For any $k, l \in \mathbb{N}$ each $(k+l)$-sector form on $T^{k+l}M$ is an l-sector form on T^kM. Particularly, a sector form on $T^{k+l}M$ can be the first differential of a sector form on T^kM.*

Proposition 6. *For $k > 1$ each exterior Cartan k-form, as a k-sector form on T^kM, vanishes on the osculating bundle $Osc^{k-1}M$. Cartan forms define an exterior geometry of the osculating bundle $Osc^{k-1}M$ in T^kM.*

See also Proposition 3.

Proposition 7. *The embedding $\zeta : \mathrm{Osc}^{k-1} \hookrightarrow \mathrm{T^kM}$ turns each k-sector form into a differential form of order k with symmetric coefficients.*

If $k = 2$, for example, we have:

$$\Phi = \varphi_{ij} u_1^i u_2^j + \varphi_i u_{12}^i \;\rightsquigarrow\; \Phi \circ \zeta = \varphi_{(ij)} du^i du^j + \varphi_i d^2 u^i .$$

Proposition 8. *Each $(k-1)$-sector form can be lifted from $T^{k-1}M$ in k different ways onto T^kM where these images may be seen as k-sector forms.*

If $k = 3$, for instance, there are three different lifts onto T^3M:

$$\begin{array}{ccc} \Phi \circ \rho_1 = \varphi_{ij} u_2^i u_3^j + \varphi_i u_{23}^i & \Phi \circ \rho_2 = \varphi_{ij} u_1^i u_3^j + \varphi_i u_{13}^i & \Phi \circ \rho_3 = \varphi_{ij} u_1^i u_2^j + \varphi_i u_{12}^i \\ \nwarrow \rho_1^* & \rho_2^* \uparrow & \rho_3^* \nearrow \\ & \Phi = \varphi_{ij} u_1^i u_2^j + \varphi_i u_{12}^i & \end{array}$$

4. Bundle connections

Definition 4.1. A connection on the bundle $\pi : M_1 \to M$ is defined by the structure $\triangle_h \oplus \triangle_v$ on a manifold M_1 where $\triangle_v = \ker T\pi$ is vertical distribution tangent to the fibers and $\triangle_h$ is horizontal distribution complementary to the distribution $\triangle_v$. The transport of the fibers along the path $\gamma \subset M$ is realized by the horizontal lifts given by the distribution $\triangle_h$ on the surface $\pi^{-1}(\gamma)$. If the bundle is a vector one and the transport of fibers along an arbitrary path is linear, then the connection is called linear.

We will assume that the base manifold M is of dimension n and the fibers are of dimension r. Then

$$\dim \triangle_h = n\,,\ \dim \triangle_v = r\,.$$

On the neighborhood $U \subset M_1$, let us consider local base and fiber coordinates:

$$(u^i, u^\alpha)\,,\ i = 1, 2, \ldots, n\,;\ \alpha = n+1, \ldots, n+r.$$

Base coordinates (u^i) are determined by the projection π and the coordinates $(\bar{u}^i)$ on a neighborhood $\bar{U} = \pi(U)$, $u^i = \bar{u}^i \circ \pi$.

Proposition 9. *On a neighborhood $U \subset M_1$ we define a local (adapted) basis of the structure $\triangle_h \oplus \triangle_v$,*

$$(X_i\ X_\alpha) = \begin{pmatrix} \dfrac{\partial}{\partial u^j} & \dfrac{\partial}{\partial u^\beta} \end{pmatrix} \cdot \begin{pmatrix} \delta_i^j & 0 \\ \Gamma_i^\beta & \delta_\alpha^\beta \end{pmatrix},\quad \begin{pmatrix} \omega^i \\ \omega^\alpha \end{pmatrix} = \begin{pmatrix} \delta_j^i & 0 \\ -\Gamma_j^\alpha & \delta_\beta^\alpha \end{pmatrix} \cdot \begin{pmatrix} du^j \\ du^\beta \end{pmatrix}.$$

The horizontal distribution $\triangle_h$ is the linear span of the vector fields (X_i) and annihilator of the forms (ω^α),

$$X_i = \partial_i + \Gamma_i^\beta \partial_\beta,\ \ \omega^\alpha = du^\alpha - \Gamma_i^\alpha du^i.$$

Proposition 10. *Any tensor field of type (p,q) on a manifold M_1 is decomposed in $\triangle_h \oplus \triangle_v$ into 2^{p+q} invariant components.*

For example, a vector field X and a 1-form Φ on M_1 can be decomposed both in natural and adapted bases:

$$X = \bar{x}^i \partial_i + \bar{x}^\alpha \partial_\alpha = x^i X_i + x^\alpha \partial_\alpha\,, \qquad x^i = \bar{x}^i,\ x^\alpha = \bar{x}^\alpha - \Gamma_i^\alpha \bar{x}^i,$$
$$\Phi = \bar{\varphi}_i du^i + \bar{\varphi}_\alpha du^\alpha = \varphi_i du^i + \varphi_\alpha \omega^\alpha\,, \qquad \varphi_i = \bar{\varphi}_i + \bar{\varphi}_\alpha \Gamma_i^\alpha,\ \varphi_\alpha = \bar{\varphi}_\alpha\,.$$

While the adapted components are invariant, the natural components are not. The components x^α and φ_i can be seen as the prototypes of covariant derivatives.

The appropriate affinor decomposes into four blocks with two of them being equal to zero:

$$d := \partial_i \otimes du^i + \partial_\alpha \otimes du^\alpha = X_i \otimes du^i + \partial_\alpha \otimes \omega^\alpha.$$

The invariant form is suitable for the description of differentials of functions in an adapted basis:

$$df = X_i f du^i + \partial_\alpha f \omega^\alpha.$$

Definition 4.2. Generally, Γ_i^α are functions of base and fiber coordinates (u^i, u^α). The curvature of a connection can be defined by $K_{ij}^\alpha = X_{[i}\Gamma_{j]}^\alpha$ appearing in the formulae

$$[X_i, X_j] = K_{ij}^\alpha \partial_\alpha\,,$$
$$\mathrm{d}\omega^\alpha = -K_{ij}^\alpha du^i \wedge du^j + \partial_\beta \Gamma_i^\alpha du^i \wedge \omega^\beta.$$

In the case of a linear connection the functions Γ_i^α and K_{ij}^α are linear on the fibers:

$$\Gamma_i^\alpha = \Gamma_{i\beta}^\alpha u^\beta,$$
$$K_{ij}^\alpha = K_{ij\beta}^\alpha u^\beta, \quad K_{ij\beta}^\alpha := \partial_{[i}\Gamma_{j]\beta}^\alpha - \Gamma_{[i|\gamma|}^\alpha \Gamma_{j]\beta}^\gamma.$$

Proposition 11. *The Pfaff system $\omega^\alpha = 0$ is equivalent to a system of differential equations with unknown functions u^α and independent variables u^i. In the case of a linear connection we have the following system of linear differential equations*

$$\omega^\alpha = 0 \quad \Leftrightarrow \quad \frac{\partial u^\alpha}{\partial u^i} = \Gamma_i^\alpha \quad \Rightarrow \quad \Big(\frac{\partial u^\alpha}{\partial u^i} = \Gamma_{i\beta}^\alpha u^\beta\Big).$$

In the case that the curvature of a connection equals to zero, $K_{ij}^\alpha = 0$, the Pfaff system $\omega^\alpha = 0$ is fully integrable.

Remark. If the system of first order differential equations is not of the form $\omega^\alpha = 0$, the following substitution can be made $\dfrac{\partial u^\alpha}{\partial u^i} \rightsquigarrow \Gamma_i^\alpha$ which translates the problem into the structure $\triangle_h \oplus \triangle_v$, see, e.g., [1], p. 85.

Proposition 12. *A classical affine connection on manifold M is seen as a linear connection on the bundle $\pi_1 : TM \to M$. On the tangent bundle $TM \to M$ one can define the structure $\triangle_h \oplus \triangle_v$. The indices in the formulae are denoted by Latin letters all of them ranging from 1 to n. The functions Γ_i^α, X_i, ω^α are of the form (in Γ_i^α the sign is changed to comply with the classical theory):*

$$\Gamma_i^\alpha \quad \rightsquigarrow \quad -\Gamma_{jk}^i u_1^k,$$
$$X_i = \partial_i + \Gamma_i^\alpha \partial_\alpha \quad \rightsquigarrow \quad X_i = \partial_i - \Gamma_{ij}^k u_1^i \partial_i^1,$$
$$\omega^\alpha = du^\alpha - \Gamma_i^\alpha du^i \quad \rightsquigarrow \quad U_{12}^i = u_{12}^i + \Gamma_{jk}^i u_1^k u_2^j.$$

The forms ω^α become 2-sector forms on a neighborhood T^2U.

Definition 4.3. Higher order connections are defined as follows:
on the tangent bundle TM the structure $\triangle \oplus \triangle_1$ is defined where $\ker T\rho_1 = \triangle_1$,
on T^2M the structure $\Delta \oplus \Delta_1 \oplus \Delta_2 \oplus \Delta_{12}$ is defined where $\ker T\rho_s = \Delta_s \oplus \Delta_{12}$,
$s = 1, 2$, etc.
See [1, 8].

Even if the theory of invariants can only be applied to tensors, algebraic methods can also be used for smooth mappings provided that the components of the jets that are not tensors can be replaced by covariant derivatives.

Let M_1, M_2 be two smooth manifolds. Let us consider the jet of the mapping $f : M_1 \to M_2$,

$$f^\alpha, f_i^\alpha, f_{ij}^\alpha, \ldots$$

We assume that the neighborhoods $U \subset M_1$ and $V = f(U) \subset M_2$ are equipped with coordinates

$$(u^i),\ i = 1, \ldots, \dim M_1,\ (v^\alpha),\ \alpha = 1, \ldots, \dim M_2.$$

The tangent mapping Tf is defined by the vector form

$$\mathcal{F}_1 = \partial_\alpha \otimes v_1^\alpha, \text{ where } v_1^\alpha := df^\alpha = f_i^\alpha u_1^i,$$

Journal of Physics: Conference Series **346** (2012) 012023 doi:10.1088/1742-6596/346/1/012023

the second order tangent mapping $T^2 f$ is defined by the vector form

$$\mathcal{F}_2 = (\,\partial_\alpha\ \partial^1_\alpha) \otimes \begin{pmatrix} v^\alpha_2 \\ v^\alpha_{12} \end{pmatrix}, \text{ where } \begin{pmatrix} v^\alpha_2 \\ v^\alpha_{12} \end{pmatrix} = \begin{pmatrix} f^\alpha_i & 0 \\ df^\alpha_i & f^\alpha_i \end{pmatrix} \cdot \begin{pmatrix} u^i_2 \\ u^i_{12} \end{pmatrix},$$

noting that , $df^\alpha_i = f^\alpha_{ij} u^j_1$, $v^\alpha_{12} = f^\alpha_{ij} u^j_1 u^i_2 + f^\alpha_i u^i_{12}$. If we define on the manifolds M_1 and M_2 affine connections with coefficients Γ^k_{ij} and $\Lambda^\alpha_{\beta\gamma}$ on the neighborhoods U and V, then the vector form $\mathcal{F}_2$ can be written, rather than in the natural, in the adapted bases as follows :

$$\mathcal{F}_2 = (\,X_\alpha\ \partial^1_\alpha) \otimes \begin{pmatrix} v^\alpha_2 \\ V^\alpha_{12} \end{pmatrix}, \text{ where } \begin{pmatrix} v^\alpha_2 \\ V^\alpha_{12} \end{pmatrix} = \begin{pmatrix} f^\alpha_i & 0 \\ F^\alpha_{ij} u^i_1 & f^\alpha_i \end{pmatrix} \cdot \begin{pmatrix} u^i_2 \\ U^i_{12} \end{pmatrix}.$$

The following invariant functions are considered:

$$\begin{aligned} u^i_{12} &\rightsquigarrow U^i_{12} = \Gamma^i_{jk} u^k_1 u^j_2 + u^i_{12}, \\ v^\alpha_{12} &\rightsquigarrow V^\alpha_{12} = \Lambda^\alpha_{\beta\gamma} v^\gamma_1 v^\beta_2 + v^i_{12}, \\ v^\alpha_{12} = f^\alpha_{ij} u^j_1 u^i_2 + f^\alpha_i u^i_{12} &\rightsquigarrow V^\alpha_{12} = F^\alpha_{ij} u^j_1 u^i_2 + f^\alpha_i U^i_{12}, \\ \text{where} &\quad \boxed{F^\alpha_{ij} = f^\alpha_{ij} - f^\alpha_k \Gamma^k_{ij} + (\Lambda^\alpha_{\beta\gamma} \circ f) f^\beta_i f^\gamma_j\,.} \end{aligned}$$

Remark. The move to adapted bases transforms the Jacobi matrix (on the neighborhood TU):

$$\begin{pmatrix} \delta^\alpha_\beta & 0 \\ \Lambda^\alpha_{\beta\gamma} v^\gamma_1 & \delta^\alpha_\beta \end{pmatrix}_{\circ f} \cdot \begin{pmatrix} f^\beta_k & 0 \\ f^\beta_{kj} u^j_1 & f^\beta_k \end{pmatrix} \cdot \begin{pmatrix} \delta^k_i & 0 \\ -\Gamma^k_{ij} u^j_1 & \delta^k_i \end{pmatrix} = \begin{pmatrix} f^\alpha_i & 0 \\ F^\alpha_{ij} u^j_1 & f^\alpha_i \end{pmatrix}.$$

The functions F^α_{ij}, as the covariant derivatives of the Jacobi matrix (f^α_i), replace the partial derivatives f^α_{ij}. The tensors $F^\alpha_{ijk}, \ldots$ are constructed in the same way, but this time using higher order connections and thus higher order tangent bundles, see [8], p. 155-168.

Let us list some cases in which (f^α_i) and (F^α_{ij}) play a major role, see [1], p. 101-110:
- Riemannian and co-Riemannian geometry,
- geodesic curves and co-geodesic fields,
- symmetries and motions in spaces with connection,
- Cartan-Laptev method.

5. Analytical mechanics

The special role played by the tangent bundles TM and T^2M in analytical mechanics was pointed out by Godbillon in [4].

To the function $H = H(u, u_1)$ on TM, the following vector field is associated:

$$X = \sum_i H_{u^i_1} \partial_i - \sum_i H_{u^i} \partial^1_i, \quad H_i := \frac{\partial H}{\partial u^i}, \quad H_{u^i_1} := \frac{\partial H}{\partial u^i_1},$$

Proposition 13. *A flow a_t of a field X in TM is defined by the following differential equations*

$$\boxed{\begin{cases} \dot u^i = H_{u^i_1} \\ \dot u^i_1 = -H_{u^i} \end{cases}, \quad \dot u^i := \frac{du^i}{dt}, \quad \dot u^i_1 := \frac{du^i_1}{dt}\,.}$$

This system is called a *Hamiltonian system* and the function H a *Hamiltonian*. In view of

$$(u^i, u_1^i, u_2^i, u_{12}^i) \rightsquigarrow (u^i, u_1^i, \dot{u}^i, \dot{u}_1^i),$$

the Hamiltonian system defines a section of $\pi_2 : T^2M \to TM$ of dimension $2n$.

Proposition 14. *In the flow of the vector field X, the Hamiltonian H and the symplectic form $\Omega = du^i \wedge du_1^i$, see [6], are invariant:*

$$XH = 0, \quad \mathcal{L}_X\Omega = 0.$$

Proposition 15. *The Poisson bracket of two Hamiltonians H and G coincides with the derivative of the function G with respect to the field X,*

$$XG = \sum_i H_{u_1^i} G_{u_i} - \sum_i H_{u^i} G_{u_1^i} = \{H, G\}.$$

Proposition 16. *If a Hamiltonian system defined on T^2M is restricted to the osculating bundle $OscM$, this leads to the following Lagrange equations*

$$\boxed{\frac{d}{dt}\Big(\frac{\partial L}{\partial \dot{u}^i}\Big) - \frac{\partial L}{\partial u^i} = 0.}$$

Proof. By definition, a Hamiltonian $H = H(u, u_1)$ is connected to a Lagrangian $L = L(u, u_2)$ by the (*Legendre transform*):

$$H(u, u_1) - \sum_i u_1^i u_2^i + L(u, u_2) = 0.$$

However, on T^2M, this equation generally makes no sense:

$$d(H - \sum_i u_1^i u_2^i + L) \equiv 0 \;\Rightarrow\; H_{u^i} + L_{u^i} = 0,\; H_{u_1^i} = u_2^i,\; L_{u_2^i} = u_1^i.$$

and only on the osculating bundle Osc$M$ where $u_1^i = u_2^i = \dot{u}$, the transformation $H \rightsquigarrow L$ is possible. On Osc$M$ the Hamiltonian system yields a Lagrange system. □

A Lagrange system defines in Osc$M \to TM$ a section of the same dimension $2n$ as a Hamiltonian system in $T^2M \to TM$.

Hamiltonian geometry on T^kM and Lagrange geometry on osculating bundles Osc^{k-1}M with $k > 2$ are related to each other in a similar way.

Acknowledgement
The second author was supported by a grant of the Czech Science Foundation (GAČR, No. 201/08/P230).

References

[1] Atanasiu G, Balan V, Brînzei N and Rahula M 2010 *Differential Geometric Structures: Tangent Bundles, Connections in Bundles, Exponential Law in the Jet Space* (in Russian, Moscow: Librokom)
[2] Atanasiu G, Balan V, Brînzei N and Rahula M 2010 *Second Order Differential Geometry and Applications: Miron-Atanasiu Theory* (in Russian, Moscow: Librokom)
[3] Bertram W 2008 Differential Geometry, Lie Groups and Symmetric Spaces over General Base Fields and Rings *Memoirs of AMS* **900**
[4] Godbillon C 1969 *Géométrie Différentielle et Mécanique Analytique* (Paris: Hermann)

[5] Ehresmann Ch 1958 Catégories doubles et catégories structurées *C. R. Acad. Sci.* **256** 1198
[6] Kushner A, Lychagin V and Rubtsov V 2007 Contact Geometry and Nonlinear Differential Equations *Ser. Encyclopedia of Mathematics and its Applications* **101** (Cambridge: University Press)
[7] Pradines J 1974 Suites exactes vectorielles doubles et connexions *C. R. Acad. Sci.* **278** 1587
[8] Rahula M 1993 *New Problems in Differential Geometry* (Singapore: World Scientific)
[9] Vagner V V 1949 *Theory of differential objects and foundations of differential geometry* (in Russian), Appendix in: Veblen O, Whitehead J H C, *The Foundations of Differential Geometry* (Moscow: Inostr. Liter.)
[10] White E J 1982 *The Method of Iterated Tangents with Applications in Local Riemannian Geometry* Monographs and Studies in Mathematics (Boston, Mass.-London: Pitman Advanced Publishing Program)

AUTHOR INDEX